Climate Change 2001: Synthesis Report

Climate Change 2001: Synthesis Report forms the fourth volume of the International Panel on Climate Change (IPCC) Third Assessment Report, and is composed of the Synthesis Report itself, the Summaries for Policymakers and Technical Summaries of the three IPCC Working Group volumes, and supporting annexes.

IPCC assessments attempt to answer such general questions as:
- Has the Earth's climate changed as a result of human activities?
- In what ways is climate projected to change in the future?
- How vulnerable are agriculture, water supply, ecosystems, coastal infrastructure, and human health to different levels of change in climate and sea level?
- What is the technical, economic, and market potential of options to adapt to climate change or reduce emissions of the gases that influence climate?

Climate Change 2001: Synthesis Report provides a policy-relevant, but not policy-prescriptive, synthesis and integration of information contained within the Third Assessment Report and also draws upon all previously approved and accepted IPCC reports that address a broad range of key policy-relevant questions. For this reason it will be especially useful for policymakers and researchers, and as a main or supplementary student textbook for courses in environmental studies, meteorology, climatology, biology, ecology, and atmospheric chemistry.

Climate Change 2001: Synthesis Report

Edited by:

Robert T. Watson
The World Bank

and the
Core Writing Team

Based on a draft prepared by:

Core Writing Team
Daniel L. Albritton, Terry Barker, Igor A. Bashmakov, Osvaldo Canziani, Renate Christ, Ulrich Cubasch, Ogunlade Davidson, Habiba Gitay, David Griggs, Kirsten Halsnaes, John Houghton, Joanna House, Zbigniew Kundzewicz, Murari Lal, Neil Leary, Christopher Magadza, James J. McCarthy, John F.B. Mitchell, Jose Roberto Moreira, Mohan Munasinghe, Ian Noble, Rajendra Pachauri, Barrie Pittock, Michael Prather, Richard G. Richels, John B. Robinson, Jayant Sathaye, Stephen Schneider, Robert Scholes, Thomas Stocker, Narasimhan Sundararaman, Rob Swart, Tomihiro Taniguchi, and D. Zhou

All IPCC Authors

Editorial Team
David J. Dokken, Maria Noguer, Paul van der Linden, Cathy Johnson, Jiahua Pan, and the GRID-Arendal Design Studio

Contribution of Working Groups I, II, and III to the Third Assessment Report of the Intergovernmental Panel on Climate Change

Published for the Intergovernmental Panel on Climate Change

CAMBRIDGE
UNIVERSITY PRESS

PUBLISHED BY THE PRESS SYNDICATE OF THE UNIVERSITY OF CAMBRIDGE
The Pitt Building, Trumpington Street, Cambridge, United Kingdom

CAMBRIDGE UNIVERSITY PRESS
The Edinburgh Building, Cambridge CB2 2RU, UK
40 West 20th Street, New York, NY 10011-4211, USA
477 Williamstown Road, Port Melbourne, VIC 3207, Australia
Ruiz de Alarcón 13, 28014 Madrid, Spain
Dock House, The Waterfront, Cape Town 8001, South Africa

http://www.cambridge.org

First published 2001

Printed in the United States of America

Typeface Times New Roman 10/12.5 pt. *System* Adobe PageMaker 6.5 [AU]

A catalog record for this book is available from the British Library.

Library of Congress Cataloging in Publication Data available

ISBN 0 521 80770 0 hardback
ISBN 0 521 01507 3 paperback

Referencing the Volume

IPCC, 2001: *Climate Change 2001: Synthesis Report. A Contribution of Working Groups I, II, and III to the Third Assessment Report of the Integovernmental Panel on Climate Change* [Watson, R.T. and the Core Writing Team (eds.)]. Cambridge University Press, Cambridge, United Kingdom, and New York, NY, USA, 398 pp.

Cover Image Credits

Center: Earth—shown for a projection centered on Asia—as seen by the Moderate-Resolution Imaging Spectroradiometer (MODIS) on board the National Aeronautics and Space Administration (NASA) EOS-Terra satellite. Land-surface data composited spatially at 1 km and temporally during May and June 2001; cloud layer derived from EOS-Terra, GOES 8/10, GMS-5, and Meteosat 5/7 sensor data; sea-ice composited over an 8-day period using MODIS data; and U.S. Geological Survey topography data overlain to visualize the terrain. Image by Reto Stöckli, Science Systems and Applications, Inc., and the Visualization and Analysis Laboratory at NASA Goddard Space Flight Center.
Right: The Lena Delta, Sakha Republic (Yakutia), Russia, as imaged from two Landsat-7 scenes taken at noon, 27 July 2000. Generated by the Norwegian Mapping Authority and GRID-Arendal, with palette derived from infrared channels to yield "natural colors" for the various landscape elements.
Lower Left: "One Way Water" (Thailand). Photograph provided by Topham/UNEP/Waranun Chutchawantipakorn.
Upper Left: "In Search of Water" (India). Photograph provided by Topham/UNEP/P.K. De.

Climate Change 2001: Synthesis Report

Contents

Foreword

The Intergovernmental Panel on Climate Change (IPCC) was jointly established in 1988, by the World Meteorological Organization (WMO) and the United Nations Environment Programme (UNEP). Its present terms of reference are to:

· Assess available information on the science, the impacts, and the economics of—and the options for mitigating and/or adapting to—climate change.
· Provide, on request, scientific/technical/socio-economic advice to the Conference of the Parties (COP) to the United Nations Framework Convention on Climate Change (UNFCCC).

Since its establishment, the IPCC has produced a series of Assessment Reports (1990, 1995, and 2001), Special Reports, Technical Papers, and methodologies, such as the *Guidelines for National Greenhouse Gas Inventories*, which have become standard works of reference, widely used by policymakers, scientists, and other experts and students.

This Synthesis Report completes the four-volume Third Assessment Report (TAR). It addresses specifically the issues of concern to the policymaker, in the context of Article 2 of the UNFCCC—issues such as the extent to which human activities have influenced and will in the future influence the global climate, the impacts of a changed climate on ecological and socio-economic systems, and existing and projected technical and policy capacity to address anthropogenic climate change. It explores briefly the linked nature of a number of multilateral environmental conventions. It draws on the work of hundreds of experts from all regions of the world who have in the past and at present participated in the IPCC process. As is customary in the IPCC, success in producing this report has depended first and foremost on the dedication, enthusiasm, and cooperation of these experts in many different but related disciplines.

We take this opportunity to express our heart-felt gratitude to the authors and reviewers of all the IPCC reports and Technical Papers, particularly the TAR. We thank likewise the IPCC Bureau; Dr. Sundararaman, Secretary of IPCC, and his Secretariat staff; and those staffing the Technical Support Units of the three Working Groups. We acknowledge with gratitude the governments and organizations that contribute to the IPCC Trust Fund, and provide support to the experts and in other ways. The IPCC has been especially successful in engaging in its work a large number of experts from the developing countries and countries with their economies in transition; the Trust Fund enables extending financial assistance for their travel to IPCC meetings.

We thank the Chairman of the IPCC, Dr. Robert T. Watson, for guiding the effort in completing the TAR.

G.O.P. Obasi
Secretary General
World Meteorological Organization

K. Töpfer
Executive Director
United Nations Environment Programme
and Director-General
United Nations Office in Nairobi

Preface

This Synthesis Report with its Summary for Policymakers is the fourth and final part of the Third Assessment Report (TAR) of the Intergovernmental Panel on Climate Change (IPCC). It draws together and integrates for the benefit of policy makers, and others, and in response to questions identified by governments and subsequently agreed by the IPCC, information that has been approved and/or accepted by the IPCC.[i] It is intended to assist governments, individually and collectively, in formulating appropriate adaptation and mitigation responses to the threat of human-induced climate change.

The Synthesis Report is based mainly on the contributions of the three IPCC Working Groups to the TAR, but also uses information from earlier IPCC assessments, Special Reports, and Technical Papers. It follows the question and answer format, and is in two parts: a Summary for Policymakers and a longer document that contains expanded responses to each of the questions posed by governments. The Summary for Policymakers references the appropriate paragraphs in the longer report, while the longer report contains references to the source of the material on which the response is based— that is, the Summaries for Policymakers and chapters from previously approved and accepted Working Group contributions to the TAR and earlier IPCC reports and Technical Papers (see the accompanying box for cross-referencing nomenclature).

The procedures for approving the Summary for Policymakers and adopting the balance of the Synthesis Report were formalized by the IPCC at its Fifteenth Session (San Jose, Costa Rica, 15–18 April 1999). A draft of the Synthesis Report and its Summary for Policymakers was prepared by a team of lead authors, who were involved in preparation of the TAR, and submitted for simultaneous government/technical and expert review. The revised drafts were circulated to governments in a final distribution before approval/adoption at the IPCC's Eighteenth Session (Wembley, United Kingdom, 24–29 September 2001).

The Synthesis Report consists of nine policy-relevant questions:

- Question 1 addresses the ultimate objective of the United Nations Framework Convention on Climate Change, which is found in Article 2 (i.e., what constitutes "dangerous anthropogenic interference in the climate system") and provides a framework for placing the issue of climate change in the context of sustainable development.
- Question 2 assesses and, where possible, attributes observed changes in climate and ecological systems since the pre-industrial era.
- Questions 3 and 4 assess the impact of future emissions of greenhouse gases and sulfate aerosol precursors (without specific policies to mitigate climate change) on climate, including changes in variability and extreme events and in ecological and socio-economic systems.
- Question 5 discusses inertia in the climate, ecological systems, and socio-economic sectors, and implications for mitigation and adaptation.
- Question 6 assesses the near- and long-term implications of stabilizing atmospheric concentrations of greenhouse gases on climate, ecological systems, and socio-economic sectors.
- Question 7 assesses the technologies, policies, and costs of near- and long-term actions to mitigate greenhouse gas emissions.
- Question 8 identifies the interactions between climate change, other environmental issues, and development.
- Question 9 summarizes the most robust findings and key uncertainties.

We take this opportunity to thank:

- The Core Writing Team who drafted this report and, with their meticulous and painstaking attention to detail, finalized it
- Other members of the IPCC Bureau who acted as Review Editors
- The members of the Working Groups' teams of Coordinating Lead Authors and Lead Authors who helped with the initial drafting
- The Heads and the staff of the Technical Support Units of the three Working Groups, particularly David Dokken, Maria Noguer, and Paul van der Linden for logistical and editorial support
- The Head and the staff of the GRID office at Arendal, Norway—Philippe Rekacewicz in particular—for working with the author team on the graphics contained in the Synthesis Report
- The staff of the IPCC Secretariat for innumerable administrative tasks performed.

---

[i] See *Procedures for the Preparation, Review, Approval, Acceptance, Adoption, and Publication of the IPCC Reports* in http://www.ipcc.ch for descriptions of terms.

The Synthesis Report with its Summary for Policymakers is published here in a single volume together with the Summaries for Policymakers and Technical Summaries of the Working Group contributions to the TAR, as well as a comprehensive, consolidated glossary. The Synthesis Report is also available in Arabic, Chinese, French, Russian, and Spanish—the other official languages of the IPCC. The Synthesis Report is also available as a stand-alone publication, as are discrete brochures consisting of the Summaries for Policymakers, Technical Summaries, and glossaries of the respective Working Group reports. The full English text of all

four volumes comprising the Third Assessment Report has been published in both print and digital form, with searchable versions available on CD-ROM and at http://www.ipcc.ch.

R.T. Watson
IPCC Chair

N. Sundararaman
IPCC Secretary

IPCC Assessments Cited in the Synthesis Report

Qx.x	Relevant paragraph in the underlying Synthesis Report
WGI TAR	Working Group I contribution to the Third Assessment Report
WGII TAR	Working Group II contribution to the Third Assessment Report
WGIII TAR	Working Group III contribution to the Third Assessment Report
SRES	Special Report on Emissions Scenarios
SRLULUCF	Special Report on Land Use, Land-Use Change, and Forestry
SRTT	Special Report on the Methodological and Technological Issues in Technology Transfer
SRAGA	Special Report on Aviation and the Global Atmosphere
DES GP	Guidance Paper on Development, Equity, and Sustainability
IPCC TP4	Technical Paper on Implications of Proposed CO_2 Emissions Limitations
IPCC TP3	Technical Paper on Stabilization of Atmospheric Greenhouse Gases: Physical, Biological, and Socio-Economic Implications
WGII SAR	Working Group II contribution to the Second Assessment Report

SPM	Summary for Policymakers
TS	Technical Summary
ES	Executive Summary
GP	Guidance Paper
TP	Technical Paper

Climate Change 2001: Synthesis Report

Summary for Policymakers

An Assessment of the Intergovernmental Panel on Climate Change

This summary, approved in detail at IPCC Plenary XVIII (Wembley, United Kingdom, 24-29 September 2001), represents the formally agreed statement of the IPCC concerning key findings and uncertainties contained in the Working Group contributions to the Third Assessment Report.

Based on a draft prepared by:

Robert T. Watson, Daniel L. Albritton, Terry Barker, Igor A. Bashmakov, Osvaldo Canziani, Renate Christ, Ulrich Cubasch, Ogunlade Davidson, Habiba Gitay, David Griggs, Kirsten Halsnaes, John Houghton, Joanna House, Zbigniew Kundzewicz, Murari Lal, Neil Leary, Christopher Magadza, James J. McCarthy, John F.B. Mitchell, Jose Roberto Moreira, Mohan Munasinghe, Ian Noble, Rajendra Pachauri, Barrie Pittock, Michael Prather, Richard G. Richels, John B. Robinson, Jayant Sathaye, Stephen Schneider, Robert Scholes, Thomas Stocker, Narasimhan Sundararaman, Rob Swart, Tomihiro Taniguchi, D. Zhou, and many IPCC authors and reviewers

Introduction

In accordance with a decision taken at its Thirteenth Session (Maldives, 22 and 25-28 September 1997) and other subsequent decisions, the IPCC decided:

- To include a Synthesis Report as part of its Third Assessment Report
- That the Synthesis Report would provide a policy-relevant, but not policy-prescriptive, synthesis and integration of information contained within the Third Assessment Report and also drawing upon all previously approved and accepted IPCC reports that would address a broad range of key policy-relevant, but not policy-prescriptive, questions
- That the questions would be developed in consultation with the Conference of the Parties (COP) to the United Nations Framework Convention on Climate Change (UNFCCC).

The following nine questions were based on submissions by governments and were approved by the IPCC at its Fifteenth Session (San José, Costa Rica, 15-18 April 1999).

Question 1

What can scientific, technical, and socio-economic analyses contribute to the determination of what constitutes dangerous anthropogenic interference with the climate system as referred to in Article 2 of the Framework Convention on Climate Change?

Q1

Natural, technical, and social sciences can provide essential information and evidence needed for decisions on what constitutes "dangerous anthropogenic interference with the climate system." At the same time, such decisions are value judgments determined through socio-political processes, taking into account considerations such as development, equity, and sustainability, as well as uncertainties and risk.   → Q1.1

The basis for determining what constitutes "dangerous anthropogenic interference" will vary among regions—depending both on the local nature and consequences of climate change impacts, and also on the adaptive capacity available to cope with climate change—and depends upon mitigative capacity, since the magnitude and the rate of change are both important. There is no universally applicable best set of policies; rather, it is important to consider both the robustness of different policy measures against a range of possible future worlds, and the degree to which such climate-specific policies can be integrated with broader sustainable development policies.   → Q1.2

The Third Assessment Report (TAR) provides an assessment of new scientific information and evidence as an input for policymakers in their determination of what constitutes "dangerous anthropogenic interference with the climate system." It provides, first, new projections of future concentrations of greenhouse gases in the atmosphere, global and regional patterns of changes and rates of change in temperature, precipitation, and sea level, and changes in extreme climate events. It also examines possibilities for abrupt and irreversible changes in ocean circulation and the major ice sheets. Second, it provides an assessment of the biophysical and socio-economic impacts of climate change, with regard to risks to unique and threatened systems, risks associated with extreme weather events, the distribution of impacts, aggregate impacts, and risks of large-scale, high-impact events. Third, it provides an assessment of the potential for achieving a broad range of levels of greenhouse gas concentrations in the atmosphere through mitigation, and information about how adaptation can reduce vulnerability.   → Q1.3-6

An integrated view of climate change considers the dynamics of the complete cycle of interlinked causes and effects across all sectors concerned (see Figure SPM-1). The TAR provides new policy-relevant information and evidence with regard to all quadrants of Figure SPM-1. A major new contribution of the *Special Report on Emissions Scenarios* (SRES) was to explore alternative development paths and related greenhouse gas emissions, and the TAR assessed preliminary work on the linkage between adaptation, mitigation, and development paths. However, the TAR does not achieve a fully integrated assessment of climate change because of the incomplete state of knowledge.

 Q1.7

Climate change decision making is essentially a sequential process under general uncertainty. Decision making has to deal with uncertainties including the risk of non-linear and/or irreversible changes, entails balancing the risks of either insufficient or excessive action, and involves careful consideration of the consequences (both environmental and economic), their likelihood, and society's attitude towards risk.

 Q1.8

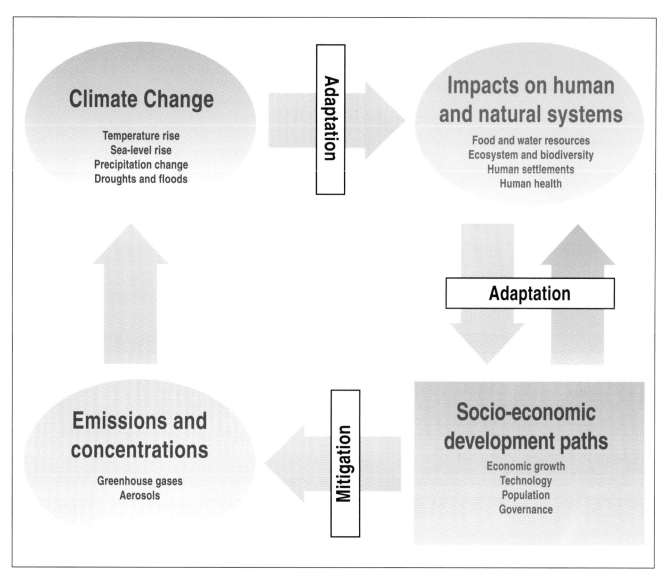

Figure SPM-1: Climate change – an integrated framework. Schematic and simplified representation of an integrated assessment framework for considering anthropogenic climate change. The yellow arrows show the cycle of cause and effect among the four quadrants shown in the figure, while the blue arrow indicates the societal response to climate change impacts. See the caption for Figure 1-1 for an expanded description of this framework.

 Q1 Figure 1-1

The climate change issue is part of the larger challenge of sustainable development. As a result, climate policies can be more effective when consistently embedded within broader strategies designed to make national and regional development paths more sustainable. This occurs because the impact of climate variability and change, climate policy responses, and associated socio-economic development will affect the ability of countries to achieve sustainable development goals. Conversely, the pursuit of those goals will in turn affect the opportunities for, and success of, climate policies. In particular, the socio-economic and technological characteristics of different development paths will strongly affect emissions, the rate and magnitude of climate change, climate change impacts, the capability to adapt, and the capacity to mitigate.

 Q1.9-10

The TAR assesses available information on the timing, opportunities, costs, benefits, and impacts of various mitigation and adaptation options. It indicates that there are opportunities for countries acting individually, and in cooperation with others, to reduce costs of mitigation and adaptation and to realize benefits associated with achieving sustainable development.

Q1.11

Question 2

What is the evidence for, causes of, and consequences of changes in the Earth's climate since the pre-industrial era?

Q2

(a) Has the Earth's climate changed since the pre-industrial era at the regional and/or global scale? If so, what part, if any, of the observed changes can be attributed to human influence and what part, if any, can be attributed to natural phenomena? What is the basis for that attribution?

(b) What is known about the environmental, social, and economic consequences of climate changes since the pre-industrial era with an emphasis on the last 50 years?

The Earth's climate system has demonstrably changed on both global and regional scales since the pre-industrial era, with some of these changes attributable to human activities.

 Q2.2

Human activities have increased the atmospheric concentrations of greenhouse gases and aerosols since the pre-industrial era. The atmospheric concentrations of key anthropogenic greenhouse gases (i.e., carbon dioxide (CO_2), methane (CH_4), nitrous oxide (N_2O), and tropospheric ozone (O_3)) reached their highest recorded levels in the 1990s, primarily due to the combustion of fossil fuels, agriculture, and land-use changes (see Table SPM-1). The radiative forcing from anthropogenic greenhouse gases is positive with a small uncertainty range; that from the direct aerosol effects is negative and smaller; whereas the negative forcing from the indirect effects of aerosols on clouds might be large but is not well quantified.

 Q2.4-5

An increasing body of observations gives a collective picture of a warming world and other changes in the climate system (see Table SPM-1).

 Q2.6

Globally it is very likely that the 1990s was the warmest decade, and 1998 the warmest year, in the instrumental record (1861–2000) (see Box SPM-1). The increase in surface temperature over the 20th century for the Northern Hemisphere is likely to have been greater than that for any other century in the last thousand years (see Table SPM-1). Insufficient data are available prior to the year 1860 in the Southern Hemisphere to compare the recent warming with changes over the last 1,000 years. Temperature changes have not been uniform globally but have varied over regions and different parts of the lower atmosphere.

 Q2.7

Table SPM-1	20th century changes in the Earth's atmosphere, climate, and biophysical system.[a]
Indicator	**Observed Changes**
Concentration indicators	
Atmospheric concentration of CO_2	280 ppm for the period 1000–1750 to 368 ppm in year 2000 (31±4% increase).
Terrestrial biospheric CO_2 exchange	Cumulative source of about 30 Gt C between the years 1800 and 2000; but during the 1990s, a net sink of about 14±7 Gt C.
Atmospheric concentration of CH_4	700 ppb for the period 1000–1750 to 1,750 ppb in year 2000 (151±25% increase).
Atmospheric concentration of N_2O	270 ppb for the period 1000–1750 to 316 ppb in year 2000 (17±5% increase).
Tropospheric concentration of O_3	Increased by 35±15% from the years 1750 to 2000, varies with region.
Stratospheric concentration of O_3	Decreased over the years 1970 to 2000, varies with altitude and latitude.
Atmospheric concentrations of HFCs, PFCs, and SF_6	Increased globally over the last 50 years.
Weather indicators	
Global mean surface temperature	Increased by 0.6±0.2°C over the 20th century; land areas warmed more than the oceans (*very likely*).
Northern Hemisphere surface temperature	Increase over the 20th century greater than during any other century in the last 1,000 years; 1990s warmest decade of the millennium (*likely*).
Diurnal surface temperature range	Decreased over the years 1950 to 2000 over land: nighttime minimum temperatures increased at twice the rate of daytime maximum temperatures (*likely*).
Hot days / heat index	Increased (*likely*).
Cold / frost days	Decreased for nearly all land areas during the 20th century (*very likely*).
Continental precipitation	Increased by 5–10% over the 20th century in the Northern Hemisphere (*very likely*), although decreased in some regions (e.g., north and west Africa and parts of the Mediterranean).
Heavy precipitation events	Increased at mid- and high northern latitudes (*likely*).
Frequency and severity of drought	Increased summer drying and associated incidence of drought in a few areas (*likely*). In some regions, such as parts of Asia and Africa, the frequency and intensity of droughts have been observed to increase in recent decades.

Box SPM-1	Confidence and likelihood statements.

Q2 Box 2-1

Where appropriate, the authors of the Third Assessment Report assigned confidence levels that represent their collective judgment in the validity of a conclusion based on observational evidence, modeling results, and theory that they have examined. The following words have been used throughout the text of the Synthesis Report to the TAR relating to WGI findings: *virtually certain* (greater than 99% chance that a result is true); *very likely* (90–99% chance); *likely* (66–90% chance); *medium likelihood* (33–66% chance); *unlikely* (10–33% chance); *very unlikely* (1–10% chance); and *exceptionally unlikely* (less than 1% chance). An explicit uncertainty range (±) is a *likely* range. Estimates of confidence relating to WGII findings are: *very high* (95% or greater), *high* (67–95%), *medium* (33–67%), *low* (5–33%), and *very low* (5% or less). No confidence levels were assigned in WGIII.

Q2.9-11

There is new and stronger evidence that most of the warming observed over the last 50 years is attributable to human activities. Detection and attribution studies consistently find evidence for an anthropogenic signal in the climate record of the last 35 to 50 years. These studies include uncertainties in forcing due to anthropogenic sulfate aerosols and natural factors (volcanoes and solar irradiance), but do not account for the effects of other types of anthropogenic aerosols and land-use changes. The sulfate and natural forcings are negative over this period and cannot explain the warming; whereas most of these studies find that, over the last 50 years, the estimated rate and magnitude of warming due to increasing greenhouse gases alone

Table SPM-1	20th century changes in the Earth's atmosphere, climate, and biophysical system.[a] (continued)
Indicator	**Observed Changes**
Biological and physical indicators	
Global mean sea level	Increased at an average annual rate of 1 to 2 mm during the 20th century.
Duration of ice cover of rivers and lakes	Decreased by about 2 weeks over the 20th century in mid- and high latitudes of the Northern Hemisphere (*very likely*).
Arctic sea-ice extent and thickness	Thinned by 40% in recent decades in late summer to early autumn (*likely*) and decreased in extent by 10–15% since the 1950s in spring and summer.
Non-polar glaciers	Widespread retreat during the 20th century.
Snow cover	Decreased in area by 10% since global observations became available from satellites in the 1960s (*very likely*).
Permafrost	Thawed, warmed, and degraded in parts of the polar, sub-polar, and mountainous regions.
El Niño events	Became more frequent, persistent, and intense during the last 20 to 30 years compared to the previous 100 years.
Growing season	Lengthened by about 1 to 4 days per decade during the last 40 years in the Northern Hemisphere, especially at higher latitudes.
Plant and animal ranges	Shifted poleward and up in elevation for plants, insects, birds, and fish.
Breeding, flowering, and migration	Earlier plant flowering, earlier bird arrival, earlier dates of breeding season, and earlier emergence of insects in the Northern Hemisphere.
Coral reef bleaching	Increased frequency, especially during El Niño events.
Economic indicators	
Weather-related economic losses	Global inflation-adjusted losses rose an order of magnitude over the last 40 years (see Q2 Figure 2-7). Part of the observed upward trend is linked to socio-economic factors and part is linked to climatic factors.

[a] This table provides examples of key observed changes and is not an exhaustive list. It includes both changes attributable to anthropogenic climate change and those that may be caused by natural variations or anthropogenic climate change. Confidence levels are reported where they are explicitly assessed by the relevant Working Group. An identical table in the Synthesis Report contains cross-references to the WGI and WGII reports.

are comparable with, or larger than, the observed warming. The best agreement between model simulations and observations over the last 140 years has been found when all the above anthropogenic and natural forcing factors are combined, as shown in Figure SPM-2.

Changes in sea level, snow cover, ice extent, and precipitation are consistent with a warming climate near the Earth's surface. Examples of these include a more active hydrological cycle with more heavy precipitation events and shifts in precipitation, widespread retreat of non-polar glaciers, increases in sea level and ocean-heat content, and decreases in snow cover and sea-ice extent and thickness (see Table SPM-1). For instance, it is very likely that the 20th century warming has contributed significantly to the observed sea-level rise, through thermal expansion of seawater and widespread loss of land ice. Within present uncertainties, observations and models are both consistent with a lack of significant acceleration of sea-level rise during the 20th century. There are no demonstrated changes in overall Antarctic sea-ice extent from the years 1978 to 2000. In addition, there are conflicting analyses and insufficient data to assess changes in intensities of tropical and extra-tropical cyclones and severe local storm activity in the mid-latitudes. Some of the observed changes are regional and some may be due to internal climate variations, natural forcings, or regional human activities rather than attributed solely to global human influence.

Q2.12-19

Observed changes in regional climate have affected many physical and biological systems, and there are preliminary indications that social and economic systems have been affected.

Q2.20 & Q2.25

Comparison between modeled and observations of temperature rise since the year 1860

Temperature anomalies in °C

(a) Natural forcing only

Model results
— Observations

Temperature anomalies in °C

(b) Anthropogenic forcing only

Model results
— Observations

Temperature anomalies in °C

(c) Natural + Anthropogenic forcing

Model results
— Observations

Figure SPM-2: Simulating the Earth's temperature variations (°C) and comparing the results to the measured changes can provide insight to the underlying causes of the major changes. A climate model can be used to simulate the temperature changes that occur from both natural and anthropogenic causes. The simulations represented by the band in (a) were done with only natural forcings: solar variation and volcanic activity. Those encompassed by the band in (b) were done with anthropogenic forcings: greenhouse gases and an estimate of sulfate aerosols. And those encompassed by the band in (c) were done with both natural and anthropogenic forcings included. From (b), it can be seen that the inclusion of anthropogenic forcings provides a plausible explanation for a substantial part of the observed temperature changes over the past century, but the best match with observations is obtained in (c) when both natural and anthropogenic factors are included. These results show that the forcings included are sufficient to explain the observed changes, but do not exclude the possibility that other forcings may also have contributed.

 Q2 Figure 2-4

Recent regional changes in climate, particularly increases in temperature, have already affected hydrological systems and terrestrial and marine ecosystems in many parts of the world (see Table SPM-1). The observed changes in these systems[1] are coherent across diverse localities and/or regions and are consistent in direction with the expected effects of regional changes in temperature. The probability that the observed changes in the expected direction (with no reference to magnitude) could occur by chance alone is negligible.

Q2.21-24

[1] There are 44 regional studies of over 400 plants and animals, which varied in length from about 20 to 50 years, mainly from North America, Europe, and the southern polar region. There are 16 regional studies covering about 100 physical processes over most regions of the world, which varied in length from about 20 to 150 years.

The rising socio-economic costs related to weather damage and to regional variations in climate suggest increasing vulnerability to climate change. Preliminary indications suggest that some social and economic systems have been affected by recent increases in floods and droughts, with increases in economic losses for catastrophic weather events. However, because these systems are also affected by changes in socio-economic factors such as demographic shifts and land-use changes, quantifying the relative impact of climate change (either anthropogenic or natural) and socio-economic factors is difficult.

 Q2.25-26

Question 3

 Q3

What is known about the regional and global climatic, environmental, and socio-economic consequences in the next 25, 50, and 100 years associated with a range of greenhouse gas emissions arising from scenarios used in the TAR (projections which involve no climate policy intervention)?

To the extent possible evaluate the:
- Projected changes in atmospheric concentrations, climate, and sea level
- Impacts and economic costs and benefits of changes in climate and atmospheric composition on human health, diversity and productivity of ecological systems, and socio-economic sectors (particularly agriculture and water)
- The range of options for adaptation, including the costs, benefits, and challenges
- Development, sustainability, and equity issues associated with impacts and adaptation at a regional and global level.

Carbon dioxide concentrations, globally averaged surface temperature, and sea level are projected to increase under all IPCC emissions scenarios during the 21st century.[2]

 Q3.2

For the six illustrative SRES emissions scenarios, the projected concentration of CO_2 in the year 2100 ranges from 540 to 970 ppm, compared to about 280 ppm in the pre-industrial era and about 368 ppm in the year 2000. The different socio-economic assumptions (demographic, social, economic, and technological) result in the different levels of future greenhouse gases and aerosols. Further uncertainties, especially regarding the persistence of the present removal processes (carbon sinks) and the magnitude of the climate feedback on the terrestrial biosphere, cause a variation of about −10 to +30% in the year 2100 concentration, around each scenario. Therefore, the total range is 490 to 1,250 ppm (75 to 350% above the year 1750 (pre-industrial) concentration). Concentrations of the primary non-CO_2 greenhouse gases by year 2100 are projected to vary considerably across the six illustrative SRES scenarios (see Figure SPM-3).

 Q3.3-5

Projections using the SRES emissions scenarios in a range of climate models result in an increase in globally averaged surface temperature of 1.4 to 5.8°C over the period 1990 to 2100. This is about two to ten times larger than the central value of observed warming over the 20th century and the projected rate of warming is very likely to be without precedent during at least the last 10,000 years, based on paleoclimate data. Temperature increases are projected to be greater than those in the Second Assessment Report (SAR), which were about 1.0 to 3.5°C based on six IS92 scenarios. The higher projected temperatures and the wider range are due primarily to lower projected sulfur dioxide (SO_2) emissions in the SRES scenarios relative to the IS92 scenarios. For the periods 1990 to 2025 and 1990 to 2050, the projected increases are 0.4 to 1.1°C and 0.8 to 2.6°C, respectively. By

 Q3.6-7 & Q3.11

---

[2] Projections of changes in climate variability, extreme events, and abrupt/non-linear changes are covered in Question 4.

the year 2100, the range in the surface temperature response across different climate models for the same emissions scenario is comparable to the range across different SRES emissions scenarios for a single climate model. Figure SPM-3 shows that the SRES scenarios with the highest emissions result in the largest projected temperature increases. Nearly all land areas will very likely warm more than these global averages, particularly those at northern high latitudes in winter.

Globally averaged annual precipitation is projected to increase during the 21st century, though at regional scales both increases and decreases are projected of typically 5 to 20%. It is likely that precipitation will increase over high-latitude regions in both summer and winter. Increases are also projected over northern mid-latitudes, tropical Africa, and Antarctica in winter, and in southern and eastern Asia in summer. Australia, Central America, and southern Africa show consistent decreases in winter rainfall. Larger year-to-year variations in precipitation are very likely over most areas where an increase in mean precipitation is projected.

→ Q3.8 & Q3.12

Glaciers are projected to continue their widespread retreat during the 21st century. Northern Hemisphere snow cover, permafrost, and sea-ice extent are projected to decrease further. The Antarctic ice sheet is likely to gain mass, while the Greenland ice sheet is likely to lose mass (see Question 4).

→ Q3.14

Global mean sea level is projected to rise by 0.09 to 0.88 m between the years 1990 and 2100, for the full range of SRES scenarios, but with significant regional variations. This rise is due primarily to thermal expansion of the oceans and melting of glaciers and ice caps. For the periods 1990 to 2025 and 1990 to 2050, the projected rises are 0.03 to 0.14 m and 0.05 to 0.32 m, respectively.

→ Q3.9 & Q3.13

Projected climate change will have beneficial and adverse effects on both environmental and socio-economic systems, but the larger the changes and rate of change in climate, the more the adverse effects predominate.

→ Q3.15

The severity of the adverse impacts will be larger for greater cumulative emissions of greenhouse gases and associated changes in climate (*medium confidence*). While beneficial effects can be identified for some regions and sectors for small amounts of climate change, these are expected to diminish as the magnitude of climate change increases. In contrast many identified adverse effects are expected to increase in both extent and severity with the degree of climate change. When considered by region, adverse effects are projected to predominate for much of the world, particularly in the tropics and subtropics.

→ Q3.16

Overall, climate change is projected to increase threats to human health, particularly in lower income populations, predominantly within tropical/subtropical countries. Climate change can affect human health directly (e.g., reduced cold stress in temperate countries but increased heat stress, loss of life in floods and storms) and indirectly through changes in the ranges of disease vectors (e.g., mosquitoes),[3] water-borne pathogens, water quality, air quality, and food availability and quality (*medium to high confidence*). The actual health impacts will be strongly influenced by local environmental conditions and socio-economic circumstances, and by the range of social, institutional, technological, and behavioral adaptations taken to reduce the full range of threats to health.

→ Q3.17

Ecological productivity and biodiversity will be altered by climate change and sea-level rise, with an increased risk of extinction of some vulnerable species (*high to medium confidence*). Significant disruptions of ecosystems from disturbances such as fire, drought, pest infestation, invasion of species, storms, and coral bleaching events are expected to

→ Q3.18-20

---

[3] Eight studies have modeled the effects of climate change on these diseases—five on malaria and three on dengue. Seven use a biological or process-based approach, and one uses an empirical, statistical approach.

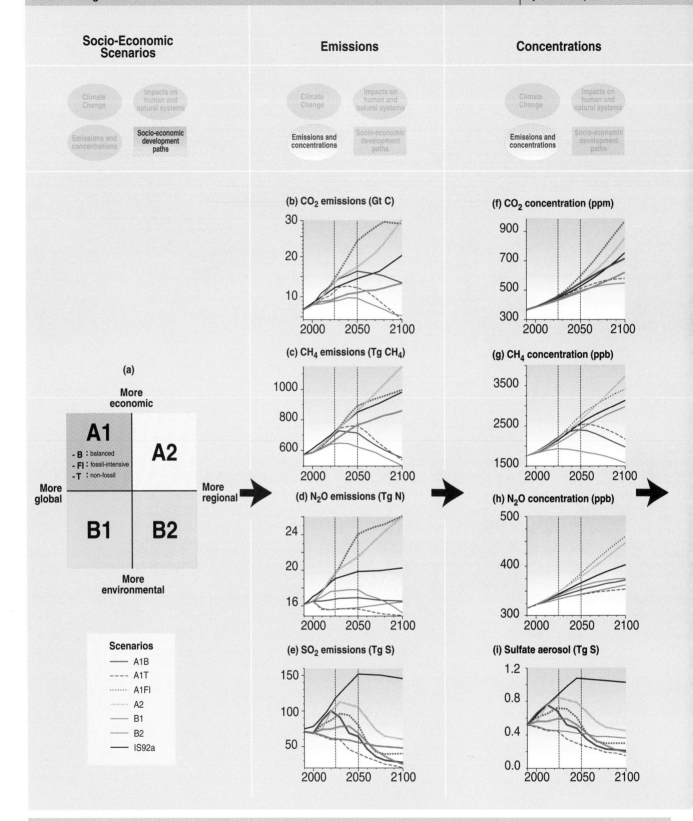

Socio-Economic Scenarios

Emissions

Concentrations

(a)

More economic

A1
- B : balanced
- FI : fossil-intensive
- T : non-fossil

A2

More global → More regional

B1

B2

More environmental

Scenarios
— A1B
--- A1T
····· A1FI
— A2
— B1
— B2
— IS92a

(b) CO₂ emissions (Gt C)

(c) CH₄ emissions (Tg CH₄)

(d) N₂O emissions (Tg N)

(e) SO₂ emissions (Tg S)

(f) CO₂ concentration (ppm)

(g) CH₄ concentration (ppb)

(h) N₂O concentration (ppb)

(i) Sulfate aerosol (Tg S)

A1FI, A1T, and A1B

The A1 storyline and scenario family describes a future world of very rapid economic growth, global population that peaks in mid-century and declines thereafter, and the rapid introduction of new and more efficient technologies. Major underlying themes are convergence among regions, capacity-building, and increased cultural and social interactions, with a substantial reduction in regional differences in per capita income. The A1 scenario family develops into three groups that describe alternative directions of technological change in the energy system. The three A1 groups are distinguished by their technological emphasis: fossil intensive (A1FI), non-fossil energy sources (A1T), or a balance across all sources (A1B) (where balanced is defined as not relying too heavily on one particular energy source, on the assumption that similar improvment rates apply to all energy supply and end use technologies).

Radiative Forcing

Temperature and Sea-Level Change

Reasons for Concern

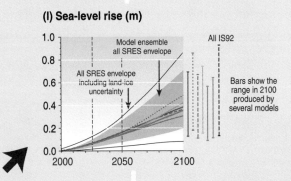

(l) Sea-level rise (m)

Scenarios
— A1B
--- A1T
········ A1FI
— A2
— B1
— B2
— IS92a

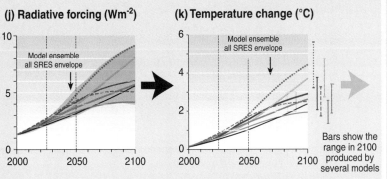

(j) Radiative forcing (Wm⁻²)

(k) Temperature change (°C)

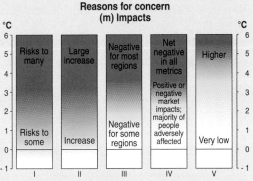

Reasons for concern (m) Impacts

I   Risks to unique and threatened systems
II Risks from extreme climate events
III Distribution of impacts
IV Aggregate impacts
V   Risks from future large-scale discontinuities

A2

The A2 storyline and scenario family describes a very heterogeneous world. The underlying theme is self-reliance and preservation of local identities. Fertility patterns across regions converge very slowly, which results in continuously increasing population. Economic development is primarily regionally oriented and per capita economic growth and technological change more fragmented and slower than other storylines.

B1

The B1 storyline and scenario family describes a convergent world with the same global population that peaks in mid-century and declines thereafter, as in the A1 storyline, but with rapid change in economic structures toward a service and information economy, with reductions in material intensity and the introduction of clean and resource-efficient technologies. The emphasis is on global solutions to economic, social, and environmental sustainability, including improved equity, but without additional climate initiatives.

B2

The B2 storyline and scenario family describes a world in which the emphasis is on local solutions to economic, social, and environmental sustainability. It is a world with continuously increasing global population, at a rate lower than A2, intermediate levels of economic development, and less rapid and more diverse technological change than in the B1 and A1 storylines. While the scenario is also oriented towards environmental protection and social equity, it focuses on local and regional levels.

Figure SPM-3: The different socio-economic assumptions underlying the SRES scenarios result in different levels of future emissions of greenhouse gases and aerosols. These emissions in turn change the concentration of these gases and Q3 Figure 3-1
aerosols in the atmosphere, leading to changed radiative forcing of the climate system. Radiative forcing due to the SRES scenarios results in projected increases in temperature and sea level, which in turn will cause impacts. The SRES scenarios do not include additional climate initiatives and no probabilities of occurrence are assigned. Because the SRES scenarios had only been available for a very short time prior to production of the TAR, the impacts assessments here use climate model results that tend to be based on equilibrium climate change scenarios (e.g., 2xCO₂), a relatively small number of experiments using a 1% per year CO₂ increase transient scenario, or the scenarios used in the SAR (i.e., the IS92 series). Impacts in turn can affect socio-economic development paths through, for example, adaptation and mitigation. The highlighted boxes along the top of the figure illustrate how the various aspects relate to the integrated assessment framework for considering climate change (see Figure SPM-1).

increase. The stresses caused by climate change, when added to other stresses on ecological systems, threaten substantial damage to or complete loss of some unique systems and extinction of some endangered species. The effect of increasing CO_2 concentrations will increase net primary productivity of plants, but climate changes, and the changes in disturbance regimes associated with them, may lead to either increased or decreased net ecosystem productivity (*medium confidence*). Some global models project that the net uptake of carbon by terrestrial ecosystems will increase during the first half of the 21st century but then level off or decline.

Models of cereal crops indicate that in some temperate areas potential yields increase with small increases in temperature but decrease with larger temperature changes (*medium to low confidence*). In most tropical and subtropical regions, potential yields are projected to decrease for most projected increases in temperature (*medium confidence*). Where there is also a large decrease in rainfall in subtropical and tropical dryland/rainfed systems, crop yields would be even more adversely affected. These estimates include some adaptive responses by farmers and the beneficial effects of CO_2 fertilization, but not the impact of projected increases in pest infestations and changes in climate extremes. The ability of livestock producers to adapt their herds to the physiological stresses associated with climate change is poorly known. Warming of a few ° C or more is projected to increase food prices globally, and may increase the risk of hunger in vulnerable populations.

→ Q3.21

Climate change will exacerbate water shortages in many water-scarce areas of the world. Demand for water is generally increasing due to population growth and economic development, but is falling in some countries because of increased efficiency of use. Climate change is projected to substantially reduce available water (as reflected by projected runoff) in many of the water-scarce areas of the world, but to increase it in some other areas (*medium confidence*) (see Figure SPM-4). Freshwater quality generally would be degraded by higher water temperatures (*high confidence*), but this may be offset in some regions by increased flows.

→ Q3.22

The aggregated market sector effects, measured as changes in gross domestic product (GDP), are estimated to be negative for many developing countries for all magnitudes of global mean temperature increases studied (*low confidence*), and are estimated to be mixed for developed countries for up to a few °C warming (*low confidence*) and negative for warming beyond a few degrees (*medium to low confidence*). The estimates generally exclude the effects of changes in climate variability and extremes, do not account for the effects of different rates of climate change, only partially account for impacts on goods and services that are not traded in markets, and treat gains for some as canceling out losses for others.

→ Q3.25

Populations that inhabit small islands and/or low-lying coastal areas are at particular risk of severe social and economic effects from sea-level rise and storm surges. Many human settlements will face increased risk of coastal flooding and erosion, and tens of millions of people living in deltas, in low-lying coastal areas, and on small islands will face risk of displacement. Resources critical to island and coastal populations such as beaches, freshwater, fisheries, coral reefs and atolls, and wildlife habitat would also be at risk.

→ Q3.23

The impacts of climate change will fall disproportionately upon developing countries and the poor persons within all countries, and thereby exacerbate inequities in health status and access to adequate food, clean water, and other resources. Populations in developing countries are generally exposed to relatively high risks of adverse impacts from climate change. In addition, poverty and other factors create conditions of low adaptive capacity in most developing countries.

→ Q3.33

Adaptation has the potential to reduce adverse effects of climate change and can often produce immediate ancillary benefits, but will not prevent all damages.

→ Q3.26

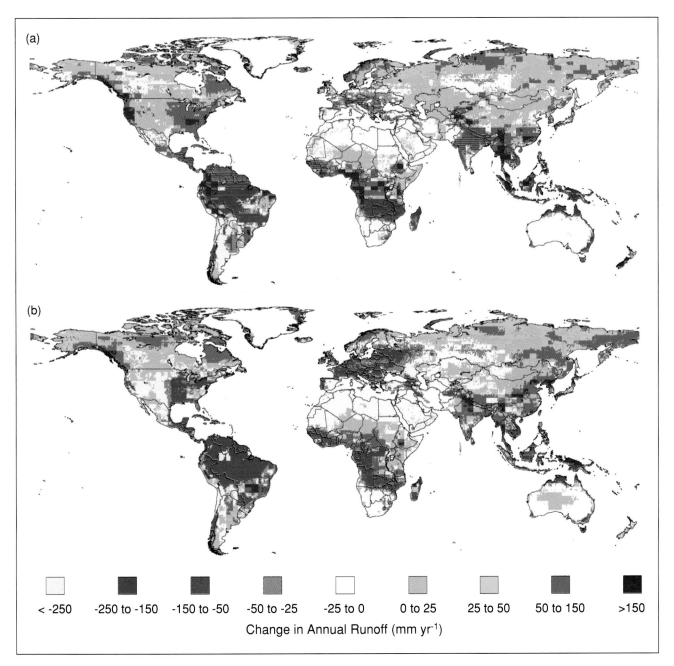

(a)

(b)

< -250	-250 to -150	-150 to -50	-50 to -25	-25 to 0	0 to 25	25 to 50	50 to 150	>150

Change in Annual Runoff (mm yr⁻¹)

Figure SPM-4: Projected changes in average annual water runoff by the year 2050, relative to average runoff for the years 1961 to 1990, largely follow projected changes in precipitation. Changes in runoff are calculated with a hydrologic model using as inputs climate projections from two versions of the Hadley Centre atmosphere-ocean general circulation model (AOGCM) for a scenario of 1% per annum increase in effective CO_2 concentration in the atmosphere: (a) HadCM2 ensemble mean and (b) HadCM3. Projected increases in runoff in high latitudes and southeast Asia and decreases in central Asia, the area around the Mediterranean, southern Africa, and Australia are broadly consistent across the Hadley Centre experiments, and with the precipitation projections of other AOGCM experiments. For other areas of the world, changes in precipitation and runoff are scenario- and model-dependent.

 Q3 Figure 3-5

Numerous possible adaptation options for responding to climate change have been identified that can reduce adverse and enhance beneficial impacts of climate change, but will incur costs. Quantitative evaluation of their benefits and costs and how they vary across regions and entities is incomplete.

 Q3.27

Greater and more rapid climate change would pose greater challenges for adaptation and greater risks of damages than would lesser and slower change. Natural and human systems have evolved capabilities to cope with a range of climate variability within which the risks of damage are relatively low and ability to recover is high. However, changes in climate that result in increased frequency of events that fall outside the historic range with which systems have coped increase the risk of severe damages and incomplete recovery or collapse of the system.

 Q3.28

Question 4

What is known about the influence of the increasing atmospheric concentrations of greenhouse gases and aerosols, and the projected human-induced change in climate regionally and globally on:

Q4

a. The frequency and magnitude of climate fluctuations, including daily, seasonal, inter-annual, and decadal variability, such as the El Niño Southern Oscillation cycles and others?
b. The duration, location, frequency, and intensity of extreme events such as heat waves, droughts, floods, heavy precipitation, avalanches, storms, tornadoes, and tropical cyclones?
c. The risk of abrupt/non-linear changes in, among others, the sources and sinks of greenhouse gases, ocean circulation, and the extent of polar ice and permafrost? If so, can the risk be quantified?
d. The risk of abrupt or non-linear changes in ecological systems?

An increase in climate variability and some extreme events is projected.

Q4.2-8

Models project that increasing atmospheric concentrations of greenhouse gases will result in changes in daily, seasonal, inter-annual, and decadal variability. There is projected to be a decrease in diurnal temperature range in many areas, decrease of daily variability of surface air temperature in winter, and increased daily variability in summer in the Northern Hemisphere land areas. Many models project more El Niño-like mean conditions in the tropical Pacific. There is no clear agreement concerning changes in frequency or structure of naturally occurring atmosphere-ocean circulation patterns such as that of the North Atlantic Oscillation (NAO).

Q4.3-8

Models project that increasing atmospheric concentrations of greenhouse gases result in changes in frequency, intensity, and duration of extreme events, such as more hot days, heat waves, heavy precipitation events, and fewer cold days. Many of these projected changes would lead to increased risks of floods and droughts in many regions, and predominantly adverse impacts on ecological systems, socio-economic sectors, and human health (see Table SPM-2 for details). High resolution modeling studies suggest that peak wind and precipitation intensity of tropical cyclones are likely to increase over some areas. There is insufficient information on how very small-scale extreme weather phenomena (e.g., thunderstorms, tornadoes, hail, hailstorms, and lightning) may change.

Q4.2-7

Greenhouse gas forcing in the 21st century could set in motion large-scale, high-impact, non-linear, and potentially abrupt changes in physical and biological systems over the coming decades to millennia, with a wide range of associated likelihoods.

Q4.9

Some of the projected abrupt/non-linear changes in physical systems and in the natural sources and sinks of greenhouse gases could be irreversible, but there is an incomplete understanding of some of the underlying processes. The likelihood of

Q4.10-16

Table SPM-2	Examples of climate variability and extreme climate events and examples of their impacts (WGII TAR Table SPM-1).
Projected Changes during the 21st Century in Extreme Climate Phenomena and their Likelihood	**Representative Examples of Projected Impacts[a] (all high confidence of occurrence in some areas)**
Higher maximum temperatures, more hot days and heat waves[b] over nearly all land areas (*very likely*)	Increased incidence of death and serious illness in older age groups and urban poor. Increased heat stress in livestock and wildlife. Shift in tourist destinations. Increased risk of damage to a number of crops. Increased electric cooling demand and reduced energy supply reliability.
Higher (increasing) minimum temperatures, fewer cold days, frost days and cold waves[b] over nearly all land areas (*very likely*)	Decreased cold-related human morbidity and mortality. Decreased risk of damage to a number of crops, and increased risk to others. Extended range and activity of some pest and disease vectors. Reduced heating energy demand.
More intense precipitation events (*very likely,* over many areas)	Increased flood, landslide, avalanche, and mudslide damage. Increased soil erosion. Increased flood runoff could increase recharge of some floodplain aquifers. Increased pressure on government and private flood insurance systems and disaster relief.
Increased summer drying over most mid-latitude continental interiors and associated risk of drought (*likely*)	Decreased crop yields. Increased damage to building foundations caused by ground shrinkage. Decreased water resource quantity and quality. Increased risk of forest fire.
Increase in tropical cyclone peak wind intensities, mean and peak precipitation intensities (*likely,* over some areas)[c]	Increased risks to human life, risk of infectious disease epidemics and many other risks. Increased coastal erosion and damage to coastal buildings and infrastructure. Increased damage to coastal ecosystems such as coral reefs and mangroves.
Intensified droughts and floods associated with El Niño events in many different regions (*likely*) (see also under droughts and intense precipitation events)	Decreased agricultural and rangeland productivity in drought- and flood-prone regions. Decreased hydro-power potential in drought-prone regions.
Increased Asian summer monsoon precipitation variability (*likely*)	Increase in flood and drought magnitude and damages in temperate and tropical Asia.
Increased intensity of mid-latitude storms (little agreement between current models)[b]	Increased risks to human life and health. Increased property and infrastructure losses. Increased damage to coastal ecosystems.

[a] These impacts can be lessened by appropriate response measures.
[b] Information from WGI TAR Technical Summary (Section F.5).
[c] Changes in regional distribution of tropical cyclones are possible but have not been established.

the projected changes is expected to increase with the rate, magnitude, and duration of climate change. Examples of these types of changes include:

· Large climate-induced changes in soils and vegetation may be possible and could induce further climate change through increased emissions of greenhouse gases from plants and soil, and changes in surface properties (e.g., albedo).

· Most models project a weakening of the thermohaline circulation of the oceans resulting in a reduction of heat transport into high latitudes of Europe, but none show an abrupt shutdown by the end of the 21st century. However, beyond the year 2100, some models suggest that the thermohaline circulation could completely, and possibly irreversibly, shut down in either hemisphere if the change in radiative forcing is large enough and applied long enough.

· The Antarctic ice sheet is likely to increase in mass during the 21st century, but after sustained warming the ice sheet could lose significant mass and contribute several meters to the projected sea-level rise over the next 1,000 years.

· In contrast to the Antarctic ice sheet, the Greenland ice sheet is likely to lose mass during the 21st century and contribute a few cm to sea-level rise. Ice sheets will continue to react to climate warming and contribute to sea-level rise for thousands of years after climate has been stabilized. Climate models indicate that the local warming over Greenland is likely to be one to three times the global average. Ice sheet models project that a local warming of larger than

3°C, if sustained for millennia, would lead to virtually a complete melting of the Greenland ice sheet with a resulting sea-level rise of about 7 m. A local warming of 5.5°C, if sustained for 1,000 years, would likely result in a contribution from Greenland of about 3 m to sea-level rise.

· Continued warming would increase melting of permafrost in polar, sub-polar, and mountain regions and would make much of this terrain vulnerable to subsidence and landslides which affect infrastructure, water courses, and wetland ecosystems.

Changes in climate could increase the risk of abrupt and non-linear changes in many ecosystems, which would affect their function, biodiversity, and productivity. The greater the magnitude and rate of the change, the greater the risk of adverse impacts. For example: Q4.17-19

· Changes in disturbance regimes and shifts in the location of suitable climatically defined habitats may lead to abrupt breakdown of terrestrial and marine ecosystems with significant changes in composition and function and increased risk of extinctions.

· Sustained increases in water temperatures of as little as 1°C, alone or in combination with any of several stresses (e.g., excessive pollution and siltation), can lead to corals ejecting their algae (coral bleaching) and the eventual death of some corals.

· Temperature increase beyond a threshold, which varies by crop and variety, can affect key development stages of some crops (e.g., spikelet sterility in rice, loss of pollen viability in maize, tubers' development in potatoes) and thus the crop yields. Yield losses in these crops can be severe if temperatures exceed critical limits for even short periods.

Question 5

What is known about the inertia and time scales associated with the changes in the climate system, ecological systems, and socio-economic sectors and their interactions?

Q5

Inertia is a widespread inherent characteristic of the interacting climate, ecological, and socio-economic systems. Thus some impacts of anthropogenic climate change may be slow to become apparent, and some could be irreversible if climate change is not limited in both rate and magnitude before associated thresholds, whose positions may be poorly known, are crossed.  Q5.1-4, Q5.8, Q5.10-12, & Q5.14-17

Inertia in Climate Systems

Stabilization of CO_2 emissions at near-current levels will not lead to stabilization of CO_2 atmospheric concentration, whereas stabilization of emissions of shorter lived greenhouse gases such as CH_4 leads, within decades, to stabilization of their atmospheric concentrations. Stabilization of CO_2 concentrations at any level requires eventual reduction of global CO_2 net emissions to a small fraction of the current emission level. The lower the chosen level for stabilization, the sooner the decline in global net CO_2 emissions needs to begin (see Figure SPM-5). Q5.3 & Q5.5

After stabilization of the atmospheric concentration of CO_2 and other greenhouse gases, surface air temperature is projected to continue to rise by a few tenths of a degree per century for a century or more, while sea level is projected to continue to rise for many centuries (see Figure SPM-5). The slow transport of heat into the oceans and slow response of ice sheets means that long periods are required to reach a new climate system equilibrium. Q5.4

Some changes in the climate system, plausible beyond the 21st century, would be effectively irreversible. For example, major melting of the ice sheets (see Question 4) and fundamental changes in the ocean circulation pattern (see Question 4) could not be reversed over  Q5.4 & Q5.14-16

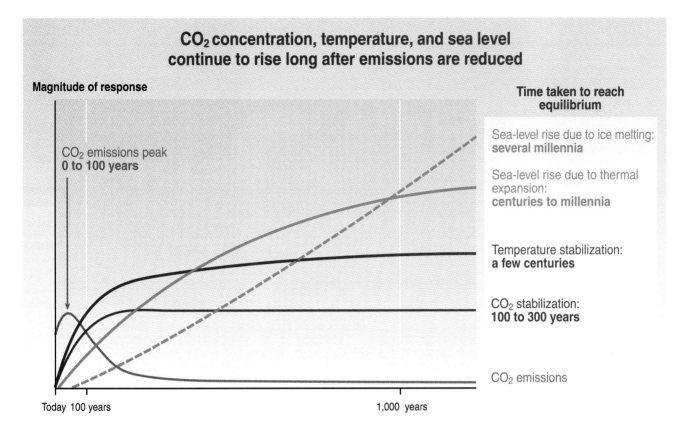

CO₂ concentration, temperature, and sea level continue to rise long after emissions are reduced

Magnitude of response

CO₂ emissions peak
0 to 100 years

Time taken to reach equilibrium

Sea-level rise due to ice melting:
several millennia

Sea-level rise due to thermal expansion:
centuries to millennia

Temperature stabilization:
a few centuries

CO₂ stabilization:
100 to 300 years

CO₂ emissions

Today 100 years   1,000 years

Figure SPM-5: After CO₂ emissions are reduced and atmospheric concentrations stabilize, surface air temperature continues to rise slowly for a century or more. Thermal expansion of the ocean continues long after CO₂ emissions have been reduced, and melting of ice sheets continues to contribute to sea-level rise for many centuries. This figure is a generic illustration for stabilization at any level between 450 and 1,000 ppm, and therefore has no units on the response axis. Responses to stabilization trajectories in this range show broadly similar time courses, but the impacts become progressively larger at higher concentrations of CO₂.

Q5 Figure 5-2

a period of many human generations. The threshold for fundamental changes in the ocean circulation may be reached at a lower degree of warming if the warming is rapid rather than gradual.

Inertia in Ecological Systems

Some ecosystems show the effects of climate change quickly, while others do so more slowly. For example, coral bleaching can occur in a single exceptionally warm season, while long-lived organisms such as trees may be able to persist for decades under a changed climate, but be unable to regenerate. When subjected to climate change, including changes in the frequency of extreme events, ecosystems may be disrupted as a consequence of differences in response times of species.

Q5.8 & Q3 Table 3-2

Some carbon cycle models project the global terrestrial carbon net uptake peaks during the 21st century, then levels off or declines. The recent global net uptake of CO₂ by terrestrial ecosystems is partly the result of time lags between enhanced plant growth and plant death and decay. Current enhanced plant growth is partly due to fertilization effects of elevated CO₂ and nitrogen deposition, and changes in climate and land-use practices. The uptake will decline as forests reach maturity, fertilization effects saturate, and decomposition catches up with growth. Climate change is likely to further reduce net terrestrial carbon uptake globally. Although warming reduces the uptake of CO₂ by the ocean, the oceanic carbon sink is projected to persist under rising atmospheric CO₂, at least for the 21st century. Movement of carbon from the surface to the deep ocean takes centuries, and its equilibration there with ocean sediments takes millennia.

Q5.6-7

Inertia in Socio-Economic Systems

Unlike the climate and ecological systems, inertia in human systems is not fixed; it can be changed by policies and the choices made by individuals. The capacity for implementing climate change policies depends on the interaction between social and economic structures and values, institutions, technologies, and established infrastructure. The combined system generally evolves relatively slowly. It can respond quickly under pressure, although sometimes at high cost (e.g., if capital equipment is prematurely retired). If change is slower, there may be lower costs due to technological advancement or because capital equipment value is fully depreciated. There is typically a delay of years to decades between perceiving a need to respond to a major challenge, planning, researching and developing a solution, and implementing it. Anticipatory action, based on informed judgment, can improve the chance that appropriate technology is available when needed.

 Q5.10-13

The development and adoption of new technologies can be accelerated by technology transfer and supportive fiscal and research policies. Technology replacement can be delayed by "locked-in" systems that have market advantages arising from existing institutions, services, infrastructure, and available resources. Early deployment of rapidly improving technologies allows learning-curve cost reductions.

 Q5.10 & Q5.22

Policy Implications of Inertia

Inertia and uncertainty in the climate, ecological, and socio-economic systems imply that safety margins should be considered in setting strategies, targets, and time tables for avoiding dangerous levels of interference in the climate system. Stabilization target levels of, for instance, atmospheric CO_2 concentration, temperature, or sea level may be affected by:
- The inertia of the climate system, which will cause climate change to continue for a period after mitigation actions are implemented
- Uncertainty regarding the location of possible thresholds of irreversible change and the behavior of the system in their vicinity
- The time lags between adoption of mitigation goals and their achievement.

Similarly, adaptation is affected by the time lags involved in identifying climate change impacts, developing effective adaptation strategies, and implementing adaptive measures.

 Q5.18-20 & Q5.23

Inertia in the climate, ecological, and socio-economic systems makes adaptation inevitable and already necessary in some cases, and inertia affects the optimal mix of adaptation and mitigation strategies. Inertia has different consequences for adaptation than for mitigation—with adaptation being primarily oriented to address localized impacts of climate change, while mitigation aims to address the impacts on the climate system. These consequences have bearing on the most cost-effective and equitable mix of policy options. Hedging strategies and sequential decision making (iterative action, assessment, and revised action) may be appropriate responses to the combination of inertia and uncertainty. In the presence of inertia, well-founded actions to adapt to or mitigate climate change are more effective, and in some circumstances may be cheaper, if taken earlier rather than later.

Q5.18-21

The pervasiveness of inertia and the possibility of irreversibility in the interacting climate, ecological, and socio-economic systems are major reasons why anticipatory adaptation and mitigation actions are beneficial. A number of opportunities to exercise adaptation and mitigation options may be lost if action is delayed.

 Q5.24

Question 6

a) How does the extent and timing of the introduction of a range of emissions reduction actions determine and affect the rate, magnitude, and impacts of climate change, and affect the global and regional economy, taking into account the historical and current emissions?

b) What is known from sensitivity studies about regional and global climatic, environmental, and socio-economic consequences of stabilizing the atmospheric concentrations of greenhouse gases (in carbon dioxide equivalents), at a range of levels from today's to double that level or more, taking into account to the extent possible the effects of aerosols? For each stabilization scenario, including different pathways to stabilization, evaluate the range of costs and benefits, relative to the range of scenarios considered in Question 3, in terms of:
- Projected changes in atmospheric concentrations, climate, and sea level, including changes beyond 100 years
- Impacts and economic costs and benefits of changes in climate and atmospheric composition on human health, diversity and productivity of ecological systems, and socio-economic sectors (particularly agriculture and water)
- The range of options for adaptation, including the costs, benefits, and challenges
- The range of technologies, policies, and practices that could be used to achieve each of the stabilization levels, with an evaluation of the national and global costs and benefits, and an assessment of how these costs and benefits would compare, either qualitatively or quantitatively, to the avoided environmental harm that would be achieved by the emissions reductions
- Development, sustainability, and equity issues associated with impacts, adaptation, and mitigation at a regional and global level.

The projected rate and magnitude of warming and sea-level rise can be lessened by reducing greenhouse gas emissions. Q6.2

The greater the reductions in emissions and the earlier they are introduced, the smaller and slower the projected warming and the rise in sea levels. Future climate change is determined by historic, current, and future emissions. Differences in projected temperature changes between scenarios that include greenhouse gas emission reductions and those that do not tend to be small for the first few decades but grow with time if the reductions are sustained. Q6.3

Reductions in greenhouse gas emissions and the gases that control their concentration would be necessary to stabilize radiative forcing. For example, for the most important anthropogenic greenhouse gas, carbon cycle models indicate that stabilization of atmospheric CO_2 concentrations at 450, 650, or 1,000 ppm would require global anthropogenic CO_2 emissions to drop below the year 1990 levels, within a few decades, about a century, or about 2 centuries, respectively, and continue to decrease steadily thereafter (see Figure SPM-6). These models illustrate that emissions would peak in about 1 to 2 decades (450 ppm) and roughly a century (1,000 ppm) from the present. Eventually CO_2 emissions would need to decline to a very small fraction of current emissions. The benefits of different stabilization levels are discussed later in Question 6 and the costs of these stabilization levels are discussed in Question 7. Q6.4

There is a wide band of uncertainty in the amount of warming that would result from any stabilized greenhouse gas concentration. This results from the factor of three Q6.5

uncertainty in the sensitivity of climate to increases in greenhouse gases.[4] Figure SPM-7 shows eventual CO_2 stabilization levels and the corresponding range of temperature change estimated to be realized in 2100 and at equilibrium.

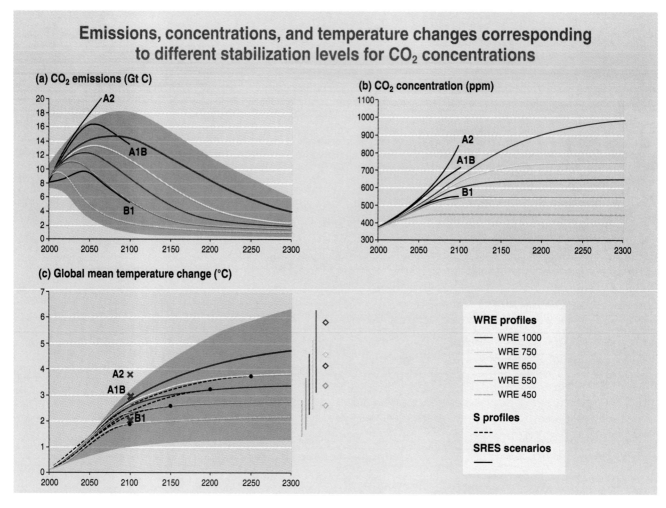

Emissions, concentrations, and temperature changes corresponding to different stabilization levels for CO_2 concentrations

Figure SPM-6: Stabilizing CO_2 concentrations would require substantial reductions of emissions below current levels and would slow the rate of warming. Q6 Figure 6-1

a) *CO_2 emissions*: The time paths of CO_2 emissions that would lead to stabilization of the concentration of CO_2 in the atmosphere at various levels are estimated for the WRE stabilization profiles using carbon cycle models. The shaded area illustrates the range of uncertainty.

b) *CO_2 concentrations*: The CO_2 concentrations specified for the WRE profiles are shown.

c) *Global mean temperature changes*: Temperature changes are estimated using a simple climate model for the WRE stabilization profiles. Warming continues after the time at which the CO_2 concentration is stabilized (indicated by black spots), but at a much diminished rate. It is assumed that emissions of gases other than CO_2 follow the SRES A1B projection until the year 2100 and are constant thereafter. This scenario was chosen as it is in the middle of the range of SRES scenarios. The dashed lines show the temperature changes projected for the S profiles (not shown in panels (a) or (b)). The shaded area illustrates the effect of a range of climate sensitivity across the five stabilization cases. The colored bars on the righthand side show uncertainty for each stabilization case at the year 2300. The diamonds on the righthand side show the average equilibrium (very long-term) warming for each CO_2 stabilization level. Also shown for comparison are CO_2 emissions, concentrations, and temperature changes for three of the SRES scenarios.

[4] The equilibrium global mean temperature response to doubling atmospheric CO_2 is often used as a measure of climate sensitivity. The temperatures shown in Figures SPM-6 and SPM-7 are derived from a simple model calibrated to give the same response as a number of complex models that have climate sensitivities ranging from 1.7 to 4.2°C. This range is comparable to the commonly accepted range of 1.5 to 4.5°C.

Emission reductions that would eventually stabilize the atmospheric concentration of CO_2 at a level below 1,000 ppm, based on profiles shown in Figure SPM-6, and assuming that emissions of gases other than CO_2 follow the SRES A1B projection until the year 2100 and are constant thereafter, are estimated to limit global mean temperature increase to 3.5°C or less through the year 2100. Global average surface temperature is estimated to increase 1.2 to 3.5°C by the year 2100 for profiles that eventually stabilize the concentration of CO_2 at levels from 450 to 1,000 ppm. Thus, although all of the CO_2 concentration stabilization profiles analyzed would prevent, during the 21st century, much of the upper end of the SRES projections of warming (1.4 to 5.8°C by the year 2100), it should be noted that for most of the profiles the concentration of CO_2 would continue to rise beyond the year 2100. The equilibrium temperature rise would take many centuries to reach, and ranges from 1.5 to 3.9°C above the year 1990 levels for stabilization at 450 ppm, and 3.5 to 8.7°C above the year 1990 levels for stabilization at 1,000 ppm.[5] Furthermore, for a specific temperature stabilization target there is a very wide range of uncertainty associated with the required stabilization level of greenhouse gas concentrations (see Figure SPM-7). The level at which CO_2 concentration is required to be stabilized for a given temperature target also depends on the levels of the non-CO_2 gases.

Q6.6

Sea level and ice sheets would continue to respond to warming for many centuries after greenhouse gas concentrations have been stabilized. The projected range of sea-level rise due to thermal expansion at equilibrium is 0.5 to 2 m for an increase in CO_2 concentration from the pre-industrial level of 280 to 560 ppm and 1 to 4 m for an increase in CO_2 concentration from 280 to 1,120 ppm. The observed rise over the 20th century was 0.1 to 0.2 m. The projected rise would be larger if the effect of increases in other greenhouse gas concentrations were to be taken into account. There are other contributions to sea-level rise over time scales of centuries to millennia. Models assessed in the TAR project sea-level rise of several meters from polar ice sheets (see Question 4) and land ice even for stablization levels of 550 ppm CO_2-equivalent.

Q6.8

Reducing emissions of greenhouse gases to stabilize their atmospheric concentrations would delay and reduce damages caused by climate change.

Q6.9

Greenhouse gas emission reduction (mitigation) actions would lessen the pressures on natural and human systems from climate change. Slower rates of increase in global mean temperature and sea level would allow more time for adaptation. Consequently, mitigation actions are expected to delay and reduce damages caused by climate change and thereby generate environmental and socio-economic benefits. Mitigation actions and their associated costs are assessed in the response to Question 7.

Q6.10

Mitigation actions to stabilize atmospheric concentrations of greenhouse gases at lower levels would generate greater benefits in terms of less damage. Stabilization at lower levels reduces the risk of exceeding temperature thresholds in biophysical systems where these exist. Stabilization of CO_2 at, for example, 450 ppm is estimated to yield an increase in global mean temperature in the year 2100 that is about 0.75 to 1.25°C less than is estimated for stabilization at 1,000 ppm (see Figure SPM-7). At equilibrium the difference is about 2 to 5°C. The geographical extent of the damage to or loss of natural systems, and the number of systems affected, which increase with the magnitude and rate of climate change, would be lower for a lower stabilization level. Similarly, for a lower stabilization level the severity of impacts from climate extremes is expected to be less, fewer regions would suffer adverse net market sector impacts, global aggregate impacts would be smaller, and risks of large-scale, high-impact events would be reduced.

Q6.11

---

[5] For all these scenarios, the contribution to the equilibrium warming from other greenhouse gases and aerosols is 0.6°C for a low climate sensitivity and 1.4°C for a high climate sensitivity. The accompanying increase in radiative forcing is equivalent to that occurring with an additional 28% in the final CO_2 concentrations.

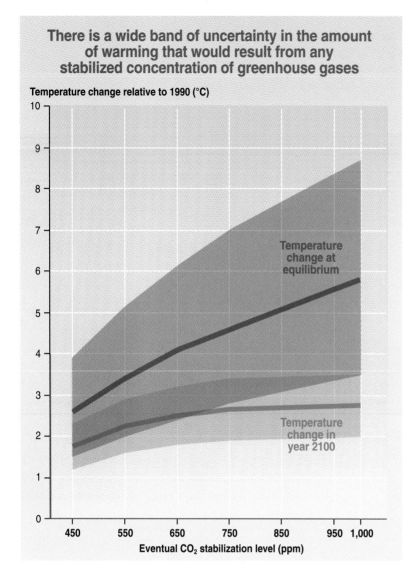

There is a wide band of uncertainty in the amount of warming that would result from any stabilized concentration of greenhouse gases

Temperature change relative to 1990 (°C)

Temperature change at equilibrium

Temperature change in year 2100

Eventual CO$_2$ stabilization level (ppm)

Figure SPM-7: Stabilizing CO$_2$ concentrations would lessen warming but by an uncertain amount. Temperature changes compared to year 1990 in (a) year 2100 and (b) at equilibrium are estimated using a simple climate model for the WRE profiles as in Figure SPM-6. The lowest and highest estimates for each stabilization level assume a climate sensitivity of 1.7 and 4.2°C, respectively. The center line is an average of the lowest and highest estimates.

 Q6 Figure 6-2

Comprehensive, quantitative estimates of the benefits of stabilization at various levels of atmospheric concentrations of greenhouse gases do not yet exist. Advances have been made in understanding the qualitative character of the impacts of climate change. Because of uncertainty in climate sensitivity, and uncertainty about the geographic and seasonal patterns of projected changes in temperatures, precipitation, and other climate variables and phenomena, the impacts of climate change cannot be uniquely determined for individual emission scenarios. There are also uncertainties about key processes and sensitivities and adaptive capacities of systems to changes in climate. In addition, impacts such as the changes in the composition and function of ecological systems, species extinction, and changes in human health, and disparity in the distribution of impacts across different populations, are not readily expressed in monetary or other common units. Because of these limitations, the benefits of different greenhouse gas emission reduction actions, including actions to stabilize greenhouse gas concentrations at selected levels, are incompletely characterized and cannot be compared directly to mitigation costs for the purpose of estimating the net economic effects of mitigation.

Q6.12

Adaptation is a necessary strategy at all scales to complement climate change mitigation efforts. Together they can contribute to sustainable development objectives.

 Q6.13

Adaptation can complement mitigation in a cost-effective strategy to reduce climate change risks. Reductions of greenhouse gas emissions, even stabilization of their concentrations in the atmosphere at a low level, will neither altogether prevent climate change or sea-level rise nor altogether prevent their impacts. Many reactive adaptations will occur in response to the changing climate and rising seas and some have already occurred. In addition, the development of planned adaptation strategies to address risks and utilize opportunities can complement mitigation actions to lessen climate change impacts. However, adaptation would entail costs and cannot prevent all damages. The costs of adaptation can be lessened by mitigation actions that will reduce and slow the climate changes to which systems would otherwise be exposed.

 Q6.14-15

The impact of climate change is projected to have different effects within and between countries. The challenge of addressing climate change raises an important issue of equity. Mitigation and adaptation actions can, if appropriately designed, advance sustainable development and equity both within and across countries and between generations. Reducing the projected increase in climate extremes is expected to benefit all countries, particularly developing countries, which are considered to be more vulnerable to climate change than developed countries. Mitigating climate change would also lessen the risks to future generations from the actions of the present generation.

Q6.16-18

Question 7

Q7

What is known about the potential for, and costs and benefits of, and time frame for reducing greenhouse gas emissions?
- What would be the economic and social costs and benefits and equity implications of options for policies and measures, and the mechanisms of the Kyoto Protocol, that might be considered to address climate change regionally and globally?
- What portfolios of options of research and development, investments, and other policies might be considered that would be most effective to enhance the development and deployment of technologies that address climate change?
- What kind of economic and other policy options might be considered to remove existing and potential barriers and to stimulate private- and public-sector technology transfer and deployment among countries, and what effect might these have on projected emissions?
- How does the timing of the options contained in the above affect associated economic costs and benefits, and the atmospheric concentrations of greenhouse gases over the next century and beyond?

There are many opportunities, including technological options, to reduce near-term emissions, but barriers to their deployment exist.

 Q7.2-7

Significant technical progress relevant to the potential for greenhouse gas emission reductions has been made since the SAR in 1995, and has been faster than anticipated. Net emissions reductions could be achieved through a portfolio of technologies (e.g., more efficient conversion in production and use of energy, shift to low- or no-greenhouse gas-emitting technologies, carbon removal and storage, and improved land use, land-use change, and forestry practices). Advances are taking place in a wide range of technologies at different stages of development, ranging from the market introduction of wind turbines and the rapid elimination of industrial by-product gases, to the advancement of fuel cell technology and the demonstration of underground CO_2 storage.

 Q7.3

The successful implementation of greenhouse gas mitigation options would need to overcome technical, economic, political, cultural, social, behavioral, and/or institutional barriers that prevent the full exploitation of the technological, economic, and social opportunities of these options. The potential mitigation opportunities and types of barriers vary by region and sector, and over time. This is caused by the wide variation in mitigative capacity. Most countries could benefit from innovative financing, social learning and innovation, institutional reforms, removing barriers to trade, and poverty eradication. In addition, in industrialized countries, future opportunities lie primarily in removing social and behavioral barriers; in countries with economies in transition, in price rationalization; and in developing countries, in price rationalization, increased access to data and information, availability of advanced technologies, financial resources, and training and capacity building. Opportunities for any given country, however, might be found in the removal of any combination of barriers.

 Q7.6

National responses to climate change can be more effective if deployed as a portfolio of policy instruments to limit or reduce net greenhouse gas emissions. The portfolio may include—according to national circumstances—emissions/carbon/energy taxes, tradable or non-tradable permits, land-use policies, provision and/or removal of subsidies, deposit/refund systems, technology or performance standards, energy mix requirement, product bans, voluntary agreements, government spending and investment, and support for research and development.

 Q7.7

Cost estimates by different models and studies vary for many reasons.

Q7.14-19

For a variety of reasons, significant differences and uncertainties surround specific quantitative estimates of mitigation costs. Cost estimates differ because of the (a) methodology[6] used in the analysis, and (b) underlying factors and assumptions built into the analysis. The inclusion of some factors will lead to lower estimates and others to higher estimates. Incorporating multiple greenhouse gases, sinks, induced technical change, and emissions trading[7] can lower estimated costs. Further, studies suggest that some sources of greenhouse gas emissions can be limited at no, or negative, net social cost to the extent that policies can exploit no-regrets opportunities such as correcting market imperfections, inclusion of ancillary benefits, and efficient tax revenue recycling. International cooperation that facilitates cost-effective emissions reductions can lower mitigation costs. On the other hand, accounting for potential short-term macro shocks to the economy, constraints on the use of domestic and international market mechanisms, high transaction costs, inclusion of ancillary costs, and ineffective tax recycling measures can increase estimated costs. Since no analysis incorporates all relevant factors affecting mitigation costs, estimated costs may not reflect the actual costs of implementing mitigation actions.

 Q7.14 & Q7.20

Studies examined in the TAR suggest substantial opportunities for lowering mitigation costs.

Q7.15-16

Bottom-up studies indicate that substantial low cost mitigation opportunities exist. According to bottom-up studies, global emissions reductions of $1.9-2.6\,\mathrm{Gt}\,C_{eq}$ (gigatonnes of carbon equivalent), and $3.6-5.0\,\mathrm{Gt}\,C_{eq}$ per year[8] could be achieved by the years 2010 and 2020, respectively. Half of these potential emissions reductions could be achieved by the year 2020 with direct benefits (energy saved) exceeding direct costs (net capital, operating, and maintenance costs), and the other half at a net direct cost of up to US$100 per t C_{eq} (at 1998 prices). These net direct cost estimates

 Q7.15 & Q7 Table 7-1

[6] The SAR described two categories of approaches to estimating costs: bottom-up approaches, which build up from assessments of specific technologies and sectors, and top-down modeling studies, which proceed from macro-economic relationships. See Box 7-1 in the underlying report.

[7] A market-based approach to achieving environmental objectives that allows those reducing greenhouse gas emissions, below what is required, to use or trade the excess reductions to offset emissions at another source inside or outside the country. Here the term is broadly used to include trade in emission allowances, and project-based collaboration.

[8] The emissions reduction estimates are with reference to a baseline trend that is similar in magnitude to the SRES B2 scenario.

are derived using discount rates in the range of 5 to 12%, consistent with public sector discount rates. Private internal rates of return vary greatly, and are often significantly higher, affecting the rate of adoption of these technologies by private entities. Depending on the emissions scenario this could allow global emissions to be reduced below year 2000 levels in 2010–2020 at these net direct cost estimates. Realizing these reductions involves additional implementation costs, which in some cases may be substantial, the possible need for supporting policies, increased research and development, effective technology transfer, and overcoming other barriers. The various global, regional, national, sector, and project studies assessed in the WGIII TAR have different scopes and assumptions. Studies do not exist for every sector and region.

Forests, agricultural lands, and other terrestrial ecosystems offer significant carbon mitigation potential. Conservation and sequestration of carbon, although not necessarily permanent, may allow time for other options to be further developed and implemented. Biological mitigation can occur by three strategies: (a) conservation of existing carbon pools, (b) sequestration by increasing the size of carbon pools,[9] and (c) substitution of sustainably produced biological products. The estimated global potential of biological mitigation options is on the order of 100 Gt C (cumulative) by year 2050, equivalent to about 10 to 20% of projected fossil-fuel emissions during that period, although there are substantial uncertainties associated with this estimate. Realization of this potential depends upon land and water availability as well as the rates of adoption of land management practices. The largest biological potential for atmospheric carbon mitigation is in subtropical and tropical regions. Cost estimates reported to date for biological mitigation vary significantly from US$0.1 to about US$20 per t C in several tropical countries and from US$20 to US$100 per t C in non-tropical countries. Methods of financial analyses and carbon accounting have not been comparable. Moreover, the cost calculations do not cover, in many instances, *inter alia*, costs for infrastructure, appropriate discounting, monitoring, data collection and implementation costs, opportunity costs of land and maintenance, or other recurring costs, which are often excluded or overlooked. The lower end of the range is assessed to be biased downwards, but understanding and treatment of costs is improving over time. Biological mitigation options may reduce or increase non-CO_2 greenhouse gas emissions.

 Q7.4 & Q7.16

The cost estimates for Annex B countries to implement the Kyoto Protocol vary between studies and regions, and depend strongly, among others, upon the assumptions regarding the use of the Kyoto mechanisms, and their interactions with domestic measures (see Figure SPM-8 for comparison of regional Annex II mitigation costs). The great majority of global studies reporting and comparing these costs use international energy-economic models. Nine of these studies suggest the following GDP impacts. In the absence of emissions trade between Annex B countries, these studies show reductions in projected GDP[10] of about 0.2 to 2% in the year 2010 for different Annex II regions. With full emissions trading between Annex B countries, the estimated reductions in the year 2010 are between 0.1 and 1.1% of projected GDP. The global modeling studies reported above show national marginal costs to meet the Kyoto targets from about US$20 up to US$600 per t C without trading, and a range from about US$15 up to US$150 per t C with Annex B trading. For most economies-in-transition countries, GDP effects range from negligible to a several percent increase. However, for some economies-in-transition countries, implementing the Kyoto Protocol will have similar impact on GDP as for Annex II countries. At the time of these studies, most models did not include sinks, non-CO_2 greenhouse gases, the Clean Development Mechanism (CDM), negative cost options,

 Q7.17-18

---

[9] Changing land use could influence atmospheric CO_2 concentration. Hypothetically, if all of the carbon released by historical land-use changes could be restored to the terrestrial biosphere over the course of the century (e.g., by reforestation), CO_2 concentration would be reduced by 40 to 70 ppm.

[10] The calculated GDP reductions are relative to each model's projected GDP baseline. The models evaluated only reductions in CO_2. In contrast, the estimates cited from the bottom-up analyses above included all greenhouse gases. Many metrics can be used to present costs. For example, if the annual costs to developed countries associated with meeting Kyoto targets with full Annex B trading are in the order of 0.5% of GDP, this represents US$125 billion (1,000 million) per year, or US$125 per person per year by 2010 in Annex II (SRES assumptions). This corresponds to an impact on economic growth *rates* over 10 years of less than 0.1 percentage point.

Projections of GDP losses and marginal cost in Annex II countries in the year 2010 from global models

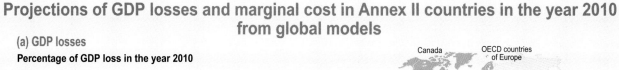

(a) GDP losses

Percentage of GDP loss in the year 2010

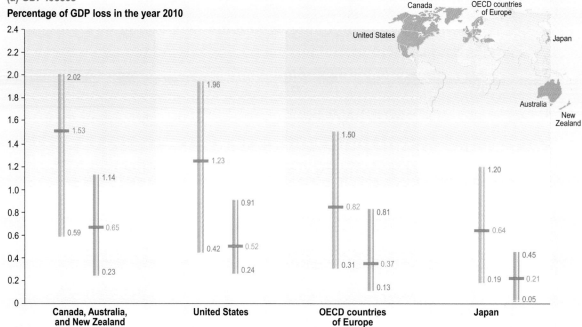

(b) Marginal cost

1990 US$ per t C

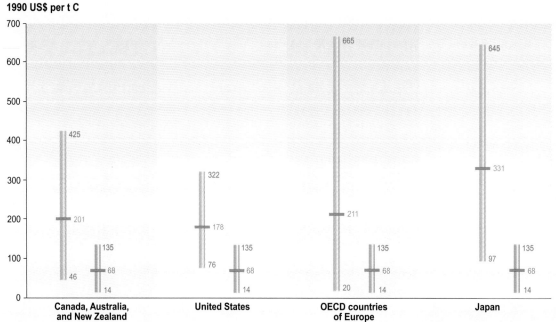

Range of outcomes for two scenarios

Absence of international trade in carbon emissions rights: each region must take the prescribed reduction

Full Annex B trading of carbon emissions rights permitted

The three numbers on each bar represent the highest, median, and lowest projections from the set of models.

Figure SPM-8: Projections of GDP losses and marginal costs in Annex II countries in the year 2010 from global models: (a) GDP losses and (b) marginal costs. The reductions in projected GDP are for the year 2010 relative to the models' reference case GDP. These estimates are based on results from nine modeling teams that participated in an Energy Modeling Forum study. The projections reported in the figure are for four regions that constitute Annex II. The models examined two scenarios. In the first, each region makes the prescribed reduction with only domestic trading in carbon emissions. In the second, Annex B trading is permitted, and thereby marginal costs are equal across regions. For the key factors, assumptions, and uncertainties underlying the studies, see Table 7-3 and Box 7-1 in the underlying report.

Q7.18-19

ancillary benefits, or targeted revenue recycling, the inclusion of which will reduce estimated costs. On the other hand, these models make assumptions which underestimate costs because they assume full use of emissions trading without transaction costs, both within and among Annex B countries, that mitigation responses would be perfectly efficient and that economies begin to adjust to the need to meet Kyoto targets between 1990 and 2000. The cost reductions from Kyoto mechanisms may depend on the details of implementation, including the compatibility of domestic and international mechanisms, constraints, and transaction costs.

Emission constraints on Annex I countries have well-established, albeit varied, "spill-over" effects[11] on non-Annex I countries. Analyses report reductions in both projected GDP and reductions in projected oil revenues for oil-exporting, non-Annex I countries. The study reporting the lowest costs shows reductions of 0.2% of projected GDP with no emissions trading, and less than 0.05% of projected GDP with Annex B emissions trading in the year 2010.[12] The study reporting the highest costs shows reductions of 25% of projected oil revenues with no emissions trading, and 13% of projected oil revenues with Annex B emissions trading in the year 2010. These studies do not consider policies and measures other than Annex B emissions trading, that could lessen the impacts on non-Annex I, oil-exporting countries. The effects on these countries can be further reduced by removal of subsidies for fossil fuels, energy tax restructuring according to carbon content, increased use of natural gas, and diversification of the economies of non-Annex I, oil-exporting countries. Other non-Annex I countries may be adversely affected by reductions in demand for their exports to Organisation for Economic Cooperation and Development (OECD) nations and by the price increase of those carbon-intensive and other products they continue to import. These other non-Annex I countries may benefit from the reduction in fuel prices, increased exports of carbon-intensive products, and the transfer of environmentally sound technologies and know-how. The possible relocation of some carbon-intensive industries to non-Annex I countries and wider impacts on trade flows in response to changing prices may lead to carbon leakage[13] on the order of 5–20%.

 Q7.19

Technology development and diffusion are important components of cost-effective stabilization.

 Q7.9-12 & Q7.23

Development and transfer of environmentally sound technologies could play a critical role in reducing the cost of stabilizing greenhouse gas concentrations. Transfer of technologies between countries and regions could widen the choice of options at the regional level. Economies of scale and learning will lower the costs of their adoption. Through sound economic policy and regulatory frameworks, transparency, and political stability, governments could create an enabling environment for private- and public-sector technology transfers. Adequate human and organizational capacity is essential at every stage to increase the flow, and improve the quality, of technology transfer. In addition, networking among private and public stakeholders, and focusing on products and techniques with multiple ancillary benefits, that meet or adapt to local development needs and priorities, is essential for most effective technology transfers.

 Q7.9-12 & Q7.23

Lower emissions scenarios require different patterns of energy resource development and an increase in energy research and development to assist accelerating the development and deployment of advanced environmentally sound energy technologies. Emissions of CO_2 due to fossil-fuel burning are virtually certain to be the dominant influence on the trend of atmospheric CO_2 concentration during the 21st century. Resource data assessed in the TAR may imply a change in the energy mix and the introduction of new sources of energy during the 21st century. The choice of energy mix and associated technologies and investments—either more in the direction of exploitation of unconventional oil and gas resources, or in the direction of

 Q7.27

---

[11] These spill-over effects incorporate only economic effects, not environmental effects.

[12] These estimated costs can be expressed as differences in GDP growth rates over the period 2000–2010. With no emissions trading, GDP growth rate is reduced by 0.02 percentage points per year; with Annex B emissions trading, growth rate is reduced by less than 0.005 percentage points per year.

[13] Carbon leakage is defined here as the increase in emissions in non-Annex B countries due to implementation of reductions in Annex B, expressed as a percentage of Annex B reductions.

non-fossil energy sources or fossil energy technology with carbon capture and storage—will determine whether, and if so, at what level and cost, greenhouse concentrations can be stabilized.

Both the pathway to stabilization and the stabilization level itself are key determinants of mitigation costs.[14]

Q7.24-25

The pathway to meeting a particular stabilization target will have an impact on mitigation cost (see Figure SPM-9). A gradual transition away from the world's present energy system towards a less carbon-emitting economy minimizes costs associated with premature retirement of existing capital stock and provides time for technology development, and avoids premature lock-in to early versions of rapidly developing low-emission technology. On the other hand, more rapid near-term action would increase flexibility in moving towards stabilization, decrease environmental and human risks and the costs associated with projected changes in climate, may stimulate more rapid deployment of existing low-emission technologies, and provide strong near-term incentives to future technological changes.

Q7.24

Studies show that the costs of stabilizing CO_2 concentrations in the atmosphere increase as the concentration stabilization level declines. Different baselines can have a strong influence on absolute costs (see Figure SPM-9). While there is a moderate increase in the costs when passing from a 750 to a 550 ppm concentration stabilization level, there is a larger increase in costs passing from 550 to 450 ppm unless the emissions in the baseline scenario are very low. Although model projections indicate long-term global growth paths of GDP are not significantly affected by mitigation actions towards stabilization, these do not show the larger variations that occur over some shorter time periods, sectors, or regions. These studies did not incorporate carbon sequestration and did not examine the possible effect of more ambitious targets on induced technological change. Also, the issue of uncertainty takes on increasing importance as the time frame is expanded.

Q7.25

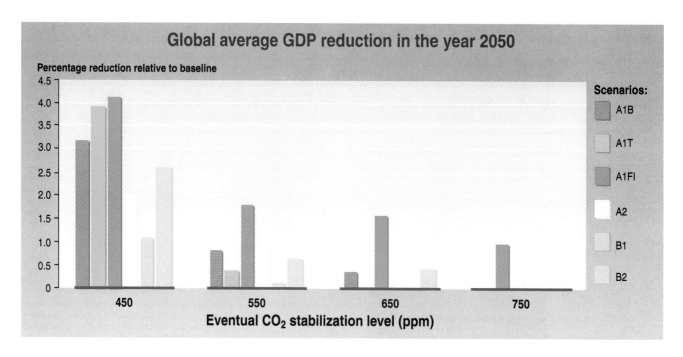

Figure SPM-9: Indicative relationship in the year 2050 between the relative GDP reduction caused by mitigation activities, the SRES scenarios, and the stabilization level. The reduction in GDP tends to increase with the stringency of the stabilization level, but the costs are very sensitive to the choice of the baseline scenario. These projected mitigation costs do not take into account potential benefits of avoided climate change.

Q7.25

[14] See Question 6 for discussion of impacts of climate change.

Question 8

What is known about the interactions between projected human-induced changes in climate and other environmental issues (e.g., urban air pollution, regional acid deposition, loss of biological diversity, stratospheric ozone depletion, and desertification and land degradation)? What is known about environmental, social, and economic costs and benefits and implications of these interactions for integrating climate change response strategies in an equitable manner into broad sustainable development strategies at the local, regional, and global scales?

Local, regional, and global environmental issues are inextricably linked and affect sustainable development. Therefore, there are synergistic opportunities to develop more effective response options to these environmental issues that enhance benefits, reduce costs, and more sustainably meet human needs.

Q8.1-2

Meeting human needs in many instances is causing environmental degradation, which in turn threatens the ability to meet present and future needs. For example, increased agricultural production can be achieved through increased use of nitrogenous fertilizers, irrigation, or the conversion of natural grasslands and forests to croplands. However, these changes can affect the Earth's climate through the release of greenhouse gases, lead to land degradation through erosion and salinization of soils, and contribute to the loss of biodiversity and reduction of carbon sequestration through the conversion and fragmentation of natural ecological systems. Agricultural productivity can in turn be adversely affected by changes in climate, especially in the tropics and subtropics, loss of biodiversity and changes at the genetic and species level, and land degradation through loss of soil fertility. Many of these changes adversely affect food security and disproportionately impact the poor.

Q8.3 & Q8.15

The primary factors underlying anthropogenic climate change are similar to those for most environmental and socio-economic issues—that is, economic growth, broad technological changes, life style patterns, demographic shifts (population size, age structure, and migration), and governance structures. These can give rise to:
- Increased demand for natural resources and energy
- Market imperfections, including subsidies that lead to the inefficient use of resources and act as a barrier to the market penetration of environmentally sound technologies; the lack of recognition of the true value of natural resources; failure to appropriate for the global values of natural resources at the local level; and failure to internalize the costs of environmental degradation into the market price of a resource
- Limited availability and transfer of technology, inefficient use of technologies, and inadequate investment in research and development for the technologies of the future
- Failure to manage adequately the use of natural resources and energy.

Q8.4

Climate change affects environmental issues such as loss of biodiversity, desertification, stratospheric ozone depletion, freshwater availability, and air quality, and in turn climate change is affected by many of these issues. For example, climate change is projected to exacerbate local and regional air pollution and delay the recovery of the stratospheric ozone layer. In addition, climate change could also affect the productivity and composition of terrestrial and aquatic ecological systems, with a potential loss in both genetic and species diversity; could accelerate the rate of land degradation; and could exacerbate problems related to freshwater quantity and quality in many areas. Conversely, local and regional air pollution, stratospheric ozone depletion, changes in ecological systems, and land degradation would affect the Earth's climate by changing the sources and sinks of greenhouse gases, radiative balance of the atmosphere, and surface albedo.

Q8.5-20

The linkages among local, regional, and global environmental issues, and their relationship to meeting human needs, offer opportunities to capture synergies in developing response options and reducing vulnerabilities to climate change, although trade-offs between issues may exist. Multiple environmental and development goals can be achieved by adopting a broad range of technologies, policies, and measures that explicitly recognize the inextricable linkages among environmental problems and human needs. Addressing the need for energy, while reducing local and regional air pollution and global climate change cost-effectively, requires an interdisciplinary assessment of the synergies and trade-offs of meeting energy requirements in the most economically, environmentally, and socially sustainable manner. Greenhouse gas emissions, as well as local and regional pollutants, could be reduced through more efficient use of energy and increasing the share of lower carbon-emitting fossil fuels, advanced fossil-fuel technologies (e.g., highly efficient combined cycle gas turbines, fuel cells, and combined heat and power) and renewable energy technologies (e.g., increased use of environmentally sound biofuels, hydropower, solar, wind- and wave-power). Further, the increase of greenhouse gas concentrations in the atmosphere can be reduced also by enhanced uptake of carbon through, for example, afforestation, reforestation, slowing deforestation, and improved forest, rangeland, wetland, and cropland management, which can have favorable effects on biodiversity, food production, land, and water resources. Reducing vulnerability to climate change can often reduce vulnerability to other environmental stresses and *vice versa*. In some cases there will be trade-offs. For example, in some implementations, monoculture plantations could decrease local biodiversity.

→ Q8.21-25

The capacity of countries to adapt and mitigate can be enhanced when climate policies are integrated with national development policies including economic, social, and other environmental dimensions. Climate mitigation and adaptation options can yield ancillary benefits that meet human needs, improve well-being, and bring other environmental benefits. Countries with limited economic resources and low level of technology are often highly vulnerable to climate change and other environmental problems.

→ Q8.26-27

A great deal of interaction exists among the environmental issues that multilateral environmental agreements address, and synergies can be exploited in their implementation. Global environmental problems are addressed in a range of individual conventions and agreements, as well as a range of regional agreements. They may contain, *inter alia*, matters of common interest and similar requirements for enacting general objectives—for example, implementation plans, data collection and processing, strengthening human and infrastructural capacity, and reporting obligations. For example, although different, the Vienna Convention for the Protection of the Ozone Layer and the United Nations Framework Convention on Climate Change are scientifically interrelated because many of the compounds that cause depletion of the ozone layer are also important greenhouse gases and because some of the substitutes for the now banned ozone-depleting substances are greenhouse gases.

→ Q8.11 & Q8.28

Question 9

What are the most robust findings and key uncertainties regarding attribution of climate change and regarding model projections of:
- Future emissions of greenhouse gases and aerosols?
- Future concentrations of greenhouse gases and aerosols?
- Future changes in regional and global climate?
- Regional and global impacts of climate change?
- Costs and benefits of mitigation and adaptation options?

In this report, a ***robust finding*** for climate change is defined as one that holds under a variety of approaches, methods, models, and assumptions and one that is expected to be relatively unaffected by uncertainties. ***Key uncertainties*** in this context are those that, if reduced, may lead to new and robust findings in relation to the questions of this report. In the examples in Table SPM-3, many of the robust findings relate to the *existence* of a climate response to human activities and the sign of the response. Many of the key uncertainties are concerned with the *quantification* of the magnitude and/or timing of the response. After addressing the attribution of climate change, the table deals in order with the issues illustrated in Figure SPM-1. Figure SPM-10 illustrates some of the main robust findings regarding climate change. Table SPM-3 provides examples and is not an exhaustive list.

Table SPM-3	Robust findings and key uncertainties.[a]	
Robust Findings		*Key Uncertainties*
Observations show Earth's surface is warming. Globally, 1990s very likely warmest decade in instrumental record (Figure SPM-10b). [Q9.8] Atmospheric concentrations of main anthropogenic greenhouse gases (CO_2 (Figure SPM-10a), CH_4, N_2O, and tropospheric O_3) increased substantially since the year 1750. [Q9.10] Some greenhouse gases have long lifetimes (e.g., CO_2, N_2O, and PFCs). [Q9.10] Most of observed warming over last 50 years likely due to increases in greenhouse gas concentrations due to human activities. [Q9.8]	**Climate change and attribution**	Magnitude and character of natural climate variability. [Q9.8] Climate forcings due to natural factors and anthropogenic aerosols (particularly indirect effects). [Q9.8] Relating regional trends to anthropogenic climate change. [Q9.8 & Q9.22]
CO_2 concentrations increasing over 21st century virtually certain to be mainly due to fossil-fuel emissions (Figure SPM-10a). [Q9.11] Stabilization of atmospheric CO_2 concentrations at 450, 650, or 1,000 ppm would require global anthropogenic CO_2 emissions to drop below year 1990 levels, within a few decades, about a century, or about 2 centuries, respectively, and continue to decrease steadily thereafter to a small fraction of current emissions. Emissions would peak in about 1 to 2 decades (450 ppm) and roughly a century (1,000 ppm) from the present. [Q9.30] For most SRES scenarios, SO_2 emissions (precursor for sulfate aerosols) are lower in the year 2100 compared with year 2000. [Q9.10]	**Future emissions and concentrations of greenhouse gases and aerosols based on models and projections with the SRES and stabilization scenarios**	Assumptions underlying the wide range[b] of SRES emissions scenarios relating to economic growth, technological progress, population growth, and governance structures (lead to largest uncertainties in projections). Inadequate emission scenarios for ozone and aerosol precursors. [Q9.10] Factors in modeling of carbon cycle including effects of climate feedbacks.[b] [Q9.10]
Global average surface temperature during 21st century rising at rates very likely without precedent during last 10,000 years (Figure SPM-10b). [Q9.13] Nearly all land areas very likely to warm more than the global average, with more hot days and heat waves and fewer cold days and cold waves. [Q9.13] Rise in sea level during 21st century that will continue for further centuries. [Q9.15] Hydrological cycle more intense. Increase in globally averaged precipitation and more intense precipitation events very likely over many areas. [Q9.14] Increased summer drying and associated risk of drought likely over most mid-latitude continental interiors. [Q9.14]	**Future changes in global and regional climate based on model projections with SRES scenarios**	Assumptions associated with a wide range[c] of SRES scenarios, as above. [Q9.10] Factors associated with model projections[c], in particular climate sensitivity, climate forcing, and feedback processes especially those involving water vapor, clouds, and aerosols (including aerosol indirect effects). [Q9.16] Understanding the probability distribution associated with temperature and sea-level projections. [Q9.16] The mechanisms, quantification, time scales, and likelihoods associated with large-scale abrupt/non-linear changes (e.g., ocean thermohaline circulation). [Q9.16] Capabilities of models on regional scales (especially regarding precipitation) leading to inconsistencies in model projections and difficulties in quantification on local and regional scales. [Q9.16]

Table SPM-3	Robust findings and key uncertainties.[a] (continued)	
Robust Findings		**Key Uncertainties**
Projected climate change will have beneficial and adverse effects on both environmental and socio-economic systems, but the larger the changes and the rate of change in climate, the more the adverse effects predominate. [Q9.17] The adverse impacts of climate change are expected to fall disproportionately upon developing countries and the poor persons within countries. [Q9.20] Ecosystems and species are vulnerable to climate change and other stresses (as illustrated by observed impacts of recent regional temperature changes) and some will be irreversibly damaged or lost. [Q9.19] In some mid- to high latitudes, plant productivity (trees and some agricultural crops) would increase with small increases in temperature. Plant productivity would decrease in most regions of the world for warming beyond a few °C. [Q9.18] Many physical systems are vulnerable to climate change (e.g., the impact of coastal storm surges will be exacerbated by sea-level rise, and glaciers and permafrost will continue to retreat). [Q9.18]	**Regional and global impacts of changes in mean climate and extremes**	Reliability of local or regional detail in projections of climate change, especially climate extremes. [Q9.22] Assessing and predicting response of ecological, social (e.g., impact of vector- and water-borne diseases), and economic systems to the combined effect of climate change and other stresses such as land-use change, local pollution, etc. [Q9.22] Identification, quantification, and valuation of damages associated with climate change. [Q9.16, Q9.22, & Q9.26]
Greenhouse gas emission reduction (mitigation) actions would lessen the pressures on natural and human systems from climate change. [Q9.28] Mitigation has costs that vary between regions and sectors. Substantial technological and other opportunities exist for lowering these costs. Efficient emissions trading also reduces costs for those participating in the trading. [Q9.31 & Q9.35-36] Emissions constraints on Annex I countries have well-established, albeit varied, "spill-over" effects on non-Annex I countries. [Q9.32] National mitigation responses to climate change can be more effective if deployed as a portfolio of policies to limit or reduce net greenhouse gas emissions. [Q9.35] Adaptation has the potential to reduce adverse effects of climate change and can often produce immediate ancillary benefits, but will not prevent all damages. [Q9.24] Adaptation can complement mitigation in a cost-effective strategy to reduce climate change risks; together they can contribute to sustainable development objectives. [Q9.40] Inertia in the interacting climate, ecological, and socio-economic systems is a major reason why anticipatory adaptation and mitigation actions are beneficial. [Q9.39]	**Costs and benefits of mitigation and adaptation options**	Understanding the interactions between climate change and other environmental issues and the related socio-economic implications. [Q9.40] The future price of energy, and the cost and availability of low-emissions technology. [Q9.33-34] Identification of means to remove barriers that impede adoption of low-emission technologies, and estimation of the costs of overcoming such barriers. [Q9.35] Quantification of costs of unplanned and unexpected mitigation actions with sudden short-term effects. [Q9.38] Quantification of mitigation cost estimates generated by different approaches (e.g., bottom-up vs. top-down), including ancillary benefits, technological change, and effects on sectors and regions. [Q9.35] Quantification of adaptation costs. [Q9.25]

[a] In this report, a ***robust finding*** for climate change is defined as one that holds under a variety of approaches, methods, models, and assumptions and one that is expected to be relatively unaffected by uncertainties. ***Key uncertainties*** in this context are those that, if reduced, may lead to new and robust findings in relation to the questions of this report. This table provides examples and is not an exhaustive list.

[b] Accounting for these above uncertainties leads to a range of CO_2 concentrations in the year 2100 between about 490 and 1,250 ppm.

[c] Accounting for these above uncertainties leads to a range for globally averaged surface temperature increase, 1990-2100, of 1.4 to 5.8°C (Figure SPM-10b) and of globally averaged sea-level rise of 0.09 to 0.88 m.

Significant progress has been made in the TAR in many aspects of the knowledge required to understand climate change and the human response to it. However, there remain important areas where further work is required, in particular:

· The detection and attribution of climate change
· The understanding and prediction of regional changes in climate and climate extremes
· The quantification of climate change impacts at the global, regional, and local levels
· The analysis of adaptation and mitigation activities
· The integration of all aspects of the climate change issue into strategies for sustainable development
· Comprehensive and integrated investigations to support the judgment as to what constitutes "dangerous anthropogenic interference with the climate system."

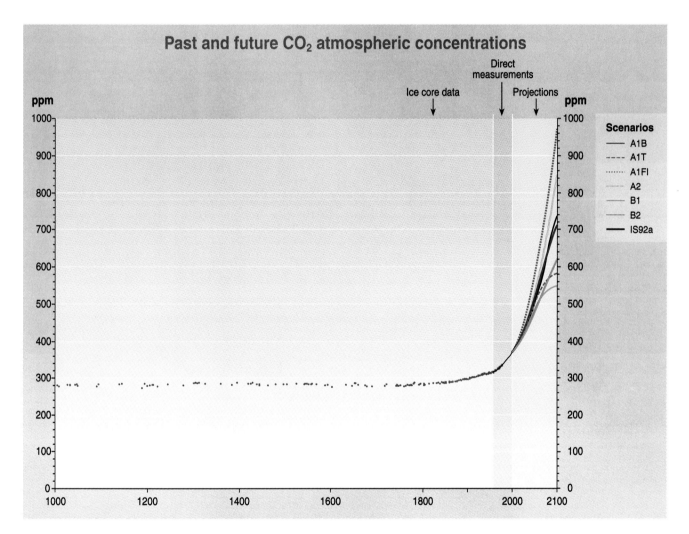

Figure SPM-10a: Atmospheric CO₂ concentration from year 1000 to year 2000 from ice core data and from direct atmospheric measurements over the past few decades. Projections of CO₂ concentrations for the period 2000 to 2100 are based on the six illustrative SRES scenarios and IS92a (for comparison with the SAR).

 Q9 Figure 9-1a

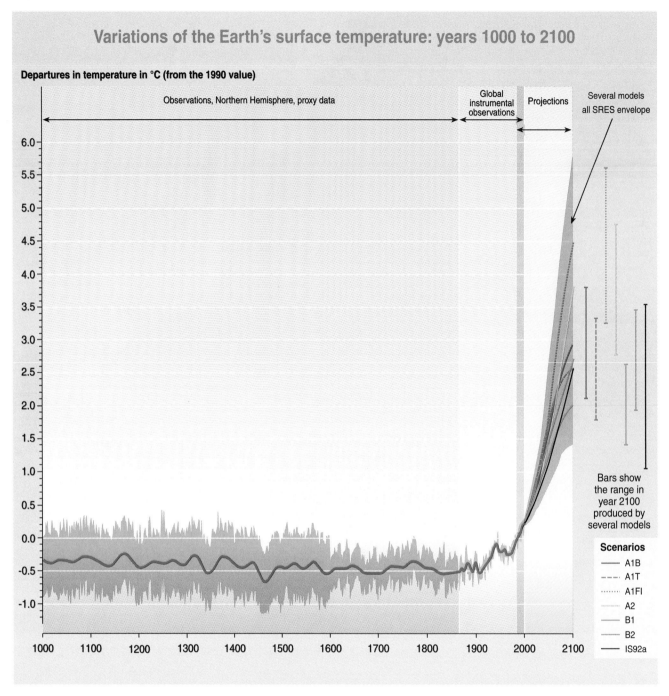

Variations of the Earth's surface temperature: years 1000 to 2100

Departures in temperature in °C (from the 1990 value)

Observations, Northern Hemisphere, proxy data

Global instrumental observations

Projections

Several models all SRES envelope

Bars show the range in year 2100 produced by several models

Scenarios
- A1B
- A1T
- A1FI
- A2
- B1
- B2
- IS92a

Figure SPM-10b: Variations of the Earth's surface temperature: years 1000 to 2100. From year 1000 to year 1860 variations in average surface temperature of the Northern Hemisphere are shown (corresponding data from the Southern Hemisphere not available) reconstructed from proxy data (tree rings, corals, ice cores, and historical records). The line shows the 50-year average, the grey region the 95% confidence limit in the annual data. From years 1860 to 2000 are shown variations in observations of globally and annually averaged surface temperature from the instrumental record; the line shows the decadal average. From years 2000 to 2100 projections of globally averaged surface temperature are shown for the six illustrative SRES scenarios and IS92a using a model with average climate sensitivity. The grey region marked "several models all SRES envelope" shows the range of results from the full range of 35 SRES scenarios in addition to those from a range of models with different climate sensitivities. The temperature scale is departure from the 1990 value; the scale is different from that used in Figure SPM-2.

 Q9 Figure 9-1b

Climate Change 2001: Synthesis Report

Synthesis Report

An Assessment of the Intergovernmental Panel on Climate Change

This underlying report, approved paragraph by paragraph at IPCC Plenary XVIII (Wembley, United Kingdom, 24-29 September 2001), represents the formally agreed statement of the IPCC concerning key findings and uncertainties contained in the Working Group contributions to the Third Assessment Report.

Based on a draft prepared by:

Core Writing Team
Robert T. Watson, Daniel L. Albritton, Terry Barker, Igor A. Bashmakov, Osvaldo Canziani, Renate Christ, Ulrich Cubasch, Ogunlade Davidson, Habiba Gitay, David Griggs, Kirsten Halsnaes, John Houghton, Joanna House, Zbigniew Kundzewicz, Murari Lal, Neil Leary, Christopher Magadza, James J. McCarthy, John F.B. Mitchell, Jose Roberto Moreira, Mohan Munasinghe, Ian Noble, Rajendra Pachauri, Barrie Pittock, Michael Prather, Richard G. Richels, John B. Robinson, Jayant Sathaye, Stephen Schneider, Robert Scholes, Thomas Stocker, Narasimhan Sundararaman, Rob Swart, Tomihiro Taniguchi, and D. Zhou

Extended Team
Q.K. Ahmad, Oleg Anisimov, Nigel Arnell, Fons Baede, Tariq Banuri, Leonard Bernstein, Daniel H. Bouille, Timothy Carter, Catrinus J. Jepma, Liu Chunzhen, John Church, Stewart Cohen, Paul Desanker, William Easterling, Chris Folland, Filippo Giorgi, Jonathan Gregory, Joanna Haigh, Hideo Harasawa, Bruce Hewitson, Jean-Charles Hourcade, Mike Hulme, Tom Karl, Pekka E. Kauppi, Rik Leemans, Anil Markandya, Luis Jose Mata, Bryant McAvaney, Anthony McMichael, Linda Mearns, Jerry Meehl, Gylvan Meira-Filho, Evan Mills, William R. Moomaw, Berrien Moore, Tsuneyuki Morita, M.J. Mwandosya, Leonard Nurse, Martin Parry, Joyce Penner, Colin Prentice, Venkatachalam Ramaswamy, Sarah Raper, Jim Salinger, Michael Scott, Roger A. Sedjo, Priyaradshi R. Shukla, Barry Smit, Joel Smith, Leena Srivastava, Ron Stouffer, Kanako Tanaka, Ferenc L. Toth, Alla Tsyban, John P. Weyant, Tom Wilbanks, Francis Zwiers, and many IPCC authors

Review Editors
Susan Barrell, Rick Bradley, Eduardo Calvo, Ian Carruthers, Oyvind Christophersen, Yuri Izrael, Eberhard Jochem, Fortunat Joos, Martin Manning, Bert Metz, Alionne Ndiaye, Buruhani Nyenzi, Ramon Pichs-Madruga, Richard Odingo, Michel Petit, Jan Pretel, Armando Ramirez, Jose Romero, John Stone, R.T.M. Sutamihardja, David Warrilow, Ding Yihui, and John Zillman

Question 1

Q1

What can scientific, technical, and socio-economic analyses contribute to the determination of what constitutes dangerous anthropogenic interference with the climate system as referred to in Article 2 of the Framework Convention on Climate Change?

Framework Convention on Climate Change, Article 2

"The ultimate objective of this Convention and any related legal instruments that the Conference of the Parties may adopt is to achieve, in accordance with the relevant provisions of the Convention, stabilization of greenhouse gas concentrations in the atmosphere at a level that would prevent dangerous anthropogenic interference with the climate system. Such a level should be achieved within a time-frame sufficient to allow ecosystems to adapt naturally to climate change, to ensure that food production is not threatened and to enable economic development to proceed in a sustainable manner."

1.1 **Natural, technical, and social sciences can provide essential information and evidence needed for decisions on what constitutes "dangerous anthropogenic interference" with the climate system. At the same time, such decisions are value judgments determined through socio-political processes, taking into account considerations such as development, equity, and sustainability, as well as uncertainties and risk.** Scientific evidence helps to reduce uncertainty and increase knowledge, and can serve as an input for considering precautionary measures.[1] Decisions are based on risk assessment, and lead to risk management choices by decision makers, about actions and policies.[2]

WGII TAR Section 2.7 & WGIII TAR Chapter 10

1.2 **The basis for determining what constitutes "dangerous anthropogenic interference" will vary among regions, depending both on the local nature and consequences of climate change impacts, and also on the adaptive capacity available to cope with climate change. It also depends upon mitigative capacity, since the magnitude and the rate of change are both important.** The consequent types of adaptation responses that will be selected depend on the effectiveness of various adaptation or mitigation responses in reducing vulnerabilities and improving the sustainability of life-support systems. There is no universally applicable best set of policies; rather, it is important to consider both the robustness of different policy measures against a range of possible future worlds, and the degree to which such climate-specific policies can be integrated with broader sustainable development policies.

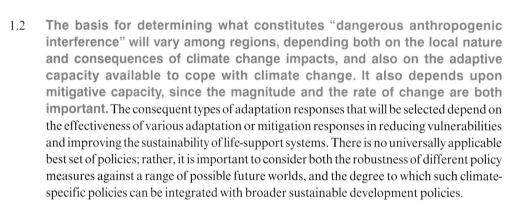
WGII TAR Chapter 18 & WGIII TAR Chapter 10

1.3 **The Third Assessment Report (TAR) provides an assessment of new scientific information and evidence as an input for policy makers in their determination of what constitutes "dangerous anthropogenic interference with the climate system"** with regard to: (1) the magnitudes and rates of changes in the climate system, (2) the ecological and socio-economic impacts of climate change, and (3) the potential for achieving a broad range of levels of concentrations through mitigation and information about how adaptation can reduce vulnerability.

WGI TAR, WGII TAR, & WGIII TAR

1.4 **With regard to the magnitudes and rates of changes in the climate system, the TAR provides scenario-based projections of future concentrations of greenhouse gases in the atmosphere, global and regional patterns of changes and rates of change in temperature, precipitation, and sea level, and changes in extreme climate events.** It also examines possibilities for abrupt and irreversible changes in ocean circulation and the major ice sheets.

WGI TAR

1.5 **The TAR reviews the biophysical and socio-economic impacts of climate change.** The TAR articulates five reasons for concern, regarding:
  · Risks to unique and threatened systems
  · Risks associated with extreme weather events
  · The distribution of impacts
  · Aggregate impacts
  · Risks of large-scale, high-impact events.
Of great significance here is an assessment of the likelihood of the critical thresholds at which natural and human systems exhibit large-scale, abrupt, or irreversible changes in their response to a changing climate. Since no single indicator (e.g., a monetary unit) captures

WGII TAR Chapter 19

---

[1] Conditions that justify the adoption of precautionary measures are described in Article 3.3 of the United Nations Framework Convention on Climate Change (UNFCCC).

[2] The risk associated with an event is most simply defined as the probability of that event, multiplied by the magnitude of its consequence. Various decision frameworks can facilitate climate risk assessment and management. These include, among others, cost-benefit analysis, cost-effectiveness analysis, multi-attribute analysis, and tolerable windows. Such techniques help to differentiate the risk levels associated with alternative futures, but in all cases the analyses are marked by considerable uncertainties.

the range of relevant risks presented by climate change, a variety of analytical approaches and criteria are required to assess impacts and facilitate decisions about risk management.

1.6   **With regard to strategies for addressing climate change, the TAR provides an assessment of the potential for achieving different levels of concentrations through mitigation and information about how adaptation can reduce vulnerability.** The causality works in both directions. Different stabilization levels result from different emission scenarios, which are connected to underlying development paths. In turn, these development paths strongly affect adaptive capacity in any region. In this way adaptation and mitigation strategies are dynamically connected with changes in the climate system and the prospects for ecosystem adaptation, food production, and sustainable economic development.

WGII TAR Chapter 18 & WGIII TAR Chapter 2

1.7   An integrated view of climate change considers the dynamics of the complete cycle of interlinked causes and effects across all sectors concerned. Figure 1-1 shows the cycle, from the underlying driving forces of population, economy, technology, and governance, through greenhouse gas and other emissions, changes in the physical climate system, biophysical and human impacts, to adaptation and mitigation, and back to the driving forces. The figure presents a schematic view of an ideal "integrated assessment" framework, in which all the parts of the climate change problem interact mutually. Changes in one part of the cycle influence other components in a dynamic manner, through multiple paths. The TAR assesses new policy-relevant information and evidence with regard to all quadrants of Figure 1-1. In particular, a new contribution has been to fill in the bottom righthand quadrant of the figure by exploring alternative development paths and their relationship to greenhouse gas emissions, and by undertaking preliminary work on the linkage between adaptation, mitigation, and development paths. However, the TAR does not achieve a fully integrated assessment of climate change, because of the incomplete state of knowledge.

WGII TAR Chapters 1 & 19, WGIII TAR Chapter 1, & SRES

1.8   **Climate change decision making is essentially a sequential process under general uncertainties.** Decision making has to deal with uncertainties including the risk of non-linear and/or irreversible changes and entails balancing the risk of either insufficient or excessive action, and involves careful consideration of the consequences (both environmental and economic), their likelihood, and society's attitude towards risk. The latter is likely to vary from country to country and from generation to generation. The relevant question is "what is the best course for the near term given the expected long-term climate change and accompanying uncertainties."

WGI TAR, WGII TAR, & WGIII TAR Section 10.1.4

1.9   **Climate change impacts are part of the larger question of how complex social, economic, and environmental subsystems interact and shape prospects for sustainable development.** There are multiple links. Economic development affects ecosystem balance and, in turn, is affected by the state of the ecosystem; poverty can be both a result and a cause of environmental degradation; material- and energy-intensive life styles and continued high levels of consumption supported by non-renewable resources and rapid population growth are not likely to be consistent with sustainable development paths; and extreme socio-economic inequality within communities and between nations may undermine the social cohesion that would promote sustainability and make policy responses more effective. At the same time, socio-economic and technology policy decisions made for non-climate-related reasons have significant implications for climate policy and climate change impacts, as well as for other environmental issues (see Question 8). In addition, critical impact thresholds and vulnerability to climate change impacts are directly connected to environmental, social, and economic conditions and institutional capacity.

WGII TAR

1.10   **As a result, the effectiveness of climate policies can be enhanced when they are integrated with broader strategies designed to make national and regional development paths more sustainable.** This occurs because of the impacts of natural

WGIII TAR Section 10.3.2

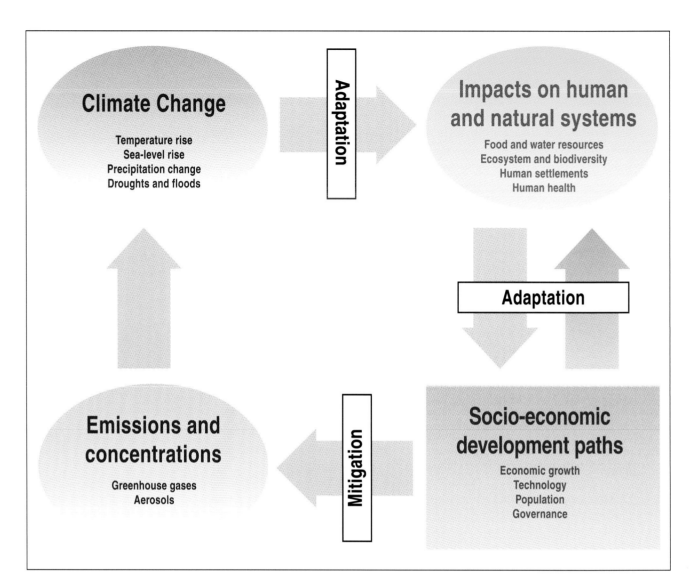

Figure 1-1: Climate change – an integrated framework. Schematic and simplified representation of an integrated assessment framework for considering anthropogenic climate change. The yellow arrows show a full clockwise cycle of cause and effect among the four quadrants shown in the figure, while the blue arrow indicates the societal response to climate change impacts. For both developed and developing countries, each **socio-economic development path** explored in the *Special Report on Emissions Scenarios* has driving forces which give rise to emissions of greenhouse gases, aerosols, and precursors—with carbon dioxide (CO_2) being the most important. The greenhouse gas **emissions accumulate in the atmosphere, changing concentrations** and disturbing the natural balances, depending on physical processes such as solar radiation, cloud formation, and rainfall. The aerosols also give rise to air pollution (e.g., acid rain) that damage human and the natural systems (not shown). The enhanced greenhouse effect will initiate **climate changes** well into the future with associated **impacts on the natural and human systems**. There is a possibility of some feedback between the changes in these systems and the climate (not shown), such as albedo effects from changing land use, and other, perhaps larger, interactions between the systems and atmospheric emissions (e.g., effects of changes in land use (again not shown)). These changes will ultimately have effects on socio-economic development paths. The development paths also have direct effects on the natural systems (shown by the anti-clockwise arrow from the development box) such as changes in land use leading to deforestation. This figure illustrates that the various dimensions of the climate change issue exist in a dynamic cycle, characterized by significant time delays. Both emissions and impacts, for example, are linked in complex ways to underlying socio-economic and technological development paths. A major contribution of the TAR has been to explicitly consider the bottom righthand domain (shown as a rectangle) by examining the relationships between greenhouse gas emissions and development paths (in SRES), and by assessing preliminary work on the linkage between adaptation, mitigation, and development paths (WGII and WGIII). However, the TAR does not achieve a fully integrated assessment of climate change, since not all components of the cycle were able to be linked dynamically. Adaptation and mitigation are shown as modifying the effects shown in the figure.

climate variation and changes, climate policy responses, and associated socio-economic development will affect the ability of countries to achieve sustainable development goals, while the pursuit of those goals will in turn affect the opportunities for, and success of, climate policies. In particular, the socio-economic and technological characteristics of different development paths will strongly affect emissions, the rate and magnitude of climate change, climate change impacts, the capability to adapt, and the capacity to mitigate climate. The *Special Report on Emissions Scenarios* (SRES, see Box 3-1) outlined multiple plausible future worlds with different characteristics, each having very different implications for the future climate and for climate policy.

1.11   **The TAR assesses available information on the timing, opportunities, costs, benefits, and impacts of various mitigation and adaptation options.** It indicates that there are opportunities for countries acting individually, or in cooperation with others, to reduce costs of mitigation and adaptation and realize benefits associated with achieving sustainable development.

WGII TAR Chapter 18, WGIII TAR Chapters 8, 9, & 10, & SRES

Question 2

Q2

What is the evidence for, causes of, and consequences of changes in the Earth's climate since the pre-industrial era?

(a) Has the Earth's climate changed since the pre-industrial era at the regional and/or global scale? If so, what part, if any, of the observed changes can be attributed to human influence and what part, if any, can be attributed to natural phenomena? What is the basis for that attribution?

(b) What is known about the environmental, social, and economic consequences of climate changes since the pre-industrial era with an emphasis on the last 50 years?

2.1   This answer focuses on classical measures of climate (e.g., temperature, precipitation, sea level, plus extreme events including floods, droughts, and storms), on other components of the Earth's climate system (e.g., greenhouse gases and aerosols, ecological systems), and on human health and socio-economic sectors. Climate *change* as defined in IPCC refers to statistically significant variations that persist for an extended period, typically decades or longer. It includes shifts in the frequency and magnitude of sporadic weather events as well as the slow continuous rise in global mean surface temperature. Thus the discussion here includes climate-weather variations on all temporal and spatial scales, ranging from brief-lived severe storms to seasonal El Niño events, decadal droughts, and century shifts in temperature and ice cover. Although short-term climate variations are considered predominantly natural at present, their impacts are discussed in this question because they represent a class of changes that may become more prevalent in a future climate perturbed by human activities (see Question 4). Attribution is used here as the process of establishing the most likely causes for the detected change with some defined level of confidence. The discussion includes both climate change that is attributable to human influence and climate change that may at present be natural but might in the future be modified through human influence (see Box 3-1).

2.2   **The Earth's climate system has demonstrably changed on both global and regional scales since the pre-industrial era, with some of these changes attributable to human activities.**

2.3   **Emissions of greenhouse gases and aerosols due to human activities continue to alter the atmosphere in ways that are expected to affect the climate (see Table 2-1).**

2.4   **Concentrations of atmospheric greenhouse gases and their radiative forcings have generally increased over the 20th century as a result of human activities.** Almost all greenhouse gases reached their highest recorded levels in the 1990s and continue to increase (see Figure 2-1). Atmospheric carbon dioxide (CO_2) and methane (CH_4) have varied substantially during glacial-interglacial cycles over the past 420,000 years, but even the largest of these earlier values are much less than their current atmospheric concentrations. In terms of radiative forcing by greenhouse gases emitted through human activity, CO_2 and CH_4 are the first and second most important, respectively. From the years 1750 to 2000, the concentration of CO_2 increased by 31±4%, and that of CH_4 rose by 151±25% (see Box 2-1 and Figure 2-1). These rates of increase are unprecedented. Fossil-fuel burning released on average 5.4 Gt C yr^{-1} during the 1980s, increasing to 6.3 Gt C yr^{-1} during the 1990s. About three-quarters of the increase in atmospheric CO_2 during the 1990s was caused by fossil-fuel burning, with land-use change including deforestation responsible for the rest. Over the 19th and much of the 20th century the terrestrial biosphere has been a net source of atmospheric CO_2, but before the end of the 20th century it had become a net sink. The increase in CH_4 can be identified with emissions from energy use, livestock, rice agriculture, and landfills. Increases in the concentrations of other greenhouse gases—particularly tropospheric ozone (O_3), the third most important—are directly attributable to fossil-fuel combustion as well as other industrial and agricultural emissions.

WGI TAR Chapters 3 & 4, & SRAGA

Box 2-1	Confidence and likelihood statements.

Where appropriate, the authors of the Third Assessment Report assigned confidence levels that represent their collective judgment in the validity of a conclusion based on observational evidence, modeling results, and theory that they have examined. The following words have been used throughout the text of the Synthesis Report to the TAR relating to WGI findings: *virtually certain* (greater than 99% chance that a result is true); *very likely* (90–99% chance); *likely* (66–90% chance); *medium likelihood* (33–66% chance); *unlikely* (10–33% chance); *very unlikely* (1–10% chance); and *exceptionally unlikely* (less than 1% chance). An explicit uncertainty range (±) is a *likely* range. Estimates of confidence relating to WGII findings are: *very high* (95% or greater), *high* (67–95%), *medium* (33–67%), *low* (5–33%), and *very low* (5% or less). No confidence levels were assigned in WGIII.

WGI TAR SPM & WGII TAR SPM

Table 2-1	20th century changes in the Earth's atmosphere, climate, and biophysical system.[a]
Indicator	**Observed Changes**
Concentration indicators	
Atmospheric concentration of CO_2	280 ppm for the period 1000–1750 to 368 ppm in year 2000 (31±4% increase). [WGI TAR Chapter 3]
Terrestrial biospheric CO_2 exchange	Cumulative source of about 30 Gt C between the years 1800 and 2000; but during the 1990s, a net sink of about 14±7 Gt C. [WG1 TAR Chapter 3 & SRLULUCF]
Atmospheric concentration of CH_4	700 ppb for the period 1000–1750 to 1,750 ppb in year 2000 (151±25% increase). [WGI TAR Chapter 4]
Atmospheric concentration of N_2O	270 ppb for the period 1000–1750 to 316 ppb in year 2000 (17±5% increase). [WGI TAR Chapter 4]
Tropospheric concentration of O_3	Increased by 35±15% from the years 1750 to 2000, varies with region. [WGI TAR Chapter 4]
Stratospheric concentration of O_3	Decreased over the years 1970 to 2000, varies with altitude and latitude. [WGI TAR Chapters 4 & 6]
Atmospheric concentrations of HFCs, PFCs, and SF_6	Increased globally over the last 50 years. [WGI TAR Chapter 4]
Weather indicators	
Global mean surface temperature	Increased by 0.6±0.2°C over the 20th century; land areas warmed more than the oceans (*very likely*). [WGI TAR Section 2.2.2.3]
Northern Hemisphere surface temperature	Increase over the 20th century greater than during any other century in the last 1,000 years; 1990s warmest decade of the millennium (*likely*). [WGI TAR Chapter 2 ES & Section 2.3.2.2]
Diurnal surface temperature range	Decreased over the years 1950 to 2000 over land: nighttime minimum temperatures increased at twice the rate of daytime maximum temperatures (*likely*). [WGI TAR Section 2.2.2.1]
Hot days / heat index	Increased (*likely*). [WGI TAR Section 2.7.2.1]
Cold / frost days	Decreased for nearly all land areas during the 20th century (*very likely*). [WGI TAR Section 2.7.2.1]
Continental precipitation	Increased by 5–10% over the 20th century in the Northern Hemisphere (*very likely*), although decreased in some regions (e.g., north and west Africa and parts of the Mediterranean). [WGI TAR Chapter 2 ES & Section 2.5.2]
Heavy precipitation events	Increased at mid- and high northern latitudes (*likely*). [WGI TAR Section 2.7.2.2]
Frequency and severity of drought	Increased summer drying and associated incidence of drought in a few areas (*likely*). In some regions, such as parts of Asia and Africa, the frequency and intensity of droughts have been observed to increase in recent decades. [WGII TAR Sections 10.1.3 & 11.1.2]

WGI TAR Chapter 5 & 6, & SRAGA Chapter 6

2.5 **The radiative forcing from the increase in anthropogenic greenhouse gases since the pre-industrial era is positive (warming) with a small uncertainty range; that from the direct effects of aerosols is negative (cooling) and smaller; whereas the negative forcing from the indirect effects of aerosols (on clouds and the hydrologic cycle) might be large but is not well quantified.** Key anthropogenic and natural factors causing a change in radiative forcing from year 1750 to year 2000 are shown in Figure 2-2, where the factors whose radiative forcing can be quantified are marked by wide, colored bars. Only some of the aerosol effects are estimated here and denoted as ranges. Other factors besides atmospheric constituents—solar irradiance and land-use change—are also shown. Stratospheric aerosols from large volcanic eruptions have led to important, but brief-lived, negative forcings (particularly the periods 1880–1920 and 1960–1994), which are not important over the time scale since the pre-industrial era and not shown. The sum of

Table 2-1	20th century changes in the Earth's atmosphere, climate, and biophysical system.[a] (continued)
Indicator	***Observed Changes***
Biological and physical indicators	
Global mean sea level	Increased at an average annual rate of 1 to 2 mm during the 20th century. [WGI TAR Chapter 11]
Duration of ice cover of rivers and lakes	Decreased by about 2 weeks over the 20th century in mid- and high latitudes of the Northern Hemisphere (*very likely*). [WGI TAR Chapter 2 ES & Section 2.2.5.5, & WGII TAR Sections 5.7 & 16.1.3.1]
Arctic sea-ice extent and thickness	Thinned by 40% in recent decades in late summer to early autumn (*likely*) and decreased in extent by 10-15% since the 1950s in spring and summer. [WGI TAR Section 2.2.5.2 & WGII TAR Section 16.1.3.1]
Non-polar glaciers	Widespread retreat during the 20th century. [WGI TAR Section 2.2.5.4 & WGII TAR Section 4.3.11]
Snow cover	Decreased in area by 10% since global observations became available from satellites in the 1960s (*very likely*). [WGI TAR Section 2.2.5.1]
Permafrost	Thawed, warmed, and degraded in parts of the polar, sub-polar, and mountainous regions. [WGI TAR Sections 2.2.5.3 & 11.2.5, & WGII TAR Section 16.1.3.1]
El Niño events	Became more frequent, persistent, and intense during the last 20 to 30 years compared to the previous 100 years. [WGI TAR Section 7.6.5]
Growing season	Lengthened by about 1 to 4 days per decade during the last 40 years in the Northern Hemisphere, especially at higher latitudes. [WGII TAR Section 5.2.1]
Plant and animal ranges	Shifted poleward and up in elevation for plants, insects, birds, and fish. [WGII TAR Sections 5.2, 5.4, 5.9, & 16.1.3.1]
Breeding, flowering, and migration	Earlier plant flowering, earlier bird arrival, earlier dates of breeding season, and earlier emergence of insects in the Northern Hemisphere. [WGII TAR Sections 5.2.1 & 5.4.3]
Coral reef bleaching	Increased frequency, especially during El Niño events. [WGII TAR Section 6.3.8]
Economic indicators	
Weather-related economic losses	Global inflation-adjusted losses rose an order of magnitude over the last 40 years (see Figure 2-7). Part of the observed upward trend is linked to socio-economic factors and part is linked to climatic factors. [WGII TAR Sections 8.2.1 & 8.2.2]

[a] This table provides examples of key observed changes and is not an exhaustive list. It includes both changes attributable to anthropogenic climate change and those that may be caused by natural variations or anthropogenic climate change. Confidence levels are reported where they are explicitly assessed by the relevant Working Group.

quantified factors in Figure 2-2 (greenhouse gases, aerosols and clouds, land-use (albedo), and solar irradiance) is positive, but this does not include the potentially large, negative forcing from aerosol indirect effects. The total change in radiative forcing since the pre-industrial era continues to be a useful tool to estimate, to a first order, the global mean surface temperature response to human and natural perturbations; however, the sum of forcings is not necessarily an indicator of the detailed aspects of the potential climate responses such as regional climate change. For the last half of the 20th century (not shown), the positive forcing due to well-mixed greenhouse gases has increased rapidly over the past 4 decades, while in contrast the sum of natural forcings has been negative over the past 2 and possibly even 4 decades.

2.6 An increasing body of observations gives a collective picture of a warming world and other changes in the climate system (see Table 2-1).

2.7 The global average surface temperature has increased from the 1860s to the year 2000, the period of instrumental record. Over the 20th century this increase was 0.6°C with a *very likely* (see Box 2-1) confidence range of 0.4–0.8°C (see Figure 2-3).

WGI TAR SPM & WGI TAR Sections 2.2.2, 2.3.2, & 2.7.2

Indicators of the human influence on the atmosphere during the industrial era

Global atmospheric concentrations of three well-mixed greenhouse gases

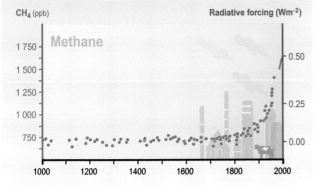

CO_2 (ppm) — Radiative forcing (Wm^{-2})

Carbon Dioxide

CH_4 (ppb) — Radiative forcing (Wm^{-2})

Methane

N_2O (ppb) — Radiative forcing (Wm^{-2})

Nitrous Oxide

Sulfate aerosols deposited in Greenland ice

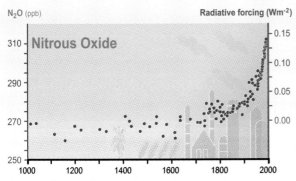

Sulfate concentration
mg SO_4^{2-} per tonne of ice

Sulfur

SO_2 emissions
from United States
and Europe
(Mt S yr^{-1})

WGI TAR Figures SPM-2, 3-2b, 4-1a, 4-1b, 4-2, & 5-4a

Figure 2-1: Records of past changes in atmospheric composition over the last millennium demonstrate the rapid rise in greenhouse gases and sulfate aerosols that is attributable primarily to industrial growth since 1750. The top three panels show increasing atmospheric concentrations of carbon dioxide (CO_2), methane (CH_4), and nitrous oxide (N_2O) over the past 1,000 years. Early sporadic data taken from air trapped in ice (symbols) matches up with continuous atmospheric observations from recent decades (solid lines). These gases are well mixed in the atmosphere, and their concentrations reflect emissions from sources throughout the globe. The estimated positive radiative forcing from these gases is indicated on the righthand scale. The lowest panel shows the concentration of sulfate in ice cores from Greenland (shown by lines for three different cores) from which the episodic effects of volcanic eruptions have been removed. Sulfate aerosols form from sulfur dioxide (SO_2) emissions, deposit readily at the surface, and are not well mixed in the atmosphere. Specifically, the increase in sulfate deposited at Greenland is attributed to SO_2 emissions from the U.S. and Europe (shown as symbols), and both show a decline in recent decades. Sulfate aerosols produce negative radiative forcing.

It is very likely that the 1990s was the warmest decade, and 1998 the warmest year, of the instrumental record. Extending the instrumental record with proxy data for the Northern Hemisphere indicates that over the past 1,000 years the 20th century increase in temperature is likely to have been the largest of any century, and the 1990s was likely the warmest decade (see Figure 2-3). Insufficient data are available in the Southern Hemisphere prior to the year 1860 to compare the recent warming with changes over the last 1,000 years. Since the year 1950, the increase in sea surface temperature is about half that of the mean land surface air temperature. During this period the nighttime daily minimum temperatures over land have increased on average by about 0.2° C per decade, about twice the corresponding

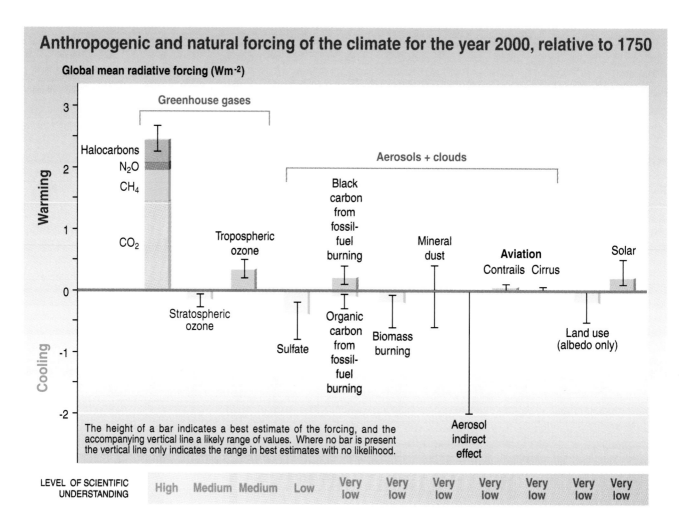

Figure 2-2: The influence of external factors on climate can be broadly compared using the concept of radiative forcing. These radiative forcings arise from changes in the atmospheric composition, alteration of surface reflectance by land use, and variation in the output of the sun. Except for solar variation, some form of human activity is linked to each. The rectangular bars represent estimates of the contributions of these forcings, some of which yield warming and some cooling. Forcing due to episodic volcanic events, which lead to a negative forcing lasting only for a few years, is not shown. The indirect effect of aerosols shown is their effect on the size and number of cloud droplets. A second indirect effect of aerosols on clouds, namely their effect on cloud lifetime, which would also lead to a negative forcing, is not shown. Effects of aviation on greenhouse gases are included in the individual bars. The vertical line about the rectangular bars indicates a range of estimates, guided by the spread in the published values of the forcings and physical understanding. Some of the forcings possess a much greater degree of certainty than others. A vertical line without a rectangular bar denotes a forcing for which no best estimate can be given owing to large uncertainties. The overall level of scientific understanding for each forcing varies considerably, as noted. Some of the radiative forcing agents are well mixed over the globe, such as CO_2, thereby perturbing the global heat balance. Others represent perturbations with stronger regional signatures because of their spatial distribution, such as aerosols. Radiative forcing continues to be a useful tool to estimate, to a first order, the relative climate impacts such as the relative global mean surface temperature response due to radiatively induced perturbations, but these global mean forcing estimates are not necessarily indicators of the detailed aspects of the potential climate responses (e.g., regional climate change).

WGI TAR SPM, WGI TAR Chapter 6 ES, & WGI TAR Figures SPM-3 & 6-6

rate of increase in daytime maximum air temperatures. These climate changes have lengthened the frost-free season in many mid- and high-latitude regions.

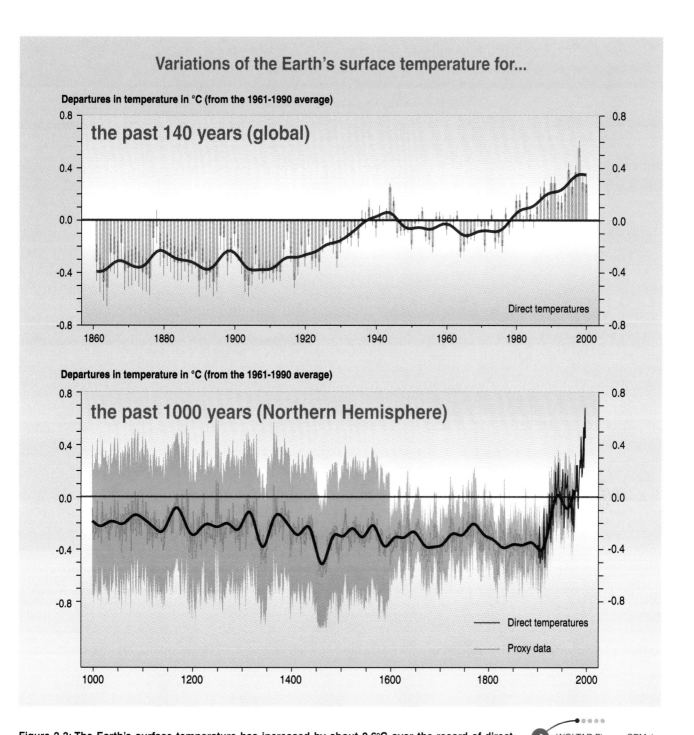

Figure 2-3: The Earth's surface temperature has increased by about 0.6°C over the record of direct temperature measurements (1860–2000, top panel)—a rise that is unprecedented, at least based on proxy temperature data for the Northern Hemisphere, over the last millennium (bottom panel). In the top panel the global mean surface temperature is shown year-by-year (red bars with *very likely* ranges as thin black whiskers) and approximately decade-by-decade (continuous red line). Analyses take into account data gaps, random instrumental errors and uncertainties, uncertainties in bias corrections in the ocean surface temperature data, and also in adjustments for urbanization over the land. The lower panel merges proxy data (year-by-year blue line with *very likely* ranges as grey band, 50-year-average purple line) and the direct temperature measurements (red line) for the Northern Hemisphere. The proxy data consist of tree rings, corals, ice cores, and historical records that have been calibrated against thermometer data. Insufficient data are available to assess such changes in the Southern Hemisphere.

WGI TAR Figures SPM-1, 2-7c, & 2-20

2.8 **In the lowest 8 km of the atmosphere the global temperature increase from the 1950s to the year 2000, about 0.1°C per decade, has been similar to that at the surface.** For the period 1979–2000 both satellite and weather balloon measurements show nearly identical warming over North America (0.3°C per decade) and Europe (0.4°C per decade) for both surface and lower atmosphere, but distinct differences over some land areas and particularly in the tropical regions ($0.10\pm0.10°$C per decade for surface versus $0.06\pm0.16°$C per decade for the lower atmosphere). Temperatures of the surface and lower atmosphere are influenced differently by factors such as stratospheric ozone depletion, atmospheric aerosols, and the El Niño phenomenon. In addition, spatial sampling techniques can also explain some of the differences in trends, but these differences are not fully resolved.

 WGI TAR SPM & WGI TAR Section 2.2.4

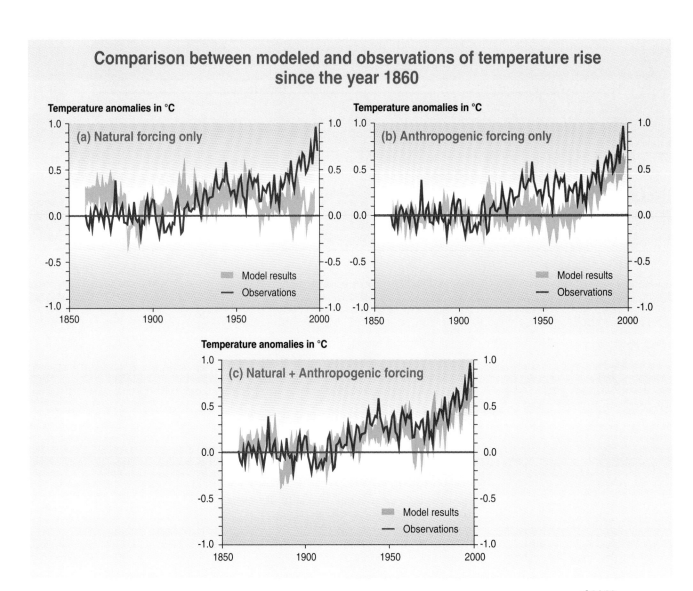

Figure 2-4: Simulating the Earth's temperature variations and comparing the results to the measured changes can provide insight into the underlying causes of the major changes. A climate model can be used to simulate the temperature changes that occur from both natural and anthropogenic causes. The simulations represented by the band in (a) were done with only natural forcings: solar variation and volcanic activity. Those encompassed by the band in (b) were done with anthropogenic forcings: greenhouse gases and an estimate of sulfate aerosols. Those encompassed by the band in (c) were done with both natural and anthropogenic forcings included. From (b), it can be seen that the inclusion of anthropogenic forcings provides a plausible explanation for a substantial part of the observed temperature changes over the past century, but the best match with observations is obtained in (c) when both natural and anthropogenic factors are included. These results show that the forcings included are sufficient to explain the observed changes, but do not exclude the possibility that other forcings may also have contributed. Similar results to those in (b) are obtained with other models with anthropogenic forcing.

 WGI TAR Figure 12-7

2.9 There is new and stronger evidence that most of the warming observed over the last 50 years is attributable to human activities.

2.10 The observed warming over the 20th century is unlikely to be entirely natural in origin. The increase in surface temperatures over the last 100 years is very unlikely to be due to internal variability alone. Reconstructions of climate data for the last 1,000 years also indicate that this 20th century warming was unusual and unlikely to be the response to natural forcing alone: That is, volcanic eruptions and variation in solar irradiance do not explain the warming in the latter half of the 20th century (see Figure 2-4a), but they may have contributed to the observed warming in the first half.

WGI TAR SPM & WGI TAR Chapter 12

2.11 In the light of new evidence and taking into account the remaining uncertainties, most of the observed warming over the last 50 years is likely to have been due to the increase in greenhouse gas concentrations. Detection and attribution studies (including greenhouse gases and sulfate aerosols as anthropogenic forcing) consistently find evidence for an anthropogenic signal in the climate record of the last 35 to 50 years, despite uncertainties in forcing due to anthropogenic sulfate aerosols and natural factors (volcanoes and solar irradiance). The sulfate and natural forcings are negative over this period and cannot explain the warming (see Figure 2-4a); whereas most of these studies find that, over the last 50 years, the estimated rate and magnitude of warming due to increasing greenhouse gases alone are comparable with, or larger than, the observed warming (Figure 2-4b). The best agreement for the 1860–2000 record is found when the above anthropogenic and natural forcing factors are combined (see Figure 2-4c). This result does not exclude the possibility that other forcings may also contribute, and some known anthropogenic factors (e.g., organic carbon, black carbon (soot), biomass aerosols, and some changes in land use) have not been used in these detection and attribution studies. Estimates of the magnitude and geographic distribution of these additional anthropogenic forcings vary considerably.

WGI TAR SPM & WGI TAR Chapter 12

2.12 Changes in sea level, snow cover, ice extent, and precipitation are consistent with a warming climate near the Earth's surface (see Table 2-1). Some of these changes are regional and some may be due to internal climate variations, natural forcings, or regional human activities rather than attributed solely to global human influence.

WGI TAR SPM & WGII TAR Section 4.3.11

2.13 It is very likely that the 20th century warming has contributed significantly to the observed rise in global average sea level and increase in ocean-heat content. Warming drives sea-level rise through thermal expansion of seawater and widespread loss of land ice. Based on tide gauge records, after correcting for land movements, the average annual rise was between 1 and 2 mm during the 20th century. The very few long records show that it was less during the 19th century (see Figure 2-5). Within present uncertainties, observations and models are both consistent with a lack of significant acceleration of sea-level rise during the 20th century. The observed rate of sea-level rise during the 20th century is consistent with models. Global ocean-heat content has increased since the late 1950s, the period with adequate observations of subsurface ocean temperatures.

WGI TAR Sections 2.2.2.5, 11.2, & 11.3.2

2.14 Snow cover and ice extent have decreased. It is very likely that the extent of snow cover has decreased by about 10% on average in the Northern Hemisphere since the late 1960s (mainly through springtime changes over America and Eurasia) and that the annual duration of lake- and river-ice cover in the mid- and high latitudes of the Northern Hemisphere has been reduced by about 2 weeks over the 20th century. There has also been a widespread retreat of mountain glaciers in non-polar regions during the 20th century. It is likely that Northern Hemisphere spring and summer sea-ice extent has decreased by about 10 to 15% from the 1950s to the year 2000 and that Arctic sea-ice thickness has declined by about 40% during late summer and early autumn in the last 3 decades of the 20th century. While there is no change in overall Antarctic sea-ice extent from 1978 to 2000 in

WGI TAR Section 2.2.5

parallel with global mean surface temperature increase, regional warming in the Antarctic Peninsula coincided with the collapse of the Prince Gustav and parts of the Larsen ice shelves during the 1990s, but the loss of these ice shelves has had little direct impact.

2.15 **Precipitation has very likely increased during the 20th century by 5 to 10% over most mid- and high latitudes of the Northern Hemisphere continents,** but in contrast, rainfall has likely decreased by 3% on average over much of the subtropical land areas (see Figure 2-6a). Increasing global mean surface temperature is very likely to lead to changes in precipitation and atmospheric moisture because of changes in atmospheric circulation, a more active hydrologic cycle, and increases in the water-holding capacity throughout the atmosphere. There has likely been a 2 to 4% increase in the frequency of heavy precipitation events in the mid- and high latitudes of the Northern Hemisphere over the latter half of the 20th century. There were relatively small long-term increases over the 20th century in land areas experiencing severe drought or severe wetness, but in many regions these changes are dominated by inter-decadal and multi-decadal climate variability with no significant trends evident over the 20th century.

WGI TAR Sections 2.5, 2.7.2.2, & 2.7.3

2.16 **Changes have also occurred in other important aspects of climate (see Table 2-1).**

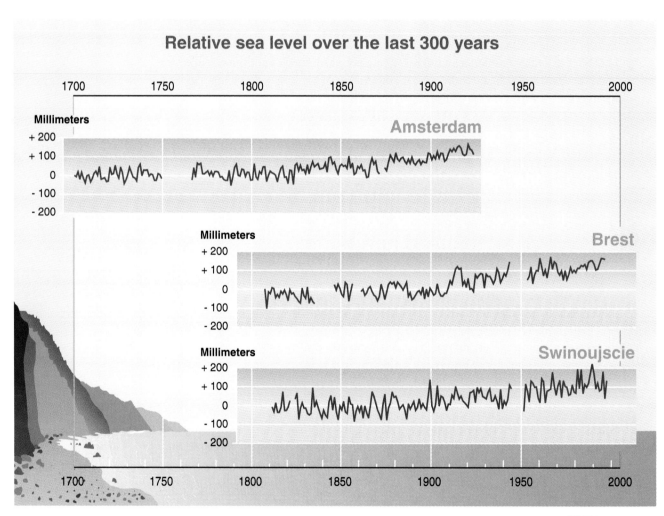

Relative sea level over the last 300 years

Figure 2-5: A limited number of sites in Europe have nearly continuous records of sea level spanning 300 years and show the greatest rise in sea level over the 20th century. Records shown from Amsterdam, The Netherlands, Brest, France, and Swinoujscie, Poland, as well as other sites, confirm the accelerated rise in sea level over the 20th century as compared to the 19th.

WGI TAR Figure 11-7

2.17 **Over the 20th century there has been a consistent, large-scale warming of both the land and ocean surface, with largest increases in temperature over the mid- and high latitudes of northern continents.** The warming of land surface faster than ocean surface from the years 1976 to 2000 (see Figure 2-6b) is consistent both with the observed changes in natural climate variations, such as the North Atlantic and Arctic Oscillations, and with the modeled pattern of greenhouse gas warming. As described below, statistically significant associations between regional warming and observed changes in biological systems have been documented in freshwater, terrestrial, and marine environments on all continents.

WGI TAR Sections 2.2.2, 2.6.3, & 2.6.5, & WGII TAR Section 6.3

2.18 **Warm episodes of the El Niño Southern Oscillation (ENSO) phenomenon have been more frequent, persistent, and intense since the mid-1970s, compared with the previous 100 years.** ENSO consistently affects regional variations of precipitation and temperature over much of the tropics, subtropics, and some mid-latitude areas. It is not obvious from models, however, that a warmer world would have a greater frequency of occurrence of El Niño events.

WGI TAR Section 2.6.2

2.19 **Some important aspects of climate appear *not* to have changed.** A few areas of the globe have not warmed in recent decades, mainly over some parts of the Southern Hemisphere oceans and parts of Antarctica (see Figure 2-6b). Antarctic sea-ice extent has stayed almost stable or even increased since 1978, the period of reliable satellite measurements. Current analyses are unable to draw conclusions about the likelihood of

WGI TAR Sections 2.2.2, 2.2.5, & 2.7.3

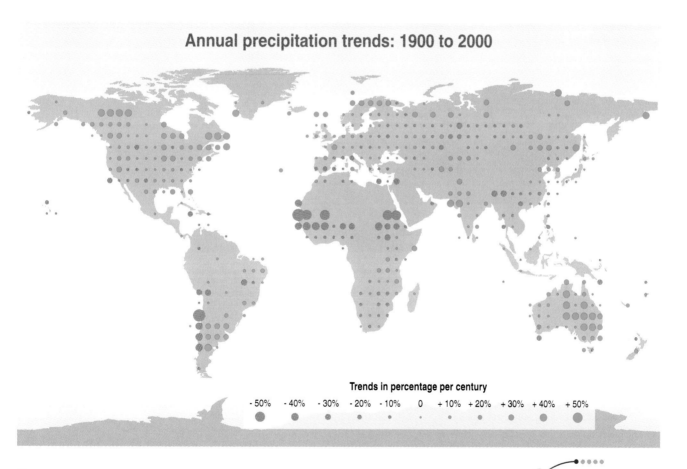

Annual precipitation trends: 1900 to 2000

Trends in percentage per century

-50%   -40%   -30%   -20%   -10%   0   +10%   +20%   +30%   +40%   +50%

Figure 2-6a: Precipitation during the 20th century has on average increased over continents outside the tropics but decreased in the desert regions of Africa and South America. While the record shows an overall increase consistent with warmer temperatures and more atmospheric moisture, trends in precipitation vary greatly from region to region and are only available over the 20th century for some continental regions. Over this period, there were relatively small long-term trends in land areas experiencing severe drought or severe wetness, but in many regions these changes are dominated by inter-decadal and multi-decadal climate variability that has no trends evident over the 20th century.

WGI TAR Figure 2-25

Annual temperature trends: 1976 to 2000

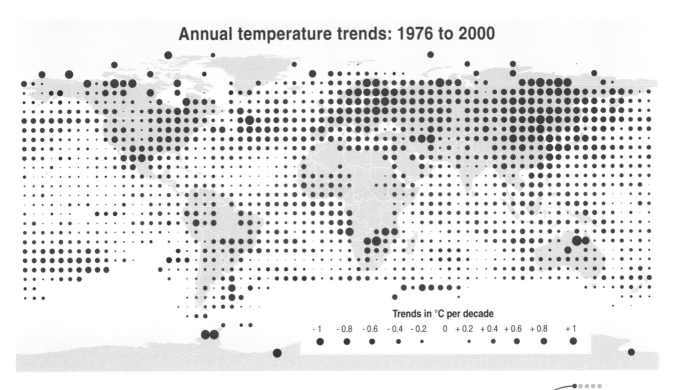

Trends in °C per decade

-1   -0.8   -0.6   -0.4   -0.2   0   +0.2   +0.4   +0.6   +0.8   +1

Figure 2-6b: A consistent, large-scale warming of both the land and ocean surface occurred over the last quarter of the 20th century, with largest temperature increases over the mid- and high latitudes of North America, Europe, and Asia. Large regions of cooling occurred only in parts of the Pacific and Southern Oceans and Antarctica. The warming of land faster than ocean surface is consistent both with the observed changes in natural climate variations such as the North Atlantic and Arctic Oscillations and with the modeled pattern of greenhouse-gas warming. As described in the text, warming in some regions is linked with observed changes in biological systems on all continents.

WGI TAR Figure 2-9d

changes in the frequency of tornadoes, thunder days, or hail events for the limited regions that have been studied. In addition, insufficient data and conflicting analyses prevent an assessment of changes in intensities of tropical and extra-tropical cyclones and severe local storm activity in the mid-latitudes.

2.20 **Observed changes in regional climate over the past 50 years have affected biological and hydrological systems in many parts of the world (see Table 2-1).**

2.21 **There has been a discernible impact of regional climate change, particularly increases in temperature, on biological systems in the 20th century.** In many parts of the world the observed changes in these systems[3], either anthropogenic or natural, are coherent across diverse localities and are consistent in direction with the expected effects of regional changes in temperature. The probability that the observed changes in the expected direction (with no reference to magnitude) could occur by chance alone is negligible. Such systems include, for example, species distributions, population sizes, and the timing of reproduction or migration events. These observations implicate regional climate change as a prominent contributing causal factor. There have been observed changes in the types (e.g., fires, droughts, blowdowns), intensity, and frequency of disturbances that are affected by regional climatic change (either anthropogenic or natural) and land-use practices, and they in turn affect the productivity of and species composition within an ecosystem,

WGII TAR Sections 5.4, 5.6.2, 10.1.3.2, 11.2, 13.1.3.1, & 13.2.4.1, & WGII TAR Figure SPM-1

[3] There are 44 regional studies of over 400 plants and animals, which varied in length from about 20 to 50 years, mainly from North America, Europe, and the southern polar region. There are 16 regional studies covering about 100 physical processes over most regions of the world, which varied in length from about 20 to 150 years.

particularly at high latitudes and high altitudes. Frequency of pests and disease outbreaks have also changed, especially in forested systems, and can be linked to changes in climate. In some regions of Africa, the combination of regional climate changes (Sahelian drought) and anthropogenic stresses has led to decreased cereal crop production since the year 1970. There are some positive aspects of warming: For example, the growing season across Europe has lengthened by about 11 days from the years 1959 to 1993, and energy consumption for heating in winter has decreased.

2.22 **Coral reefs are adversely affected by rising sea surface temperatures.** Increasing sea surface temperatures have been recorded in much of the tropical oceans over the past several decades. Many corals have undergone major, although often partially reversible, bleaching episodes when sea surface temperatures rise by 1°C in any one season, and extensive mortality occurs for a 3°C rise. This typically occurs during El Niño events and is exacerbated by rising sea surface temperatures. These bleaching events are often associated with other stresses such as pollution.

WGI TAR Section 2.2.2.2 & WGII TAR Sections 6.4.5 & 17.2.4.1

2.23 **Changes in marine systems, particularly fish populations, have been linked to large-scale climate oscillations.** The El Niño affects fisheries off the coasts of South America and Africa and the decadal oscillations in the Pacific are linked to decline of fisheries off the west coast of North America.

WGI TAR Section 2.6.3 & WGII TAR Sections 10.2.2.2, 14.1.3, & 15.2.3.3

2.24 **Changes in stream flow, floods, and droughts have been observed.** Evidence of regional climate change impacts on elements of the hydrological cycle suggest that warmer temperatures lead to intensification of the hydrological cycle. Peak stream flow has shifted back from spring to late winter in large parts of eastern Europe, European Russia, and North America in the last decades. The increasing frequency of droughts and floods in some areas is related to variations in climate—for example, droughts in Sahel and in northeast and southern Brazil, and floods in Colombia and northwest Peru.

WGI TAR Section 2.7.3.3, WGII TAR SPM, WGII TAR Sections 4.3.6, 10.2.1.2, 14.3, & 19.2.2.1, & WGII TAR Table 4-1

2.25 **There are preliminary indications that some human systems have been affected by recent increases in floods and droughts. The rising socio-economic costs related to weather damage and to regional variations in climate suggest increasing vulnerability to climate change (see Table 2-1).**

2.26 **Extreme weather or climatic events cause substantial, and increasing, damage.** Extreme events are currently a major source of climate-related impacts. For example, heavy losses of human life, property damage, and other environmental damages were recorded during the El Niño event of the years 1997–1998. The impacts of climatic extremes and variability are a major concern. Preliminary indications suggest that some social and economic systems have been affected by recent increases in floods and droughts, with increases in economic losses for catastrophic weather events. Because these systems are also affected by changes in socio-economic factors such as demographic shifts and land-use changes, quantifying the relative impacts of climate change (either anthropogenic or natural) and of socio-economic factors is difficult. For example, direct costs of global catastrophic weather-related losses, corrected for inflation, have risen an order of magnitude from the 1950s to the 1990s (see Figure 2-7), and costs for non-catastrophic weather events have grown similarly. The number of weather-related catastrophic events has risen three times faster than the number of non-weather-related events, despite generally enhanced disaster preparedness. Part of this observed upward trend in weather-related losses over the past 50 years is linked to socio-economic factors (e.g., population growth, increased wealth, urbanization in vulnerable areas), and part is linked to regional climatic factors (e.g., changes in precipitation, flooding events).

WGII TAR SPM & WGII TAR Sections 8.2 & 14.3

2.27 **The fraction of weather-related losses covered by insurance varies considerably by region,** and the uneven impacts of climatic hazards raise issues for development and equity. Insurers pay only 5% of total economic losses today in Asia and South America, 10% in Africa, and about 30% in Australia, Europe, and North and Central America. The fraction covered is typically much higher when just storm losses are considered, but flood- and crop-related losses have much lower coverage. The balance of the losses are absorbed by governments and affected individuals and organizations.

WGII TAR Sections
8.3.3.1 & 8.5.4

2.28 **Climate-related health effects are observed.** Many vector-, food-, and water-borne infectious diseases are known to be sensitive to changes in climatic conditions. Extensive experience makes clear that any increase in floods will increase the risk of drowning, diarrheal and respiratory diseases, water-contamination diseases, and—in developing countries—hunger and malnutrition (*high confidence*). Heat waves in Europe and North America are associated with a significant increase in urban mortality, but warmer wintertime temperatures also result in reduced wintertime mortality. In some cases health effects are clearly related to recent climate changes, such as in Sweden where tick-borne encephalitis

WGII TAR SPM & WGII
TAR Sections 9.5.1, 9.7.8,
10.2.4, & 13.2.5

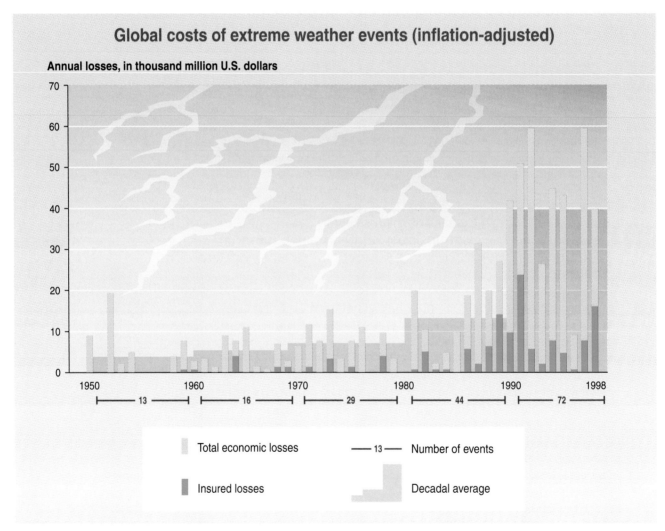

Global costs of extreme weather events (inflation-adjusted)

Annual losses, in thousand million U.S. dollars

Legend:
- Total economic losses
- Insured losses
- —13— Number of events
- Decadal average

Figure 2-7: The economic losses from catastrophic weather events have risen globally 10-fold (inflation-adjusted) from the 1950s to the 1990s, much faster than can be accounted for with simple inflation. The insured portion of these losses rose from a negligible level to about 23% in the 1990s. The total losses from small, non-catastrophic weather-related events (not included here) are similar. Part of this observed upward trend in weather-related disaster losses over the past 50 years is linked to socio-economic factors (e.g., population growth, increased wealth, urbanization in vulnerable areas), and part is linked to regional climatic factors (e.g., changes in precipitation, flooding events).

WGII TAR Figure 8-1

incidence increased after milder winters and moved northward following the increased frequency of milder winters over the years 1980 to 1994.

2.29 **The recognition and anticipation of adverse impacts of climate change has led to both public and governmental responses.**

2.30 **As a consequence of observed and anticipated climate change, socio-economic and policy responses have occurred in the last decade.** These have included stimulation of the renewable energy market, development of energy-efficiency improvement programs enhanced by climate change concerns, integration of climate policies into broader national policies, carbon taxes in several countries, domestic greenhouse gases trading regimes in some countries, national and international voluntary agreements with industries to increase energy efficiency or otherwise decrease greenhouse gas emissions, creation of carbon exchange markets, public and political pressures for utilities to reduce or offset carbon emissions from new energy projects, industry reconnaissance into approaches to offset carbon emissions, and establishment of programs to assist the developing and least developed countries reduce vulnerabilities and adapt to climate change and engage in mitigation activities.

WGIII TAR Sections 3.2, 3.4-5, 3.8.4, 6.2.2, 6.3.2, & 9.2.1

Question 3

What is known about the regional and global climatic, environmental, and socio-economic consequences in the next 25, 50, and 100 years associated with a range of greenhouse gas emissions arising from scenarios used in the TAR (projections which involve no climate policy intervention)?

To the extent possible evaluate the:
- Projected changes in atmospheric concentrations, climate, and sea level
- Impacts and economic costs and benefits of changes in climate and atmospheric composition on human health, diversity and productivity of ecological systems, and socio-economic sectors (particularly agriculture and water)
- The range of options for adaptation, including the costs, benefits, and challenges
- Development, sustainability, and equity issues associated with impacts and adaptation at a regional and global level.

Q3

3.1   The greenhouse gas emissions scenarios used as the basis for the climate projections in the TAR are those contained in the IPCC *Special Report on Emissions Scenarios* (see Box 3-1). Because the SRES scenarios had only been available for a very short time prior to production of the TAR, it was not possible to include impact assessments based on these scenarios. Hence, the impacts assessments in the TAR use climate model results that tend to be based on equilibrium climate change scenarios (e.g., $2xCO_2$), a relatively small number of experiments using a 1% per year CO_2 increase transient scenario, or the scenarios used in the Second Assessment Report (i.e., the IS92 series). The challenge in answering this question therefore is to try and map these impact results onto the climate change results, which have used the SRES scenarios. This, by necessity, requires various approximations to be made and in many cases only qualitative conclusions can be drawn. Projections of changes in climate variability, extreme events, and abrupt/non-linear changes are covered in Question 4.

Box 3-1	Future emissions of greenhouse gases and aerosols due to human activities will alter the atmosphere in ways that are expected to affect the climate.

WGI TAR Chapters 3, 4, 5, & 6

Changes in climate occur as a result of internal variability of the climate system and external factors (both natural and as a result of human activities). Emissions of greenhouse gases and aerosols due to human activities change the composition of the atmosphere. Future emissions of greenhouse gases and aerosols are determined by driving forces such as population, socio-economic development, and technological change, and hence are highly uncertain. Scenarios are alternative images of how the future might unfold and are an appropriate tool with which to analyze how driving forces may influence future emission outcomes and to assess the associated uncertainties. The SRES scenarios, developed to update the IS92 series, consist of six scenario groups, based on narrative storylines, which span a wide range of these driving forces (see Figure 3-1). They are all plausible and internally consistent, and no probabilities of occurrence are assigned. They encompass four combinations of demographic change, social and economic development, and broad technological developments (A1B, A2, B1, B2). Two further scenario groups, A1FI and A1T, explicitly explore alternative energy technology developments to A1B (see Figure 3-1a). The resulting emissions of the greenhouse gases CO_2, CH_4, and N_2O, along with SO_2 which leads to the production of sulfate aerosols, are shown in Figures 3-1b to 3-1e; other gases and particles are also important. These emissions cause changes in the concentrations of these gases and aerosols in the atmosphere. The changes in the concentrations for the SRES scenarios are shown in Figures 3-1f to 3-1i. Note that for gases which stay in the atmosphere for a long period, such as CO_2 shown in panel (f), the atmospheric concentration responds to changes in emissions relatively slowly (e.g., see Figure 5-3); whereas for short-lived gases and aerosols, such as sulfate aerosols shown in panel (i), the atmospheric concentration responds much more quickly. The influence of changes in the concentrations of greenhouse gases and aerosols in the atmosphere on the climate system can broadly be compared using the concept of radiative forcing, which is a measure of the influence a factor has in altering the balance of incoming and outgoing energy in the Earth-atmosphere system. A positive radiative forcing, such as that produced by increasing concentrations of greenhouse gases, tends to warm the surface; conversely a negative radiative forcing, which can arise from an increase in some types of aerosols such as sulfate aerosols, tends to cool the surface. The radiative forcing resulting from the increasing concentrations in panels (f) to (i) is shown in panel (j). Note that, as with the IS92 scenarios, all combinations of emissions of greenhouse gases and aerosols in the SRES scenarios result in increased radiative forcing.

3.2   Carbon dioxide concentrations, globally averaged surface temperature, and sea level are projected to increase under all IPCC emissions scenarios during the 21st century.

WGI TAR Section 3.7.3.3

3.3   **All SRES emissions scenarios result in an increase in the atmospheric concentration of** CO_2. For the six illustrative SRES scenarios, the projected concentrations of CO_2—the primary anthropogenic greenhouse gas—in the year 2100 range from 540 to 970 ppm, compared to about 280 ppm in the pre-industrial era and about 368 ppm in the year 2000 (see Figure 3-1f). These projections include the land and ocean climate feedbacks. The different socio-economic assumptions (demographic, social, economic, and technological) result in different levels of future greenhouse gases and aerosols. Further uncertainties, especially regarding the persistence of the present removal processes (carbon sinks) and the magnitude of the climate feedback on the terrestrial biosphere, cause a variation of about −10 to +30% in the year 2100 concentration, around each scenario. The total range is 490 to 1,260 ppm (75 to 350% above the year 1750 (pre-industrial) concentration).

3.4 Model calculations of the concentrations of the primary non-CO_2 greenhouse gases by year 2100 vary considerably across the six illustrative SRES scenarios. For most cases, A1B, A1T, and B1 have the smallest increases, and A1FI and A2 the largest (see Figures 3-1g and 3-1h).

WGI TAR Section 4.4.5 &
WGI TAR Box 9.1

3.5 The SRES scenarios include the possibility of either increases or decreases in anthropogenic aerosols, depending on the extent of fossil-fuel use and policies to abate polluting emissions. As seen in Figure 3-1i, sulfate aerosol concentrations are projected to fall below present levels by 2100 in all six illustrative SRES scenarios. This would result in warming relative to present day. In addition, natural aerosols (e.g., sea salt, dust, and emissions leading to sulfate and carbon aerosols) are projected to increase as a result of changes in climate.

WGI TAR Section 5.5 &
SRES Section 3.6.4

3.6 The globally averaged surface temperature is projected to increase by 1.4 to 5.8°C over the period 1990 to 2100 (see Figure 3-1k). This is about two to ten times larger than the central value of observed warming over the 20th century and the projected rate of warming is very likely to be without precedent during at least the last 10,000 years, based on paleoclimate data (see Figure 9-1). For the periods 1990 to 2025 and 1990 to 2050, the projected increases are 0.4 to 1.1°C and 0.8 to 2.6°C, respectively. These results are for the full range of 35 SRES scenarios, based on a number of climate models.[4] Temperature increases are projected to be greater than those in the SAR, which were about 1.0 to 3.5°C based on six IS92 scenarios. The higher projected temperatures and the wider range are due primarily to lower projected SO_2 emissions in the SRES scenarios relative to the IS92 scenarios, because of structural changes in the energy system as well as concerns about local and regional air pollution.

WGI TAR Section 9.3.3

3.7 By 2100, the range in the surface temperature response across different climate models for the same emissions scenario is comparable to the range across different SRES emissions scenarios for a single climate model. Figure 3-1 shows that the SRES scenarios with the highest emissions result in the largest projected temperature increases. Further uncertainties arise due to uncertainties in the radiative forcing. The largest forcing uncertainty is that due to the sulfate aerosols.

WGI TAR Section 9.3.3

→ **Figure 3-1: The different socio-economic assumptions underlying the SRES scenarios result in different levels of future emissions of greenhouse gases and aerosols.** These emissions in turn change the concentration of these gases and aerosols in the atmosphere, leading to changed radiative forcing of the climate system. Radiative forcing due to the SRES scenarios results in projected increases in temperature and sea level, which in turn will cause impacts. The SRES scenarios do not include additional climate initiatives and no probabilities of occurrence are assigned. Because the SRES scenarios had only been available for a very short time prior to production of the TAR, the impacts assessments here use climate model results which tend to be based on equilibrium climate change scenarios (e.g., $2xCO_2$), a relatively small number of experiments using a 1% per year CO_2 increase transient scenario, or the scenarios used in the Second Assessment Report (i.e., the IS92 series). Impacts in turn can affect socio-economic development paths through, for example, adaptation and mitigation. The highlighted boxes along the top of the figure illustrate how the various aspects relate to the integrated assessment framework for considering climate change (see Figure 1-1).

WGI TAR Figures 3.12, 4.14, 5.13, 9.13, 9.14, & 11.12, WGII TAR Figure 19-7, & SRES Figures SPM-2, SPM-5, SPM-6, & TS-10

---

[4] Complex physically based climate models are the main tool for projecting future climate change. In order to explore the range of scenarios, these are complemented by simple climate models calibrated to yield an equivalent response in temperature and sea level to complex climate models. These projections are obtained using a simple climate model whose climate sensitivity and ocean heat uptake are calibrated to each of seven complex climate models. The climate sensitivity used in the simple model ranges from 1.7 to 4.2°C, which is comparable to the commonly accepted range of 1.5 to 4.5°C. For the atmosphere-ocean general circulation model (AOGCM) experiments for the end of the 21st century (years 2071 to 2100) compared with the period 1961 to 1990, the mean warming for SRES scenario A2 is 3.0°C with a range of 1.3 to 4.5°C, while for SRES scenario B2 the mean warming is 2.2°C with a range of 0.9 to 3.4°C.

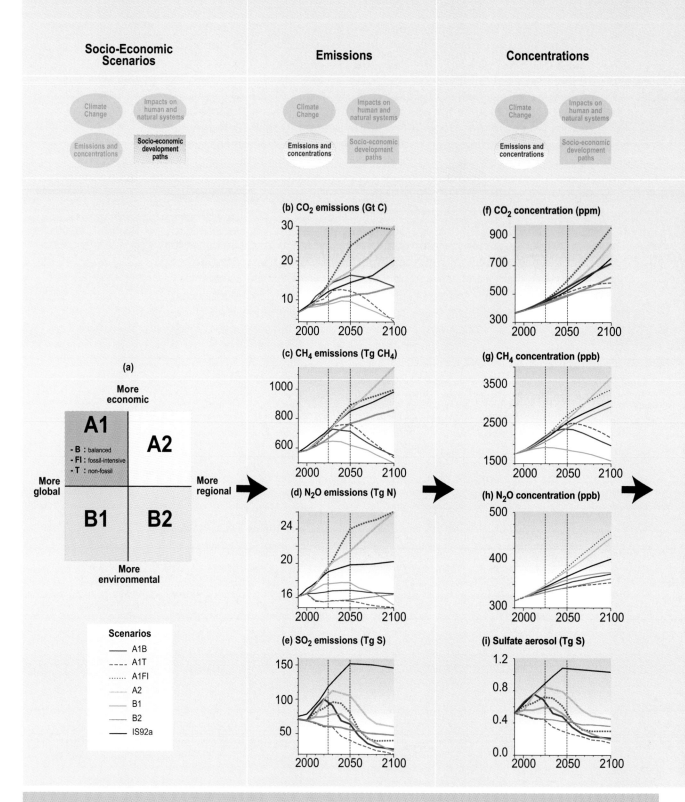

Socio-Economic Scenarios

Emissions

Concentrations

(a)

More economic

A1
- **B** : balanced
- **FI** : fossil-intensive
- **T** : non-fossil

A2

More global

More regional

B1

B2

More environmental

Scenarios

—— A1B
- - - - A1T
········· A1FI
—— A2
—— B1
—— B2
—— IS92a

(b) CO₂ emissions (Gt C)

(c) CH₄ emissions (Tg CH₄)

(d) N₂O emissions (Tg N)

(e) SO₂ emissions (Tg S)

(f) CO₂ concentration (ppm)

(g) CH₄ concentration (ppb)

(h) N₂O concentration (ppb)

(i) Sulfate aerosol (Tg S)

A1FI, A1T, and A1B

The A1 storyline and scenario family describes a future world of very rapid economic growth, global population that peaks in mid-century and declines thereafter, and the rapid introduction of new and more efficient technologies. Major underlying themes are convergence among regions, capacity-building, and increased cultural and social interactions, with a substantial reduction in regional differences in per capita income. The A1 scenario family develops into three groups that describe alternative directions of technological change in the energy system. The three A1 groups are distinguished by their technological emphasis: fossil intensive (A1FI), non-fossil energy sources (A1T), or a balance across all sources (A1B) (where balanced is defined as not relying too heavily on one particular energy source, on the assumption that similar improvment rates apply to all energy supply and end use technologies).

Radiative Forcing

Temperature and Sea-Level Change

Reasons for Concern

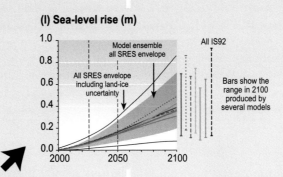

(l) Sea-level rise (m)

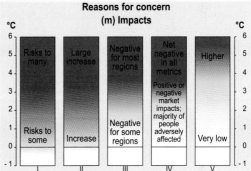

(j) Radiative forcing (Wm⁻²)

(k) Temperature change (°C)

Bars show the range in 2100 produced by several models

Reasons for concern

(m) Impacts

I Risks to unique and threatened systems
II Risks from extreme climate events
III Distribution of impacts
IV Aggregate impacts
V Risks from future large-scale discontinuities

Scenarios

— A1B
---- A1T
······ A1FI
——— A2
——— B1
——— B2
— IS92a

A2

The A2 storyline and scenario family describes a very heterogeneous world. The underlying theme is self-reliance and preservation of local identities. Fertility patterns across regions converge very slowly, which results in continuously increasing population. Economic development is primarily regionally oriented and per capita economic growth and technological change more fragmented and slower than other storylines.

B1

The B1 storyline and scenario family describes a convergent world with the same global population that peaks in mid-century and declines thereafter, as in the A1 storyline, but with rapid change in economic structures toward a service and information economy, with reductions in material intensity and the introduction of clean and resource-efficient technologies. The emphasis is on global solutions to economic, social, and environmental sustainability, including improved equity, but without additional climate initiatives.

B2

The B2 storyline and scenario family describes a world in which the emphasis is on local solutions to economic, social, and environmental sustainability. It is a world with continuously increasing global population, at a rate lower than A2, intermediate levels of economic development, and less rapid and more diverse technological change than in the B1 and A1 storylines. While the scenario is also oriented towards environmental protection and social equity, it focuses on local and regional levels.

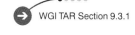

3.8   **Globally averaged annual precipitation is projected to increase during the 21st century.** Globally averaged water vapor and evaporation are also projected to increase.

WGI TAR Section 9.3.1

3.9   **Global mean sea level is projected to rise by 0.09 to 0.88 m between the years 1990 and 2100, for the full range of SRES scenarios (see Figure 3-1I).** For the periods 1990 to 2025 and 1990 to 2050, the projected rises are 0.03 to 0.14 m and 0.05 to 0.32 m, respectively. This is due primarily to thermal expansion and loss of mass from glaciers and ice caps. The range of sea-level rise presented in the SAR was 0.13 to 0.94 m, based on the IS92 scenarios. Despite the higher temperature change projections in this assessment, the sea-level projections are slightly lower, primarily due to the use of improved models, which give a smaller contribution from glaciers and ice sheets.

WGI TAR Section 11.5.1

3.10   **Substantial differences are projected in regional changes in climate and sea level, compared to the global mean change.**

3.11   **It is very likely that nearly all land areas will warm more rapidly than the global average, particularly those at northern high latitudes in winter.** Most notable of these is the warming in the northern regions of North America, and northern and central Asia, which exceeds global mean warming in each model by more than 40%. In contrast, the warming is less than the global mean change in south and southeast Asia in summer and in southern South America in winter (see Figure 3-2).

WGI TAR Section 10.3.2

3.12   **At the regional scale, both increases and decreases in precipitation are projected, typically of 5 to 20%.** It is likely that precipitation will increase over high latitude regions in both summer and winter. Increases are also projected over northern mid-latitudes, tropical Africa and Antarctica in winter, and in southern and eastern Asia in summer. Australia, Central America, and southern Africa show consistent decreases in winter rainfall. Larger year-to-year variations in precipitation are very likely over most areas where an increase in mean precipitation is projected (see Figure 3-3).

WGI TAR Section 10.3.2

3.13   **The projected range of regional variation in sea-level change is substantial compared to projected global average sea-level rise, because the level of the sea at the shoreline is determined by many factors (see Figure 3-4).** Confidence in the regional distribution of sea-level change from complex models is low because there is little similarity between model results, although nearly all models project greater than average rise in the Arctic Ocean and less than average rise in the Southern Ocean.

WGI TAR Section 11.5.2

3.14   **Glaciers and ice caps are projected to continue their widespread retreat during the 21st century.** Northern Hemisphere snow cover, permafrost, and sea-ice extent are projected to decrease further. The Antarctic ice sheet is likely to gain mass because of greater precipitation, while the Greenland ice sheet is likely to lose mass because the increase in runoff will exceed the precipitation increase. Concerns that have been expressed about the stability of the West Antarctic ice sheet are covered in Question 4.

WGI TAR Section 11.5.4

→ **Figure 3-2: The background shows the annual mean change of temperature (color shading) for (a) the SRES scenario A2 and (b) the SRES scenario B2.** Both SRES scenarios show the period 2071 to 2100 relative to the period 1961 to 1990, and were performed by AOGCMs. Scenarios A2 and B2 are shown as no AOGCM runs were available for the other SRES scenarios. The boxes show an analysis of inter-model consistency in regional relative warming (i.e., warming relative to each model's global average warming) for the same scenarios. Regions are classified as showing either agreement on warming in excess of 40% above the global mean annual average (*much greater than average warming*), agreement on warming greater than the global mean annual average (*greater than average warming*), agreement on warming less than the global mean annual average (*less than average warming*), or disagreement amongst models on the magnitude of regional relative warming (*inconsistent magnitude of warming*). There is also a category for agreement on cooling (this category never occurs). A consistent result from at least seven of the nine models is defined as being necessary for agreement. The global mean annual average warming of the models used span 1.2 to 4.5°C for A2 and 0.9 to 3.4°C for B2, and therefore a regional 40% amplification represents warming ranges of 1.7 to 6.3°C for A2 and 1.3 to 4.7°C for B2.

WGI TAR Figures 9.10d & 9.10e, & WGI TAR Box 10.1 (Figure 1)

Change in temperature for scenarios A2 and B2

a) Scenario A2

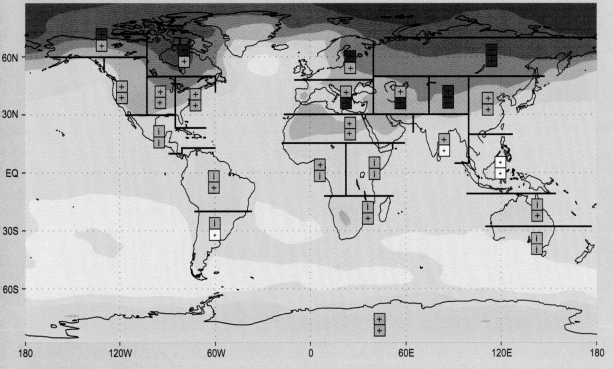

b) Scenario B2

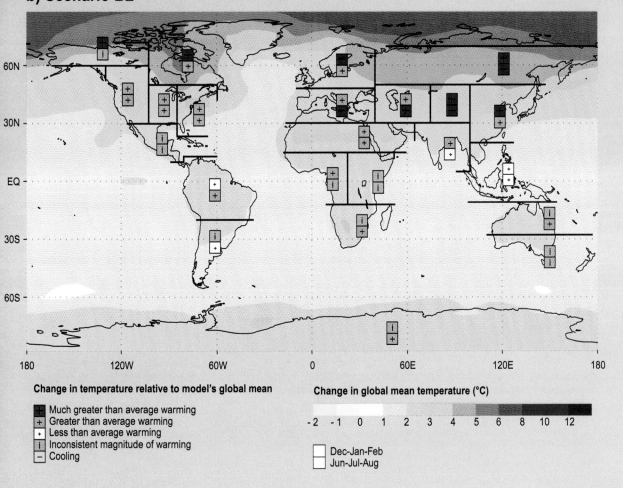

Change in temperature relative to model's global mean

- ■ Much greater than average warming
- ⊞ Greater than average warming
- ⊡ Less than average warming
- ⊟ Inconsistent magnitude of warming
- – Cooling

Change in global mean temperature (°C)

-2 -1 0 1 2 3 4 5 6 8 10 12

□ Dec-Jan-Feb
□ Jun-Jul-Aug

Change in precipitation for scenarios A2 and B2

a) Scenario A2

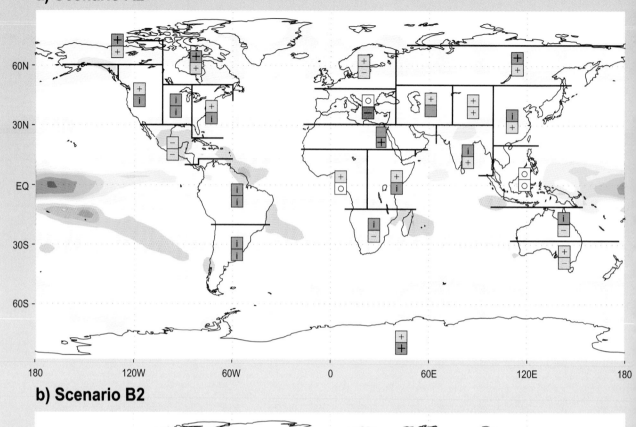

b) Scenario B2

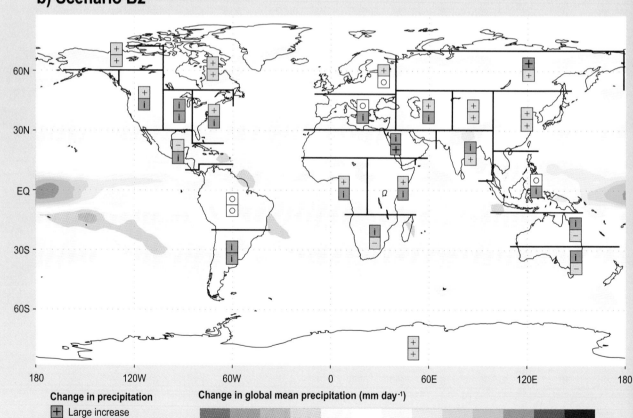

Change in precipitation

+	Large increase
+	Small increase
○	No change
–	Small decrease
■	Large decrease
i	Inconsistent sign

Change in global mean precipitation (mm day⁻¹)

- 1   - 0.75   - 0.50   - 0.25   0   0.25   0.50   0.75   1   1.5   2   3

Dec-Jan-Feb
Jun-Jul-Aug

WGI TAR Box 10.1 (Figure 2)

Figure 3-3: The background shows the annual mean change of rainfall (color shading) for (a) the SRES scenario A2 and (b) the SRES scenario B2. Both SRES scenarios show the period 2071 to 2100 relative to the period 1961 to 1990, and were performed by AOGCMs. Scenarios A2 and B2 are shown as no AOGCM runs were available for the other SRES scenarios. The boxes show an analysis of inter-model consistency in regional precipitation change. Regions are classified as showing either agreement on increase with an average change of greater than 20% (*large increase*), agreement on increase with an average change between 5 and 20% (*small increase*), agreement on a change between –5 and +5% or agreement with an average change between –5 and +5% (*no change*), agreement on decrease with an average change between –5 and -20% (*small decrease*), agreement on decrease with an average change of more than –20% (*large decrease*), or disagreement (*inconsistent sign*). A consistent result from at least seven of the nine models is defined as being necessary for agreement.

3.15 **Projected climate change will have beneficial and adverse environmental and socio-economic effects, but the larger the changes and rate of change in climate, the more the adverse effects predominate.**

3.16 **The impacts of climate change will be more severe the greater the cumulative emissions of greenhouse gases (*medium confidence*).** Climate change can have beneficial as well as adverse effects, but adverse effects are projected to predominate for much of the world. The various effects of climate change pose risks that increase with global mean temperature. Many of these risks have been organized into five reasons for concern: threats to endangered species and unique systems, damages from extreme climate events, effects that fall most heavily on developing countries and the poor within countries, global aggregate impacts, and large-scale high-impact events (see Box 3-2 and Figure 3-1). The effects of climate change on human health, ecosystems, food production, water resources, small islands and low-lying coastal regions, and aggregate market activities are summarized below. However, note that future changes in the frequency or intensity of extreme events have not been taken into account in most of these studies (see also Question 4).

WGII TAR Sections 1.2, 19.3, 19.5, & 19.8

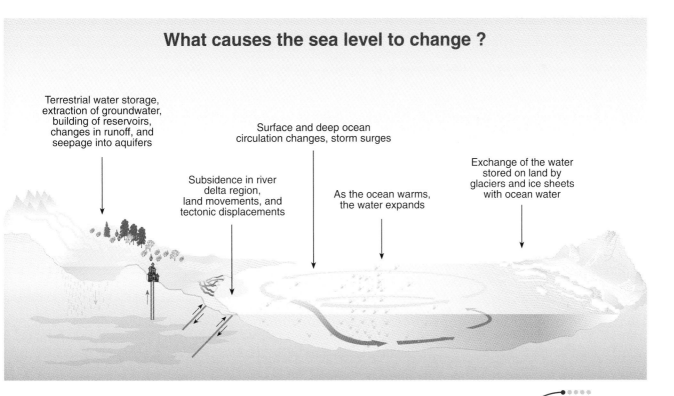

What causes the sea level to change ?

Terrestrial water storage, extraction of groundwater, building of reservoirs, changes in runoff, and seepage into aquifers

Surface and deep ocean circulation changes, storm surges

Exchange of the water stored on land by glaciers and ice sheets with ocean water

Subsidence in river delta region, land movements, and tectonic displacements

As the ocean warms, the water expands

Figure 3-4: The level of the sea at the shoreline is determined by many factors in the global environment that operate on a great range of time scales, from hours (tidal) to millions of years (ocean basin changes due to tectonics and sedimentation). On the time scale of decades to centuries, some of the largest influences on the average levels of the sea are linked to climate and climate change processes.

WGI TAR Box TS-2

Box 3-2 | Concerns about the risks from climate change rise with temperature.

- *Unique and threatened systems:* Some changes in species and systems have already been associated with observed changes in climate, and some highly vulnerable species and systems may be at risk of damage or even loss for very small changes in climate. Greater warming would intensify the risks to these species and systems, and place additional ones at risk.

  WGII TAR Sections 5.2, 5.4, & 19.3

- *Extreme climate events:* Increased frequencies and intensities of some extreme events have already been observed (see Question 2) and are likely to increase with further warming, as would the risks to human life, property, crops, livestock, and ecosystems. These risks increase where development is occurring in inherently dynamic and unstable zones (e.g., river floodplains and low-lying coastal regions) (see also Question 4).

  WGII TAR Sections 15.2 & 19.6

- *Uneven distribution of impacts:* In general, developing countries are at greater risk of adverse impacts from climate change than are developed countries, of which some of the latter may experience market sector benefits for warming less than a few °C. For greater warming, most regions are at risk of predominantly negative effects from climate change. But developing countries generally would continue to be more severely impacted than developed countries. Within countries, vulnerability varies and the poorest populations often have higher exposure to impacts that threaten their lives and livelihoods.

  WGII TAR Section 19.4

- *Global aggregate impacts:* Globally aggregated market sector impacts may be positive or negative up to a few °C, though the majority of people may be negatively affected. With greater warming, the risk of negative global market sector impacts increases, and impacts would be predominantly negative for most people.

  WGII TAR Section 19.5

- *Large-scale, high-impact events:* The probability of large-scale, high-impact events within a 100-year time horizon such as shutdown of the thermohaline circulation or collapse of the West Antarctic ice sheet is very low for warming less than a few °C. The risk, which is a product of the probabilities of these events and the magnitude of their consequences, is largely unquantified. For greater warming, and over a time horizon longer than 100 years, the probabilities and the risks increase, but by an amount that cannot now be estimated. See also Question 4.

  WGII TAR Section 19.6

Human Health

3.17 **Overall climate change is projected to increase threats to human health, particularly in lower income populations predominantly within tropical/ subtropical countries.** Climate change can affect human health through multiple pathways, including direct effects (e.g., reduced cold stress in temperate countries but increased heat stress, loss of life in floods and storms) and indirect effects that operate through changes in the ranges of disease vectors (e.g., mosquitoes)[5], water-borne pathogens, water quality, air quality, food availability and quality (e.g., decreased protein content in some cereals), population displacement, and economic disruption (*medium to high confidence*). Some effects may be beneficial (e.g., reduced cold stress, and reduced disease transmission in some cases), but the predominant effect is anticipated to be adverse (see Table 3-1). Actual impacts will be strongly influenced by local environmental conditions and socio-economic circumstances, and for each anticipated adverse health impact there is a range of social, institutional, technological, and behavioral adaptation options to lessen that impact. Adaptations could, for example, encompass strengthening of the public health infrastructure, health-oriented management of the environment (including air and water quality, food safety, urban and housing design, and surface water management), and the provision of appropriate medical care.

WGII TAR Sections 5.3, 9.1, 9.5, & 9.11

Biodiversity and Productivity of Ecological Systems

3.18 **Diversity in ecological systems is expected to be affected by climate change and sea-level rise, with an increased risk of extinction of some vulnerable species (*high confidence*).** Significant disruptions of ecosystems from disturbances such as fire, drought, pest infestation, invasion of species, storms, and coral bleaching events are expected to increase (see Table 3-2). The stresses caused by climate change, added to other stresses on ecological systems (e.g., land conversion, land degradation, harvesting, and pollution), threaten substantial damage to or complete loss of some unique ecosystems, and extinction of some critically endangered and endangered species. Coral

WGII TAR Sections 5.2.3, 5.4.1, 16.2, 17.2, & 19.3.2-3

---

[5] Eight studies have modeled the effects of climate change on these diseases, five on malaria and three on dengue. Seven use a biological or process-based approach, and one uses an empirical, statistical approach.

Table 3-1 Human health consequences of climate change if no climate policy interventions are made.

	2025	2050	2100
CO_2 concentration[a]	405–460 ppm	445–640 ppm	540–970 ppm
Global mean temperature change from the year 1990[b]	0.4–1.1°C	0.8–2.6°C	1.4–5.8°C
Global mean sea-level rise from the year 1990[b]	3–14 cm	5–32 cm	9–88 cm
Human Health Effects[c]			
Heat stress and winter mortality [WGII TAR Section 9.4]	Increase in heat-related deaths and illness (*high confidence*[d]). Decrease in winter deaths in some temperate regions (*high confidence*[d]).	Thermal stress effects amplified (*high confidence*[d]).	Thermal stress effects amplified (*high confidence*[d]).
Vector- and water-borne diseases [WGII TAR Section 9.7]		Expansion of areas of potential transmission of malaria and dengue (*medium to high confidence*[d]).	Further expansion of areas of potential transmission (*medium to high confidence*[d]).
Floods and storms [WGII TAR Sections 3.8.5 & 9.5]	Increase in deaths, injuries, and infections associated with extreme weather (*medium confidence*[d]).	Greater increases in deaths, injuries, and infections (*medium confidence*[d]).	Greater increases in deaths, injuries, and infections (*medium confidence*[d]).
Nutrition [WGII TAR Sections 5.3.6 & 9.9]	Poor are vulnerable to increased risk of hunger, but state of science very incomplete.	Poor remain vulnerable to increased risk of hunger.	Poor remain vulnerable to increased risk of hunger.

[a] The reported ranges for CO_2 concentration are estimated with fast carbon cycle models for the six illustrative SRES scenarios and correspond to the minimum and maximum values estimated with a fast carbon cycle model for the 35 SRES projections of greenhouse gas emissions. See WGI TAR Section 3.7.3.
[b] The reported ranges for global mean temperature change and global mean sea-level rise correspond to the minimum and maximum values estimated with a simple climate model for the 35 SRES projections of greenhouse gas and SO_2 emissions. See WGI TAR Sections 9.3.3 and 11.5.1.
[c] Summary statements about effects of climate change in the years 2025, 2050, and 2100 are inferred from Working Group II's assessment of studies that investigate the impacts of scenarios other than the SRES projections, as studies that use the SRES projections have not been published yet. Estimates of the impacts of climate change vary by region and are highly sensitive to estimates of regional and seasonal patterns of temperature and precipitation changes, changes in the frequencies or intensities of climate extremes, and rates of change. Estimates of impacts are also highly sensitive to assumptions about characteristics of future societies and the extent and effectiveness of future adaptations to climate change. In consequence, summary statements about the impacts of climate change in the years 2025, 2050, and 2100 must necessarily be general and qualitative. The statements in the table are considered to be valid for a broad range of scenarios. Note, however, that few studies have investigated the effects of climate changes that would accompany global temperature increases near the upper end of the range reported for the year 2100.
[d] Judgments of confidence use the following scale: *very high* (95% or greater), *high* (67–95%), *medium* (33–67%), *low* (5–33%), and *very low* (5% or less). See WGII TAR Box 1-1.

reefs and atolls, mangroves, boreal and tropical forests, polar and alpine ecosystems, prairie wetlands, and remnant native grasslands are examples of systems threatened by climate change. In some cases the threatened ecosystems are those that could mitigate against some climate change impacts (e.g., coastal systems that buffer the impacts of storms). Possible adaptation methods to reduce the loss of biodiversity include the establishment of refuges, parks and reserves with corridors to allow migration of species, and the use of captive breeding and translocation of species.

3.19 **The productivity of ecological systems is highly sensitive to climate change and projections of change in productivity range from increases to decreases (*medium confidence*).** Increasing CO_2 concentrations would increase net primary productivity (CO_2 fertilization) and net ecosystem productivity in most vegetation systems, causing carbon to accumulate in vegetation and soils over time. Climate change may either augment or reduce the direct effects of CO_2 on productivity, depending on the type of vegetation, the region, and the scenario of climate change.

WGI TAR Section 3.7 & WGII TAR Sections 5.2.2 & 5.6.3

Table 3-2	Ecosystem effects of climate change if no climate policy interventions are made.[*]		
	2025	*2050*	*2100*
CO₂ concentration[a]	405–460 ppm	445–640 ppm	540–970 ppm
Global mean temperature change from the year 1990[b]	0.4–1.1°C	0.8–2.6°C	1.4–5.8°C
Global mean sea-level rise from the year 1990[b]	3–14 cm	5–32 cm	9–88 cm

Ecosystem Effects[c]

	2025	*2050*	*2100*
Corals [WGII TAR Sections 6.4.5, 12.4.7, & 17.2.4]	Increase in frequency of coral bleaching and death of corals (*high confidence[d]*).	More extensive coral bleaching and death (*high confidence[d]*).	More extensive coral bleaching and death (*high confidence[d]*). Reduced species biodiversity and fish yields from reefs (*medium confidence[d]*).
Coastal wetlands and shorelines [WGII TAR Sections 6.4.2 & 6.4.4]	Loss of some coastal wetlands to sea-level rise (*medium confidence[d]*). Increased erosion of shorelines (*medium confidence[d]*).	More extensive loss of coastal wetlands (*medium confidence[d]*). Further erosion of shorelines (*medium confidence[d]*).	Further loss of coastal wetlands (*medium confidence[d]*). Further erosion of shorelines (*medium confidence[d]*).
Terrestrial ecosystems [WGII TAR Sections 5.2.1, 5.4.1, 5.4.3, 5.6.2, 16.1.3, & 19.2]	Lengthening of growing season in mid- and high latitudes; shifts in ranges of plant and animal species (*high confidence[d]*).[e,f] Increase in net primary productivity of many mid- and high-latitude forests (*medium confidence[d]*). Increase in frequency of ecosystem disturbance by fire and insect pests (*high confidence[d]*).	Extinction of some endangered species; many others pushed closer to extinction (*high confidence[d]*). Increase in net primary productivity may or may not continue. Increase in frequency of ecosystem disturbance by fire and insect pests (*high confidence[d]*).	Loss of unique habitats and their endemic species (e.g., vegetation of Cape region of South Africa and some cloud forests) (*medium confidence[d]*). Increase in frequency of ecosystem disturbance by fire and insect pests (*high confidence[d]*).
Ice environments [WGI TAR Sections 2.2.5 & 11.5; WGII TAR Sections 4.3.11, 11.2.1, 16.1.3, 16.2.1, 16.2.4, & 16.2.7]	Retreat of glaciers, decreased sea-ice extent, thawing of some permafrost, longer ice-free seasons on rivers and lakes (*high confidence[d]*).[f]	Extensive Arctic sea-ice reduction, benefiting shipping but harming wildlife (e.g., seals, polar bears, walrus) (*medium confidence[d]*). Ground subsidence leading to infrastructure damage (*high confidence[d]*).	Substantial loss of ice volume from glaciers, particularly tropical glaciers (*high confidence[d]*).

[*] Refer to footnotes a–d accompanying Table 3-1.
[e] Aggregate market effects represent the net effects of estimated economic gains and losses summed across market sectors such as agriculture, commercial forestry, energy, water, and construction. The estimates generally exclude the effects of changes in climate variability and extremes, do not account for the effects of different rates of change, and only partially account for impacts on goods and services that are not traded in markets. These omissions are likely to result in underestimates of economic losses and overestimates of economic gains. Estimates of aggregate impacts are controversial because they treat gains for some as canceling out losses for others and because the weights that are used to aggregate across individuals are necessarily subjective.
[f] These effects have already been observed and are expected to continue [TAR WGII Sections 5.2.1, 5.4.3, 16.1.3, & 19.2].

3.20 **The terrestrial ecosystems at present are a carbon sink which may diminish with increased warming by the end of the 21st century (see Table 3-2) (*medium confidence*).** The terrestrial ecosystems at present are a sink for carbon. This is partly a result of delays between enhanced plant growth and plant death and decay. Current enhanced plant growth is partly due to fertilization effects of elevated CO₂ on plant photosynthesis (either directly via increased carbon assimilation, or indirectly through higher water-use efficiency), nitrogen deposition (especially in the Northern Hemisphere), climate change, and land-use practices over past decades. The uptake will decline as forests reach maturity, fertilization effects saturate and decomposition catches up with growth, and possibly through changes in disturbance regimes (e.g., fire and insect outbreaks) mediated through climate change. Some global models project that the net uptake of carbon by terrestrial ecosystems

WGI TAR Section 3.2.2, WGII TAR Sections 5.2, 5.5-6, & 5.9, & SRLULUCF Section 1.4

will increase during the first half of the 21st century but may diminish and even become a source with increased warming towards the end of the 21st century.

Agriculture

3.21 **Models of cereal crops indicate that in some temperate areas potential yields increase for small increases in temperature but decrease with larger temperature changes (*medium to low confidence*). In most tropical and subtropical regions potential yields are projected to decrease for most projected increases in temperature (*medium confidence*) (see Table 3-3).** In mid-latitudes, crop models indicate that warming of less than a few °C and the associated increase in CO_2 concentrations will lead to generally positive responses and generally negative responses with greater warming. In tropical agricultural areas, similar assessments indicate that yields of some crops would decrease with even minimal increases in temperature because they are near their maximum temperature tolerance. Where there is also a large decrease in rainfall in subtropical and tropical dryland/rainfed systems, crop yields would be even more adversely affected. Assessments that include autonomous agronomic adaptation (e.g., changes in planting times and crop varieties) tend to project yields less adversely affected by climate change than without adaptation. These assessments include the effects of CO_2 fertilization but not technological innovations or changes in the impacts of pests and diseases, degradation of soil and water resources, or climate extremes. The ability of livestock producers to adapt their herds to the physiological stresses associated with climate change is poorly known. Warming of a few °C or more is projected to increase food prices globally, and may increase the risk of hunger in vulnerable populations (*low confidence*).

WGII TAR Sections 5.3.4-6, & 9.9

Table 3-3	Agricultural effects of climate change if no climate policy interventions are made.[*]		
	2025	*2050*	*2100*
CO_2 concentration[a]	405–460 ppm	445–640 ppm	540–970 ppm
Global mean temperature change from the year 1990[b]	0.4–1.1°C	0.8–2.6°C	1.4–5.8°C
Global mean sea-level rise from the year 1990[b]	3–14 cm	5–32 cm	9–88 cm
Agricultural Effects[c]			
Average crop yields[g] [WGII TAR Sections 5.3.6, 10.2.2, 11.2.2, 12.5, 13.2.3, 14.2.2, & 15.2.3]	Cereal crop yields increase in many mid- and high-latitude regions (*low to medium confidence*[d]). Cereal crop yields decrease in most tropical and subtropical regions (*low to medium confidence*[d]).	Mixed effects on cereal yields in mid-latitude regions. More pronounced cereal yield decreases in tropical and subtropical regions (*low to medium confidence*[d]).	General reduction in cereal yields in most mid-latitude regions for warming of more than a few °C (*low to medium confidence*[d]).
Extreme low and high temperatures [WGII TAR Section 5.3.3]	Reduced frost damage to some crops (*high confidence*[d]). Increased heat stress damage to some crops (*high confidence*[d]). Increased heat stress in livestock (*high confidence*[d]).	Effects of changes in extreme temperatures amplified (*high confidence*[d]).	Effects of changes in extreme temperatures amplified (*high confidence*[d]).
Incomes and prices [WGII TAR Sections 5.3.5-6]		Incomes of poor farmers in developing countries decrease (*low to medium confidence*[d]).	Food prices increase relative to projections that exclude climate change (*low to medium confidence*[d]).

[*] Refer to footnotes a-d accompanying Table 3-1.
[g] These estimates are based on the sensitivity of the present agricultural practices to climate change, allowing (in most cases) for adaptations based on shifting use of only existing technologies.

Water

3.22 **Projected climate change would exacerbate water shortage and quality problems in many water-scarce areas of the world, but alleviate it in some other areas.** Demand for water is generally increasing due to population growth and economic development, but is falling in some countries because of increased efficiency of use. Climate change is projected to reduce streamflow and groundwater recharge in many parts of the world but to increase it in some other areas (*medium confidence*). The amount of change varies among scenarios partly because of differences in projected rainfall (especially rainfall intensity) and partly because of differences in projected evaporation. Projected streamflow changes under two climate change scenarios are shown in Figure 3-5. Several hundred million to a few billion people are projected to suffer a supply reduction of 10% or more by the year 2050 for climate change projections corresponding to 1% per year increase in CO_2 emissions (see Table 3-4). Freshwater quality generally would be degraded by higher water temperatures (*high confidence*), but this may be offset by increased flows in some regions. The effects of climate changes on water scarcity, water quality, and the frequency and intensity of floods and droughts will intensify challenges for water and flood management. Unmanaged and poorly managed water systems are the most vulnerable to adverse effects of climate change.

WGI TAR Section 9.3.6 &
WGII TAR Sections 4.3-4,
4.5.2, & 4.6

Table 3-4	Water resource effects of climate change if no climate policy interventions are made.*		
	2025	*2050*	*2100*
CO_2 concentration[a]	405–460 ppm	445–640 ppm	540–970 ppm
Global mean temperature change from the year 1990[b]	0.4–1.1°C	0.8–2.6°C	1.4–5.8°C
Global mean sea-level rise from the year 1990[b]	3–14 cm	5–32 cm	9–88 cm
Water Resource Effects[c]			
Water supply [WGII TAR Sections 4.3.6 & 4.5.2]	Peak river flow shifts from spring toward winter in basins where snowfall is an important source of water (*high confidence[d]*).	Water supply decreased in many water-stressed countries, increased in some other water-stressed countries (*high confidence[d]*).	Water supply effects amplified (*high confidence[d]*).
Water quality [WGII TAR Section 4.3.10]	Water quality degraded by higher temperatures. Water quality changes modified by changes in water flow volume. Increase in saltwater intrusion into coastal aquifers due to sea-level rise (*medium confidence[d]*).	Water quality degraded by higher temperatures(*high confidence[d]*). Water quality changes modified by changes in water flow volume (*high confidence[d]*).	Water quality effects amplified (*high confidence[d]*).
Water demand [WGII TAR Section 4.4.3]	Water demand for irrigation will respond to changes in climate; higher temperatures will tend to increase demand (*high confidence[d]*).	Water demand effects amplified (*high confidence[d]*).	Water demand effects amplified (*high confidence[d]*).
Extreme events [WGI TAR SPM; WGII TAR SPM]	Increased flood damage due to more intense precipitation events (*high confidence[d]*). Increased drought frequency (*high confidence[d]*).	Further increase in flood damage (*high confidence[d]*). Further increase in drought events and their impacts.	Flood damage several-fold higher than "no climate change scenarios."

* Refer to footnotes a-d accompanying Table 3-1.

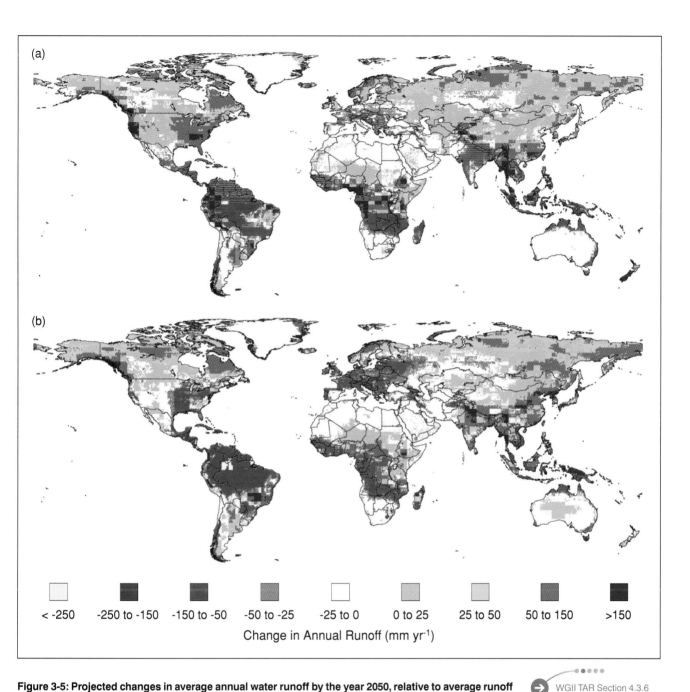

Figure 3-5: Projected changes in average annual water runoff by the year 2050, relative to average runoff for the period 1961–1990, largely follow projected changes in precipitation. Changes in runoff are calculated with a hydrologic model using as inputs climate projections from two versions of the Hadley Centre AOGCM for a scenario of 1% per year increase in effective CO_2 concentration in the atmosphere: (a) HadCM2 ensemble mean and (b) HadCM3. Projected increases in runoff in high latitudes and southeast Asia, and decreases in central Asia, the area around the Mediterranean, southern Africa, and Australia are broadly consistent across the Hadley Centre experiments, and with the precipitation projections of other AOGCM experiments. For other areas of the world, changes in precipitation and runoff are scenario- and model-dependent.

WGII TAR Section 4.3.6

Small Islands and Low-Lying Coasts

3.23 **Populations that inhabit small islands and/or low-lying coastal areas are at particular risk of severe social and economic effects from sea-level rise and storm surges.** Many human settlements will face increased risk of coastal flooding and erosion, and tens of millions of people living in deltas, low-lying coastal areas, and on small islands will face the risk of displacement of populations and loss of infrastructure and/or substantial efforts and costs to protect vulnerable coastal areas. Resources critical

WGII TAR Sections 7.2.2, 17.2, & 19.3.4

to island and coastal populations such as freshwater, fisheries, coral reefs and atolls, beaches, and wildlife habitat would also be at risk.

3.24 **Projected sea-level rise will increase the average annual number of people flooded in coastal storm surges (*high confidence*).** The areas of greatest absolute increase in populations at risk are southern Asia and southeast Asia, with lesser but significant increases in eastern Africa, western Africa, and the Mediterranean from Turkey to Algeria. Significant portions of many highly populated coastal cities are also vulnerable to permanent land submergence and especially to more frequent coastal flooding superimposed on surge heights, due to sea-level rise. These estimates assume no change in the frequency or intensity of storms, which could exacerbate the effects of sea-level rise on flooding risks in some areas.

WGII TAR Sections 6.5.1, 7.2.2, & 17.2.2

Market Effects

3.25 **The aggregated market sector effects, measured as changes in gross domestic product (GDP), are estimated to be negative for many developing countries for all magnitudes of global mean temperature increases studied (*low confidence*), and are estimated to be mixed for developed countries for up to a few °C warming (*low confidence*) and negative for warming beyond a few °C (*medium to low confidence*).** The effects of climate change will have market sector effects by changing the abundance, quality, and prices of food, fiber, water, and other goods and services (see Table 3-5). In addition, climate change can have market effects through changes in energy demand, hydropower supply, transportation, tourism and construction, damages to property and insurance losses from extreme climate events, loss of coastal land from sea-level rise, location and relocation decisions for development and populations, and the resource needs and costs of adapting to climate change. Estimates of net market effects from a few published studies, aggregated across sectors and to national or regional scales, indicate losses for most developing countries and regions studied. Both gains and losses are estimated for developed

WGII TAR Sections 6.5, 7.2-3, 8.3, 18.3.4, 18.4.3, 19.4.1-3, & 19.5

Table 3-5	Other market sector effects of climate change if no climate policy interventions are made.*		
	2025	*2050*	*2100*
CO_2 concentration[a]	405–460 ppm	445–640 ppm	540–970 ppm
Global mean temperature change from the year 1990[b]	0.4–1.1°C	0.8–2.6°C	1.4–5.8°C
Global mean sea-level rise from the year 1990[b]	3–14 cm	5–32 cm	9–88 cm
Other Market Sector Effects[c]			
Energy [WGII TAR Section 7.3]	Decreased energy demand for heating buildings (*high confidence*[d]). Increased energy demand for cooling buildings (*high confidence*[d]).	Energy demand effects amplified (*high confidence*[d]).	Energy demand effects amplified (*high confidence*[d]).
Financial sector [WGII TAR Section 8.3]		Increased insurance prices and reduced insurance availability (*high confidence*[d]).	Effects on financial sector amplified.
Aggregate market effects[e] [WGII TAR Sections 19.4-5]	Net market sector losses in many developing countries (*low confidence*[d]). Mixture of market gains and losses in developed countries (*low confidence*[d]).	Losses in developing countries amplified (*medium confidence*[d]). Gains diminished and losses amplified in developed countries (*medium confidence*[d]).	Losses in developing countries amplified (*medium confidence*[d]). Net market sector losses in developed countries from warming of more than a few °C (*medium confidence*[d]).
* Refer to footnotes a-d accompanying Table 3-1 and footnote e accompanying Table 3-2.			

countries and regions for increases in global mean temperature of up to a few °C. Economic losses are estimated for developed countries at larger temperature increases. When aggregated to a global scale, world GDP would change by plus or minus a few percent for global mean temperature increases of up to a few °C, with increasing net losses for larger increases in temperature. The estimates generally exclude the effects of changes in climate variability and extremes, do not account for the effects of different rates of climate change, only partially account for impacts on goods and services that are not traded in markets, and treat gains for some as canceling out losses for others. Therefore, confidence in estimates of market effects for individual countries is generally *low*, and the various omissions are likely to result in underestimates of economic losses and overestimates of economic gains.

3.26 Adaptation has the potential to reduce adverse effects of climate change and can often produce immediate ancillary benefits, but will not prevent all damages.

3.27 **Numerous possible adaptation options for responding to climate change have been identified that can reduce adverse and enhance beneficial impacts of climate change, but will incur costs.** Quantitative evaluation of their benefits and costs and how they vary across regions and entities is incomplete. Adaptation to climate change can take many forms, including actions taken by people with the intent of lessening impacts or utilizing new opportunities, and structural and functional changes in natural systems made in response to changes in pressures. The focus in this report is on the adaptive actions of people. The range of options includes reactive adaptations (actions taken concurrent with changed conditions and without prior preparation) and planned adaptations (actions taken either concurrent with or in anticipation of changed conditions, but with prior preparation). Adaptations can be taken by private entities (e.g., individuals, households, or business firms) or by public entities (e.g., local, state, or national government agencies). Examples of identified options are listed in Table 3-6. The benefits and costs of adaptation options, evaluation of which is incomplete, will also vary across regions and entities. Despite the incomplete and evolving state of knowledge about adaptation, a number of robust findings have been derived and summarized.

WGII TAR Sections 18.2.3 & 18.3.5

3.28 **Greater and more rapid climate change would pose greater challenges for adaptation and greater risks of damages than would lesser and slower change.** Key features of climate change to be adapted to include the magnitudes and rates of changes in climate extremes, variability, and mean conditions. Natural and human systems have evolved capabilities to cope with a range of climate variability within which the risks of damage are relatively low and ability to recover is high. Changes in climate that result in increased frequency of events that fall outside the historic range with which systems have coped, however, increase the risk of severe damages and incomplete recovery or collapse of the system. Changes in mean conditions (e.g., increases in average temperature), even in the absence of changes in variance, can lead to increases in the frequencies of some events (e.g., more frequent heat waves) that exceed the coping range, and decreases in the frequencies of others (e.g., less frequent cold spells) (see Question 4 and Figure 4-1).

WGII TAR Sections 18.2.2, 18.3.3, & 18.3.5

3.29 **Enhancement of adaptive capacity can extend or shift ranges for coping with variability and extremes to generate benefits in the present and future.** Many of the adaptation options listed in Table 3-6 are presently employed to cope with current climate variability and extremes, and their expanded use can enhance both current and future capacity to cope. But such efforts may not be as effective in the future as the amount and rate of climate change increase.

WGII TAR Sections 18.2.2 & 18.3.5

3.30 **The potential direct benefits of adaptation are substantial and take the form of reduced adverse and enhanced beneficial impacts of climate change.** Results of studies of future impacts of climate change indicate the potential for adaptation to

WGII TAR Sections 5.3.4, 6.5.1, & 18.3.2

Table 3-6	Examples of adaptation options for selected sectors.
Sector/System	**Adaptation Options**
Water [WGII TAR Sections 4.6 & 7.5.4; WGII SAR Sections 10.6.4 & 14.4]	Increase water-use efficiency with "demand-side" management (e.g., pricing incentives, regulations, technology standards). Increase water supply, or reliability of water supply, with "supply-side" management (e.g., construct new water storage and diversion infrastructure). Change institutional and legal framework to facilitate transfer of water among users (e.g., establish water markets). Reduce nutrient loadings of rivers and protect/augment streamside vegetation to offset eutrophying effects of higher water temperatures. Reform flood management plans to reduce downstream flood peaks; reduce paved surfaces and use vegetation to reduce storm runoff and increase water infiltration. Reevaluate design criteria of dams, levees, and other infrastructure for flood protection.
Food and fiber [WGII TAR Sections 5.3.4-5; WGII SAR Sections 2.9, 4.4.4, 13.9, & 15.6; SRTT Section 11.2.1]	Change timing of planting, harvesting, and other management activities. Use minimum tillage and other practices to improve nutrient and moisture retention in soils and to prevent soil erosion. Alter animal stocking rates on rangelands. Switch to crops or crop cultivars that are less water-demanding and more tolerant of heat, drought, and pests. Conduct research to develop new cultivars. Promote agroforestry in dryland areas, including establishment of village woodlots and use of shrubs and trees for fodder. Replant with mix of tree species to increase diversity and flexibility. Promote revegetation and reforestation initiatives. Assist natural migration of tree species with connected protected areas and transplanting. Improve training and education of rural work forces. Establish or expand programs to provide secure food supplies as insurance against local supply disruptions. Reform policies that encourage inefficient, non-sustainable, or risky farming, grazing, and forestry practices (e.g., subsidies for crops, crop insurance, water).
Coastal areas and marine fisheries [WGII TAR Sections 6.6 & 7.5.4; WGII SAR Section 16.3; SRTT Section 15.4]	Prevent or phase-out development in coastal areas vulnerable to erosion, inundation, and storm-surge flooding. Use "hard" (dikes, levees, seawalls) or "soft" (beach nourishment, dune and wetland restoration, afforestation) structures to protect coasts. Implement storm warning systems and evacuation plans. Protect and restore wetlands, estuaries, and floodplains to preserve essential habitat for fisheries. Modify and strengthen fisheries management institutions and policies to promote conservation of fisheries. Conduct research and monitoring to better support integrated management of fisheries.
Human health [WGII TAR Sections 7.5.4 & 9.11; WGII SAR Section 12.5; SRTT Section 14.4]	Rebuild and improve public health infrastructure. Improve epidemic preparedness and develop capacities for epidemic forecasting and early warning. Monitor environmental, biological, and health status. Improve housing, sanitation, and water quality. Integrate urban designs to reduce heat island effect (e.g., use of vegetation and light colored surfaces). Conduct public education to promote behaviors that reduce health risks.
Financial services [WGII TAR Section 8.3.4]	Risk spreading through private and public insurance and reinsurance. Risk reduction through building codes and other standards set or influenced by financial sector as requirements for insurance or credit.

substantially reduce many of the adverse impacts and enhance beneficial impacts. For example, analyses of coastal flood risks from storm surges estimate that climate change-driven sea-level rise would increase the average annual number of people flooded many-fold if coastal flood protection is unchanged from the present. But if coastal flood protection is enhanced in proportion to future GDP growth, the projected increase is cut by as much as two-thirds (see Figure 3-6). However, estimates such as these indicate only potential benefits from adaptation, not the likely benefits—as analyses generally use arbitrary assumptions about adaptation options and obstacles, often omit consideration of changes in climate extremes and variability, and do not account for imperfect foresight.

3.31 **Estimates of the costs of adaptation are few; the available estimates indicate that costs are highly sensitive to decision criteria for the selection and timing of specific adaptation measures.** The costs of measures to protect coastal areas from sea-level rise are perhaps the best studied to date. Evaluated measures include construction

WGII TAR Sections 6.5.2 & 18.4.3

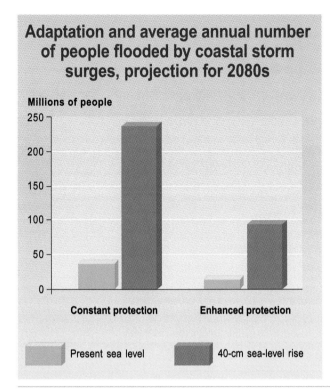

Adaptation and average annual number of people flooded by coastal storm surges, projection for 2080s

Millions of people

WGII TAR Section 6.5.1

Figure 3-6: Adaptation and the average annual number of people flooded by coastal storm surges, projection for the 2080s. The left two bars show the average annual number of people projected to be flooded by coastal storm surges in the year 2080 for present sea level and for a rise in sea level of ~40 cm, assuming that coastal protection is unchanged from the present and a moderate population increase. The right two bars show the same, but assuming that coastal protection is enhanced in proportion to GDP growth.

of "hard structures" such as dikes, levees, and seawalls, and the use of "soft structures" such as nourishment of beaches with sand and dune restoration. Estimates of the costs of protecting coasts vary depending on assumptions about what decisions will be made regarding the extent of the coastline to be protected, the types of structures to be used, the timing of their implementation (which is influenced by the rate of sea-level rise), and discount rates. Different assumptions about these factors yield estimates for protection of U.S. coasts from 0.5-m sea-level rise by the year 2100 that range from US$20 billion to US$150 billion in present value.

3.32 **Climate change is expected to negatively impact development, sustainability, and equity.**

3.33 **The impacts of climate change will fall disproportionately upon developing countries and the poor persons within all countries, and thereby exacerbate inequities in health status and access to adequate food, clean water, and other resources.** As already noted, populations in developing countries are generally expected to be exposed to relatively high risks of adverse impacts from climate change on human health, water supplies, agricultural productivity, property, and other resources. Poverty, lack of training and education, lack of infrastructure, lack of access to technologies, lack of diversity in income opportunities, degraded natural resource base, misplaced incentives, inadequate legal framework, and struggling public and private institutions create conditions of low adaptive capacity in most developing countries. The exposures and low capacity to adapt combine to make populations in developing countries generally more vulnerable than populations in developed countries.

WGII TAR Sections 18.5.1-3

3.34 **Non-sustainable resource use adds to the vulnerability to climate change.** Conversion of natural habitat to human uses, high harvesting rates of resources from the environment, cultivation and grazing practices that fail to protect soils from degradation, and pollution of air and water can reduce the robustness of systems to cope with variations or change in climate, and the resilience of systems to recover from declines. Such pressures

WGII TAR Sections 1.2.2, 4.7, 5.1, 6.3.4, & 6.4.4

make systems, and the populations that derive goods, services, and livelihoods from them, highly vulnerable to climate change. These pressures are present in developed as well as developing countries, but satisfying development goals in ways that do not place non-sustainable pressures on systems pose a particular dilemma for developing countries.

3.35 **Hazards associated with climate change can undermine progress toward sustainable development.** More frequent and intensified droughts can exacerbate land degradation. Increases in heavy precipitation events can increase flooding, landslides, and mudslides, the destruction from which can set back development efforts by years in some instances. Advances in health and nutritional status could be set back in some areas by climate change impacts on human health and agriculture. Hazards such as these can also be exacerbated by further development in inherently dynamic and unstable zones (e.g., floodplains, barrier beaches, low-lying coasts, and deforested steep slopes).

3.36 **Climate change can detract from the effectiveness of development projects if not taken into account.** Development projects often involve investments in infrastructure, institutions, and human capital for the management of climate-sensitive resources such as water, hydropower, agricultural lands, and forests. The performance of these projects can be affected by climate change and increased climate variability, yet these factors are given little consideration in the design of projects. Analyses have shown that flexibility to perform well under a wider range of climate conditions can be built into projects at modest incremental costs in some instances, and that greater flexibility has immediate value because of risks from present climate variability.

3.37 **Many of the requirements for enhancing capacity to adapt to climate change are similar to those for promoting sustainable development.** Examples of common requirements for enhancing adaptive capacity and sustainable development include increasing access to resources and lowering inequities in access, reducing poverty, improving education and training, investing in infrastructure, involving concerned parties in managing local resources, and raising institutional capacities and efficiencies. Additionally, initiatives to slow habitat conversion, manage harvesting practices to better protect the resource, adopt cultivation and grazing practices that protect soils, and better regulate the discharge of pollutants can reduce vulnerabilities to climate change while moving toward more sustainable use of resources.

WGII TAR Section 18.6.1

Question 4

Q4

What is known about the influence of the increasing atmospheric concentrations of greenhouse gases and aerosols, and the projected human-induced change in climate regionally and globally on:

a. The frequency and magnitude of climate fluctuations, including daily, seasonal, inter-annual, and decadal variability, such as the El Niño Southern Oscillation cycles and others?
b. The duration, location, frequency, and intensity of extreme events such as heat waves, droughts, floods, heavy precipitation, avalanches, storms, tornadoes, and tropical cyclones?
c. The risk of abrupt/non-linear changes in, among others, the sources and sinks of greenhouse gases, ocean circulation, and the extent of polar ice and permafrost? If so, can the risk be quantified?
d. The risk of abrupt or non-linear changes in ecological systems?

4.1   This answer focuses on projected changes in the frequency and magnitude of climate fluctuations as a result of increasing concentrations of greenhouse gases and aerosols. Particular emphasis is placed on changes in the frequency, magnitude, and duration of climatic extremes, which represent important climate change risks for ecological systems and socio-economic sectors. Projected abrupt or other non-linear changes in the biophysical system are discussed here; the gradual changes in the physical, biological, and social systems are discussed in Question 3.

4.2   **Models project that increasing atmospheric concentrations of greenhouse gases will result in changes in daily, seasonal, inter-annual, and decadal variability.** There is projected to be a decrease in diurnal temperature range in many areas, with nighttime lows increasing more than daytime highs. A number of models show a general decrease of daily variability of surface air temperature in winter and increased daily variability in summer in the Northern Hemisphere land areas. Current projections show little change or a small increase in amplitude for El Niño events over the next 100 years. Many models show a more El Niño-like mean response in the tropical Pacific, with the central and eastern equatorial Pacific sea surface temperatures projected to warm more than the western equatorial Pacific and with a corresponding mean eastward shift of precipitation. Even with little or no change in El Niño strength, global warming is likely to lead to greater extremes of drying and heavy rainfall and increase the risk of droughts and floods that occur with El Niño events in many different regions. There is no clear agreement between models concerning the changes in frequency or structure of other naturally occurring atmosphere-ocean circulation pattern such as the North Atlantic Oscillation (NAO).

WGI TAR Sections 9.3.5-6, & WGII TAR Section 14.1.3

4.3   **The duration, location, frequency, and intensity of extreme weather and climate events are likely to very likely to change, and would result in mostly adverse impacts on biophysical systems.**

4.4   Natural circulation patterns, such as ENSO and NAO, play a fundamental role in global climate and its short-term (daily, intra- and inter-annual) and longer term (decadal to multi-decadal) variability. Climate change may manifest itself as a shift in means as well as a change in preference of specific climate circulation patterns that could result in changes in the variance and frequency of extremes of climatic variables (see Figure 4-1).

WGI TAR Sections 1.2 & 2.7

4.5   **More hot days and heat waves and fewer cold and frost days are very likely over nearly all land areas.** Increases in mean temperature will lead to increases in hot weather and record hot weather, with fewer frost days and cold waves (see Figure 4-1a,b). A number of models show a generally decreased daily variability of surface air temperature in winter and increased daily variability in summer in Northern Hemisphere land areas. The changes in temperature extremes are likely to result in increased crop and livestock losses, higher energy use for cooling and lower for heating, and increased human morbidity and heat-stress-related mortality (see Table 4-1). Fewer frost days will result in decreased cold-related human morbidity and mortality, and decreased risk of damage to a number of crops, though the risk to other crops may increase. Benefits to agriculture from a small temperature increase could result in small increases in the GDP of temperate zone countries.

WGI TAR Sections 9.3.6 & 10.3.2, & WGII TAR Sections 5.3, 9.4.2, & 19.5

4.6   **The amplitude and frequency of extreme precipitation events is very likely to increase over many areas** and the return period for extreme precipitation events are projected to decrease. This would lead to more frequent floods and landslides with attendant loss of life, health impacts (e.g., epidemics, infectious diseases, food poisoning), property damage, loss to infrastructure and settlements, soil erosion, pollution loads, insurance and agriculture losses, amongst others. A general drying of the mid-continental areas during summer is likely to lead to increases in summer droughts and could increase the risk of wild fires. This general drying is due to a combination of increased temperature and potential evaporation that is not balanced by increases in precipitation. It is likely that global warming will lead to an increase in the variability of Asian summer monsoon precipitation.

WGI TAR Section 9.3.6 & WGII TAR Sections 4.3.8, 9.5.3, 9.7.10, & 9.8

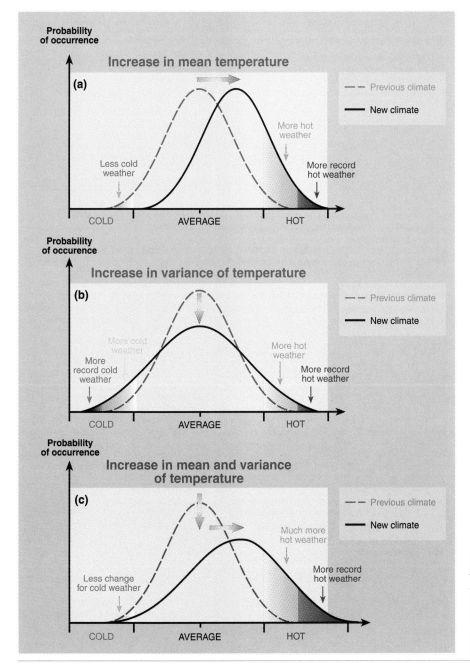

WGI TAR Figure 2.32

Figure 4-1: Schematic diagrams showing the effects on extreme temperatures when (a) the mean increases, leading to more record hot weather, (b) the variance increases, and (c) when both the mean and variance increase, leading to much more record hot weather.

4.7 **High resolution modeling studies suggest that over some areas the peak wind intensity of tropical cyclones is likely to increase** by 5 to 10% and precipitation rates may increase by 20 to 30%, but none of the studies suggest that the locations of the tropical cyclones will change. There is little consistent modeling evidence for changes in the frequency of tropical cyclones.

WGI TAR Box 10.2

4.8 **There is insufficient information on how very small-scale phenomena may change.** Very small-scale phenomena such as thunderstorms, tornadoes, hail, hailstorms, and lightning are not simulated in global climate models.

WGI TAR Section 9.3.6

4.9 **Greenhouse gas forcing in the 21st century could set in motion large-scale, high-impact, non-linear, and potentially abrupt changes in physical and biological systems over the coming decades to millennia, with a wide range of associated likelihoods.**

Table 4-1	Examples of climate variability and extreme climate events and examples of their impacts (WGII TAR Table SPM-1).
Projected Changes during the 21st Century in Extreme Climate Phenomena and their Likelihood	**Representative Examples of Projected Impacts[a] (all high confidence of occurrence in some areas)**
Higher maximum temperatures, more hot days and heat waves[b] over nearly all land areas (*very likely*)	Increased incidence of death and serious illness in older age groups and urban poor. Increased heat stress in livestock and wildlife. Shift in tourist destinations. Increased risk of damage to a number of crops. Increased electric cooling demand and reduced energy supply reliability.
Higher (increasing) minimum temperatures, fewer cold days, frost days, and cold waves[b] over nearly all land areas (*very likely*)	Decreased cold-related human morbidity and mortality. Decreased risk of damage to a number of crops, and increased risk to others. Extended range and activity of some pest and disease vectors. Reduced heating energy demand.
More intense precipitation events (*very likely*, over many areas)	Increased flood, landslide, avalanche, and mudslide damage. Increased soil erosion. Increased flood runoff could increase recharge of some floodplain aquifers. Increased pressure on government and private flood insurance systems and disaster relief.
Increased summer drying over most mid-latitude continental interiors and associated risk of drought (*likely*)	Decreased crop yields. Increased damage to building foundations caused by ground shrinkage. Decreased water resource quantity and quality. Increased risk of forest fire.
Increase in tropical cyclone peak wind intensities, mean and peak precipitation intensities (*likely*, over some areas)[c]	Increased risks to human life, risk of infectious disease epidemics and many other risks. Increased coastal erosion and damage to coastal buildings and infrastructure. Increased damage to coastal ecosystems such as coral reefs and mangroves.
Intensified droughts and floods associated with El Niño events in many different regions (*likely*) (see also under droughts and intense precipitation events)	Decreased agricultural and rangeland productivity in drought- and flood-prone regions. Decreased hydro-power potential in drought-prone regions.
Increased Asian summer monsoon precipitation variability (*likely*)	Increase in flood and drought magnitude and damages in temperate and tropical Asia.
Increased intensity of mid-latitude storms (little agreement between current models)[b]	Increased risks to human life and health. Increased property and infrastructure losses. Increased damage to coastal ecosystems.

[a] These impacts can be lessened by appropriate response measures.
[b] Information from WGI TAR Technical Summary (Section F.5).
[c] Changes in regional distribution of tropical cyclones are possible but have not been established.

4.10 The climate system involves many processes that interact in complex non-linear ways, which can give rise to thresholds (thus potentially abrupt changes) in the climate system that could be crossed if the system were perturbed sufficiently. These abrupt and other non-linear changes include large climate-induced increase in greenhouse gas emissions from terrestrial ecosystems, a collapse of the thermohaline circulation (THC; see Figure 4-2), and disintegration of the Antarctic and the Greenland ice sheets. Some of these changes have low probability of occurrence during the 21st century; however, greenhouse gas forcing in the 21st century could set in motion changes that could lead to such transitions in subsequent centuries (see Question 5). Some of these changes (e.g., to THC) could be irreversible over centuries to millennia. There is a large degree of uncertainty about the mechanisms involved and about the likelihood or time scales of such changes; however, there is evidence from polar ice cores of atmospheric regimes changing within a few years and large-scale hemispheric changes as fast as a few decades with large consequences on the biophysical systems.

WGI TAR Sections 7.3, 9.3.4, & 11.5.4; WGII TAR Sections 5.2 & 5.8; & SRLULUCF Chapters 3 & 4

4.11 **Large climate-induced increases in greenhouse gas emissions due to large-scale changes in soils and vegetation may be possible in the 21st century.** Global warming interacting with other environmental stresses and human activity could lead to the rapid breakdown of existing ecosystems. Examples include drying of the tundra,

WGII TAR Sections 5.2, 5.8, & 5.9; & SRLULUCF Chapters 3 & 4

boreal and tropical forests, and their associated peatlands leaving them susceptible to fires. Such breakdowns could induce further climate change through increased emissions of CO_2 and other greenhouse gases from plants and soil and changes in surface properties and albedo.

4.12 **Large, rapid increases in atmospheric CH_4 either from reductions in the atmospheric chemical sink or from release of buried CH_4 reservoirs appear exceptionally unlikely.** The rapid increase in CH_4 lifetime possible with large emissions of tropospheric pollutants does not occur within the range of SRES scenarios. The CH_4 reservoir buried in solid hydrate deposits under permafrost and ocean sediments is enormous, more than 1,000-fold the current atmospheric content. A proposed climate feedback occurs when the hydrates decompose in response to warming and release large amounts of CH_4; however, most of the CH_4 gas released from the solid form is decomposed by bacteria in the sediments and water column, thus limiting the amount emitted to the atmosphere unless explosive ebullient emissions occur. The feedback has not been quantified, but there are no observations to support a rapid, massive CH_4 release in the record of atmospheric CH_4 over the past 50,000 years.

WGI TAR Section 4.2.1.1

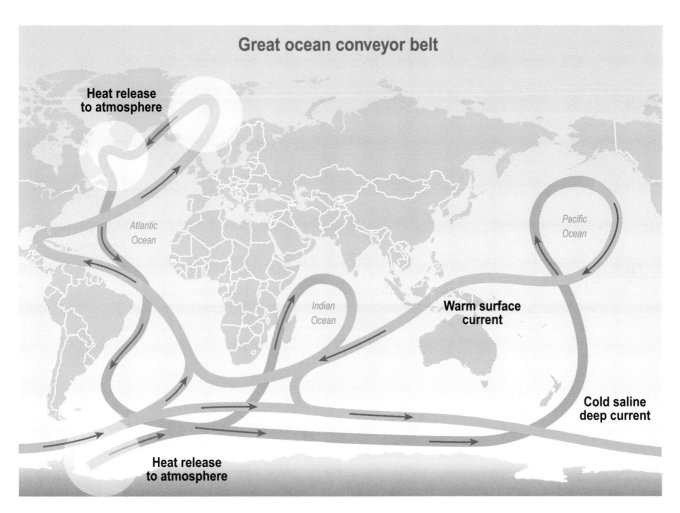

Figure 4-2: Schematic illustration of the global circulation system in the world ocean consisting of major north-south thermohaline circulation routes in each ocean basin joining in the Antarctic circumpolar circulation. Warm surface currents and cold deep currents are connected in the few areas of deepwater formation in the high latitudes of the Atlantic and around Antarctica (blue), where the major ocean-to-atmosphere heat transfer occurs. This current system contributes substantially to the transport and redistribution of heat (e.g., the poleward flowing currents in the North Atlantic warm northwestern Europe by up to 10°C). Model simulations indicate that the North Atlantic branch of this circulation system is particularly vulnerable to changes in atmospheric temperature and in the hydrological cycle. Such perturbations caused by global warming could disrupt the current system, which would have a strong impact on regional-to-hemispheric climate. Note that this is a schematic diagram and it does not give the exact locations of the water currents that form part of the THC.

4.13 **Most models project a weakening of the ocean thermohaline circulation, which leads to a reduction of the heat transport into high latitudes of Europe (see Figure 4-2).** However, even in models where THC weakens, there is still a warming over Europe due to increased concentrations of greenhouse gases. The current projections do not exhibit a complete shutdown of THC by the year 2100. Beyond the year 2100, some models suggest that THC could completely, and possibly irreversibly, shut down in either hemisphere if the change in radiative forcing is large enough and applied long enough. Models indicate that a decrease in THC reduces its resilience to perturbations (i.e., a once-reduced THC appears to be less stable and a shutdown can become more likely).

WGI TAR SPM & WGI TAR Sections 7.3 & 9.3.4

4.14 **The Antarctic ice sheet as a whole is likely to increase in mass during the 21st century. However, the West Antarctic ice sheet could lose mass over the next 1,000 years with an associated sea-level rise of several meters, but there is an incomplete understanding of some of the underlying processes.** Concerns have been expressed about the stability of the West Antarctic ice sheet because it is grounded below sea level. However, loss of grounded ice leading to substantial sea-level rise from this source is widely agreed to be very unlikely during the 21st century. Current climate and ice dynamic models project that over the next 100 years the Antarctic ice sheet as a whole is likely to gain mass because of a projected increase in precipitation, contributing to a relative decrease of several centimeters to sea level. Over the next 1,000 years, these models project that the West Antarctic ice sheet could contribute up to 3 m to sea-level rise.

WGI TAR Section 11.5.4

4.15 **The Greenland ice sheet is likely to lose mass during the 21st century and contribute a few centimeters to sea-level rise.** Over the 21st century, the Greenland ice sheet is likely to lose mass because the projected increase in runoff will exceed the increase in precipitation and contribute 10 cm maximum to the total sea-level rise. The ice sheets will continue to react to climate warming and contribute to sea-level rise for thousands of years after climate has stabilized. Climate models indicate that the local warming over Greenland is likely to be one to three times the global average. Ice sheet models project that a local warming of larger than 3°C, if sustained for thousands of years, would lead to virtually a complete melting of the Greenland ice sheet with a resulting sea-level rise of about 7 m. A local warming of 5.5°C, if sustained for 1,000 years, would likely result in a contribution from Greenland of about 3 m to sea-level rise (see Question 3).

WGI TAR Section 11.5.4

4.16 **Pronounced changes in permafrost temperature, surface morphology, and distribution are expected in the 21st century.** Permafrost currently underlies 24.5% of the exposed land area of the Northern Hemisphere. Under climatic warming, much of this terrain would be vulnerable to subsidence, particularly in areas of relatively warm, discontinuous permafrost. The area of the Northern Hemisphere occupied by permafrost could eventually be reduced by 12 to 22% of its current extent and could eventually disappear from half the present-day Canadian permafrost region. The changes on the southern limit may become obvious by the late 21st century, but some thick ice-rich permafrost could persist in relict form for centuries or millennia. Thawing of ice-rich permafrost can be accompanied by mass movements and subsidence of the surface, possibly increasing the sediment loads in water courses and causing damage to the infrastructure in developed regions. Depending on the precipitation regime and drainage conditions, degradation of permafrost could lead to emission of greenhouse gases, conversion of forest to bogs, grasslands, or wetland ecosystems and could cause major erosion problems and landslides.

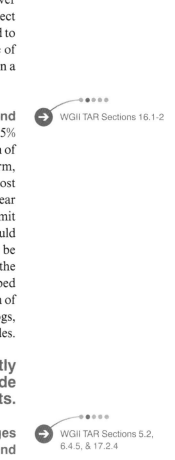

WGII TAR Sections 16.1-2

4.17 **Many natural and managed ecosystems may change abruptly or non-linearly during the 21st century. The greater the magnitude and rate of the change, the greater the risk of adverse impacts.**

4.18 **Changes in climate could increase the risk of abrupt and non-linear changes in many ecosystems, which would affect their biodiversity, productivity, and**

WGII TAR Sections 5.2, 6.4.5, & 17.2.4

function. For example, sustained increases in water temperatures of as little as 1°C, alone or in combination with any of several stresses (e.g., excessive pollution and siltation), can lead to corals ejecting their algae (coral bleaching; see Figure 4-3 and Question 2), the eventual death of the corals, and a possible loss of biodiversity. Climate change will also shift suitable habitats for many terrestrial and marine organisms polewards or terrestrial ones to higher altitudes in mountainous areas. Increased disturbances along with the shift in habitats and the more restrictive conditions needed for establishment of species could lead to abrupt and rapid breakdown of terrestrial and marine ecosystems, which could result in new plant and animal assemblages that are less diverse, that include more "weedy" species, and that increase risk of extinctions (see Question 3).

4.19 **Ecological systems have many interacting non-linear processes and are thus subject to abrupt changes and threshold effects arising from relatively small changes in driving variables, such as climate.** For example:

· Temperature increase beyond a threshold, which varies by crop and variety, can affect key development stages of some crops and result in severe losses in crop yields. Examples of key development stages and their critical thresholds include spikelet sterility in rice (e.g., temperatures greater than 35°C for more than 1 hour during the flowering and pollination process greatly reduce flower formation and eventually grain production), loss of pollen viability in maize (>35°C), reversal of cold-hardening in wheat (>30°C for more than 8 hours), and reduced formation of tubers and tuber bulking in potatoes (>20°C). Yield losses in these crops can be severe if temperatures exceed critical limits for even short periods.

WGII SAR Sections 13.2.2 & 13.6.2

· Mangroves occupy a transition zone between sea and land that is set by a balance between the erosional processes from the sea and siltation processes from land. The erosional

WGII TAR Sections 5.3, 10.2.2, 15.2, & 17.2

Figure 4-3: The diversity of corals could be affected with the branching corals (e.g., staghorn coral) decreasing or becoming locally extinct as they tend to be more severely affected by increases in sea surface temperatures, and the massive corals (e.g., brain corals) increasing.

WGII TAR Section 17.2.4

processes from the sea might be expected to increase with sea-level rise, and the siltation processes through climate change and other human activities (e.g., coastal development). Thus, the impact on the mangrove forests will be determined by the balance between these two processes, which will determine whether mangrove systems migrate landward or seaward.

4.20 **Large-scale changes in vegetation cover could affect regional climate.** Changes in land surface characteristics, such as those created by land cover, can modify energy, water, and gas fluxes and affect atmospheric composition creating changes in local/regional climate and thus changing the disturbance regime (e.g., in the Arctic). In areas without surface water (typically semi-arid or arid), evapotranspiration and albedo affect the local hydrologic cycle, thus a reduction in vegetative cover could lead to reduced precipitation at the local/regional scale and change the frequency and persistence of droughts.

WGII TAR Sections 1.3.1, 5.2, 5.9, 10.2.6.3, 13.2.2, 13.6.2, & 14.2.1

Question 5

What is known about the inertia and time scales associated with the changes in the climate system, ecological systems, and socio-economic sectors and their interactions?

Q5

| **Box 5-1** | Time scale and inertia. |

The terms "time scale" and "inertia" have no generally accepted meaning across all the disciplines involved in the TAR. The following definitions are applied for the purpose of responding to this question:
- "Time scale" is the time taken for a perturbation in a process to show at least half of its final effect. The time scales of some key Earth system processes are shown in Figure 5-1.
- "Inertia" means a delay, slowness, or resistance in the response of climate, biological, or human systems to factors that alter their rate of change, including continuation of change in the system after the cause of that change has been removed.

These are only two of several concepts used in the literature to describe the responses of complex, non-linear, adaptive systems to external forcing.

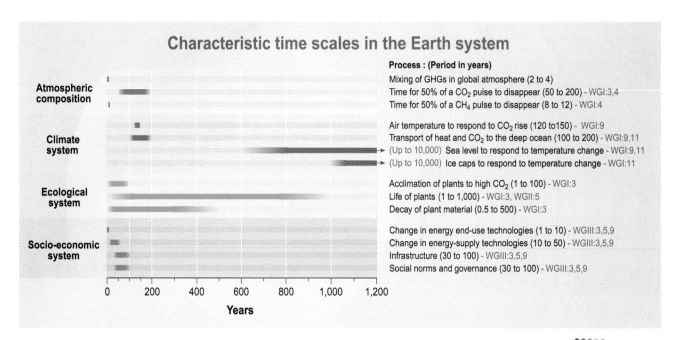

Figure 5-1: The characteristic time scales of some key processes in the Earth system: atmospheric composition (blue), climate system (red), ecological system (green), and socio-economic system (purple). "Time scale" is defined here as the time needed for at least half of the consequences of a change in a driver of the process to have been expressed. Problems of adaptation arise when response process (such as the longevity of some plants) are much slower than driving process (the change in temperature). Inter-generational equity problems arise for all processes with time scales greater than a human generation, since a large part of the consequences of activities of a given generation will be borne by future generations.

WGI TAR Chapters 3, 4, 7, & 11, WGII TAR Chapter 5, & WGIII TAR Chapters 5, 6, & 10

5.1 This response dicusses, and gives examples of, inertia and varying time scales associated with important processes in the interacting climate, ecological, and socio-economic systems. It then discusses potentially irreversible changes—that is, situations where parts of the climate, ecological, or socio-economic systems may fail to return to their former state within time scales of multiple human generations after the driving forces leading to change are reduced or removed. Finally, it explores how the effects of inertia may influence decisions regarding the mitigation of, or adaptation to, climate change.

5.2 **Inertia is a widespread inherent characteristic of the interacting climate, ecological, and socio-economic systems. Thus some impacts of anthropogenic climate change may be slow to become apparent, and some could be irreversible if climate change is not limited in both rate and magnitude before associated thresholds, whose positions may be poorly known, are crossed.**

5.3 **The combined effect of the interacting inertias of the various component processes is such that stabilization of the climate and climate-impacted**

WGI TAR Sections 3.2, 3.7, & 4.2, & WGI TAR Figure 9.16

systems will only be achieved long after anthropogenic emissions of greenhouse gases have been reduced. The perturbation of the atmosphere and oceans, resulting from CO_2 already emitted due to human activities since 1750, will persist for centuries because of the slow redistribution of carbon between large ocean and terrestrial reservoirs with slow turnover (see Figures 5-2 and 5-4). The future atmospheric concentration of CO_2 is projected to remain for centuries near the highest level reached, since natural processes can only return the concentration to pre-industrial levels over geological time scales. By contrast, stabilization of emissions of shorter lived greenhouse gases such as CH_4 leads, within decades, to stabilization of atmospheric concentrations. Inertia also implies that avoidance of emissions of long-lived greenhouse gases has long-lasting benefits.

5.4   The oceans and cryosphere (ice caps, ice sheets, glaciers, and permafrost) are the main sources of physical inertia in the climate system for time scales up to 1,000 years. Due to the great mass, thickness, and thermal capacity of the oceans and cryosphere, and the slowness of the heat transport process, linked ocean-climate models predict that the average temperature of the atmosphere near the Earth's surface will take hundreds of years to finally approach the new "equilibrium" temperature following a change in radiative forcing. Penetration of heat from the atmosphere into the upper "mixed layer" of the ocean occurs within decades, but transport of heat into the deep ocean requires centuries. An associated consequence is that human-induced sea-level rise will continue inexorably for many centuries after the atmospheric concentration of greenhouse gases has been stabilized.

WGI TAR Sections 7.3, 7.5, & 11.5.4, & WGI TAR Figures 9.1, 9.24, & 11.16

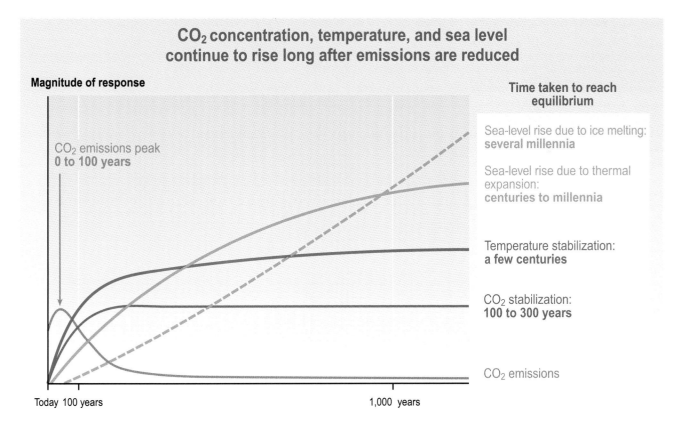

Figure 5-2: After CO_2 emissions are reduced and atmospheric concentrations stabilize, surface air temperature continues to rise by a few tenths of a degree per century for a century or more. Thermal expansion of the ocean continues long after CO_2 emissions have been reduced, and melting of ice sheets continues to contribute to sea-level rise for many centuries. This figure is a generic illustration for stabilization at any level between 450 and 1,000 ppm, and therefore has no units on the response axis. Responses to stabilization trajectories in this range show broadly similar time courses, but the impacts become progressively larger at higher concentrations of CO_2.

WGI TAR Sections 3.7, 9.3, & 11.5, & WGI TAR Figures 3.13, 9.16, 9.19, 11.15, & 11.16

5.5 **The lower the stabilization target for atmospheric CO_2, the sooner emissions of CO_2 would need to decrease to meet it.** If emissions were held at present levels, carbon cycle models indicate that the atmospheric concentration of CO_2 would continue to rise (see Figure 5-3).

WGI TAR Sections 3.2.3.2, 3.7.3, & 9.3.3.1

· Stabilization of CO_2 concentrations at any level requires ultimate reduction of global net emissions to a small fraction of the current emission level.

· Stabilization of atmospheric CO_2 concentrations at 450, 650, or 1,000 ppm would require global anthropogenic CO_2 emissions to drop below the year 1990 level, within a few decades, about a century, or about 2 centuries, respectively, and continue to decrease steadily thereafter (see Figure 6-1).

These time constraints are partly due to the rate of CO_2 uptake by the ocean, which is limited by the slow transport of carbon between the surface and deep waters. There is sufficient uptake capacity in the ocean to incorporate 70 to 80% of foreseeable anthropogenic CO_2 emissions to the atmosphere, but this would take several centuries. Chemical reaction involving ocean sediments has the potential to remove up to a further 15% over a period of 5,000 years.

5.6 **A delay between biospheric carbon uptake and carbon release is manifest as a temporary net carbon uptake.** The main flows in the global carbon cycle have widely differing characteristic time scales (see Figures 5-1 and 5-4). The net terrestrial carbon uptake that has developed over the past few decades is partly a result of the time lag between photosynthetic carbon uptake and carbon release when plants eventually die and decay. For example, the uptake resulting from regrowth of forests on agricultural lands, abandoned over the last century in the Northern Hemisphere, will decline as the forests reach their mature biomass, growth slows, and death increases. Enhancement of plant carbon uptake due to elevated CO_2 or nitrogen deposition will eventually saturate, then decomposition of the increased biomass will catch up. Climate change is likely to increase disturbance and decomposition rates in the future. Some models project that the recent global net terrestrial carbon uptake will peak, then level off or decrease. The peak could be passed within the 21st century according to several model projections. Projections of the global net terrestrial carbon exchange with the atmosphere beyond a few decades remain uncertain (see Figure 5-5).

WGI TAR Sections 3.2.2-3 & 3.7.1-2, & WGI TAR Figure 3.10

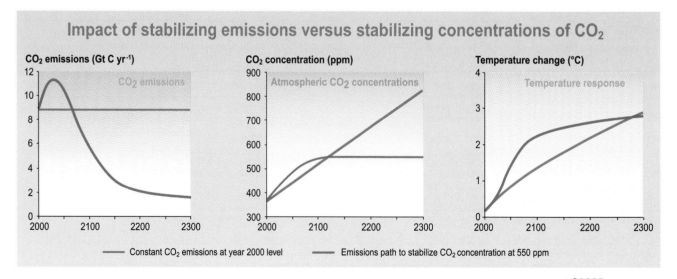

Impact of stabilizing emissions versus stabilizing concentrations of CO_2

CO_2 emissions (Gt C yr^{-1})

CO_2 emissions

CO_2 concentration (ppm)

Atmospheric CO_2 concentrations

Temperature change (°C)

Temperature response

—— Constant CO_2 emissions at year 2000 level   —— Emissions path to stabilize CO_2 concentration at 550 ppm

Figure 5-3: Stabilizing CO_2 *emissions* at current levels will result in a continuously rising atmospheric CO_2 *concentration* and temperature. Stabilization of atmospheric CO_2 and temperature change will eventually require the emissions to drop well below current levels. In all three panels the red curves illustrate the result of emissions held constant at the level prescribed by the WRE 550 profile for the year 2000 (which is slightly higher than the actual emissions for the year 2000), while the blue curves are the result of emissions following the WRE 550 stabilization profile. Both cases are illustrative only: Constant global emissions are unrealistic in the short term, and no preference is expressed for the WRE 550 profile over others. Other stabilization profiles are illustrated in Figure 6-1. Figure 5-3 was constructed using the models described in WGI TAR Chapters 3 & 9.

WGI TAR Sections 3.7 & 9.3

5.7   Although warming reduces the uptake of CO_2 by the ocean, the oceanic net carbon uptake is projected to persist under rising atmospheric CO_2, at least for the 21st century. Movement of carbon from the surface to the deep ocean takes centuries, and its equilibration there with ocean sediments takes millennia.

 WGI TAR Sections 3.2.3 & 3.7.2, & WGI TAR Figures 3.10c,d

5.8   **When subjected to rapid climate change, ecological systems are likely to be disrupted as a consequence of the differences in response times within the system.** The resulting loss of capacity by the ecosystem to supply services such as food, timber, and biodiversity maintenance on a sustainable basis may not be immediately apparent. Climate change may lead to conditions unsuitable for the establishment of key species, but the slow and delayed response of long-lived plants hides the importance of the change until the already established individuals die or are killed in a disturbance. For example, for climate change of the degree possible within the 21st century, it is likely, in some forests, that when a stand is disturbed by fire, wind, pests, or harvesting, instead of the community regenerating as in the past, species may be lost or replaced by different species.

WGII TAR Section 5.2

5.9   **Humans have shown a capacity to adapt to long-term mean climate conditions, but there is less success in adapting to extremes and to year-to-year variations in climatic conditions.** Climatic changes in the next 100 years are expected to exceed

WGII TAR SPM 2.7, WGII TAR Sections 4.6.4, 18.2-4, & 18.8, & WGIII TAR Section 10.4.2

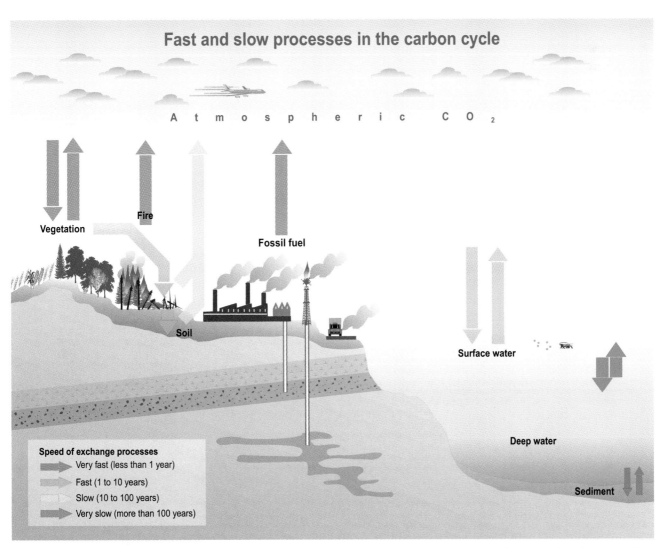

Figure 5-4: The range of time scales of major processes within the global carbon cycle leads to a range of response times for perturbations of CO_2 in the atmosphere, and contributes to the development of transient sinks, as when the atmospheric CO_2 concentration rose above its pre-1750 equilibrium level.

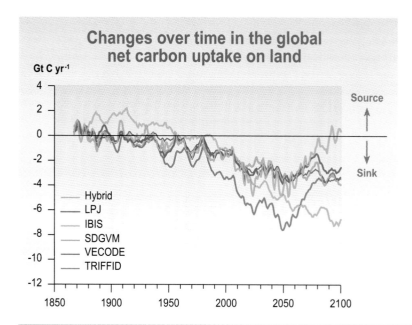

Changes over time in the global net carbon uptake on land

Gt C yr⁻¹

Hybrid
LPJ
IBIS
SDGVM
VECODE
TRIFFID

WGI TAR Figure 3.10b

Figure 5-5: The recent net uptake of carbon on the land is partly due to enhanced CO₂ uptake through plant growth, with a delay before this carbon is returned to the atmosphere via the decay of plant material and soil organic matter. Several processes contribute to the enhanced plant growth: changes in land use and management, fertilizing effects of elevated CO₂ and nitrogen, and some climate changes (such as a longer growing season at high latitudes). A range of models (identified by their acronyms in the figure) project a continued increase in the strength of the net carbon uptake on land for several decades, then a leveling off or decline late in the 21st century for reasons explained in the text. The model results illustrated here arise from the IS92a scenario, but similar conclusions are reached using other scenarios.

any experienced by human societies over at least the past 5 millennia. The magnitude and rate of these changes will pose a major challenge for humanity. The time needed for socio-economic adaptation varies from years to decades, depending on the sector and the resources available to assist the transition. There is inertia in decision making in the area of adaptation and mitigation, and in implementing those decisions, on the order of decades. The fact that adaptation and mitigation decisions are generally not made by the same entities compounds the difficulties inherent in the identification and implementation of the best possible combination of strategies, and hence contributes to the delays of climate change response.

5.10 **There is typically a delay of years to decades between perceiving a need to respond to a major challenge, planning, researching and developing a solution, and implementing it.** This delay can be shortened by anticipating needs through the application of foresight, and thus developing technologies in advance. The response of technological development to energy price changes has historically been relatively rapid (typically, less than 5 years elapses between a price shock and the response in terms of patenting activity and introduction of new model offerings) but its diffusion takes much longer. The diffusion rate often depends on the rate of retirement of previously installed equipment. Early deployment of rapidly improving technologies allows learning-curve cost reductions (learning by doing), without premature lock-in to existing, low-efficiency technology. The rate of technology diffusion is strongly dependent not only on economic feasibility but also on socio-economic pressures. For some technologies, such as the adoption of new crop varieties, the availability of, and information on, pre-existing adaptation options allows for rapid adaptation. In many regions, however, population pressures on limited land and water resources, government policies impeding change, or limited access to information or financial resources make adaptation difficult and slow. Optimal adaptation to climate change trends, such as more frequent droughts, may be delayed if they are perceived to be due to natural variability, while they might actually be related to climate change. Conversely, maladaptation can occur if climate variability is mistaken for a trend.

WGII TAR Sections 1.4.1, 12.8.4, & 18.3.5, & WGIII TAR Sections 3.2, 5.3.1, & 10.4

5.11 Social structures and personal values interact with society's physical infrastructure, institutions, and the technologies embodied within them, and the combined system evolves relatively slowly. This is obvious, for instance, in relation to the impact of urban design and infrastructure on energy consumption for heating, cooling, and transport. Markets sometimes "lock in" to technologies and practices that are sub-optimal because of the investment in supporting infrastructure, which block out alternatives. Diffusion of many

WGIII TAR Sections 3.2, 3.8.6, 5.2-3, & 10.3, SRTT SPM, & SRTT Chapter 4 ES

innovations comes up against people's traditional preferences and other social and cultural barriers. Unless advantages are very clear, social or behavioral changes on the part of technology users may require decades. Energy use and greenhouse gas mitigation are peripheral interests in most people's everyday lives. Their consumption patterns are driven not only by demographic, economic and technological change, resource availability, infrastructure, and time constraints, but also by motivation, habit, need, compulsion, social structures, and other factors.

5.12 **Social and economic time scales are not fixed: They are sensitive to social and economic forces, and could be changed by policies and the choices made by individuals.** Behavioral and technological changes can occur rapidly under severe economic conditions. For example, the oil crises of the 1970s triggered societal interest in energy conservation and alternative sources of energy, and the economy in most Organisation for Economic Cooperation and Development (OECD) countries deviated strongly from the traditional tie between energy consumption and economic development growth rates (see Figure 5-6). Another example is the observed reduction in CO_2 emissions caused by the disruption of the economy of the Former Soviet Union (FSU) countries in 1988. The response in both cases was very rapid (within a few years). The converse is also apparently true: In situations where pressure to change is small, inertia is large. This has implicitly been assumed to be the case in the SRES scenarios, since they do not consider major stresses, such as economic recession, large-scale conflict, or collapses in food stocks and associated human suffering, which are inherently difficult to forecast.

WGIII TAR Chapter 2, WGIII TAR Sections 3.2 & 10.1.4.3, & WGII SAR Section 20.1

5.13 **Stabilization of atmospheric CO_2 concentration at levels below about 600 ppm is only possible with reductions in carbon intensity and/or energy intensity greater than have been achieved historically.** This implies shifts toward alternative development pathways with new social, institutional, and technological configurations that address environmental constraints. Low historical rates of improvement in energy intensity (energy use per unit GDP) reflect the relatively low priority placed on energy efficiency by most producers and users of technology. By contrast, labor productivity increased at higher rates over the period 1980 to 1992. The historically recorded annual rates of improvement of global energy intensity (1 to 1.5% per year) would have to be increased and maintained over long time frames to achieve stabilization of CO_2 concentrations at about 600 ppm or below (see Figure 5-7). Carbon intensity (carbon per unit energy produced) reduction rates would eventually have to change by even more (e.g., up to 1.5% per year (the historical baseline is 0.3 to 0.4% per year)). In reality, both energy intensity and carbon intensity are likely to continue to improve, but greenhouse gas stabilization at levels below 600 ppm requires that at least one of them do so at a rate much higher than historically achieved. The lower the stabilization target and the higher the level of baseline emissions, the larger the CO_2 divergence from the baseline that is needed, and the earlier it would need to occur.

WGI TAR Section 3.7.3.4, WGIII TAR Section 2.5, & SRES Section 3.3.4

5.14 **Some climate, ecological, and socio-economic system changes are effectively irreversible over many human lifetimes, and others are intrinsically irreversible.**

5.15 **There are two types of apparent irreversibility.** "Effective irreversibility" derives from processes that have the potential to return to their pre-disturbance state, but take centuries to millennia to do so. An example is the partial melting of the Greenland ice sheet. Another is the projected rise in mean sea level, partly as a result of melting of the cryosphere, but primarily due to thermal expansion of the oceans. The world is already committed to some sea-level rise as a consequence of the surface atmospheric warming that has occurred over the past century. "Intrinsic irreversibility" results from crossing a threshold beyond which the system no longer spontaneously returns to the previous state. An example of an intrinsically irreversible change due to crossing a threshold is the extinction of species, resulting from a combination of climate change and habitat loss.

WGI TAR Chapter 11, WGII TAR Chapter 5, & WGII TAR Sections 16.2.1 & 17.2.5

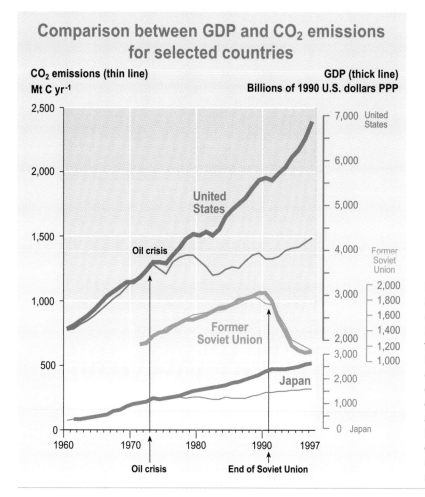

Comparison between GDP and CO₂ emissions for selected countries

CO₂ emissions (thin line)
Mt C yr⁻¹

GDP (thick line)
Billions of 1990 U.S. dollars PPP

WGIII TAR Table 3.1 & WGII
SAR Figure 20-1

Figure 5-6: The response of the energy system, as indicated by the emission of CO₂ (expressed as carbon), to economic changes, indicated by GDP (expressed in Purchasing Power Parity (PPP) terms). The response can be almost without inertia if the shock is large. The "oil crisis"—during which energy prices rose substantially over a short period of time—led to an almost immediate and sustained divergence of the formerly closely linked emissions and GDP in most developed countries: Japan and United States are shown as examples. At the breakup of the Former Soviet Union, the two indicators remained closely linked, leading the emission to drop rapidly in tandem with declining GDP.

WGI TAR Sections 2.4.3,
7.3.7, & 9.3.4.3, & WGII
TAR Section 1.4.3.5

5.16 **The location of a threshold, and the resistance to change in its vicinity, can be affected by the rate at which the threshold is approached.** Model results indicate that a threshold may exist in the ocean thermohaline circulation (see Question 4) such that a transition to a new ocean circulation, as occurred during the emergence from the last glacial period, could be induced if the world warms rapidly. While such a transition is very unlikely during the 21st century, some models suggest that it would be irreversible (i.e., the new circulation would persist even after the perturbation disappeared). For slower rates of warming, THC would likely gradually adjust and thresholds may not be crossed. This implies that the greenhouse gas emission trajectory is important in determining the evolution of THC. When a system approaches a threshold, as is the case for a weakening THC under global warming, resilience to perturbations decreases.

WGII TAR Sections 1.2.1.2,
4.7.3, & 5.2, WGIII TAR TS
2.3, SRES Box 4.2, & WGII
SAR A.4.1

5.17 **Higher rates of warming and the compounded effects of multiple stresses increase the likelihood of a threshold crossing.** An example of an ecological threshold is provided by the migration of plant species as they respond to a changing climate. Fossil records indicate that the maximum rate at which most plant species have migrated in the past is about 1 km per year. Known constraints imposed by the dispersal process (e.g., the mean period between germination and the production of seeds, and the mean distance that an individual seed can travel) suggest that, without human intervention, many species would not be able to keep up with the rate of movement of their preferred climatic niche projected for the 21st century, even if there were no barriers to their movement imposed by land use. An example of a socio-economic threshold is provided by conflicts in already stressed situations—for example, a river basin shared by several nations with competition for a limited water resource. Further pressure from an environmental stress

Acceleration of energy system change

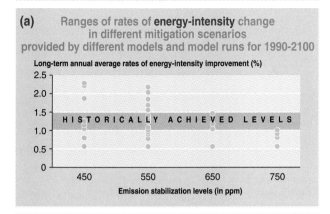

(a) Ranges of rates of **energy-intensity** change in different mitigation scenarios provided by different models and model runs for 1990-2100

Long-term annual average rates of energy-intensity improvement (%)

HISTORICALLY ACHIEVED LEVELS

Emission stabilization levels (in ppm)

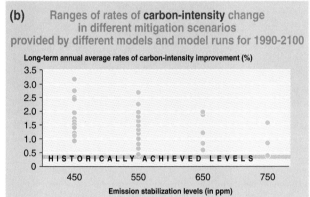

(b) Ranges of rates of **carbon-intensity** change in different mitigation scenarios provided by different models and model runs for 1990-2100

Long-term annual average rates of carbon-intensity improvement (%)

HISTORICALLY ACHIEVED LEVELS

Emission stabilization levels (in ppm)

WGIII TAR Figures 2.8 & 2.18

Figure 5-7: (a) The required rate of decrease in energy intensity (energy per unit GDP) in order to meet given CO_2 concentration stabilization targets is within the range of historically achieved rates for stabilization above 550 ppm, and possibly even at 450 ppm, but (b) the required rate of improvement in carbon intensity (carbon emissions per unit energy) to stabilize at levels below about 600 ppm is higher than the historically achieved rates. As a consequence, the cost of mitigation rises as the stabilization level decreases, and does so more steeply below a target of about 600 ppm than above (see Figure 7-3).

such as reduced stream flow could trigger more severe conflict. If impacted systems are not fully understood, the presence of a threshold may not be apparent until it is reached.

5.18 **Inertia in the climate, ecological, and socio-economic systems makes adaptation inevitable and already necessary in some cases, and inertia affects the optimal mix of adaptation and mitigation strategies.**

5.19 **As a result of the time lags and inertias inherent in the Earth system, including its social components, some of the consequences of actions taken, or not taken, will only be felt many years in the future.** For example, the differences in the initial trajectories of the various SRES and stabilization scenarios are small, but the outcomes in terms of the climate in the year 2100 are large. The choice of development path has consequences at all the affected time scales; thus, long-term total costs and benefits may differ considerably from short-term ones.

WGIII TAR Section 8.4.2

5.20 **In the presence of inertia, well-founded actions to adapt to or mitigate climate change are more effective, and under some circumstances may be cheaper, if taken earlier rather than later.** Time lags provide a breathing space between emissions and impacts, thus allowing time for planned adaptation. The inertia of technology development and capital stock replacement is an important argument for gradual mitigation. The essential point of inertia in economic structures and processes is that deviation from any given trend incurs costs, and these costs rise with the speed of deviations (e.g., the costs of early retirement of carbon-intensive facilities). Earlier mitigation action may reduce the risk of incurring severe lasting or irreversible impacts, while reducing the need for more rapid mitigation later. Accelerated action may help to drive down the costs of mitigation and

WGII TAR Sections 1.3.4 & 2.7.1, WGIII TAR Chapter 2, WGIII TAR Sections 10.1 & 10.4.2-3, & WGIII TAR Table 10.7

adaptation in the long term by accelerating technology development and the early realization of benefits currently obscured by market imperfections. Abatement over the next few years is economically valuable if there is a significant probability of having to stay below ceilings that would otherwise be reached within the characteristic time scales of the systems producing greenhouse gases. Climate change mitigation decisions depend on the interplay of inertia and uncertainty, resulting in a sequential decision-making process. Foresight and early adaptation will be most advantageous in sectors with long-lived infrastructure, such as dams and bridges, and large social inertia, such as misallocated property rights. Anticipatory adaptive action can be very cost-effective if the anticipated trend materializes.

5.21   **The existence of time lags, inertia, and irreversibility in the Earth system means that a mitigation action or technology development can have different outcomes, depending on when it is taken.** For example, in one model analysis of the hypothetical effect of reducing anthropogenic greenhouse gas emissions to zero in the year 1995, on sea-level rise during the 21st century in the Pacific, showed that the sea-level rise that would inevitably occur due to warming incurred to 1995 (5 to 12 cm) would be substantially less than if the same emission reduction occurred in the year 2020 (14 to 32 cm). This demonstrates the increasing commitment to future sea-level rise due to past and present emissions, and the effect of delaying the hypothetical emissions reduction.

WGII TAR Sections 2.7.1 & 17.2.1

5.22   **Technological inertia in less developed countries can be reduced through "leapfrogging" (i.e., adopting anticipative strategies to avoid the problems faced today by industrial societies).** It cannot be assumed that developing countries will automatically follow the past development paths of industrialized countries. For example, some developing countries have bypassed land-lines for communication, and proceeded directly to mobile phones. Developing countries could avoid the past energy-inefficient practices of developed countries by adopting technologies that use energy in a more sustainable way, recycling more wastes and products, and handling residual wastes in a more acceptable manner. This may be easier to achieve in new infrastructure and energy systems in developing countries since large investments are needed in any case. Transfer of technology between countries and regions can reduce technological inertia.

WGII TAR Chapter 2, WGIII TAR Section 10.3.3, SRES Section 3.3.4.8, & SRTT SPM

5.23   **Inertia and uncertainty in the climate, ecological, and socio-economic systems imply that safety margins should be considered in setting strategies, targets, and time tables for avoiding dangerous levels of interference in the climate system.** Stabilization target levels of, for instance, atmospheric CO_2 concentration, temperature, or sea level may be affected by:
   · The inertia of the climate system, which will cause climate change to continue for a period after mitigation actions are implemented
   · Uncertainty regarding the location of possible thresholds of irreversible change and the behavior of the system in their vicinity
   · The time lags between adoption of mitigation goals and their achievement.
Similarly, adaptation is affected by time lags involved in identifying climate change impacts, developing effective adaptation strategies, and implementing adaptive measures. Hedging strategies and sequential decision making (iterative action, assessment, and revised action) may be appropriate responses to the combination of inertia and uncertainty. Inertia has different consequences for adaptation than for mitigation, with adaptation being primarily oriented to address localized impacts of climate change, while mitigation aims to address the impacts on the climate system. Both issues involve time lags and inertia, with inertia suggesting a generally greater sense of urgency for mitigation.

WGII TAR Section 2.7.1 & WGIII TAR Sections 10.1.4.1-3

5.24   **The pervasiveness of inertia and the possibility of irreversibility in the interacting climate, ecological, and socio-economic systems are major reasons why anticipatory adaptation and mitigation actions are beneficial.** A number of opportunities to exercise adaptation and mitigation options may be lost if action is delayed.

Question 6

Q6

a) How does the extent and timing of the introduction of a range of emissions reduction actions determine and affect the rate, magnitude, and impacts of climate change, and affect the global and regional economy, taking into account the historical and current emissions?

b) What is known from sensitivity studies about regional and global climatic, environmental, and socio-economic consequences of stabilizing the atmospheric concentrations of greenhouse gases (in carbon dioxide equivalents), at a range of levels from today's to double that level or more, taking into account to the extent possible the effects of aerosols? For each stabilization scenario, including different pathways to stabilization, evaluate the range of costs and benefits, relative to the range of scenarios considered in Question 3, in terms of:
  - Projected changes in atmospheric concentrations, climate, and sea level, including changes beyond 100 years
  - Impacts and economic costs and benefits of changes in climate and atmospheric composition on human health, diversity and productivity of ecological systems, and socio-economic sectors (particularly agriculture and water)
  - The range of options for adaptation, including the costs, benefits, and challenges
  - The range of technologies, policies, and practices that could be used to achieve each of the stabilization levels, with an evaluation of the national and global costs and benefits, and an assessment of how these costs and benefits would compare, either qualitatively or quantitatively, to the avoided environmental harm that would be achieved by the emissions reductions
  - Development, sustainability, and equity issues associated with impacts, adaptation, and mitigation at a regional and global level.

6.1   The climatic, environmental, and socio-economic consequences of greenhouse gas emissions were assessed in Question 3 for scenarios that do not include any climate policy interventions. These same issues are addressed here in Question 6, but this time to assess the benefits that would result from a set of climate policy interventions. Among the emission reduction scenarios considered are scenarios that would achieve stabilization of CO_2 concentrations in the atmosphere. The role of adaptation as a complement to mitigation and the potential contributions of reducing emissions to the goals of sustainable development and equity are evaluated. The policies and technologies that might be used to implement the emission reductions and their costs are considered in Question 7.

6.2   **The projected rate and magnitude of warming and sea-level rise can be lessened by reducing greenhouse gas emissions.**

6.3   **The greater the reductions in emissions and the earlier they are introduced, the smaller and slower the projected warming and rise in sea levels.** Future climate change is determined by historic, current, and future emissions. Estimates have been made of the global mean temperature and sea-level rise effects of a 2% per year reduction in CO_2 emissions by developed countries over the period 2000 to 2100, assuming that developing countries do not reduce their emissions.[6] Under these assumptions, global emissions and the atmospheric concentration of CO_2 grow throughout the century but at a diminished rate compared to scenarios that assume no actions to reduce developed country emissions. The effects of the emission limit accrue slowly but build with time. By the year 2030, the projected concentration of CO_2 in the atmosphere is reduced roughly 20% relative to the IS92a scenario of unabated emissions, which diminishes warming and sea-level rise by a small amount within this time frame. By the year 2100, the projected CO_2 concentration is reduced by 35% relative to the IS92a scenario, projected global mean warming reduced by 25%, and projected sea-level rise reduced by 20%. Analyses of CO_2 emission reductions of 1% per year by developed countries indicate that the lesser reductions would yield smaller reductions in CO_2 concentration, temperature change, and sea-level rise. Actions such as these taken now would have a greater effect at the year 2100 than the same emissions reductions implemented at a later time.

6.4   **Reductions in greenhouse gas emissions and the gases that control their concentration would be necessary to stabilize radiative forcing.** For example, for the most important anthropogenic greenhouse gas, carbon cycle models indicate that stabilization of atmospheric CO_2 concentrations at 450, 650, or 1,000 ppm would require global anthropogenic CO_2 emissions to drop below year 1990 levels within a few decades, about a century, or about 2 centuries, respectively, and continue to decrease steadily thereafter (see Figure 6-1). These models illustrate that emissions would peak in about 1 to 2 decades (450 ppm) and roughly a century (1,000 ppm) from the present (see Table 6-1). Eventually CO_2 emissions would need to decline to a very small fraction of current emissions. The benefits of different stabilization levels are discussed later in Question 6 and the costs of these stabilization levels are discussed in Question 7.

6.5   **There is a wide band of uncertainty in the amount of warming that would result from any stabilized greenhouse gas concentration.** Estimates of global mean temperature change for scenarios that would stabilize the concentration of CO_2 at different levels, and hold them constant thereafter, are presented in Figure 6-1c. The uncertainty about climate sensitivity yields a wide range of estimates of temperature change that would result

---

[6] In these analyses, emissions by developed countries of CH_4, N_2O, and SO_2 are kept constant at their year 1990 values, and halocarbons follow a scenario consistent with the Copenhagen version of the Montreal Protocol. Developing country emissions of CO_2 and other greenhouse gases are assumed to follow the IS92 scenario projections. The temperature projections were made with a simple climate model. The IS92 scenarios are described in the IPCC *Special Report on Radiative Forcing of Climate Change*.

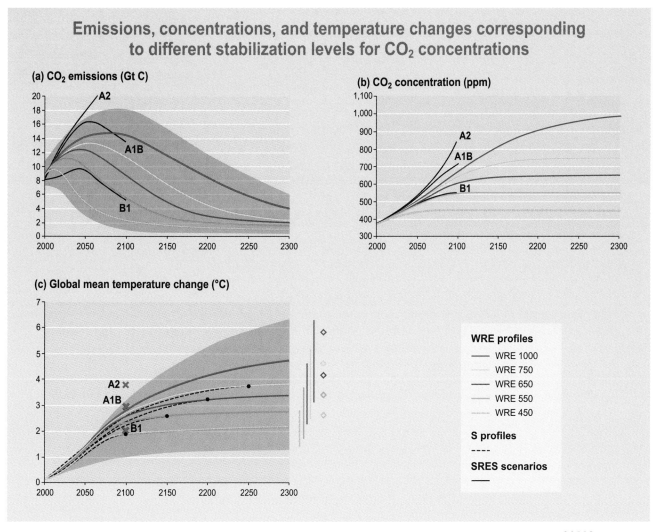

Emissions, concentrations, and temperature changes corresponding to different stabilization levels for CO_2 concentrations

(a) CO_2 emissions (Gt C)

(b) CO_2 concentration (ppm)

(c) Global mean temperature change (°C)

WRE profiles
— WRE 1000
— WRE 750
— WRE 650
— WRE 550
— WRE 450

S profiles

SRES scenarios
—

Figure 6-1: Stabilizing CO_2 concentrations would require substantial reductions of emissions below current levels and would slow the rate of warming.

 WGI TAR Sections 3.7.3 & 9.3.3, & IPCC TP3

a) **CO_2 emissions:** The time paths of CO_2 emissions that would lead to stabilization of the concentration of CO_2 in the atmosphere at 450, 550, 650, 750, and 1,000 ppm are estimated for the WRE stabilization profiles using carbon cycle models. Lower CO_2 concentration levels would require an earlier reversal of emissions growth and earlier decreases to levels below current emissions. The shaded area illustrates the range of uncertainty in estimating CO_2 emissions corresponding to specified concentration time paths, as represented in carbon cycle models. Also shown for comparison are CO_2 emissions for three of the SRES scenarios (A1B, A2, and B1), which do not include greenhouse gas emission limits.

b) **CO_2 concentrations:** The CO_2 concentrations specified for the WRE profiles gradually approach stabilized levels that range from 450 to 1,000 ppm. Also shown for comparison are estimates of CO_2 concentrations that would result from three of the SRES projections of emissions (A1B, A2, and B1).

c) **Global mean temperature changes:** Global mean temperature changes are estimated for the WRE stabilization profiles using a simple climate model tuned in turn to each of several more complex models. Estimated warming slows as growth in the atmospheric concentration of CO_2 slows and warming continues after the time at which the CO_2 concentration is stabilized (indicated by black spots) but at a much diminished rate. It is assumed that emissions of gases other than CO_2 follow the SRES A1B projection until the year 2100 and are constant thereafter. This scenario was chosen as it is in the middle of the range of the SRES scenarios. The dashed lines show the temperature changes projected for the S profiles, an alternate set of CO_2 stabilization profiles (not shown in panels (a) or (b)). The shaded area illustrates the effect of a range of climate sensitivity across the five stabilization cases. The colored bars on the righthand side show, for each WRE profile, the range at the year 2300 due to the different climate model tunings and the diamonds on the righthand side show the equilibrium (very long-term) warming for each stabilization level using average climate model results. Also shown for comparison are temperature increases in the year 2100 estimated for the SRES emission scenarios (indicated by red crosses).

| Table 6-1 | Projected CO_2 concentrations for the SRES emissions scenarios and deduced emissions for the WRE profiles leading to stabilization of atmospheric CO_2.[a] |

	CO_2 Emissions ($Gt\ C\ yr^{-1}$)		Accumulated CO_2 Emissions	Year in which Emissions		Atmospheric Concentration (ppm)		Year of Concentration Stabilization
	2050	2100	2001 to 2100 (Gt C)	Peak	Fall below 1990 Levels[b]	2050	2100	
SRES Emissions Scenarios								
A1B	16.4	13.5	1,415			490–600	615–920	
A1T	12.3	4.3	985			465–560	505–735	
A1FI	23.9	28.2	2,105			520–640	825–1,250	
A2	17.4	29.1	1,780			490–600	735–1,080	
B1	11.3	4.2	900			455–545	485–680	
B2	11.0	13.3	1,080			445–530	545–770	
WRE Stabilization Profiles								
450	3.0–6.9	1.0–3.7	365–735	2005–2015	<2000–2045	445	450	2090
550	6.4–12.6	2.7–7.7	590–1,135	2020–2030	2030–2100	485	540	2150
650	8.1–15.3	4.8–11.7	735–1,370	2030–2045	2055–2145	500	605	2200
750	8.9–16.4	6.6–14.6	820–1,500	2040–2060	2080–2180	505	640	2250
1,000	9.5–17.2	9.1–18.4	905–1,620	2065–2090	2135–2270	510	675	2375

[a] blue text = prescribed and black text = model results; both fossil-fuel and land-use change emissions are considered. Ranges from two simple carbon cycle models: ISAM model range is based on complex model results, while BERN-CC model range is based on uncertainties in system responses and feedbacks. The SRES results can be found in Appendix II.1.1 of the WGI TAR. The exact timing of the WRE emissions depends on the pathway to stabilization.

[b] 1990 emissions are taken to be 7.8 Gt C; this value is uncertain primarily due to the uncertainty in the size of the land-use change emissions, assumed here to be 1.7 Gt C, the annual average value through the 1980s.

from emissions corresponding to a selected concentration level.[7] This is shown more clearly in Figure 6-2, which shows eventual CO_2 concentration stabilization levels and the corresponding range of temperature change that is estimated to be realized in the year 2100 and at long-run equilibrium. To estimate temperature changes for these scenarios, it is assumed that emissions of greenhouse gases other than CO_2 would follow the SRES A1B scenario until the year 2100 and that emissions of these gases would be constant thereafter. Different assumptions about emissions of other greenhouse gases would result in different estimates of warming for each CO_2 stabilization level.

6.6 **Emission reductions that would eventually stabilize the atmospheric concentration of CO_2 at a level below 1,000 ppm, based on profiles shown in Figure 6-1, and assuming that emissions of gases other than CO_2 follow the SRES A1B projection until the year 2100 and are constant thereafter, are estimated to limit global mean temperature increase to 3.5°C or less through the year 2100.**
Global average surface temperature is estimated to increase 1.2 to 3.5°C by the year 2100 for profiles that would limit CO_2 emissions so as to eventually stabilize the concentration of CO_2 at a level from 450 to 1,000 ppm. Thus, although all of the CO_2 concentration stabilization profiles analyzed would prevent, during the 21st century, much of the upper end of the SRES projections of warming (1.4 to 5.8°C by the year 2100), it should be noted that for most of the profiles the concentration of CO_2 would continue to rise beyond the year 2100. Owing to the large inertia of the ocean (see Question 5), temperatures are projected to continue to rise even after stabilization of CO_2 and other greenhouse gas concentrations, though at a rate that is slower than is projected for the period prior to stabilization and that diminishes with time. The equilibrium temperature rise would take many centuries to reach, and ranges from 1.5 to 3.9°C above the year 1990 levels for

WGI TAR Section 9.3.3 & WGI TAR Table 9.3

---

[7] The equilibrium global mean temperature response to doubling atmospheric CO_2 is often used as a measure of climate sensitivity. The temperatures shown in Figures 6-1 and 6-2 are derived from a simple model calibrated to give the same response as a number of complex models that have climate sensitivities ranging from 1.7 to 4.2°C. This range is comparable to the commonly accepted range of 1.5 to 4.5°C.

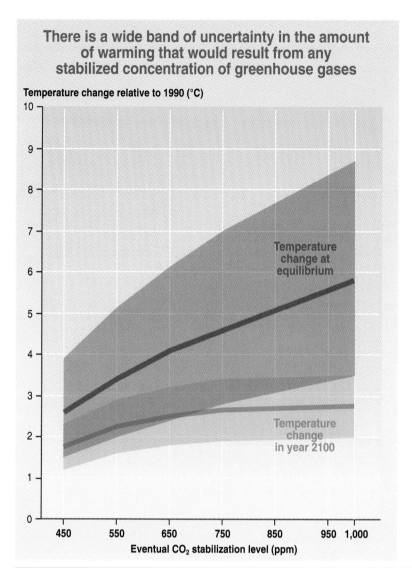

There is a wide band of uncertainty in the amount of warming that would result from any stabilized concentration of greenhouse gases

Temperature change relative to 1990 (°C)

Temperature change at equilibrium

Temperature change in year 2100

Eventual CO_2 stabilization level (ppm)

WGI TAR Section 9.3.3

Figure 6-2: Temperature changes relative to 1990 in (a) year 2100 and (b) at equilibrium are estimated using a simple climate model for the WRE profiles as in Figure 6-1. The lowest and highest estimates for each stabilization level assume a climate sensitivity of 1.7 and 4.2°C, respectively. The center line is an average of the low and high estimates.

stabilization at 450 ppm and 3.5 to 8.7°C above the year 1990 levels for stabilization at 1,000 ppm.[8] Furthermore, for a specific temperature stabilization target, there is a very wide range of uncertainty associated with the required stabilization level of greenhouse gas concentration (see Figure 6-2). The level at which CO_2 concentration is required to be stabilized for a given temperature target also depends on the levels of the non-CO_2 gases. Results from the only comprehensive climate model that has been used to analyze the regional effects of stabilizing CO_2 concentrations project that regionally averaged temperature changes would be similar in geographic pattern but less in magnitude than those projected for a baseline scenario with a 1% per year increase in CO_2 emissions from the year 1990.[9]

6.7 **Different time paths of emissions that lead to a common level for stabilization of the atmospheric concentration of greenhouse gases yield different time paths of temperature change.** For CO_2 stabilization levels of 450, 550, 650, and 750 ppm, two sets of emission time paths have been analyzed in previous IPCC reports and are

WGI TAR Section 9.3.3.1

[8] For all these scenarios, the contribution to the equilibrium warming from other greenhouse gases and aerosols is 0.6°C for a low climate sensitivity and 1.4°C for a high climate sensitivity. The accompanying increase in radiative forcing is equivalent to that occurring with an additional 28% in the final CO_2 concentrations.

[9] This rate of emission growth closely approximates the IS92a emission scenario.

referred to as the S and WRE profiles.[10] The WRE profiles allow higher emissions in early decades than do the S profiles, but then must require lower emissions in later decades to achieve a specified stabilization level. This deferment of emission reductions in the WRE profiles is estimated to reduce mitigation costs (see Question 7) but would result in a more rapid rate of warming initially. The difference in temperature projections for the two sets of pathways is 0.2°C or less in the year 2050, when the difference is most pronounced. Beyond the year 2100, the temperature changes of the WRE and S profiles converge. The temperature projections for the S and WRE profiles are compared in Figure 6-1c.

6.8 **Sea level and ice sheets would continue to respond to warming for many centuries after greenhouse gas concentrations have been stabilized (see Question 5).** The projected range of sea-level rise due to thermal expansion at equilibrium is 0.5 to 2 m for an increase in CO_2 concentration from the pre-industrial level of 280 to 560 ppm and 1 to 4 m for an increase in CO_2 concentration from 280 to 1,120 ppm. The observed rise over the 20th century was 0.1 to 0.2 m. The projected rise would be larger if the effect of increases in other greenhouse gas concentrations were to be taken into account. There are other contributions to sea-level rise over time scales of centuries to millennia (see Question 5). Models assessed in the TAR project sea-level rise of several meters from polar ice sheets (see Question 4) and land ice even for stabilization levels of 550 ppm CO_2-equivalent.

WGI TAR SPM & WGI TAR Section 11.5.4

6.9 **Reducing emissions of greenhouse gases to stabilize their atmospheric concentrations would delay and reduce damages caused by climate change.**

6.10 **Greenhouse gas emission reduction (mitigation) actions would lessen the pressures on natural and human systems from climate change.** Slower rates of increase in global mean temperature and sea level would allow more time for adaptation. Consequently, mitigation actions are expected to delay and reduce damages caused by climate change and thereby generate environmental and socio-economic benefits. Mitigation actions and their associated costs are assessed in the response to Question 7.

WGII TAR Sections 1.4.3, 18.8, & 19.5

6.11 **Mitigation actions to stabilize atmospheric concentrations of greenhouse gases at lower levels would generate greater benefits in terms of less damage.** Stabilization at lower levels reduces the risk of exceeding temperature thresholds in biophysical systems where these exist. Stabilization of CO_2 at, for example, 450 ppm is estimated to yield an increase in global mean temperature in the year 2100 that is about 0.75 to 1.25°C less than is estimated for stabilization at 1,000 ppm (see Figure 6-2). At equilibrium the difference is about 2 to 5°C. The geographical extent of the damage to or loss of natural systems, and the number of systems affected, which increase with the magnitude and rate of climate change, would be lower for a lower stabilization level. Similarly, for a lower stabilization level the severity of impacts from climate extremes is expected to be less, fewer regions would suffer adverse net market sector impacts, global aggregate impacts would be smaller, and risks of large-scale high-impact events would be reduced. Figure 6-3 presents a summary of climate change risks or reasons for concern (see Box 3-2) juxtaposed against the ranges of global mean temperature change in the year 2100 that have been estimated for different scenarios.[11]

WGI TAR Section 9.3.3 & WGII TAR Sections 1.4.3.5, 5.2, 5.4, & 19.3-6

6.12 **Comprehensive, quantitative estimates of the benefits of stabilization at various levels of atmospheric concentrations of greenhouse gases do not yet exist.**

WGII TAR Sections 19.4-5

---

[10] The S and WRE profiles are discussed in the WGI SAR and are described in more detail in IPCC Technical Paper 3.
[11] Climate change impacts will vary by region and sector or system, and the impacts will be influenced by regional and seasonal changes in mean temperature and precipitation, climate variability, the frequencies and intensities of extreme climate events, and sea-level rise. Global mean temperature change is used as a summary measure of the pressures exerted by climate change.

Risks of climate change damages would be reduced by stabilizing CO_2 concentrations

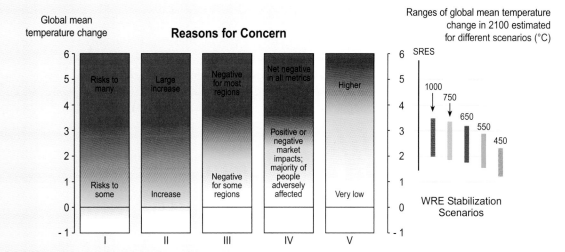

I. Unique and Threatened Systems

Extinction of species.

Loss of unique habitats, coastal wetlands.

Bleaching and death of coral.

II. Extreme Climate Events

Health, property, and environmental impacts from increased frequency and intensity of some climate extremes.

III. Distribution of Impacts

Cereal crop yield changes that vary from increases to decreases across regions but which are estimated to decrease in most tropical and subtropical regions.

Decrease in water availability in some water-stressed countries, increase in others.

Greater risks to health in developing countries than in developed countries.

Net market sector losses estimated for many developing countries; mixed effects estimated for developed countries up to a few degrees warming and negative effects for greater warming.

IV. Global Aggregate Impacts

Estimates of globally aggregated net market sector impacts are positive and negative up to a few degrees warming and negative for greater warming.

More people adversely affected than beneficially affected even for warming less than a few degrees.

V. Large-Scale, High-Impact Events

Significant slowing of thermohaline circulation possible by 2100.

Melting and collapse of ice sheets adding substantially to sea-level rise (very low likelihood before 2100; likelihood higher on multi-century time scale).

Figure 6-3: Risks of climate change damages would be reduced by stabilizing CO_2 concentration. The risks of adverse impacts from climate change are depicted for different magnitudes of global mean temperature change, where global mean temperature change is used as a proxy for the magnitude of climate change.

WGI TAR Section 9.3.3 & WGII TAR Section 19.8.2

Estimates of global mean temperature change by the year 2100 relative to the year 1990 are shown on the righthand side of the figure for scenarios that would lead to stabilization of the atmospheric concentration of CO_2, as well as for the full set of SRES projections. Many risks associated with warming above 3.5°C by the year 2100 would be avoided by stabilizing CO_2 concentration at or below 1,000 ppm. Stabilization at a lower level would reduce risks further. White indicates neutral or small negative or small positive impacts or risks; yellow indicates negative impacts for some systems or low risks; and red means negative impacts or risks that are more widespread and/or greater in magnitude. The assessment of impacts or risks takes into account only the magnitude of change and not the rate of change. Global mean annual temperature change is used as a proxy for the magnitude of climate change, but impacts would be a function of, among other factors, the magnitude and rate of global and regional changes in mean climate, climate variability and extreme climate phenomena, social and economic conditions, and adaptation.

While advances have been made in understanding the qualitative character of the impacts of future climate change, the impacts that would result under different scenarios are incompletely quantified. Because of uncertainty in climate sensitivity, and uncertainty about the geographic and seasonal patterns of changes in temperatures, precipitation, and other climate variables and phenomena, the impacts of climate change cannot be uniquely determined for individual emission scenarios. There are also uncertainties about key processes and sensitivities and adaptive capacities of systems to changes in climate. In addition, impacts such as changes in the composition and function of ecological systems, species extinction, and changes in human health, and disparity in the distribution of impacts across different populations and regions, are not readily expressed in monetary or other common units. Because of these limitations, the benefits of different greenhouse gas reduction actions, including actions to stabilize greenhouse gas concentrations at selected levels, are incompletely characterized and cannot be compared directly to mitigation costs for the purpose of estimating the net economic effects of mitigation.

6.13 Adaptation is a necessary strategy at all scales to complement climate change mitigation efforts. Together they can contribute to sustainable development objectives.

6.14 **Adaptation can complement mitigation in a cost-effective strategy to reduce climate change risks.** Reductions of greenhouse gas emissions, even stabilization of their concentrations in the atmosphere at a low level, will neither altogether prevent climate change or sea-level rise nor altogether prevent their impacts. Many reactive adaptations will occur in response to the changing climate and rising seas and some have already occurred. In addition, the development of planned adaptation strategies to address risks and utilize opportunities can complement mitigation actions to lessen climate change impacts. However, adaptation would entail costs and cannot prevent all damages. Adaptation implemented in combination with mitigation can be a more cost-effective approach to reducing the impacts of climate change than either applied alone. The potential for adaptation to substantially reduce many of the adverse impacts of climate change was assessed in Question 3. Because there are overlapping ranges of global temperature increases associated with the various stabilization levels (see Figure 6-1c), many adaptation options will be appropriate for a range of stabilization levels. Improved knowledge will narrow the uncertainties associated with particular stabilization levels and identification of appropriate adaptation strategies.

WGII TAR Sections 1.4.4.2, 18.3.5, & 18.4.1

6.15 **Adaptation costs and challenges can be lessened by mitigation of climate change.** Greenhouse gas emission reductions would reduce the magnitude and rate of changes to be adapted to, possibly including changes in the frequencies and intensities of extreme events. The smaller changes to which systems would be exposed, and slower pace at which stresses would increase, would allow more time for adaptation and lessen the degree to which current practices for coping with climate variability and extremes might need to be modified (see Question 3). More aggressive mitigation efforts will therefore reduce adaptation costs to attain a specified level of effectiveness.

WGII TAR Sections 18.2.2, 18.3, & 18.8

6.16 **Mitigation and adaptation actions can, if appropriately designed, advance sustainable development objectives.** As described in Question 3, risks associated with climate change have the potential to undermine progress toward sustainable development (e.g., damages from extreme climate events, water shortage and degraded water quality, food supply disruptions and hunger, land degradation, and diminished human health). By reducing these risks, climate change mitigation and adaptation policies can improve the prospects for sustainable development.[12]

WGII TAR Section 18.6.1, & WGIII TAR Sections 2.2.3 & 10.3.2

---

[12] The relationships between mitigation actions themselves and sustainable development and equity are addressed in Question 7. The relationships among adaptation, sustainable development, and equity are covered in Question 3.

6.17   **The impact of climate change is projected to have different effects within and between countries. The challenge of addressing climate change raises an important issue of equity.** Climate change pressures can exacerbate inequities between developing and developed countries; lessening these pressures through mitigation and enhancement of adaptive capacity can reduce these inequities. People in developing countries, particularly the poorest people in these countries, are considered to be more vulnerable to climate change than people in developed countries (see Question 3). Reducing the rate of warming and sea-level rise and increasing the capacity to adapt to climate change would benefit all countries, particularly developing countries.

WGII TAR Sections 18.5.3 & 19.4

6.18   **Reducing and slowing climate change can also promote inter-generational equity.** Emissions of the present generation will affect many future generations because of inertia in the atmosphere-ocean-climate system and the long-lived and sometimes irreversible effects of climate change on the environment. Future generations are generally anticipated to be wealthier, better educated and informed, and technologically more advanced than the present generation and consequently better able to adapt in many respects. But the changes set in motion in coming decades will accumulate and some could reach magnitudes that would severely test the abilities of many societies to cope. For irreversible impacts, such as the extinction of species or loss of unique ecosystems, there are no adaptation responses that can fully remedy the losses. Mitigating climate change would lessen the risks to future generations from the actions of the present generation.

WGII TAR Sections 1.2 & 18.5.2, & WGIII TAR Section 10.4.3

Question 7

Q7

What is known about the potential for, and costs and benefits of, and time frame for reducing greenhouse gas emissions?
- What would be the economic and social costs and benefits and equity implications of options for policies and measures, and the mechanisms of the Kyoto Protocol, that might be considered to address climate change regionally and globally?
- What portfolios of options of research and development, investments, and other policies might be considered that would be most effective to enhance the development and deployment of technologies that address climate change?
- What kind of economic and other policy options might be considered to remove existing and potential barriers and to stimulate private- and public-sector technology transfer and deployment among countries, and what effect might these have on projected emissions?
- How does the timing of the options contained in the above affect associated economic costs and benefits, and the atmospheric concentrations of greenhouse gases over the next century and beyond?

7.1 This question focuses on the potential for, and costs of, mitigation both in the near and long term. The issue of the primary mitigation benefits (the avoided costs and damages of slowing climate change) is addressed in Questions 5 and 6, and that of ancillary mitigation benefits is addressed in this response and the one to Question 8. This response describes a variety of factors that contribute to significant differences and uncertainties in the quantitative estimates of the costs of mitigation options. The SAR described two categories of approaches to estimating costs: bottom-up approaches, which often assess near-term cost and potential, and are built up from assessments of specific technologies and sectors; and top-down approaches, which proceed from macro-economic relationships. These two approaches lead to differences in the estimates of costs, which have been narrowed since the SAR. The response below reports on cost estimates from both approaches for the near term, and from the top-down approach for the long term. Mitigation options and their potential to reduce greenhouse gas emissions and sequester carbon are discussed first. This is followed by a discussion of the costs for achieving emissions reductions to meet near-term emissions constraints, and long-term stabilization goals, and the timing of reductions to achieve such goals. This response concludes with a discussion of equity as it relates to climate change mitigation.

Potential, Barriers, Opportunities, Policies, and Costs of Reducing Greenhouse Gas Emissions in the Near Term

7.2 **Significant technological and biological potential exists for near-term mitigation.**

7.3 **Significant technical progress relevant to greenhouse gas emissions reduction has been made since the SAR, and has been faster than anticipated.** Advances are taking place in a wide range of technologies at different stages of development—for example, the market introduction of wind turbines; the rapid elimination of industrial by-product gases, such as N_2O from adipic acid production and perfluorocarbons from aluminum production; efficient hybrid engine cars; the advancement of fuel cell technology; and the demonstration of underground CO_2 storage. Technological options for emissions reduction include improved efficiency of end-use devices and energy conversion technologies, shift to zero- and low-carbon energy technologies, improved energy management, reduction of industrial by-product and process gas emissions, and carbon removal and storage. Table 7-1 summarizes the results from many sectoral studies, largely at the project, national, and regional level with some at the global level, providing estimates of potential greenhouse gas emissions reductions to the 2010 and 2020 time frame.

WGIII TAR Sections 3.3-8, & WGIII TAR Chapter 3 Appendix

7.4 **Forests, agricultural lands, and other terrestrial ecosystems offer significant carbon mitigation potential. Conservation and sequestration of carbon, although not necessarily permanent, may allow time for other options to be further developed and implemented (see Table 7-2).** Biological mitigation can occur by three strategies: a) conservation of existing carbon pools, b) sequestration by increasing the size of carbon pools,[13] and c) substitution of sustainably produced biological products (e.g., wood for energy-intensive construction products and biomass for fossil fuels). Conservation of threatened carbon pools may help to avoid emissions, if leakage can be prevented, and can only become sustainable if the socio-economic drivers for deforestation and other losses of carbon pools can be addressed. Sequestration reflects the biological dynamics of growth, often starting slowly, passing through a maximum, and then declining over decades to centuries. The potential of biological mitigation options is on the order of 100 Gt C (cumulative) by the year 2050, equivalent to about 10 to 20% of projected fossil-fuel emissions during that period, although there are substantial uncertainties associated with

WGIII TAR Sections 3.6.4 & 4.2-4, & SRLULUCF

---

[13] Changing land use could influence atmospheric CO_2 concentration. Hypothetically, if all of the carbon released by historical land-use changes could be restored to the terrestrial biosphere over the course of the century (e.g., by reforestation), CO_2 concentration would be reduced by 40 to 70 ppm.

| **Table 7-1** | Estimates of potential global greenhouse gas emission reductions in 2010 and in 2020 (WGIII SPM Table SPM-1). |

Sector	*Historic Emissions in 1990 [Mt C_{eq} yr^{-1}]*	*Historic C_{eq} Annual Growth Rate over 1990-1995 [%]*	*Potential Emission Reductions in 2010 [Mt C_{eq} yr^{-1}]*	*Potential Emission Reductions in 2020 [Mt C_{eq} yr^{-1}]*	*Net Direct Costs per Tonne of Carbon Avoided*
Buildings[a] CO$_2$ only	1,650	1.0	700–750	1,000–1,100	Most reductions are available at negative net direct costs.
Transport CO$_2$ only	1,080	2.4	100–300	300–700	Most studies indicate net direct costs less than US$25 per t C but two suggest net direct costs will exceed US$50 per t C.
Industry CO$_2$ only – Energy efficiency – Material efficiency	2,300	0.4	300–500 ~200	700–900 ~600	More than half available at net negative direct costs. Costs are uncertain.
Industry Non-CO$_2$ gases	170		~100	~100	N$_2$O emissions reduction costs are US$0–10 per t C_{eq}.
Agriculture[b] CO$_2$ only Non-CO$_2$ gases	210 1,250–2,800	n/a	150–300	350–750	Most reductions will cost between US$0–100 per t C_{eq} with limited opportunities for negative net direct cost options.
Waste[b] CH$_4$ only	240	1.0	~200	~200	About 75% of the savings as CH$_4$ recovery from landfills at net negative direct cost; 25% at a cost of US$20 per t C_{eq}.
Montreal Protocol replacement applications Non-CO$_2$ gases	0	n/a	~100	n/a	About half of reductions due to difference in study baseline and SRES baseline values. Remaining half of the reductions available at net direct costs below US$200 per t C_{eq}.
Energy supply and conversion[c] CO$_2$ only	(1,620)	1.5	50–150	350–700	Limited net negative direct cost options exist; many options are available for less than US$100 per t C_{eq}.
Total	6,900–8,400[d]		1,900–2,600[e]	3,600–5,050[e]	

[a] Buildings include appliances, buildings, and the building shell.

[b] The range for agriculture is mainly caused by large uncertainties about CH$_4$, N$_2$O, and soil-related emissions of CO$_2$. Waste is dominated by methane landfill and the other sectors could be estimated with more precision as they are dominated by fossil CO$_2$.

[c] Included in sector values above. Reductions include electricity generation options only (fuel switching to gas/nuclear, CO$_2$ capture and storage, improved power station efficiencies, and renewables).

[d] Total includes all sectors reviewed in WGIII TAR Chapter 3 for all six gases. It excludes non-energy related sources of CO$_2$ (cement production, 160 Mt C; gas flaring, 60 Mt C; and land-use change, 600–1,400 Mt C) and energy used for conversion of fuels in the end-use sector totals (630 Mt C). If petroleum refining and coke oven gas were added, global year 1990 CO$_2$ emissions of 7,100 Mt C would increase by 12%. Note that forestry emissions and their carbon sink mitigation options are not included.

[e] The baseline SRES scenarios (for six gases included in the Kyoto Protocol) project a range of emissions of 11,500–14,000 Mt C_{eq} for the year 2010 and of 12,000–16,000 Mt C_{eq} for the year 2020. The emissions reduction estimates are most compatible with baseline emissions trends in the SRES B2 scenario. The potential reductions take into account regular turnover of capital stock. They are not limited to cost-effective options, but exclude options with costs above US$100 t C_{eq} (except for Montreal Protocol gases) or options that will not be adopted through the use of generally accepted policies.

this estimate. Realization of this potential depends upon land and water availability as well as the rates of adoption of land management practices. The largest biological potential for atmospheric carbon mitigation is in subtropical and tropical regions.

7.5 Adoption of opportunities including greenhouse gas-reducing technologies and measures may require overcoming barriers through the implementation of policy measures.

Table 7-2	Estimates of potential global greenhouse gas emission reductions in the year 2010: land use, land-use change, and forestry.

Categories of Mitigation Options	*Potential Emission Reductions in 2010 [Mt C yr⁻¹]*	*Potential Emission Reductions [Mt C]*	
Afforestation/reforestation (AR)[a]	197–584		Includes carbon in above- and below-ground biomass. Excludes carbon in soils and in dead organic matter.
Reducing deforestation (D)[b]		1,788	Potential for reducing deforestation is very uncertain for the tropics and could be in error by as much as ±50%.
Improved management within a land use (IM)[c]	570		Assumed to be the best available suite of management practices for each land use and climatic zone.
Land-use change (LC)[c]	435		
Total	1,202–1,589	1,788	

[a] Source: SRLULUCF Table SPM-3. Based on IPCC definitional scenario. Information is not available for other definitional scenarios. Potential refers to the estimated range of accounted average stock change for the period 2008–2012 (Mt C yr⁻¹).
[b] Source: SRLULUCF Table SPM-3. Based on IPCC definitional scenario. Information is not available for other definitional scenarios. Potential refers to the estimated average stock change (Mt C).
[c] Source: SRLULUCF Table SPM-4. Potential refers to the estimated net change in carbon stocks in the year 2010 (Mt C yr⁻¹). The list of activities is not exclusive or complete, and it is unlikely that all countries will apply all activities. Some of these estimates reflect considerable uncertainty.

7.6 **The successful implementation of greenhouse gas mitigation options would need to overcome technical, economic, political, cultural, social, behavioral, and/or institutional barriers that prevent the full exploitation of the technological, economic, and social opportunities of these mitigation options (see Figure 7-1).** The potential mitigation opportunities and types of barriers vary by region and sector, and over time. Most countries could benefit from innovative financing, social learning and innovation, and institutional reforms, removing barriers to trade, and poverty eradication. This is caused by a wide variation in mitigation capacity. The poor in any country are faced with limited opportunities to adopt technologies or change their social behavior, particularly if they are not part of a cash economy. Most countries could benefit from innovative financing and institutional reform and removing barriers to trade. In the industrialized countries, future opportunities lie primarily in removing social and behavioral barriers; in countries with economies in transition, in price rationalization; and in developing countries, in price rationalization, increased access to data and information, availability of advanced technologies, financial resources, and training and capacity building. Opportunities for any given country, however, might be found in the removal of any combination of barriers.

WGIII TAR Sections 1.5 & 5.3-5

7.7 **National responses to climate change can be more effective if deployed as a portfolio of policy instruments to limit or reduce net greenhouse gas emissions.** The portfolio of national climate policy instruments may include—according to national circumstances—emissions/carbon/energy taxes, tradable or non-tradable permits, provision and/or removal of subsidies, land-use policies, deposit/refund systems, technology or performance standards, energy mix requirements, product bans, voluntary agreements, information campaigns, environmental labeling, government spending and investment, and support for research and development (R&D). The literature in general gives no preference for any particular policy instrument.

WGIII TAR Sections 1.5.3, 5.3-4, & 6.2

7.8 **Coordinated actions among countries and sectors may help to reduce mitigation cost by addressing competitiveness concerns, potential conflicts with international trade rules, and carbon leakage. A group of countries that wants to limit its collective greenhouse gas emissions could agree to implement well-designed international instruments.** Instruments assessed in the WGIII TAR, and being developed in the Kyoto Protocol, are emissions trading, Joint Implementation

WGIII TAR Sections 6.3-4 & 10.2

Concepts of mitigation potentials

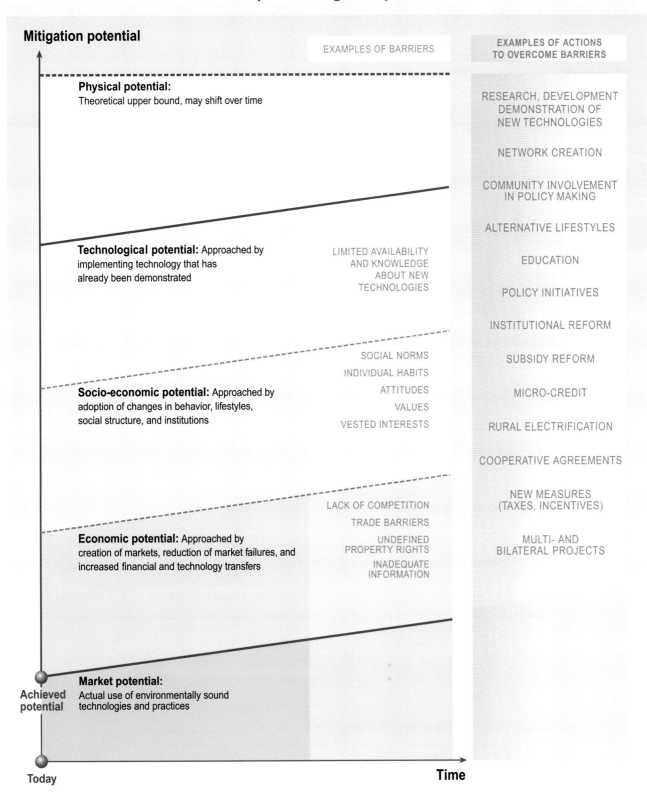

Figure 7-1: Penetration of environmentally sound technologies (including practices): a conceptual framework. Various barriers prevent the different potentials from being realized. Opportunities exist to overcome barriers through innovative projects, programs, and financing arrangements. An action can address more than one barrier. Actions may be pursued to address barriers at all levels simultaneously. Their implementation may require public policies, measures, and instruments. The socio-economic potential may lie anywhere in the space between the economic and technological potential.

WGIII TAR Section 5.2

(JI), and the Clean Development Mechanism (CDM). Other international instruments also assessed in the WGIII TAR include coordinated or harmonized emission/carbon/energy taxes, an emission/carbon/energy tax, technology and product standards, voluntary agreements with industries, direct transfers of financial resources and technology, and coordinated creation of enabling environments such as reduction of fossil-fuel subsidies. Some of these have been considered only in some regions to date.

7.9   Transfer of technologies between countries and regions would widen the choice of options at the regional level, and economies of scale and learning will lower the costs of their adoption.

7.10   **Adequate human and organizational capacity at every stage can increase the flow, and improve the quality, of technologies transferred within and across countries.** The transfer of environmentally sound technologies has come to be seen as a major element of global strategies to achieve sustainable development and climate change mitigation. The local availability of technical, business, management, and regulatory skills can enhance the flow of international capital, helping to promote technology transfer. Technical skills are enhanced by the creation of competence in associated services, organizational know-how, and capacity improvement to formulate and enforce regulations. Capacity building is a continuous process that needs to keep up with the evolution of mitigation options as they respond to technological and social changes.

WGIII TAR Sections 2.4.5 & 10.3.3, & SRTT SPM

7.11   **Governments through sound economic policy and regulatory frameworks, transparency, and political stability can create an enabling environment for private- and public-sector technology transfers.** At the macro-level, actions to consider include reform of the legal system, protection of intellectual property rights, open and competitive markets, reduced corruption, discouragement of restrictive business practices, reform of export credit, political risk insurance, reduction of tied aid, development of physical and communications infrastructure, and improvement of macro-economic stability. At the sectoral and project levels, actions include fuel and electricity price rationalization, energy industry institutional reform, improving land tenure, transparent project approval procedures, ensuring assessment of local technology needs and social impact of technologies, cross-country R&D on innovative technologies, and demonstration programs.

WGIII TAR Section 10.3.3 & SRTT SPM

7.12   **Networking among private and public stakeholders, and focusing on products and techniques with multiple ancillary benefits that meet or adapt to local development needs and priorities foster effective technology transfer.** National systems of innovation (NSI) can help achieve this through activities such as (a) strengthening educational institutions; (b) collection, assessment, and dissemination of technical, commercial, financial, and legal information; (c) technology assessment, demonstration projects, and extension services; (d) supporting market intermediary organizations; and (e) innovative financial mechanisms. Increasing flows of national and multilateral assistance can help to mobilize and multiply additional financial resources, including official development assistance, to support NSI activities.

WGIII TAR Section 10.3.3 & SRTT SPM

7.13   For participating countries, an increasing scale of international cooperation, such as emissions trading[14] and technology transfer, will lower mitigation costs.

7.14   A large number of studies using both top-down and bottom-up approaches (see Box 7-1 for definitions) report on the costs of greenhouse gas mitigation. Estimates of the costs of

---

[14] This market-based approach to achieve environmental objectives allows those reducing greenhouse gas emissions below what is required to use or trade the excess reductions to offset emissions at another source inside or outside the country. Here the term is broadly used to include trade in emission allowances and project-based collaboration.

Box 7-1	Bottom-up and top-down approaches to cost estimates: critical factors and the importance of uncertainties.

For a variety of reasons, significant differences and uncertainties surround specific quantitative estimates of mitigation costs. Cost estimates differ because of the (a) methodology used in the analysis, and (b) underlying factors and assumptions built into the analysis. Bottom-up models incorporate detailed studies of engineering costs of a wide range of available and anticipated technologies, and describe energy consumption in great detail. However, they typically incorporate relatively little detail on non-energy consumer behavior and interactions with other sectors of the economy. The costs estimated by bottom-up models can range from negative values (due to the adoption of "no-regrets" options) to positive values. Negative costs indicate that the direct energy benefits of a mitigation option exceed its direct costs (net capital, operating, and maintenance costs). Market and institutional barriers, however, can prevent, delay, or make more costly the adoption of these options. Inclusion of implementation and policy costs would add to the costs estimated by bottom-up models.

Top-down models are aggregate models of the economy that often draw on analysis of historical trends and relationships to predict the large-scale interactions between sectors of the economy, especially the interactions between the energy sector and the rest of the economy. Top-down models typically incorporate relatively little detail on energy consumption and technological change. The costs estimated by top-down models usually range from zero to positive values. This is because negative cost options estimated in bottom-up models are assumed to be adopted in both the baseline and policy scenarios. This is an important factor in the differences in the estimates from these two types of models.

The inclusion of some factors will lead to lower cost estimates and others to higher estimates. Incorporating multiple greenhouse gases, sinks, induced technical change, and emissions trading can lower costs. Further, studies suggest that some sources of greenhouse gas emissions can be limited at no or negative net social cost to the extent that policies can exploit no-regret opportunities such as correcting market imperfections, inclusion of ancillary benefits, and efficient tax revenue recycling. International cooperation that facilitates cost-effective emissions reductions can lower mitigation costs. On the other hand, accounting for potential short-term macro shocks to the economy, constraints on the use of domestic and international market mechanisms, high transaction costs, inclusion of ancillary costs, and ineffective tax recycling measures can increase estimated costs. Since no analysis incorporates all relevant factors affecting mitigation costs, estimated costs may not reflect the actual costs of implementing mitigation actions.

WGIII TAR Sections 3.3-8, 7.6.3, 8.2-3, & 9.4, & WGIII TAR Box SPM-2

limiting fossil-fuel greenhouse gas emissions vary widely and depend on choice of methodologies, underlying assumptions, emissions scenarios, policy instruments, reporting year, and other criteria.

7.15   **Bottom-up studies indicate that substantial low-cost mitigation opportunities exist.** According to bottom-up assessments (see Box 7-1) of specific technologies and sectors, half of the potential emissions reductions noted in Table 7-1 may be achieved by the year 2020 with direct benefits exceeding direct costs, and the other half at a net direct cost of up to US$100 per t C_{eq} (at 1998 prices). However, for reasons described below, the realized potential may be different. These cost estimates are derived using discount rates in the range of 5 to 12%, consistent with public-sector discount rates. Private internal rates of return vary greatly, and are often significantly higher, affecting the rate of adoption of these technologies by private entities. Depending on the emissions scenario, this could allow global emissions to be reduced below year 2000 levels in the period 2010–2020 at these net direct costs. Realizing these reductions involves additional implementation costs, which in some cases may be substantial, the possible need for supporting policies, increased R&D, effective technology transfer, and overcoming other barriers. The various global, regional, national, sector, and project studies assessed in the WGIII TAR have different scopes and assumptions. Studies do not exist for every sector and region.

WGIII TAR Sections 1.5, 3.3-8, 5.3-4, & 6.2

7.16   **Cost estimates using bottom-up analyses reported to date for biological mitigation vary significantly and do not consistently account for all significant components of cost.** Cost estimates using bottom-up analyses reported to date for biological mitigation vary significantly from US$0.1 to about US$20 per t C in several tropical countries and from US$20 to US$100 per t C in non-tropical countries. Methods of financial analyses and carbon accounting have not been comparable. Moreover, the cost calculations do not cover, in many instances, *inter alia*, costs for infrastructure, appropriate discounting, monitoring, data collection and implementation costs, opportunity costs of land and maintenance, or

WGIII TAR Sections 4.3-4

other recurring costs, which are often excluded or overlooked. The lower end of the range is assessed to be biased downwards, but understanding and treatment of costs is improving over time. Biological mitigation options may reduce or increase non-CO_2 greenhouse gas emissions.

7.17 **Projections of abatement cost of near-term policy options implemented without Annex B emissions trade for meeting a given near-term CO_2 emissions target as reported by several models[15] of the global economy (top-down models) vary within regions (as shown by the brown lines in Figure 7-2a for Annex II regions and in Table 7-3a).** Reasons for the differentiation among models within regions is due to varying assumptions about future GDP growth rates and changes in carbon and energy intensity (different socio-economic development paths). The same reasons also apply to differences across regions. These models assume that national policy instruments are efficient and consistent with international policy instruments. That is, they assume that reductions are made through the use of market mechanisms (e.g., cap and trade) within each region. To the extent that regions employ a mix of market mechanisms and command and control policies, costs will likely be higher. On the other hand, inclusion of carbon sinks, non-CO_2 greenhouse gases, induced technical change, ancillary benefits, or targeted revenue recycling could reduce costs.

WGIII TAR Sections 8.2-3

7.18 **The models used in the above study show that the Kyoto mechanisms are important in controlling risks of high costs in given countries, and thus could complement domestic policy mechanisms, and could minimize risks of inequitable international impacts.** For example, the brown and blue lines in Figure 7-2b and Table 7-3b show that the national marginal costs to meet the Kyoto targets range from about US$20 up to US$600 per t C without Annex B trading, and range from about US$15 up to US$150 per t C with Annex B trading, respectively. At the time of these studies, most models did not include sinks, non-CO_2 greenhouse gases, CDM, negative cost options, ancillary benefits, or targeted revenue recycling, which will reduce estimated costs. On the other hand, these models make assumptions which underestimate costs because they assume full use of emissions trading without transaction costs, both within and among Annex B countries, and that mitigation responses would be perfectly efficient and that economies begin to adjust to the need to meet Kyoto targets between the years 1990 and 2000. The cost reductions from Annex B trading will depend on the details of implementation, including the compatibility of domestic and international mechanisms, constraints, and transaction costs. The following is indicative of the broad variation in the change in GDP reported for Annex B countries:

WGIII TAR Sections TS 8.3, 7.3, 8.3, 9.2, & 10.2

· *For Annex II countries, the above modeling studies show reductions in GDP, compared to projected levels in the year 2010.* Figure 7-2 indicates that in the absence of Annex B trading losses range from 0.2 to 2% of GDP. With Annex B trading, losses range from 0.1 to 1% of GDP. National studies, which explore a more diverse set of policy packages and take account of specific national circumstances, vary even more widely.

· *For most economies in transition, GDP effects range from negligible to a several percent increase, reflecting opportunities for energy-efficiency improvements not available to Annex II countries.* Under assumptions of drastic energy-efficiency improvement and/or continuing economic recessions in some countries, the assigned amounts may exceed projected emissions in the first commitment period. In this case, models show increased GDP due to revenues from trading assigned amounts. However, for some economies in transition, implementing the Kyoto Protocol will have similar impact on GDP as for Annex II countries.

---

[15] The above-referenced models report results for Energy Modeling Forum scenarios examining the benefits of emissions trading. For the analyses reported here, these models exclude sinks, multiple gases, ancillary benefits, macro-economic shocks, and induced technical change, but include lump sum tax revenue recycling. In the model baseline, additional no-regrets options, which are not listed above, are included.

Projections of GDP losses and marginal cost in Annex II countries in the year 2010 from global models

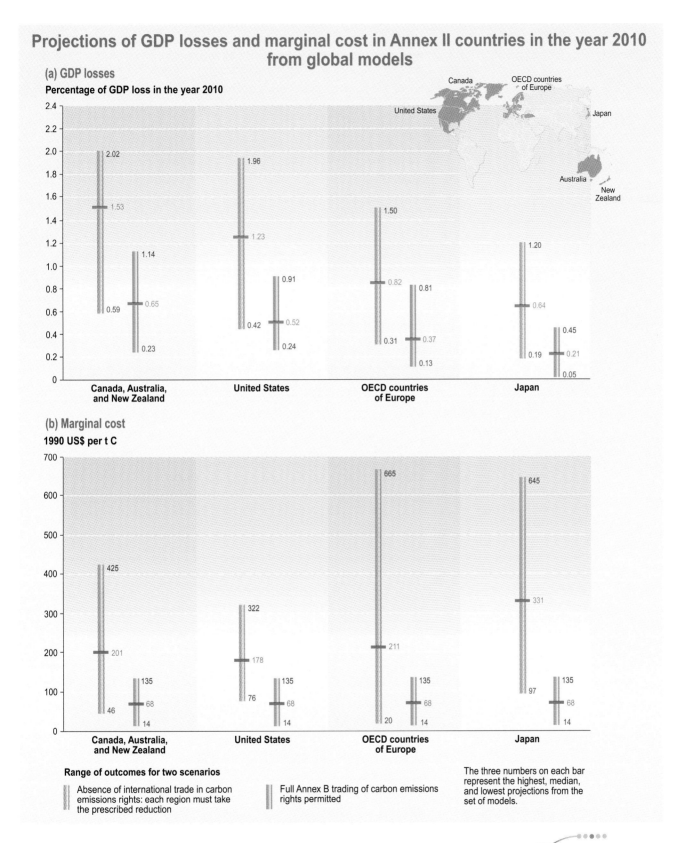

(a) GDP losses

Percentage of GDP loss in the year 2010

(b) Marginal cost

1990 US$ per t C

Range of outcomes for two scenarios

Absence of international trade in carbon emissions rights: each region must take the prescribed reduction

Full Annex B trading of carbon emissions rights permitted

The three numbers on each bar represent the highest, median, and lowest projections from the set of models.

Figure 7-2: Projections of GDP losses and marginal costs in Annex II countries in the year 2010 from global models: (a) GDP losses and (b) marginal costs. The reductions in projected GDP are for the year 2010 relative to the model reference case GDP. These estimates are based on results of an Energy Modeling Forum study. The projections reported in the figures are for four regions, which constitute Annex II. The models examined two scenarios. In the first, each region makes the prescribed reduction with only domestic trading in carbon emissions. In the second, Annex B trading is permitted and thereby marginal costs are equal across regions. For the key factors, assumptions, and uncertainties underlying the studies, see Box 7-1.

WGIII TAR Sections 8.3.1 & 10.4.4

| Table 7-3 | Results of model comparison from the Energy Modeling Forum.[a] |

(a) Calculated losses (as % of total GDP) for various postulated trading regimes associated with meeting the Kyoto targets in Annex B countries.

	No Trading				Annex I Trading			
Model	**CANZ**	**USA**	**OECD Europe**	**Japan**	**CANZ**	**USA**	**OECD Europe**	**Japan**
ABARE-GTEM	1.96	1.96	0.94	0.72	0.23	0.47	0.13	0.05
AIM	0.59	0.45	0.31	0.25	0.36	0.31	0.17	0.13
CETA		1.93				0.67		
G-Cubed	1.83	0.42	1.50	0.57	0.72	0.24	0.61	0.45
GRAPE			0.81	0.19			0.81	0.10
MERGE3	2.02	1.06	0.99	0.80	1.14	0.51	0.47	0.19
MS-MRT	1.83	1.88	0.63	1.20	0.88	0.91	0.13	0.22
RICE	0.96	0.94	0.55	0.78	0.54	0.56	0.28	0.30

(b) Marginal abatement costs (in 1990 US$ per t C; 2010 Kyoto target).

Model	**CANZ**	**USA**	**OECD Europe**	**Japan**	**Annex I Trading**
ABARE-GTEM	425	322	665	645	106
AIM	147	153	198	234	65
CETA		168			46
Fund					14
G-Cubed	157	76	227	97	53
GRAPE			204	304	70
MERGE3	250	264	218	500	135
MIT_EPPA	247	193	276	501	76
MS-MRT	213	236	179	402	77
RICE	145	132	159	251	62
SGM	201	188	407	357	84
WorldScan	46	85	20	122	20

(c) Costs of Kyoto Protocol implementation for oil-exporting countries according to various models.[b]

Model[c]	**Without Trading[d]**	**With Annex I Trading**	**With "Global Trading"**
G-Cubed	−25% oil revenue	−13% oil revenue	−7% oil revenue
GREEN	−3% real income	"substantially reduced loss"	n/a
GTEM	0.2% GDP loss	<0.05% GDP loss	n/a
MS-MRT	1.39% welfare loss	1.15% welfare loss	0.36% welfare loss
OPEC	−17% OPEC revenue	−10% OPEC revenue	−8% OPEC revenue
CLIMOX	n/a	−10% some oil exporters' revenues	n/a

[a] Table 7-3a derived from WGIII TAR Table TS-5, Table 7-3b from WGIII TAR Table TS-4, and Table 7-3c from WGIII TAR Table TS-6.
[b] The definition of oil-exporting country varies. For G-Cubed and the OPEC models, it is the OPEC countries; for GREEN, a group of oil-exporting countries; for GTEM, Mexico and Indonesia; for MS-MRT, OPEC countries plus Mexico; and for CLIMOX, west Asian and north African oil exporters.
[c] The models report impact on the global economy in the year 2010 with mitigation according to the Kyoto Protocol targets (usually in the models applied to CO_2 mitigation by the year 2010 rather than greenhouse gas emissions to the period 2008–2012) achieved by imposing a carbon tax or auctioned emission permits with revenues recycled through lump-sum payments to consumers. No ancillary benefits, such as reductions in local air pollution damages, are taken into account in the results.
[d] "Trading" denotes trading in emission permits between countries.
n/a = not available.

7.19 **Emission constraints on Annex I countries have well-established, albeit varied, "spill-over" effects[16] on non-Annex I countries.**

WGIII TAR Sections 8.3.2 & 9.3.1-2

· *Oil-exporting, non-Annex I countries: Analyses report costs differently, including,* inter alia, *reductions in projected GDP and reductions in projected oil revenues.* The study reporting the lowest costs shows reductions of 0.2% of projected GDP with no emissions trading, and less than 0.05% of projected GDP with Annex B emissions trading in the year 2010.[17]

[16] Spill-over effects incorporate only economic, not environmental, effects.
[17] These estimated costs can be expressed as differences in GDP growth rates over the period 2000–2010. With no emissions trading, GDP growth rate is reduced by 0.02 percentage points per year; with Annex B emissions trading, growth rate is reduced by less than 0.005 percentage points per year.

The study reporting the highest costs shows reductions of 25% of projected oil revenues with no emissions trading, and 13% of projected oil revenues with Annex B emissions trading in the year 2010 (see Table 7-3c). These studies do not consider policies and measures[18] other than Annex B emissions trading, which could lessen the impact on non-Annex I, oil-exporting countries, and therefore tend to overstate both the costs to these countries and overall costs. The effects on these countries can be further reduced by removal of subsidies for fossil fuels, energy tax restructuring according to carbon content, increased use of natural gas, and diversification of the economies of non-Annex I, oil-exporting countries.

· *Other non-Annex I countries: They may be adversely affected by reductions in demand for their exports to OECD nations and by the price increase of those carbon-intensive and other products they continue to import. These countries may benefit from the reduction in fuel prices, increased exports of carbon-intensive products, and the transfer of environmentally sound technologies and know-how.* The net balance for a given country depends on which of these factors dominates. Because of these complexities, the breakdown of winners and losers remains uncertain.

· *Carbon leakage: The possible relocation of some carbon-intensive industries to non-Annex I countries and wider impacts on trade flows in response to changing prices may lead to leakage on the order of 5-20%.*[19] Exemptions (e.g., for energy-intensive industries) make the higher model estimates for carbon leakage unlikely, but would raise aggregate costs. The transfer of environmentally sound technologies and know-how, not included in models, may lead to lower leakage and especially on the longer term may more than offset the leakage.

7.20 **Some sources of greenhouse gas emissions can be limited at no, or negative, net social cost to the extent that policies can exploit no-regret opportunities. This may be achieved by removal of market imperfections, accounting for ancillary benefits (see Question 8), and recycling revenues to finance reductions in distortionary taxes ("double dividend").**

WGIII TAR Sections 5.3-5, 7.3.3, 8.2.2, 8.2.4, 9.2.1-2, 9.2.4, 9.2.8, & 10.4

· *Market imperfections:* Reduction of existing market or institutional failures and other barriers that impede adoption of cost-effective emission reduction measures can lower private costs compared to current practice. This can also reduce private costs overall.

· *Ancillary benefits:* Climate change mitigation measures will have effects on other societal issues. For example, reducing carbon emissions in many cases will result in the simultaneous reduction in local and regional air pollution. It is likely that mitigation strategies will also affect transportation, agriculture, land-use practices, and waste management and will have an impact on other issues of social concern, such as employment, and energy security. However, not all of the effects will be positive; careful policy selection and design can better ensure positive effects and minimize negative impacts. In some cases, the magnitude of ancillary benefits of mitigation may be comparable to the costs of the mitigating measures, adding to the no-regret potential, although estimates are difficult to make and vary widely.

· *Double dividend:* Instruments (such as taxes or auctioned permits) provide revenues to the government. If used to finance reductions in existing distortionary taxes ("revenue recycling"), these revenues reduce the economic cost of achieving greenhouse gas reductions. The magnitude of this offset depends on the existing tax structure, type of tax cuts, labor market conditions, and method of recycling. Under some circumstances, it is possible that the economic benefits may exceed the costs of mitigation.

---

[18] These policies and measures include those for non-CO_2 gases and non-energy sources of all gases; offsets from sinks; industry restructuring (e.g., from energy producer to supplier of energy services); use of OPEC's market power; and actions (e.g., of Annex B Parties) related to funding, insurance, and the transfer of technology. In addition, the studies typically do not include the following policies and effects that can reduce the total cost of mitigation: the use of tax revenues to reduce tax burdens or finance other mitigation measures; environmental ancillary benefits of reductions in fossil-fuel use; and induced technical change from mitigation policies.

[19] Carbon leakage is defined here as the increase in emissions in non-Annex B countries due to implementation of reductions in Annex B, expressed as a percentage of Annex B reductions.

Potential, Barriers, Opportunities, Policies, and Costs of Stabilizing Atmospheric Greenhouse Gas Concentrations in the Long Term

7.21   Cost of stabilization depends on both the target and the emissions pathway.

7.22   **There is no single path to a low-emission future, and countries and regions will have to choose their own path. Most model results indicate that known technological options[20] could achieve a broad range of atmospheric CO_2 stabilization levels, such as 550 ppmv, 450 ppmv, or below over the next 100 years or more, but implementation would require associated socio-economic and institutional changes.** To achieve stabilization at these levels, the scenarios suggest that a very significant reduction in world carbon emissions per unit of GDP from year 1990 levels will be necessary. For the crucial energy sector, almost all greenhouse gas mitigation and concentration stabilization scenarios are characterized by the introduction of efficient technologies for both energy use and supply, and of low- or no-carbon energy. However, no single technology option will provide all of the emissions reductions needed for stabilization. Reduction options in non-energy sources and non-CO_2 greenhouse gases will also provide significant potential for reducing emissions.

WGIII TAR Sections 2.3.2, 2.4.5, 2.5.1-2, 3.5, & 8.4, & WGIII TAR Chapter 3 Appendix

7.23   **The development and diffusion of new economically competitive and environmentally sound technology can substantially reduce the costs of stabilizing concentrations at a given level.** A substantial body of work has considered the implication of technology development and diffusion on the cost of meeting alternative stabilization levels. The principal conclusion is that the cost of emissions mitigation depends crucially on the ability to develop and deploy new technology. The value of successful technology diffusion appears to be large and depends upon the magnitude and timing of emissions mitigation, the assumed reference scenario, and the economic competitiveness of the technology.

WGIII TAR Section 10.3.3

7.24   **The pathway to stabilization can be as important as the stabilization level itself in determining mitigation cost.** Economic modeling studies completed since the SAR indicate that a gradual near-term transition from the world's present energy system towards a less carbon-emitting economy minimizes costs associated with premature retirement of existing capital stock. It also provides time for investment in technology development and diffusion, and may reduce the risk of lock-in to early versions of rapidly developing low-emission technology. On the other hand, more rapid near-term action would increase flexibility in moving towards stabilization, decrease environmental and human risks associated with rapid climatic changes, while minimizing potential implications of inertia in climate and ecological systems (see Question 5). It may also stimulate more rapid deployment of existing low-emission technologies and provide strong near-term incentives to future technological changes that may help reduce the risks of lock-in to carbon-intensive technologies. It also would give greater scope for later tightening of targets should that be deemed desirable in light of evolving scientific understanding.

WGIII TAR Sections 2.3.2, 5.3.1, 8.4, & 10.4.2-3

7.25   **Cost-effectiveness studies with a century time scale estimate that the mitigation costs of stabilizing CO_2 concentrations in the atmosphere increase as the concentration stabilization level declines. Different baselines can have a strong influence on absolute costs.** While there is a moderate increase in the costs when passing from a 750 to a 550 ppmv concentration stabilization level, there is a larger

WGIII TAR Sections 2.5.2, 8.4.1, 8.4.3, & 10.4.6

---

[20] "Known technological options" refer to technologies that exist in operation or pilot plant stage today, as referenced in the mitigation scenarios discussed in this report. It does not include any new technologies that will require drastic technological breakthroughs. In this way it can be considered to be a conservative estimate, considering the length of the scenario period.

increase in costs passing from 550 to 450 ppmv (see Figure 7-3) unless the emissions in the baseline scenario are very low (see Figure 7-4). Although model projections indicate long-term global growth paths of GDP are not significantly affected by mitigation actions towards stabilization, these do not show the larger variations that occur over some shorter time periods, sectors, or regions. These results, however, do not incorporate carbon sequestration, and did not examine the possible effect of more ambitious targets on induced technological change. Costs associated with each concentration level depend on numerous factors including the rate of discount, distribution of emission reductions over time, policies and measures employed, and particularly the choice of the baseline scenario. For scenarios characterized by a focus on local and regional sustainable development for example, total costs of stabilizing at a particular level are significantly lower than for other scenarios. Also, the issue of uncertainty takes on increasing importance as the time frame is expanded.

7.26 Energy R&D and social learning can contribute to the flow and adoption of improved energy technologies throughout the 21st century.

7.27 Lower emissions scenarios require different patterns of energy resource development and an increase in energy R&D to assist accelerating the development and deployment of advanced environmentally sound energy technologies. Emissions of CO_2 due to fossil-fuel burning are virtually certain to be the dominant influence on the atmospheric CO_2 concentration trend during the 21st century. Resource data assessed in the TAR may imply a change in the energy mix and the introduction of new sources of energy during the 21st century. Fossil-fuel resources will not limit carbon

WGIII TAR Sections 2.5.1-2, 3.8.4, & 8.4.5

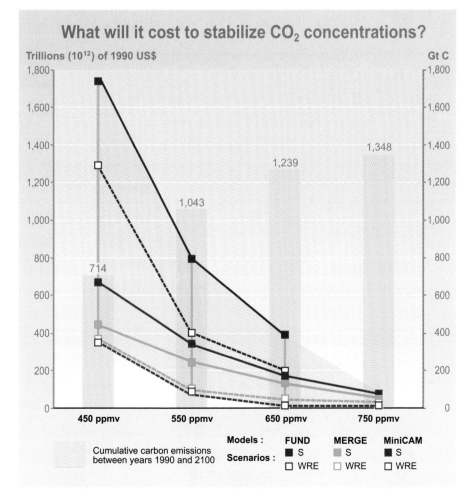

What will it cost to stabilize CO_2 concentrations?

WGIII TAR Sections 2.5.2, 8.4.1, 8.4.3, & 10.4.6

Figure 7-3: The mitigation costs (1990 US$, present value discounted at 5% per year for the period 1990 to 2100) of stabilizing CO_2 concentrations at 450 to 750 ppmv are calculated using three global models, based on different model-dependent baselines. Avoided impacts of climate change are not included. In each instance, costs were calculated based on two emission pathways for achieving the prescribed target: S (referred as WGI emissions pathways in WGIII TAR) and WRE as described in response to Question 6. The bars show cumulative carbon emissions between the years 1990 and 2100. Cumulative future emissions until carbon budget ceiling is reached are reported above the bars in Gt C.

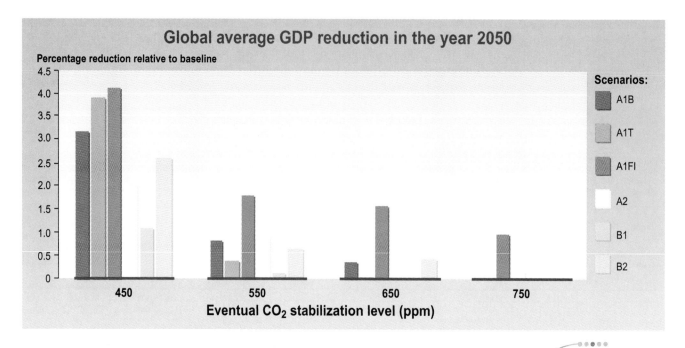

Global average GDP reduction in the year 2050

Percentage reduction relative to baseline

Eventual CO₂ stabilization level (ppm)

Scenarios:
- A1B
- A1T
- A1FI
- A2
- B1
- B2

Figure 7-4: Indicative relationship in the year 2050 between the relative GDP reduction caused by mitigation activities, the SRES scenarios, and the stabilization level. The reduction in GDP tends to increase with the stringency of the stabilization level, but the costs are very sensitive to the choice of the baseline scenario. These projected mitigation costs do not take into account potential benefits of avoided climate change.

WGIII TAR Figure 8-18

emissions during the 21st century (see Figure 7-5). The carbon in proven conventional oil and gas reserves is much less than the cumulative carbon emissions associated with stabilization of CO₂ at levels of 450 ppmv or higher.[21] These resource data may imply a change in the energy mix and the introduction of new sources of energy during the 21st century. The choice of energy mix and associated technologies and investments—either more in the direction of exploitation of unconventional oil and gas resources, or in the direction of non-fossil energy sources or fossil energy technology with carbon capture and storage—will determine whether, and if so, at what level and cost, greenhouse concentrations can be stabilized.

7.28 **The decline in energy R&D expenditure is inconsistent with the goal of accelerating the development and deployment of advanced energy technologies.** Energy-related R&D expenditure by Annex II governments increased dramatically after the 1970 oil price increases, but as a group it has decreased steadily in real terms since the early 1980s. In some countries the decrease has been as great as 75%. The support for energy conservation and renewable energy R&D has increased. However, other important energy technologies relevant to climate change, such as, for example, commercial biomass and carbon capture and storage, remain minor constituents of the energy R&D portfolio.

WGIII TAR Section 10.3.3 & SRTT Section 2.3

7.29 **Social learning and innovation and changes in institutional structure could contribute to climate change mitigation.** Changes in collective rules and individual behaviors may have significant effects on greenhouse gas emissions, but take place within a complex institutional, regulatory, and legal setting. Several studies suggest that current incentive systems can encourage resource-intensive production and consumption patterns that increase greenhouse gas emissions in all sectors (e.g., transport and housing). In the shorter term, there are opportunities to influence through social innovations individual and organizational behaviors.

WGIII TAR Sections 1.4.3, 5.3.7, 10.3.2, & 10.3.4

---

[21] The reference to a particular concentration level does not imply an agreed-upon desirability of stabilization at this level.

In the longer term, such innovations in combination with technological change may further enhance socio-economic potential, particularly if preferences and cultural norms shift towards lower emitting and sustainable behaviors. These innovations frequently meet with resistance, which may be addressed by encouraging greater public participation in the decision-making process. This can help contribute to new approaches to sustainability and equity.

Integrating Near- and Long-Term Considerations

7.30 **Climate change decision making is a sequential process under uncertainty. Decision making at any point in time entails balancing the risks of either insufficient or excessive action.**

7.31 **Development of a prudent risk management strategy involves careful consideration of the consequences (both environmental and economic), their likelihood, and society's attitude toward risk.** The latter is likely to vary from country to country and perhaps even from generation to generation. This report therefore confirms the SAR finding that the value of better information about climate change processes and impacts

 WGIII TAR Section 10.4.3

Carbon in fossil-fuel reserves and resources compared with historical fossil-fuel carbon emissions, and with cumulative carbon emissions from a range of SRES scenario and TAR stabilization scenarios until the year 2100

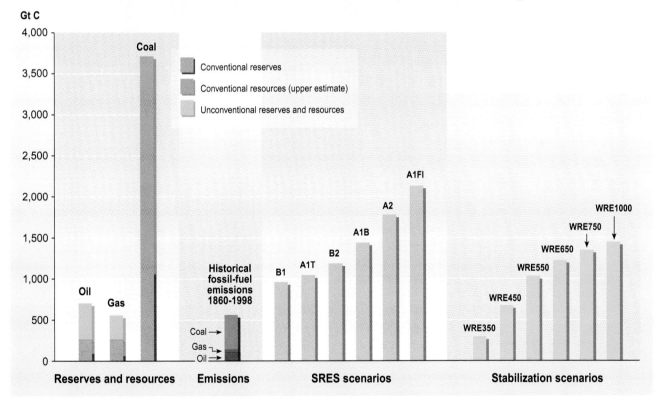

Figure 7-5: Carbon in oil, gas, and coal reserves and resources is compared with historic fossil-fuel carbon emissions over the period 1860–1998, and with cumulative carbon emissions from a range of SRES scenarios and TAR stabilization scenarios until the year 2100. Data for current reserves and resources are shown in the lefthand columns. Unconventional oil and gas includes tar sands, shale oil, other heavy oil, coal bed methane, deep geopressured gas, gas in aquifers, etc. Gas hydrates (clathrates) that amount to an estimated 12,000 Gt C are not shown. The scenario columns show both SRES reference scenarios as well as scenarios that lead to stabilization of CO_2 concentrations at a range of levels. Note that if by the year 2100 cumulative emissions associated with SRES scenarios are equal to or smaller than those for stabilization scenarios, this does not imply that these scenarios equally lead to stabilization.

 WGIII TAR Section 3.8.1

and society's responses to them is likely to be great. Decisions about near-term climate policies are in the process of being made while the concentration stabilization target is still being debated. The literature suggests a step-by-step resolution aimed at stabilizing greenhouse gas concentrations. This will also involve balancing the risks of either insufficient or excessive action. The relevant question is not "what is the best course for the next 100 years," but rather "what is the best course for the near term given the expected long-term climate change and accompanying uncertainties."

7.32 **Stabilizing atmospheric concentrations would depend upon emissions reductions beyond those agreed to in the Kyoto Protocol.** Most post-SRES scenario analyses suggest that achievement of stabilization at 450 ppmv may require emission reductions during the period 2008 to 2012 in Annex I countries that are significantly stronger than the Kyoto Protocol commitments. This analysis also suggests that achieving the aggregate Kyoto commitments may be consistent with trajectories that achieve stabilization at 550 ppmv or higher. Other analyses suggest a more gradual departure from emissions baselines even for 450 ppmv followed by sharper reductions in subsequent budget periods. The path is influenced by the representation of inertia in the system and expectations about how initial reductions by Annex I countries may relate to the strength and scope of emissions limitation in subsequent periods.

WGIII TAR Sections 2.5.2 & 8.4

7.33 **Climate change mitigation raises both inter-regional and inter-temporal equity considerations.**

7.34 **Differences in the distribution of technological, natural, and financial resources among and within nations and regions, and between generations, as well as differences in mitigation costs, are often key considerations in the analysis of climate change mitigation options.** Much of the debate about the future differentiation of contributions of countries to mitigation and related equity issues also considers these circumstances.[22] The challenge of addressing climate change raises an important issue of equity, namely the extent to which the impacts of climate change or mitigation policies ameliorate or exacerbate inequities both within and across nations and regions, and between generations. Findings with respect to these different aspects of equity include:

WGIII TAR Sections 1.3, 2.5.2, 8.2.2, 10.2, & 10.4.5

· *Equity within nations: Most studies show that the distributional effects of a carbon tax are regressive unless the tax revenues are used either directly or indirectly in favor of the low-income groups; the regressive aspect can be totally or partially compensated by a revenue-recycling policy.*

· *Equity across nations and regions: Greenhouse gas stabilization scenarios assessed in this report assume that developed countries and countries with economies in transition limit and reduce their greenhouse gas emissions first.[23]* Another aspect of equity across nations and regions is that mitigation of climate change can offset inequities that would be exacerbated by the impacts of climate change (see Question 6).

· *Equity between generations: Stabilization of concentrations depends more upon cumulative than annual emissions; emissions reductions by any generation will reduce the need for those by future generations.[24]* Inter-generational equity can be promoted by reducing climate change impacts through mitigation of climate change by any generation, since not only would impacts—which are expected to affect especially those with the fewest resources—be reduced, but also subsequent generations will have less climate change to adapt to (see Question 6).

---

[22] Approaches to equity have been classified into a variety of categories, including those based on allocation, outcome, process, rights, liability, poverty, and opportunity, reflecting the diverse expectations of fairness used to judge policy processes and the corresponding outcomes.

[23] Emissions from all regions diverge from baselines at some point. Global emissions diverge earlier and to a greater extent as stabilization levels are lower or underlying scenarios are higher. Such scenarios are uncertain, and do not provide information on equity implications and how such changes may be achieved or who may bear any costs incurred.

[24] See above for other aspects of timing of greenhouse gas emissions reductions.

Question 8

What is known about the interactions between projected human-induced changes in climate and other environmental issues (e.g., urban air pollution, regional acid deposition, loss of biological diversity, stratospheric ozone depletion, and desertification and land degradation)? What is known about environmental, social, and economic costs and benefits and implications of these interactions for integrating climate change response strategies in an equitable manner into broad sustainable development strategies at the local, regional, and global scales?

8.1   The answer to this question recognizes two major points. The first is that the human impacts on the environment are manifested in several issues, many driven by common factors associated with the meeting of human needs. The second is that many of these issues—their causes and impacts—are biogeophysically and socio-economically interrelated. With a central emphasis on climate change, this answer assesses the current understanding of the interrelations between the causes and impacts of the key environmental issues of today. To that is added a summary of the now largely separate policy approaches to these issues. In so doing, this answer frames how choices associated with one issue may positively or negatively influence another. With such knowledge, there is the prospect of efficient integrated approaches.

8.2   **Local, regional, and global environmental issues often combine in ways that jointly affect the sustainable meeting of human needs.**

8.3   **Meeting human needs is degrading the environment in many instances, and environmental degradation is hampering the meeting of human needs.** Society has a range of socio-economic paths to development; however, these will only be sustainable if due consideration is given to the environment. Environmental degradation is already evident at the local, regional, and global scale, such as air pollution, scarcity of freshwater, deforestation, desertification, acid deposition, loss of biological diversity and changes at the genetic and species level, land degradation, stratospheric ozone depletion, and climate change. Very frequently, addressing human needs causes or exacerbates several environmental problems, which may increase the vulnerability to climatic changes. For example, with the aim of higher agricultural production, there is increased use of nitrogeneous fertilizers, irrigation, and conversion of forested areas to croplands. These agricultural activities can affect the Earth's climate through release of greenhouse gases, degrade land by erosion and salinization, and reduce biodiversity. In turn, an environmental change can impact meeting human needs. For example, agricultural productivity can be adversely affected by changes in the magnitude and pattern of rainfall, and human health in an urban environment can be impacted by heat waves.

WGI TAR Sections 3.4, 4.1, & 5.2, WGII TAR Sections 4.1 & 5.1-2, & WGIII TAR Sections 3.6 & 4.2

8.4   **Just as different environmental problems are often caused by the same underlying driving forces (economic growth, broad technological changes, life-style patterns, demographic shifts (population size, age structure, and migration), and governance structures), common barriers inhibit solutions to a variety of environmental and socio-economic issues.** Approaches to the amelioration of environmental issues can be hampered by many of the same barriers, for example:
  · Increased demand for natural resources and energy
  · Market imperfections, including subsidies that lead to the inefficient use of resources and act as a barrier to the market penetration of environmentally sound technologies; the lack of recognition of the true value of natural resources; failure to appropriate for the global values of natural resources at the local level; and failure to internalize the costs of environmental degradation into the market price of a resource
  · Limited availability and transfer of technology, inefficient use of technologies, and inadequate investment in research and development for the technologies of the future
  · Failure to manage adequately the use of natural resources and energy.

WGIII TAR Chapter 5, SRES Chapter 3, & SRTT TS 1.5

8.5   **Several environmental issues that traditionally have been viewed as separate are indeed linked with climate change via common biogeochemical and socio-economic processes.**

8.6   Figure 8-1 illustrates how climate change is interlinked with several other environmental issues.

Surface Ozone Air Pollution and Climate Change

8.7   **Surface ozone air pollution and the emissions that drive it are important contributors to global climate change.** The same pollutants that generate surface

WGI TAR Sections 4.2.3-4

Linkages between climate change and other environmental issues

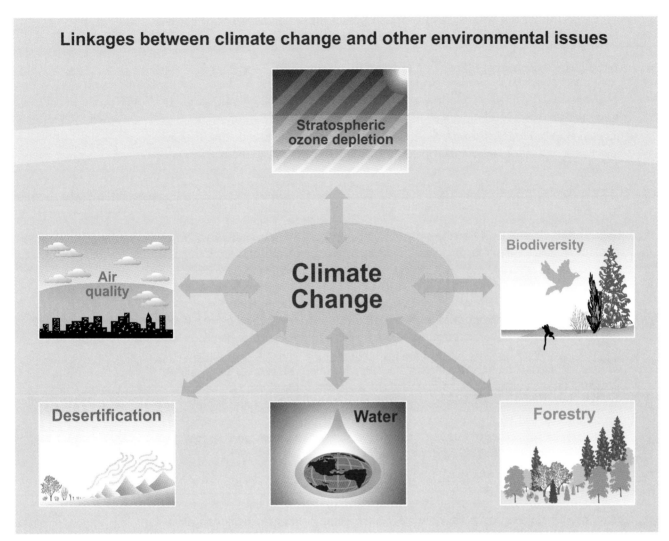

Figure 8-1: Climate is controlled by geochemical processes and cycles resulting from the interplay among the environment's components involved, as affected by human action. The scheme shows some of these issues. For simplicity, the single double-ended arrows between issues represent some of the linkages involved. For example, biological and ecological processes play an important role in modulating the Earth's climate at both regional and global scale by controlling the amounts of water vapor and other greenhouse gases that enter into or are depleted from the atmosphere. Changes in climate affect the boundaries, composition, and functioning of ecological systems, such as forests, and changes in the structure and functioning of forests affect the Earth's climate system through changes in the biogeochemical cycles, particularly cycles of carbon, nitrogen, and water. There are other linkages such as the connection between air quality and forestry, directly or through acid precipitation, which for simplicity are not shown here.

ozone pollution (nitrogen oxides, carbon monoxide, and volatile organic compounds) also contribute to the rise in global tropospheric ozone, making it the third most important contributor to radiative forcing after CO_2 and CH_4 (see Figure 2-2). In some regions emissions of ozone precursor substances are controlled by regional environmental treaties (see Table 8-3) and other regulations.

8.8 **Global climate changes and rising tropospheric ozone levels may exacerbate urban air pollution problems.** Projections based on some SRES scenarios show increases in tropospheric ozone of more than 40 ppb over most of the Northern Hemisphere mid-latitudes. Such increases would approximately double the baseline levels of ozone entering many metropolitan regions, substantially degrading air quality. Climate change would affect the meteorological conditions (regional temperature, cloud cover, and surface wind) that influence photochemistry, and the occurrence of major pollution episodes. While warmer temperatures would generally contribute to more urban ozone, the change in frequency and intensity of pollution episodes

WGI TAR Sections 4.4.4 & 4.5-6, & WGII TAR Sections 7.2.2.3 & 9.6

has not been evaluated. Adverse health effects attributable to urban air quality would be exacerbated by increases in heat waves that would accompany anthropogenic climate change.

Acid Deposition and Climate Change

8.9 **The sulfate aerosols formed from sulfur emissions from the burning of fossil fuels lead to both acid deposition and a cooling of the climate system.** Acid deposition has adverse impacts on both terrestrial and aquatic ecosystems and causes damage to human health and many materials. Some of these impacts could be exacerbated by climate change (e.g., through increase in humidity and temperature). Actions to reduce sulfur emissions have been taken in many countries, and declines in sulfate deposition have been observed in some regions in recent years (see Table 8-3). In the SRES scenarios, this situation has led to projections of future sulfate aerosol abundances that are lower than those in the SAR. This has led, in turn, to less negative projections for the radiative forcing by aerosols, hence less of a cooling effect to offset the greenhouse gas-induced warming.

WGI TAR Sections 5.2.2.6, 5.5.3, 6.7, & 6.15, WGII TAR Sections 5.6, 5.7.3, & 15.2.4.2, & SRES Section 3.6.4

Stratospheric Ozone Depletion and Climate Change

8.10 **Depletion of the stratospheric ozone layer leads to an increased penetration of UV-B radiation and to a cooling of the climate system.** Ozone depletion has allowed for increased penetration of UV-B radiation, with harmful effects on human and animal health, plants, etc. During the last 2 decades, the observed losses of stratospheric ozone have decreased the downward infrared emissions to the troposphere from the (now colder) lower stratosphere. Stratospheric ozone depletion has also altered tropospheric ozone concentrations, and, by allowing more ultraviolet sunlight into the troposphere, it has led to more rapid photochemical destruction of CH_4 thereby reducing its radiative forcing. These effects lead also to a cooling of the climate system.

WGI TAR Sections 4.2.2 & 6.4

8.11 **Many of the halocarbons that cause depletion of the ozone layer are also important greenhouse gases.** Chlorofluorocarbons, for example, add a notable fraction to the total positive radiative forcing since the pre-industrial era. The negative radiative forcing from the associated stratospheric ozone depletion (noted above) reduces this by about half. The Montreal Protocol will eventually eliminate both of these radiative-forcing contributions. However, one class of substitutes for the now-banned chlorofluorocarbons is hydrofluorocarbons, which are among the greenhouse gases listed under the Kyoto Protocol. This overlap can give rise to a potential conflict beween the goals of the two Protocols.

WGI TAR Sections 4.2.2 & 6.3.3

8.12 **Climate change will alter the temperature and wind patterns of the stratosphere, possibly enhancing chlorofluorocarbon depletion of stratospheric ozone over the next 50 years.** Increases in greenhouse gases lead in general to a colder stratosphere, which alters stratospheric chemistry. Some studies predict that current rates of climate change will result in significant increases in the depletion of the Arctic stratospheric ozone layer over the next decade before chlorofluorocarbon concentrations have declined substantially. Although many climate/ozone-layer feedbacks have been identified, no quantitative consensus is reached in this assessment.

WGI TAR Sections 4.5, 6.4, & 7.2.4.2

Biodiversity, Agriculture and Forestry, and Climate Change

8.13 **Changes in terrestrial and marine ecosystems are closely linked to changes in climate and *vice versa*.** Changes in climate and in atmospheric concentrations of CO_2 cause changes in the biodiversity and function of some ecosystems. In turn, ecosystem changes influence the land-atmosphere exchange of greenhouse gases (e.g., CO_2, CH_4, and N_2O) and of water and energy, and change surface albedo. Therefore, understanding these combined effects and feedbacks are a requisite for evaluating the future state of the atmosphere and the natural systems and their biodiversity.

WGI TAR Section 4.5.3

8.14 **Natural climate variations have illustrated the impacts of climate change on natural and managed ecosystems.** The impacts of floods, droughts, and heat waves are etched into human history. Further, the warming events associated with El Niño illustrate that changes in climate patterns adversely affect fish, marine mammals, and coastal and ocean biodiversity. Coastal ecosystems—such as coral reefs, salt marshes, and mangrove forests—are affected by sea-level rise, warming ocean temperatures, increased CO_2 concentrations, and changes in storm frequency and intensity. Table 8-1 gives main implications of climate change for natural ecosystems at the regional scale.

WGII TAR Chapters 5 & 6

8.15 **Climate change is but one of many stresses on managed and unmanaged ecosystems.** Land-use change, resource demands, deposition of nutrients and pollutants, harvesting, grazing, habitat fragmentation and loss, and invasive species are major stressors on ecosystems. They can lead to species extinction, resulting in losses of biodiversity. Therefore, climate change constitutes an additional stress and could change or endanger ecosystems and the many services they provide. As a result, the impact of climate change will be influenced by management of natural resources, adaptation, and interaction with other pressures. Figure 8-2 exemplifies the manner in which climate change interacts with other factors in food supply and demand.

WGII TAR Chapters 5 & 6, & WGIII TAR Sections 4.1-2

8.16 **Climate change can influence the distribution and migration of species in unmanaged ecosystems.** Populations of many species are already threatened with extinction and are expected to be placed at greater risk by the stresses of changing climate, rendering portions of their current habitat unsuitable. Vegetation distribution models since the SAR suggest that a mass ecosystem or biome movement is most unlikely to occur because differerent species have different climate tolerance and different migration abilities, and are affected differently by the arrival of new species. Lastly, in a related sense, climate change can enhance the spreading of pests and diseases, thereby affecting both natural ecosystems, crops, and livestock (e.g., changes in temperature and humidity thresholds allow pests and diseases to move to new areas).

WGII TAR Chapter 5

8.17 **Carbon storage capacities of managed and unmanaged ecosystems, particularly forests, influence impacts and feedbacks with climate change.** For example, forests, agricultural lands, and other terrestrial ecosystems offer a significant carbon mitigation potential. Although not necessarily permanent, conservation and sequestration may allow time for other options to be further developed and implemented. Terrestrial ecosystem degradation may be exacerbated by climate change, affecting the storage of carbon, and adding to the stresses resulting from the current deforestation practices. It should be noted that, if appropriate management practices are not carried out, CO_2 emissions in the future could be higher. For example, abandoning fire management in forests or reverting from direct seeding to intensive tillage in agriculture may result in rapid loss of part, at least, of the accumulated carbon.

WGIII TAR Section 4.3 & SRLULUCF SPM

Land Degradation and Desertification and Climate Change

8.18 **Projected levels of climate change would exacerbate the continuation of land degradation and desertification that has occurred over the past few centuries in many areas.** Land-use conversion and the intensive use of land, particularly in the world's arid and semi-arid regions, has resulted in decreased soil fertility and increased land degradation and desertification. The changes have been large enough to be apparent from satellite images. Land degradation already affects more than 900 million people in 100 countries, and one quarter of the world soil resources, most of them in the developing countries. The annual recorded losses of millions of hectares significantly undermine economies and create some irreversible situations. The TAR projections using the SRES scenarios indicate increased droughts, higher intensity of rainfall, more irregular rainfall patterns, and more frequent tropical summer drought in the mid-latitude continental interiors. The systems that

WGI TAR Sections 2.7.3.3, 9.3, & 10.3, WGII TAR Section 5.5, & WGII TAR Table SPM-1

Table 8-1	Examples for observed and projected regional implications of climate change on natural ecosystems, biodiversity, and food supply.

Region	Impacts	Reference Section in WGII TAR
Africa	Irreversible losses of biodiversity could be accelerated with climate change. Significant extinctions of plant and animal species are projected and would impact rural livelihoods, tourism, and genetic resources (*medium confidence*).	TS 5.1.3 & Section 10.2.3.2
Asia	Decreases in agricultural productivity and aquaculture due to thermal and water stress, sea-level rise, floods and droughts, and tropical cyclones would diminish food security in many countries of arid, tropical, and temperate Asia; agriculture would expand and increase in productivity in northern areas (*medium confidence*). Climate change would exacerbate threat to biodiversity due to land-use and land-cover change and population pressure (*medium confidence*). Sea-level rise would put ecological security at risk including mangroves and coral reefs (*high confidence*).	TS 5.2.1-2 & Sections 11.2.1-2
Australia and New Zealand	A warming of 1°C would threaten the survival of species currently near the upper limit of their temperature range, notably in marginal alpine regions. Some species with restricted climatic niches and that are unable to migrate due to fragmentation of the landscape soil differences or topography could become endangered or extinct (*high confidence*). Australian ecosystems that are particularly vulnerable to climate change include coral reefs, arid and semi-arid habitats in southwest and inland Australia, and Australian alpine systems. Freshwater wetlands in coastal zones in both Australia and New Zealand are vulnerable, and some New Zealand ecosystems are vulnerable to accelerated invasion by weeds.	TS 5.3.2 & Sections 12.4.2, 12.4.4-5, & 12.4.7
Europe	Natural ecosystems will change due to increasing temperature and atmospheric concentration of CO_2. Diversity in nature reserves is under threat of rapid change. Loss of important habitats (wetlands, tundra, and isolated habitats) would threaten some species, including rare/endemic species and migratory birds. There will be some broadly positive effects on agriculture in northern Europe (*medium confidence*); productivity will decrease in southern and eastern Europe (*medium confidence*).	TS 5.4.2-3 & Sections 13.2.1.4, 13.2.2.1, 13.2.2.3-5, & 13.2.3.1
Latin America	It is well-established that Latin America accounts for one of the Earth's largest concentrations of biodiversity and the impacts of climate change can be expected to increase the risk of biodiversity loss (*high confidence*). Yields of important crops are projected to decrease in many locations even when the effects of CO_2 are taken into account; subsistence farming in some regions could be threatened (*high confidence*).	TS 5.5.2 & 5.5.4, & Sections 14.2.1-2
North America	There is strong evidence that climate change can lead to the loss of specific ecosystem types (e.g., high alpine areas and specific coastal (salt marshes and inland prairie "potholes") wetlands) (*high confidence*). Some crops would benefit from modest warming accompanied by increasing CO_2, but effect would vary among crops and regions (*high confidence*), including declines due to drought in some areas of Canada's Prairies and the U.S. Great Plains, potential increased food production in areas of Canada north of current production areas, and increased warm temperate mixed forest production (*medium confidence*). However, benefits for crops would decline at an increasing rate and possibly become a net loss with further warming (*medium confidence*). Unique natural ecosystems such as prairie wetlands, alpine tundra, and coldwater ecosystems will be at risk and effective adaptation is unlikely (*medium confidence*).	TS 5.6.4-5 & Sections 15.2.2-3
Arctic	The Arctic is extremely vulnerable to climate change, and major physical, ecological, and economic impacts are expected to appear rapidly.	TS 5.7 & Sections 16.2.7-8
Antarctic	In the Antarctic projected climate change will generate impacts that will be realized slowly (*high confidence*). Warmer temperatures and reduced ice extent are likely to produce long-term changes in the physical oceanography and ecology of the Southern Ocean, with intensified biological activity and increased growth rate of fish.	TS 5.7 & Sections 16.2.3 & 16.2.4.2
Small Islands	Projected future climate change and sea-level rise will affect shifts in species composition and competition. It is estimated that one out of every three (30%) known threatened plants are island endemics, while 23% of bird species are threatened. Coral reefs, mangroves, and seagrass beds that often rely on stable environmental conditions will be adversely affected by rising air and sea temperatures and sea-level rise (*medium confidence*). Declines in coastal ecosystems would negatively impact reef fish and threaten reef fisheries (*medium confidence*).	TS 5.8 & Sections 17.2.4-5 & 17.2.8.2

likely would be impacted include those with scarce water resources, rangelands, and land subsidence (see Table 8-2).

Climate change and food

Climate Change

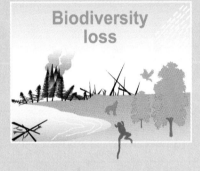

Unsustainable forestry

Nitrogen fertilization

Hydrologic and CO_2 changes

Extensification

Loss and fragmentation of habitats

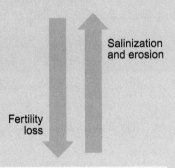

Food supply and demand

Loss of crops genetic diversity

Biodiversity loss

Salinization and erosion

Fertility loss

Land degradation

Figure 8-2: This figure shows linkages between climate change and other environmental factors in food supply and demand. Increasing food demand by a growing world population calls for larger food production. This, in turn, brings a series of implications in the use of land, such as converting wildlands to croplands (extensification), and using chemical fertilizers and/or using irrigation to increase yield (intensification) or enabling cultivation in otherwise non-usable land. Expanding the land under cultivation results in loss of biodiversity, as ecosystems are converted to fields growing only a few species (usually exotics). Change of forests to agriculture brings a net loss of carbon to the atmosphere, as forests are replaced by grassland or cropland. This clearing also increases flooding probability, as the agricultural systems retain less precipitation than forests. Intensification of crop production can involve a variety of chemical treatments, most of them being nitrogen fertilizers bringing the side effect of release of nitrogen gas compounds (some of which are strong greenhouse gases) to the atmosphere and nitrogen runoff into watersheds, with many environmental and health implications. The expansion of irrigation affects the supply of freshwater for other uses, leading to shortages and conflicts over water-use rights. Meeting the needs for increased agricultural production has the potential to increase global rates of biodiversity loss, climate change, and desertification. There are interrelations, particularly to water, that underly all these issues, but for simplicity are not shown in the figure.

Table 8-2	Examples of regional impacts of climate change on water resources, land degradation, and desertification.

Region	Projections	Reference Section in WGII TAR
Africa	Changes in rainfall and intensified land use would exacerbate the desertification processes. Desertification would be exacerbated by reduction in the average annual rainfall, runoff, and soil moisture in countries of west African Sahel, and northern and southern Africa (*medium confidence*). Increases in droughts and other extreme events would add to stresses on water resources, food security, and human health, and would constrain development in the region (*high confidence*).	TS 5.1.6, Chapter 10 ES, Sections 10.2.1 & 10.2.6, & Table SPM-2
Asia	Water shortage—already a limiting factor for ecosystems, food and fiber production, human settlements, and human health—may be exacerbated by climate change. Runoff and water availability may decrease in arid and semi-arid Asia but increase in northern Asia (*medium confidence*). Reduced soil moisture in summer would exacerbate land degradation and desertification in arid and semi-arid regions.	TS 5.2.3 & Sections 11.1.1 & 11.2.3
Australia and New Zealand	Interannual variability due to ENSO leads to major floods and droughts in Australia and New Zealand. Such variations are expected to continue under enhanced greenhouse gas conditions, but possibly with greater hydrological extremes. Water is likely to be a key issue (*high confidence*) due to projected drying trends over much of the region and change to a more El Niño-like event state. Water quality would be affected, and more intense rainfall events would increase fast runoff, soil erosion, and sediment loading. Eutrophication is a major water quality problem in Australia.	TS 5.3 & Sections 12.1.5.3 & 12.3
Europe	Summer runoff, water availability, and soil moisture are likely to decrease in southern Europe, and would widen the gap between the north and south (*high confidence*). Flood hazards will increase across much of Europe (*medium to high confidence*); risk would be substantial for coastal areas where flooding will increase erosion and result in loss of wetlands. Half of alpine glaciers and large permafrost areas could disappear by the end of the 21st century (*medium confidence*).	TS 5.4.1, Chapter 13 ES, & Section 13.2.1
Latin America	Some studies based on model experiments suggest that under climate change the hydrological cycle would be more intense, with changes in the distribution of extreme rainfall, wet spells, and dry spells. Frequent severe drought in Mexico during the last decade coincides with some of these model findings. El Niño is related to dry conditions in northeastern Brazil, northern Amazons, and the Peruvian-Bolivian altiplano. Southern Brazil and northwestern Peru exhibit anomalous wet conditions during these periods. Loss and retreat of glaciers would adversely impact runoff and water supply in areas where snowmelt is an important water resource (*high confidence*).	TS 5.5.1, Chapter 14 ES, & Section 14.2.4
North America	Snowmelt-dominated watersheds in western North America will experience earlier spring peak flows (*high confidence*) and reduction in summer flow (*medium confidence*); adaptive responses may offset some, but not all, of the impacts on water resources and aquatic ecosystems (*medium confidence*).	TS 5.6.2, Section 15.2.1, & Table SPM-2
Small Islands	Islands with very limited water supplies are highly vulnerable to the impacts of climate change on the water balance (*high confidence*).	TS 5.8.4, Section 17.2.6, & Table SPM-2

Freshwater and Climate Change

8.19 **All three classes of freshwater problems—having too little, too much, and too dirty water—may be exacerbated by climate change.** Freshwater is essential for human health, food production, and sanitation, as well as for manufacturing and other industrial uses and sustaining ecosystems. There are several indicators of water resources stress. When withdrawals are greater than 20% of the total renewable resources, water stress often is a limiting factor on development. Withdrawals of 40% or more represent high stress. Similarly, water stress may be a problem if a country or region has less than 1,700 m³ yr⁻¹ of water per capita. In the year 1990, approximately one-third of the world's population lived in countries using more than 20% of their water resources, and by the year 2025 about 60% of a larger total would be living in such a stressed country, only because of population growth. Higher temperatures could increase such stress conditions. However, adaptation through appropriate water management practices can reduce the adverse impacts. While climate change is just one of the stresses on water resources in this increasingly populated world, it is clear that it is an important one (see Table 8-2). The TAR projections using the SRES scenarios of future

WGII TAR Sections 4.1, 4.4.3, 4.5.2, & 4.6.2

climate indicate a tendency for increased flood and drought risks for many areas under most scenarios. Decreases of water availability in parts of a warmer world are projected in areas like southern Africa and countries around the Mediterranean. Because of sea-level rise, many coastal systems will experience saltwater intrusion into fresh groundwater and encroachment of tidal water into estuaries and river systems, with consequential effects on freshwater availability.

8.20 **Water managers in some countries are beginning to consider climate change explicitly, although methodologies for doing so are not yet well defined.** By its nature, water management is based around minimization of risks and adaptation to changing circumstances, now also changing climate. There has been a gradual shift from "supply-side" approaches (i.e., providing water to satisfy demands by increased capacity reservoirs or structural flood defenses) towards "demand-side" approaches (i.e., trimming demands adequately to match water availability, using water more efficiently, and non-structural means of preparedness to floods and droughts).

WGII TAR Section 4.2.4

8.21 **Interactions between climate change and other environmental problems offer opportunities to capture synergies in developing response options, enhancing benefits, and reducing costs (see Figure 1-1).**

8.22 **By capturing synergies, some greenhouse gas mitigation actions may yield extensive ancillary benefits for several other environmental problems, but also trade-offs may occur.** Examples include, *inter alia*, reduction of negative environmental impacts such as air pollution and acid deposition; protecting forests, soils, and watersheds; reducing distortionary subsidies and taxes; and inducing more efficient technological change and diffusion, contributing to wider goals of sustainable development. However, dependent on the way climate change or other environmental problems are addressed, and the degree to which interlinking issues are taken into account, significant trade-offs may occur and unanticipated costs may be incurred. For example, policy options to reduce greenhouse gas emissions from the energy and land-use sectors can have both positive and negative effects on other environmental problems:

WGIII TAR Sections 3.6.4, 4.4, 8.2.4, & 9.2.2-5

   · In the energy sector, greenhouse gas emissions as well as local and regional pollutants could be reduced through more efficient and environmentally sound use of energy and increasing the share of lower carbon emitting fossil fuels, advanced fossil-fuel technologies (e.g., highly efficient combined cycle gas turbines, fuel cells, and combined heat and power), and renewable energy technologies (e.g., increased use of environmentally sound biofuels, hydropower, solar, wind- and wave-power). Increased use of biomass as a substitute for fossil fuel could have positive or negative impacts on soils, biodiversity, and water availability depending on the land use it replaces and the management regime.

   · In the land-use sector, conservation of biological carbon pools not only prevents carbon from being emitted into the atmosphere, it also can have a favorable effect on soil productivity, prevent biodiversity loss, and reduce air pollution problems from biomass burning. Carbon sequestration by plantation forestry can enhance carbon sinks and protect soils and watersheds, but—if developed improperly—may have negative effects on biodiversity and water availability. For example, in some implementations, monoculture plantations could decrease local biodiversity.

8.23 **Conversely, addressing environmental problems other than climate change can have ancillary climate benefits, but the linkages between the various problems may also lead to trade-offs.** Examples include:

   · There are likely to be substantial greenhouse gas benefits from policies aimed at reducing air pollution. For example, increasing pollution is often associated with the rapidly growing transportation sector in all regions, involving emissions of particulate matter and precursors of ozone pollution. Addressing these emissions to reduce the impacts

WGIII TAR Sections 2.4, 9.2.8, & 10.3.2, & SRES

on human health, agriculture, and forestry through increasing energy efficiency or penetration of non-fossil-fuel energy can also reduce greenhouse gas emissions.

· Controlling sulfur emissions has positive impacts on human health and vegetation, but sulfate aerosols partly offset the warming effect of greenhouse gases and therefore control of sulfur emissions can amplify possible climate change. If sulfur emissions are controlled through desulfurization of flue gases at power plants, an energy penalty results, with associated increase of greenhouse gas emissions.

8.24 **Adopting environmentally sound technologies and practices offer particular opportunities for economically, environmentally, and socially sound development while avoiding greenhouse gas-intensive activities.** For example, the application of supply- and demand-side energy-efficient technologies simultaneously reduces various energy-related environmental impacts and can lower the pressure on energy investments, reduce public investments, improve export competitiveness, and enlarge energy reserves. The adoption of more sustainable agricultural practices (e.g., in Africa) illustrates the mutually reinforcing effects of climate change mitigation, environmental protection, and long-term economic benefits. The introduction or expansion of agroforestry and balanced fertilizer agriculture can improve food security and at the same time reduce greenhouse gas emissions. More decentralized development patterns based on a stronger role for small- and medium-sized cities can decrease the migration of rural population into urban centers, reduce needs for transportation, and allow the use of environmentally sound technologies (bio-fuel, solar energy, wind, and small-scale hydropower) to tap the large reserves of natural resources.

WGII TAR Section 7.5.4 &
WGIII TAR Section 10.3.2

8.25 **Reducing vulnerability to climate change can often reduce vulnerability to other environmental stresses and *vice versa*.** Examples include, *inter alia*:

· *Protecting threatened ecosystems:* Removing societal stresses and managing resources in a sustainable manner may help unique and threatened systems also to cope with the additional stress posed by climate change. Accounting for potential climatic changes and integration with socio-economic needs and development plans can make biodiversity conservation strategies and climate change adaptation measures more effective.

· *Land-use management:* Addressing or avoiding land degradation also decreases vulnerability to climate change, especially when response strategies consider the social and economic factors defining the land-use practices together with the additional risks imposed by climate change. In regions where deforestation is progressing and leading to carbon loss and increased peak runoff, restoring vegetation by reforestation (and when possible by afforestation) and revegetation can help to combat desertification.

· *Freshwater management:* Problems with availability, abundance, and pollution of freshwater, which are often caused by demographic and development pressures, can be exacerbated by climate change. Reducing vulnerability to water stress (e.g., by water conservation, water-demand management, and more efficient water use) also reduces vulnerability to additional stress by climate change.

WGII TAR Sections 4.1-2
& 7.5.4

8.26 **Approaches that exploit synergies between environmental policies and key national socio-economic objectives like growth and equity could help mitigate and reduce vulnerability to climate change, as well as promote sustainable development.** Sustainable development is closely linked with the environmental, social, and economic components defining the status of each community. The interconnections among those elements of sustainable development are reflected in Figure 8-3, illustrating that important issues such as climate change, sustainability, poverty, and equity can be related to all three components. Just as climate policies can yield ancillary benefits that improve well-being, non-climate socio-economic policies may bring climate benefits. Utilizing such ancillary benefits would aid in making development more sustainable. Complex interactions among environmental, social, and economic challenges exist, and therefore none of these three types of problems can be resolved in isolation.

WGIII TAR Sections 1.3.4,
2.2.3, & 10.3.2, & DES GP

Key elements of sustainable development and interconnections

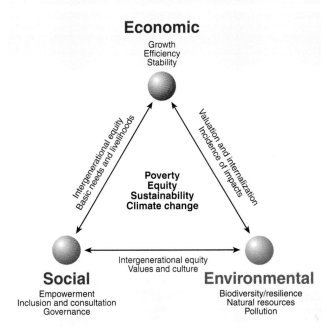

DES GP

Figure 8-3: The vertices of the triangle represent the three major dimensions or domains of sustainable development: economic, social, and environmental. The economic domain is geared mainly towards improving human welfare, primarily through increases in the consumption of goods and services. The environmental domain focuses on protection of the integrity and resilience of ecological systems. The social domain emphasizes the strengthening of human relationships and achievement of individual and group aspirations. Examples of linkages between the three domains are shown along the sides of the triangle. Important issues such as climate change, poverty, equity, and sustainability lie within the triangle and interact with all three domains.

8.27 **Countries with limited economic resources, low levels of technology, poor information systems, inadequate infrastructure, unstable and weak institutions, and inequitable empowerment and access to resources are not only highly vulnerable to climate change but also to other environmental problems, and at the same time have limited capacity to adapt to these changing circumstances and/or mitigate them.** The capacity of these countries to adapt and mitigate can be enhanced when climate policies are integrated with non-climate objectives of national policy development and turned into broad transition strategies to achieve the long-term social and technological changes required by both sustainable development and climate change mitigation.

WGII TAR Chapter 18 & WGIII TAR Sections 1.5.1, 2.4.4, 5.3, 10.3.2, & 10.3.4

8.28 **A great deal of interaction exists among the environmental issues that multilateral environmental agreements address, and synergies can be exploited in their implementation.** Global environmental problems are addressed in a range of individual conventions and agreements—the Vienna Convention and its Montreal Protocol, the United Nations Framework Convention on Climate Change, the United Nations Convention on Biological Diversity, the United Nations Convention to Combat Desertification, and the United Nations Forum on Forests—as well as a range of regional agreements, such as the Convention on Long-Range Transboundary Air Pollution. Table 8-3 provides a list of selected examples of such conventions and instruments. They may contain, *inter alia*, similar requirements concerning common shared or coordinated governmental and civil institutions to enact the general objectives—for example, formulation of strategies and action plans as a framework for country-level implementation; collection of data and processing information and new and strengthened capacities for both human resources and institutional structures; and reporting obligations. Also they provide a framework within which synergies in scientific assessment can be utilized (see Box 8-1).

WGIII TAR Section 10.3.2

Table 8-3 | Selected international environmental treaties.

Convention and Agreement	Place and Date of Adoption
The Antarctic Treaty – Protocol to the Antarctic Treaty on Environmental Protection	Washington, 1959 Madrid, 1991
Convention on Wetlands of International Importance especially as Waterfowl Habitat – Protocol to Amend the Convention on Wetlands of International Importance Especially as Waterfowl Habitat	Ramsar, 1971 Paris, 1982
International Convention for the Prevention of Pollution from Ships	London, 1973
Convention on International Trade on Endangered Species of Wild Fauna and Flora	Washington, 1973
Convention on the Prevention of Marine Pollution from Land-based Sources	Paris, 1974
Convention on the Conservation of Migratory Species of Wild Animals	Bonn, 1979

UN/ECE Convention on Long-Range Transboundary Air Pollution
  – Protocol on Long-Term Financing of the Cooperative Programme for Monitoring and Evaluation of the Long-Range Transmission of Air Pollutants in Europe (EMEP)
  – Protocol on the Reduction of Sulfur Emissions or their Transboundary Fluxes by at least 30%
  – Protocol Concerning the Control of Emissions of Nitrogen or their Transboundary Fluxes
  – Protocol Concerning the Control of Emissions of Volatile Organic Compounds or their Transboundary Fluxes
  – Protocol on Further Reduction of Sulfur Emission
  – Protocol on Heavy Metals
  – Protocol on Persistent Organic Pollutants
  – Protocol to Abate Acidification, Eutrophication, and Ground-level Ozone
	Geneva, 1979
	Geneva, 1984
	
	Helsinki, 1985
	Sofia, 1988
	Geneva, 1991
	Oslo, 1994
	Aarhus, 1998
	Aarhus, 1998
	Gothenburg, 1999

United Nations Convention on the Law of the Sea	Montego Bay, 1982
Vienna Convention for the Protection of the Ozone Layer – Montreal Protocol on Substances that Deplete the Ozone Layer	Vienna, 1985 Montreal, 1987
Basel Convention on the Control of Transboundary Movements of Hazardous Wastes and their Disposal – Amendment to the Basel Convention on the Control of Transboundary Movements of Hazardous Wastes and their Disposal	Basel, 1989 Geneva, 1995
UN/ECE Convention on the Protection and Use of Transboundary Watercourses and International Lakes	Helsinki, 1992
United Nations Framework Convention on Climate Change – Kyoto Protocol to the United Nations Framework Convention on Climate Change	New York, 1992 Kyoto, 1997
Convention on Biological Diversity – Cartagena Protocol on Biosafety to the Convention on Biological Diversity	Rio de Janeiro, 1992 Montreal, 2000
United Nations Convention to Combat Desertification in those Countries Experiencing Serious Drought and/or Desertification, Particularly in Africa	Paris, 1994
Stockholm Convention on Persistent Organic Pollutants	Stockholm, 2001
United Nations Forum on Forests[a]	New York, 2001

[a] This reference is included in view of the importance of international efforts towards a treaty on the issue of forests and their environmental value.

Box 8-1 | Assessing climate change and stratospheric ozone depletion.

The Ozone Scientific Assessment Panel of the Montreal Protocol and the IPCC have had integrated assessment activities regarding the state of understanding of the coupling of the stratospheric ozone layer and the climate system. For the past several years, the Scientific Assessments of Ozone Depletion have included the climate relevance of ozone-depleting gases. Further these assessments have included how current and future climate change and greenhouse gas abundances can influence ozone layer recovery. The IPCC has assessed the climate-cooling tendency due to ozone layer depletion. In addition, joint activities have been undertaken such as the assessment of the climate and ozone-layer impacts of aviation and how the mitigative needs of the Montreal Protocol for substitutes for ozone-depleting gases (notably hydrofluorocarbons) could be impacted by potential decisions about the global warming properties of these gases. These assessments provide information on how decisions and actions regarding one issue would influence the other, and they foster effective dialog between the policy frameworks.

WGI TAR Sections 4.2, 5.5, 6.13, & 7.2.4. WGIII TAR Chapter 3 Appendix, & SRAGA Section 4.2

Question 9

Q9

What are the most robust findings and key uncertainties regarding attribution of climate change and regarding model projections of:
  • Future emissions of greenhouse gases and aerosols?
  • Future concentrations of greenhouse gases and aerosols?
  • Future changes in regional and global climate?
  • Regional and global impacts of climate change?
  • Costs and benefits of mitigation and adaptation options?

Introduction

9.1   **The understanding of climate change, its impacts, and the options to mitigate and adapt is developed through multi- and interdisciplinary research and monitoring in an integrated assessment framework.** As understanding deepens, some findings become more robust and some uncertainties emerge as critical for informed policy formulation. Some uncertainties arise from a lack of data and a lack of understanding of key processes and from disagreement about what is known or even knowable. Other uncertainties are associated with predicting social and personal behavior in response to information and events. The uncertainties tend to escalate with the complexity of the problem, as additional elements are introduced to include a more comprehensive range of physical, technical, social, and political impacts and policy responses. The climate responds to human influence without deliberation or choice; but human society can respond to climate change deliberately, making choices between different options. An objective of the TAR and other IPCC reports is to explore, assess, quantify, and, if possible, reduce these uncertainties.

9.2   **In this report, a robust finding for climate change is defined as one that holds under a variety of approaches, methods, models, and assumptions and one that is expected to be relatively unaffected by uncertainties.** A robust finding can be expected to fall into the categories of *well-established* (high level of agreement and high amount of evidence) and *established but incomplete* (high level of agreement, but incomplete evidence) in the literature. Robustness is different from likelihood: A finding that an outcome is "exceptionally unlikely" may be just as robust as the finding that it is "virtually certain." A major development in the TAR is that of the multiple alternative pathways for emissions and concentrations of greenhouse gases as represented by the SRES. Robust findings are those that are maintained under a wide range of these possible worlds.

9.3   **Key uncertainties in this context are those which, if reduced, may lead to new and robust findings in relation to the questions of this report.** These findings may, in turn, lead to better or more of the information that underpins policy making. The uncertainties can never be fully resolved, but often they can be bounded by more evidence and understanding, particularly in the search for consistent outcomes or robust conclusions.

9.4   **Robust findings and key uncertainties can be brought together in the context of an integrated assessment framework.**

9.5   **The integrated assessment framework described in this report is used to bring together the robust findings and key uncertainties in the model projections.** Such a framework can encompass all the disciplines involved in understanding the climate, the biosphere, and human society. It emphasizes the linkages between the systems described in the different Working Group reports of the TAR as well as considers linkages between climate change and other environmental issues, and helps to identify gaps in knowledge. It suggests how key uncertainties can affect the whole picture. Figure 1-1 shows how adaptation and mitigation can be integrated into the assessment. The human and natural systems will have to adapt to climate change, and development will be affected. The adaptation will be both autonomous and via government initiatives, and adaptation actions will reduce (but cannot entirely avoid) some of the impacts of climate change on these systems and on development. Adaptation actions provide benefits but also entail costs. Mitigation is unlike adaptation in that it reduces emissions at the start of the cycle, it reduces concentrations (compared to what would otherwise occur), and it reduces climate change and the risks and uncertainties associated with climate change. It further reduces the need for adaptation, the impacts of climate change, and effects on socio-economic development. It is also different in that mitigation aims to address the impacts on the climate system, whereas adaptation is primarily oriented to address localized impacts of climate change. The primary benefit of mitigation is avoided climate change, but it also has costs. In addition, mitigation gives rise to ancillary benefits (e.g., reduced air pollution leading to improvements in human health). A fully integrated approach to climate change assessment would consider the whole cycle shown in Figure 1-1 dynamically with all the feedbacks but this could not be accomplished in the TAR.

9.6   Many of the **robust findings** as listed in Table SPM-3 are concerned with the *existence* of a climate response to human activities and the sign of the response. Many of the **key uncertainties** are concerned with the *quantification* of the magnitude and/or the timing of the response and the potential effects of improving methods and relaxing assumptions.

Attribution of Climate Change

9.7   **There is now stronger evidence for a human influence on the global climate.**

9.8   **An increasing body of observations gives a collective picture of a warming world and modeling studies indicate that most of the observed warming at the Earth's surface over the last 50 years is likely to have been due to human activities.** Globally, the 1990s were very likely to have been the warmest decade in the instrumental record (i.e., since the year 1861). For the Northern Hemisphere, the magnitude of the warming in the last 100 years is likely to be the largest of any century during the past 1,000 years. Observations, together with model simulations, provide stronger evidence that most of the warming observed over the last 50 years is attributable to the increase in greenhouse gas concentrations. The observations also provide increased confidence in the ability of models to project future climate change. Better quantification of the human influence depends on reducing the **key uncertainties** relating to the magnitude and character of natural variability and the magnitude of climate forcings due to natural factors and anthropogenic aerosols (particularly indirect effects) and the relating of regional trends to anthropogenic climate change.

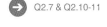 Q2.7 & Q2.10-11

Future Emissions and Concentrations of Greenhouse Gases and Aerosols

9.9   **Human activities increase the atmospheric concentrations of greenhouse gases.**

9.10   **Since the year 1750 (i.e., the beginning of the Industrial Revolution), the atmospheric concentration of CO_2 (the largest contributor to anthropogenic radiative forcing) has increased by 31% due to human activities, and all SRES scenarios project substantial increases in the future (Figure 9-1a).** Other greenhouse gases have also increased in concentrations since the year 1750 (e.g., CH_4 by 150%, N_2O by 17%). The present CO_2 concentration has not been exceeded during the past 420,000 years (the span measurable in ice cores) and likely not during the past 20 million years. The rate of increase is unprecedented relative to any sustained global changes over at least the last 20,000 years. In projections of greenhouse gas concentrations based on the set of SRES scenarios (see Box 3-1), CO_2 concentrations continue to grow to the year 2100. Most SRES scenarios show reductions in SO_2 emissions (precursor for sulfate aerosols) by the year 2100 compared with the year 2000. Some greenhouse gases (e.g., CO_2, N_2O, perfluorocarbons) have long lifetimes (a century or more) for their residence in the atmosphere, while the lifetime of aerosols is measured in days. **Key uncertainties** are inherent in the assumptions that underlie the wide range of future emissions in the SRES scenarios and therefore the quantification of future concentrations. These uncertainties relate to population growth, technological progress, economic growth, and governance structures, which are particularly difficult to quantify. Further, inadequate emission scenarios have been available of lower atmosphere ozone and aerosol precursors. Smaller uncertainties arise from lack of understanding of all the factors inherent in modeling the carbon cycle and including the effects of climate feedbacks. Accounting for all these uncertainties leads

 Q2.4, Q3.3, Q3.5, & Q5.3

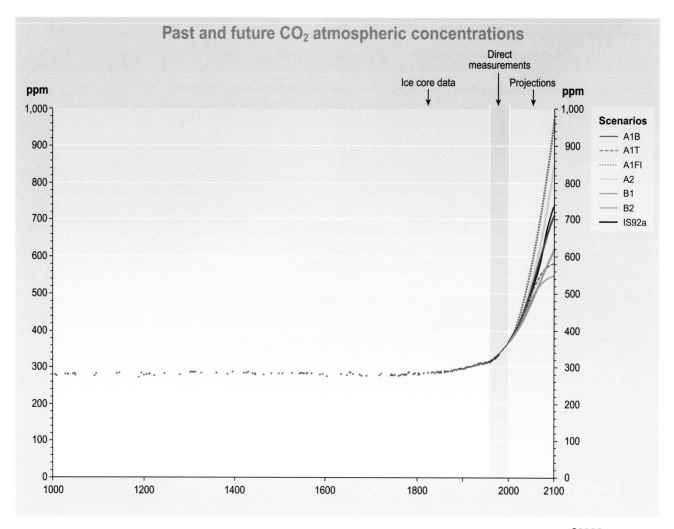

Past and future CO₂ atmospheric concentrations

Figure 9-1a: Observations of atmospheric CO₂ concentration over the years 1000 to 2000 from ice core data supplemented with data from direct atmospheric measurements over the past few decades. Over the period 2000 to 2100, projections are shown of CO₂ concentrations based on the six illustrative SRES scenarios and IS92a (for comparison with the SAR).

WGI TAR SPM Figures 2a & 5b

to a range of CO₂ concentrations in the year 2100 between about 490 and 1,260 ppm (compared to the pre-industrial concentration of about 280 ppm and of about 368 ppm in the year 2000).

9.11 **Fossil-fuel CO₂ emissions are virtually certain to remain the dominant influence on the trends in CO₂ concentrations over the 21st century.** This is implied by the range of SRES scenarios in which projected fossil-fuel emissions exceed the foreseeable biospheric sources and sinks for CO₂. It is estimated that, even if all the carbon so far released by land-use changes could be restored to the terrestrial biosphere (e.g., by reforestation), CO₂ concentration would be reduced by 40 to 70 ppm. There are *key uncertainties* in the influence of changing land use and biospheric feedbacks on the uptake, storage, and release of carbon that in turn could influence CO₂ concentrations.

Q4.11 & Q7.4

Future Changes in Regional and Global Climate

9.12 **The climate has changed during the 20th century; larger changes are projected for the 21st century.**

9.13 **Under all SRES scenarios, projections show the global average surface temperature continuing to rise during the 21st century at rates of rise that are very likely to be without precedent during the last 10,000 years, based on paleoclimate data (Figure 9-1b).** It is very likely that nearly all land areas will warm more rapidly than the global average, particularly those at high northern latitudes in the cold season. There are very likely to be more hot days; fewer cold days, cold waves, and frost days; and a reduced diurnal temperature range.

→ Q3.7, Q3.11, & Q4.5

9.14 **In a warmer world the hydrological cycle will become more intense.** Global average precipitation is projected to increase. More intense precipitation events (hence flooding) are very likely over many areas. Increased summer drying and associated risk of drought is likely over most mid-latitude continental interiors. Even with little or no change in El Niño amplitude, an increase in temperatures globally is likely to lead to greater extremes of drying and heavy rainfall, and increase the risk of droughts and floods that occur with El Niño events in many different regions.

→ Q2.24, Q3.8, Q3.12, Q4.2, & Q4.6

9.15 **In a warmer world the sea level will rise, primarily due to thermal expansion and loss of mass from glaciers and ice caps, the rise being continued for hundreds of years even after stabilization of greenhouse gas concentrations.** This is due to the long time scales on which the deep ocean adjusts to climate change. Ice sheets will continue to react to climate change for thousands of years. Models project that a local warming (annually averaged) of larger than 3°C, sustained for many millennia, would lead to virtually a complete melting of the Greenland ice sheet with a resulting sea-level rise of about 7 m.

→ Q3.9, Q3.14, Q4.15, & Q5.4

9.16 *Key uncertainties* that influence the quantification and the detail of future projections of climate change are those associated with the SRES scenarios, and also those associated with the modeling of climate change, in particular those that concern the understanding of key feedback processes in the climate system, especially those involving clouds, water vapor, and aerosols (including their indirect forcing). Allowing for these uncertainties leads to a range of projections of surface temperature increase for the period 1990 to 2100 of 1.4 to 5.8°C (see Figure 9-1b) and of sea-level rise from 0.09 to 0.88 m. Another uncertainty concerns the understanding of the probability distribution associated with temperature and sea-level projections for the range of SRES scenarios. *Key uncertainties* also affect the detail of regional climate change and its impacts because of the limited capabilities of the regional models, and the global models driving them, and inconsistencies in results between different models especially in some areas and in precipitation. A further key uncertainty concerns the mechanisms, quantification, time scales, and likelihoods associated with large-scale abrupt/non-linear changes (e.g., ocean thermohaline circulation).

→ Q3.6, Q3.9, & Q4.9-19

Regional and Global Impacts of Climate Change

9.17 **Projected climate change will have beneficial and adverse effects on both environmental and socio-economic systems, but the larger the changes and the rate of change in climate, the more the adverse effects predominate.**

9.18 **Regional changes in climate, particularly increases in temperature, have already affected and will continue to affect a diverse set of physical and biological systems in many parts of the world.** Examples of observed changes include shrinkage of glaciers, reductions in seasonal snow cover, thawing of permafrost, later freezing and earlier break-up of ice on rivers and lakes, loss of Arctic sea ice, lengthening of mid- to high-latitude growing seasons, poleward and altitudinal shifts of plant and animal ranges, changes in the seasonal progression of some plants and animals, declines in some plant and animal populations, and damage to coral reefs. These observed rates of change would be expected to increase

→ Q3.14 & Q3.18-21

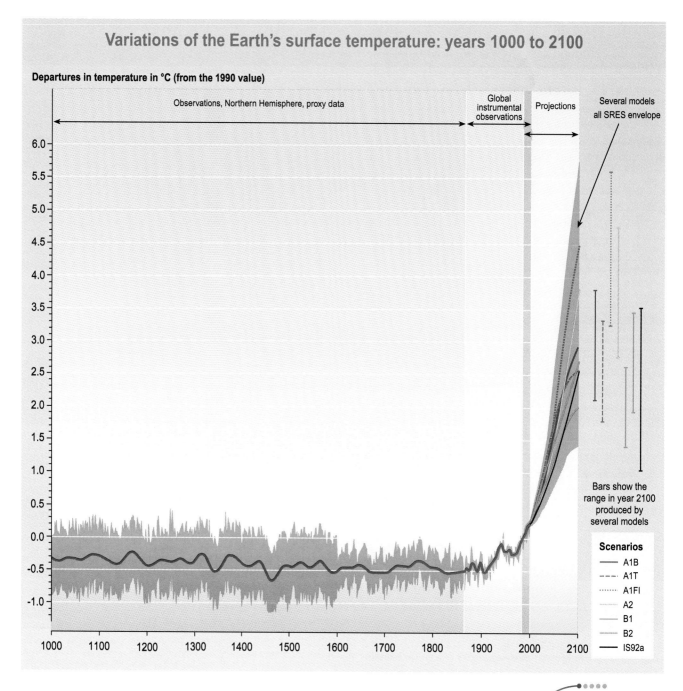

Variations of the Earth's surface temperature: years 1000 to 2100

Departures in temperature in °C (from the 1990 value)

Observations, Northern Hemisphere, proxy data

Global instrumental observations

Projections

Several models all SRES envelope

Bars show the range in year 2100 produced by several models

Scenarios
- —— A1B
- - - - A1T
- ········ A1FI
- ——— A2
- —— B1
- —— B2
- —— IS92a

Figure 9-1b: Variations of the Earth's surface temperature: years 1000 to 2100. Over the period 1000 to 1860, observations are shown of variations in average surface temperature of the Northern Hemisphere (corresponding data from the Southern Hemisphere not available) constructed from proxy data (tree rings, corals, ice cores, and historical records). The line shows the 50-year average, and the grey region the 95% confidence limit in the annual data. From the years 1860 to 2000, observations are shown of variations of global and annual averaged surface temperature from the instrumental record. The line shows the decadal average. Over the period 2000 to 2100, projections are shown of globally averaged surface temperature for the six illustrative SRES scenarios and IS92a as estimated by a model with average climate sensitivity. The grey region "several models all SRES envelope" shows the range of results from the full range of 35 SRES scenarios in addition to those from a range of models with different climate sensitivities.

WGI TAR SPM Figures 1b & 5d

in the future represented by any of the SRES scenarios, for which the warming trends for the 21st century are two to ten times those observed for the 20th century. Many physical systems are vulnerable to climate change: For example, the impact of coastal storm surges will be exacerbated by sea-level rise, and glaciers and permafrost will continue to retreat. In some mid- to high latitudes, plant productivity (trees and some agricultural crops) would increase with small increases in temperature. Plant productivity would decrease in most regions of the world for warming beyond a few °C. In most tropical and subtropical regions, yields are projected to decrease for almost any increase in temperature.

9.19 **Ecosystems and species are vulnerable to climate change and other stresses (as illustrated by observed impacts of recent regional temperature changes) and some will be irreversibly damaged or lost.** Natural systems at risk include coral reefs and atolls, mangroves, boreal and tropical forests, polar and alpine ecosystems, prairie wetlands, and remnant native grasslands. While some species may increase in abundance or range, climate change will increase existing risks of extinction of some more vulnerable species and loss of biodiversity. It is *well-established* that the geographical extent of the damage or loss, and the number of systems affected, will increase with the magnitude and rate of climate change.

→ Q3.18

9.20 **The adverse impacts of climate change are expected to fall disproportionately upon developing countries and the poor persons within countries.** Projected changes in climate extremes could have major consequences especially on water and food security and on health. The vulnerability of human societies and natural systems to climate extremes is demonstrated by the damage, hardship, and death caused by events such as droughts, floods, heat waves, avalanches, landslides, and windstorms, which have shown an increasing trend during recent decades. While overall precipitation is projected to increase, there are likely to be much larger changes in intensity and frequency, which will increase the likelihood of extremes of drying and precipitation, and thus droughts and floods during the 21st century. These increases combined with increased water stress (occurring already because of increasing demand) will affect food security and health especially in many developing countries. Conversely, the frequency and magnitude of extreme low-temperature events, such as cold spells, is projected to decrease in the future, with both positive and negative impacts.

→ Q3.17, Q3.21-22, & Q3.33

9.21 **Populations that inhabit small islands and low-lying coastal areas are at particular risk of severe social and economic effects from sea-level rise and storm surges.** Tens of millions of people living in deltas, low-lying coastal areas, and on small islands will face risk of displacement. Further negative impacts will be increased by saltwater intrusion and flooding due to storm surges and loss of coastal wetlands and slowing down of river discharges.

→ Q3.23-24

9.22 *Key uncertainties* in the identification and quantification of impacts arise from the lack of reliable local or regional detail in climate change, especially in the projection of extremes, inadequate accounting in impacts assessments for the effects of changes in extremes and disasters, limited knowledge of some non-linear processes and feedbacks, uncertainties in the costing of the damage due to climate impacts, lack of both relevant data and understanding of key processes in different regions, and uncertainties in assessing and predicting the response of ecological and social (e.g., impact of vector- and water-borne diseases), and economic systems to the combined effect of climate change and other stresses such as land-use change, local pollution, etc.

→ Q3.13, Q4.10, & Q4.18-19

Costs and Benefits of Adaptation and Mitigation Options

9.23 **Adaptation is a necessity; its cost can be reduced by anticipation, analysis, and planning.**

9.24 Adaptation is no longer an option, it is a necessity, given that climate changes and related impacts are already occurring. Anticipatory and reactive adaptation, which will vary with location and sector, has the potential to reduce adverse impacts of climate change, to enhance beneficial impacts, and to produce many immediate ancillary benefits, but will not prevent all damages.

→ Q3.26-28 & Q3.33

However, its potential is much more limited for natural systems than for human systems. The capacity of different regions to adapt to climate change depends highly upon their current and future states of socio-economic development and their exposure to climate stress. Therefore the potential for adaptation is more limited for developing countries, which are projected to be the most adversely affected. Adaptation appears to be easier if the climate changes are modest and/or gradual rather than large and/or abrupt. If climate changes more rapidly than expected in any region, especially with respect to climate extremes, then the potential of adaptation to diminish vulnerability of human systems will be lessened.

9.25 **The costs of adaptation can be reduced by anticipation and planned action, and many costs may be relatively small, especially when adaptation policies and measures contribute to other goals of sustainable development.**

Q3.31 & Q3.36-37

9.26 *Key uncertainties* regarding adaptations relate to the inadequate representation by models of local changes, lack of foresight, inadequate knowledge of benefits and costs, possible side effects including acceptability and speed of implementation, various barriers to adaptation, and more limited opportunities and capacities for adaptation in developing countries.

Q3.27

9.27 **The primary economic benefits of mitigation are the *avoided* costs associated with the adverse impacts of climate change.**

9.28 **Greenhouse gas emission reduction (mitigation) action would lessen the pressures on natural and human systems from climate change.** Comprehensive, quantitative estimates of global primary benefits of mitigating climate change do not exist. For mean temperature increases over a few °C relative to the year 1990, impacts are predominantly adverse, so net primary benefits of mitigation are positive. A *key uncertainty* is the net balance of adverse and beneficial impacts of climate change for temperature increases less than about a few °C. These averages conceal wide regional variations.

Q6.10

9.29 **Mitigation generates costs and ancillary benefits.**

9.30 **Major reductions in global greenhouse gas emissions would be necessary to achieve stabilization of their concentrations.** For example, for the most important anthropogenic greenhouse gas, carbon cycle models indicate that stabilization of atmospheric CO_2 concentrations at 450, 650, or 1,000 ppm would require global anthropogenic CO_2 emissions to drop below year 1990 levels within a few decades, about a century, or about 2 centuries, respectively, and continue to decrease steadily thereafter. Emissions would peak in about 1 to 2 decades (450 ppm) and roughly a century (1,000 ppm) from the present. Eventually stabilization would require CO_2 emissions to decline to a very small fraction of current global emissions. The *key uncertainties* here relate to the possibilities of climate change feedbacks and development pathways and how these affect the timing of emissions reductions.

Q6.4

9.31 **Mitigation costs and benefits vary widely across sectors, countries, and development paths.** In general, it is easier to identify sectors—such as coal, possibly oil and gas, and some energy-intensive industries dependent on energy produced from these fossil fuels—that are very likely to suffer an economic disadvantage from mitigation. Their economic losses are more immediate, more concentrated, and more certain. The sectors that are likely to benefit include renewable energy, services, and new industries whose development is stimulated by demand for low-emission fuels and production techniques. Different countries and development paths have widely different energy structures, so they too have different costs and benefits from mitigation. Carbon taxes can have negative income effects on low-income groups unless the tax revenues are used directly or indirectly to compensate such effects.

Q7.14, Q7.17, & Q7.34

9.32 **Emission constraints in Annex I countries have well established, albeit varied, "spill-over" effects on non-Annex I countries.** Analyses of the effects of emissions constraints on Annex I countries report reductions below what would otherwise occur in both projected GDP and in projected oil revenues for oil-exporting non-Annex I countries.

9.33 **Lower emissions scenarios require different patterns of energy resource development and an increase in energy R&D to assist accelerating the development and deployment of advanced environmentally sound energy technologies.** Emissions of CO_2 due to fossil-fuel burning are virtually certain to be to the dominant influence on the trend on the atmospheric CO_2 concentration during the 21st century. Resource data assessed in the TAR may imply a change in the energy mix and the introduction of new sources of energy during the 21st century. Fossil-fuel resources will not limit carbon emissions during the 21st century. The carbon in proven conventional oil and gas reserves is much less, however, than the cumulative carbon emissions associated with stabilization of CO_2 at levels of 450 ppm or higher.[25] These resource data may imply a change in the energy mix and the introduction of new sources of energy during the 21st century. The choice of energy mix and associated technologies and investments—either more in the direction of exploitation of unconventional oil and gas resources, or in the direction of non-fossil energy sources, or fossil energy technology with carbon capture and storage—will determine whether, and if so, at what level and cost, greenhouse concentrations can be stabilized. *Key uncertainties* are the future relative prices of energy and carbon-based fuels, and the relative technical and economic attractiveness of non-fossil-fuel energy alternatives compared with unconventional oil and gas resources.

9.34 **Significant progress in energy-saving and low-carbon technologies has been made since 1995, and the progress has been faster than anticipated in the SAR.** Net emission reductions could be achieved through, *inter alia*, improved techniques in production and use of energy, shifts to low- or no-carbon technologies, CO_2 removal and storage, improved land-use and forestry practices, and movement to more sustainable lifestyles. Significant progress is taking place in the development of wind turbines, solar energy, hybrid engine cars, fuel cells, and underground CO_2 storage. *Key uncertainties* are (a) the likelihood of technological breakthroughs leading to substantial reductions in costs and rapid take-up of low-carbon processes and products, and (b) the future scale of private and public R&D expenditures on these technologies.

9.35 **Studies examined in the TAR suggest substantial technological and other opportunities for lowering mitigation costs. National mitigation responses to climate change can be more effective if deployed as a portfolio of policy instruments to limit or reduce net greenhouse gas emissions.** The costs of mitigation are strongly affected by development paths, with those paths involving substantial increases in greenhouse gas emissions requiring more mitigation to reach a stabilization target, and hence higher costs. These costs can be substantially reduced or even turned into net benefits with a portfolio of policy instruments (including those that help to overcome barriers) to the extent that policies can exploit "no-regrets" opportunities in the following areas:

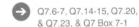

· *Technological options:* Technological options may achieve global emissions reductions of 1.9 to 2.6 Gt C_{eq} yr^{-1} by year 2010 and 3.6 to 5.0 Gt C_{eq} yr^{-1} by year 2020. Half of these reductions may be realized with one component of their economic cost (net capital, operating, and maintenance costs) with direct benefits exceeding direct costs, and the other half with that component of their economic cost ranging from US$0 to US$100 per t C_{eq}.[26] Depending on the emissions scenario, global emissions could be reduced below year 2000 levels over the 2010 to 2020 time frame. *Key uncertainties* are the identification,

---

[25] The reference to a particular concentration level does not imply an agreed-upon desirability of stabilization at this level.

[26] These cost estimates in 1998 prices are derived using discount rates in the range of 5 to 12%, consistent with public-sector discount rates. Private internal rates of return vary greatly and are often significantly higher.

extent, and nature of any barriers that impede adoption of promising low-emission technologies, and the estimation of the costs of overcoming the barriers.

- *Ancillary benefits:* Depending on factors (such as location of the greenhouse gas emissions, the prevailing local climate, and the population density, composition, and health) the magnitude of the ancillary benefits of mitigation may be comparable to the costs of the mitigating policies and measures. ***Key uncertainties*** are the magnitude and location of these benefits involving the scientific assessment and valuation of health risks of air pollution, particularly those involving fine aerosols and particles.
- *Double dividends:* Instruments (such as taxes or auctioned permits) provide revenues to the government. If used to finance reductions in existing distortionary taxes ("revenue recycling"), these revenues reduce the economic cost of achieving greenhouse gas reductions. The magnitude of this offset depends on the existing tax structure, type of tax cuts, labor market conditions, and method of recycling. Under some circumstances, it is possible that the economic benefits may exceed the costs of mitigation. ***Key uncertainties*** regarding the overall net costs of mitigation vary between countries, depending on the existing tax structure, the extent of the distortion, and the type of tax cuts that are acceptable.

9.36 **Modeling studies show that emissions trading reduces costs of mitigation for those participating in the trading.** Global modeling studies, with results depending strongly upon assumptions, project that costs of mitigation based on Kyoto targets are likely to be reduced by full carbon-permit trading within the Annex B[27] group of countries. *Annex I OECD[28] countries* may expect aggregate costs to be reduced by about half through full permit trading. *Annex I economies in transition* are projected to be unaffected or to gain several percent increase in GDP. *Oil-exporting, non-Annex I countries* may also expect similar reductions in costs under such trading. The aggregate effects of trading are expected to be positive for *other non-Annex I countries.* Those countries that may expect a loss or gain without Annex I trading may expect a smaller change with trading. A ***key uncertainty*** is the extent of the underlying costs, which vary widely across countries, and how these cost estimates will be changed (a) when methods are improved and (b) when some of the assumptions of the models are relaxed. Such assumptions are concerned with:

Q7.18-19

- Allowance for exemptions in the emission-permit trading in concert with other policies and measures
- Consideration of various market imperfections
- Allowance for induced technical change
- Inclusion of ancillary benefits
- Opportunities for double dividends
- Inclusion of policies for non-CO_2 greenhouse gases and non-energy sources of all greenhouse gases (e.g., CH_4 from agriculture)
- Offsets from sinks.

9.37 **Although model projections indicate that long-term global growth paths of GDP are not significantly affected by mitigation actions towards stabilization, these do not show the larger variations that occur over some shorter time periods, sectors, or regions.**

Q7.25

9.38 **Unexpected public policies ("quick fixes") with sudden short-term effects may cost economies much more than expected policies with gradual effects.** A ***key uncertainty*** in the magnitude of the costs lies in the existence of well-designed contingency plans in the event of policy shifts (e.g., as a result of a sudden shift in public

Q7.24 & Q7.31

---

[27] *Annex B countries:* Group of countries included in Annex B of the Kyoto Protocol that have agreed to a target for their greenhouse gas emissions, including all the Annex I countries (as amended in 1998) but Turkey and Belarus.

[28] *Annex I countries:* Group of countries included in Annex I to the United Nations Framework Convention on Climate Change, including all developed countries in the the Organisation for Economic Cooperation and Development and those with economies in transition.

perception of the climate change). Other *key uncertainties* for costs lie in the possibilities of the rapid short-term effects including, or leading to, abrupt reductions in costs of low-carbon processes and products, shifts towards low-emission technologies, and/or changes towards more sustainable lifestyles.

9.39 **Near-term action in mitigation and adaptation would reduce risks.** Because of the long time lags associated both with the climate system (e.g., ~100 years for atmospheric CO_2) and with human response, near-term action in mitigation and adaptation would reduce risks. Inertia in the interacting climate, ecological, and socio-economic systems is a major reason why anticipatory adaptation and mitigation actions are beneficial.

 Q5.19 & Q5.24

9.40 **Adaptation can complement mitigation in a cost-effective strategy to reduce climate change risks; together they can contribute to sustainable development objectives.** Some future paths that focus on the social, economic, and environmental elements of sustainable development may result in lower greenhouse gas emissions than other paths, so that the level of additional policies and measures required for a particular level of stabilization and any associated costs can also be lower. A *key uncertainty* is the lack of appropriate knowledge on the interactions between climate change and other environmental issues and the related socio-economic implications. A related issue is the pace of change in integrating the main global conventions and protocols associated with climate change (e.g., those involving world trade, transboundary pollution, biodiversity, desertification, stratospheric ozone depletion, health, and food security). It is also uncertain at which rate individual countries will integrate sustainable development concepts into policy-making processes.

Q1.9 & Q8.21-28

9.41 **Development paths that meet sustainable development objectives may result in lower levels of greenhouse gas emissions.** Key choices about future development paths and the future of the climate are being made now in both developed and developing countries. Information is available to help decision makers evaluate benefits and costs from adaptation and mitigation over a range of options and sustainable development pathways. Anticipated adaptation could be much less costly than reactive adaptation. Mitigation of climate change can reduce and postpone the impacts, lowering the damages and giving human societies as well as animals and plants more time to adapt.

 Q5.22, Q7.25, & Q8.26

Further Work

9.42 **Significant progress has been made in the TAR in many aspects of the knowledge required to understand climate change and the human response to it.** However, there remain important areas where further work is required, in particular:
- The detection and attribution of climate change
- The understanding and prediction of regional changes in climate and climate extremes
- The quantification of climate change impacts at the global, regional, and local levels
- The analysis of adaptation and mitigation activities
- The integration of all aspects of the climate change issue into strategies for sustainable development
- Comprehensive and integrated investigations to support the judgment as to what constitutes "dangerous anthropogenic interference with the climate system."

 WGI TAR SPM, WGII TAR SPM, & WGIII TAR SPM

Climate Change 2001:
Synthesis Report

Working Group Summaries

Working Group Summaries for Policymakers and Technical Summaries

Working Group I: The Scientific Basis
Working Group II: Impacts, Adaptation, and Vulnerability
Working Group III: Mitigation

Climate Change 2001:
The Scientific Basis

Working Group I Summaries

Summary for Policymakers
A Report of Working Group I of the Intergovernmental Panel on Climate Change

Technical Summary of the Working Group I Report
A Report accepted by Working Group I of the Intergovernmental Panel on Climate Change but not approved in detail

Part of the Working Group I contribution to the Third Assessment Report
of the Intergovernmental Panel on Climate Change

Contents

Climate Change 2001:
The Scientific Basis

Summary for Policymakers

A Report of Working Group I of the Intergovernmental Panel on Climate Change

Based on a draft prepared by:

Daniel L. Albritton, Myles R. Allen, Alfons P. M. Baede, John A. Church, Ulrich Cubasch, Dai Xiaosu, Ding Yihui, Dieter H. Ehhalt, Christopher K. Folland, Filippo Giorgi, Jonathan M. Gregory, David J. Griggs, Jim M. Haywood, Bruce Hewitson, John T. Houghton, Joanna I. House, Michael Hulme, Ivar Isaksen, Victor J. Jaramillo, Achuthan Jayaraman, Catherine A. Johnson, Fortunat Joos, Sylvie Joussaume, Thomas Karl, David J. Karoly, Haroon S. Kheshgi, Corrine Le Quéré, Kathy Maskell, Luis J. Mata, Bryant J. McAvaney, Mack McFarland, Linda O. Mearns, Gerald A. Meehl, L. Gylvan Meira-Filho, Valentin P. Meleshko, John F. B. Mitchell, Berrien Moore, Richard K. Mugara, Maria Noguer, Buruhani S. Nyenzi, Michael Oppenheimer, Joyce E. Penner, Steven Pollonais, Michael Prather, I. Colin Prentice, Venkatchala Ramaswamy, Armando Ramirez-Rojas, Sarah C. B. Raper, M. Jim Salinger, Robert J. Scholes, Susan Solomon, Thomas F. Stocker, John M. R. Stone, Ronald J. Stouffer, Kevin E. Trenberth, Ming-Xing Wang, Robert T. Watson, Kok S. Yap, John Zillman

with contributions from many authors and reviewers.

Introduction

The Third Assessment Report of Working Group I of the Intergovernmental Panel on Climate Change (IPCC) builds upon past assessments and incorporates new results from the past five years of research on climate change[1]. Many hundreds of scientists[2] from many countries participated in its preparation and review.

This Summary for Policymakers (SPM), which was approved by IPCC member governments in Shanghai in January 2001[3], describes the current state of understanding of the climate system and provides estimates of its projected future evolution and their uncertainties. Further details can be found in the underlying report, and the appended Source Information provides cross references to the report's chapters.

An increasing body of observations gives a collective picture of a warming world and other changes in the climate system.

Since the release of the Second Assessment Report (SAR[4]), additional data from new studies of current and palaeoclimates, improved analysis of data sets, more rigorous evaluation of their quality, and comparisons among data from different sources have led to greater understanding of climate change.

The global average surface temperature has increased over the 20th century by about 0.6°C.

- The global average surface temperature (the average of near surface air temperature over land, and sea surface temperature) has increased since 1861. Over the 20th century the increase has been $0.6 \pm 0.2°$C[5,6] (Figure 1a). This value is about 0.15°C larger than that estimated by the SAR for the period up to 1994, owing to the relatively high temperatures of the additional years (1995 to 2000) and improved methods of processing the data. These numbers take into

account various adjustments, including urban heat island effects. The record shows a great deal of variability; for example, most of the warming occurred during the 20th century, during two periods, 1910 to 1945 and 1976 to 2000.

- Globally, it is very likely[7] that the 1990s was the warmest decade and 1998 the warmest year in the instrumental record, since 1861 (see Figure 1a).
- New analyses of proxy data for the Northern Hemisphere indicate that the increase in temperature in the 20th century is likely[7] to have been the largest of any century during the past 1,000 years. It is also likely[7] that, in the Northern Hemisphere, the 1990s was the warmest decade and 1998 the warmest year (Figure 1b). Because less data are available, less is known about annual averages prior to 1,000 years before present and for conditions prevailing in most of the Southern Hemisphere prior to 1861.
- On average, between 1950 and 1993, night-time daily minimum air temperatures over land increased by about 0.2°C per decade. This is about twice the rate of increase in daytime daily maximum air temperatures (0.1°C per decade). This has lengthened the freeze-free season in many mid- and high latitude regions. The increase in sea surface temperature over this period is about half that of the mean land surface air temperature.

Temperatures have risen during the past four decades in the lowest 8 kilometres of the atmosphere.

- Since the late 1950s (the period of adequate observations from weather balloons), the overall global temperature increases in the lowest 8 kilometres of the atmosphere and in surface temperature have been similar at 0.1°C per decade.
- Since the start of the satellite record in 1979, both satellite and weather balloon measurements show that the global average temperature of the lowest 8 kilometres of the atmosphere has changed by $+0.05 \pm 0.10°$C per decade, but the global average surface temperature has increased significantly by $+0.15 \pm 0.05°$C per decade. The difference in the warming rates is statistically significant. This difference occurs primarily over the tropical and sub-tropical regions.
- The lowest 8 kilometres of the atmosphere and the surface are influenced differently by factors such as stratospheric ozone depletion, atmospheric aerosols, and the El Niño phenomenon. Hence, it is physically plausible to expect that over a short time period (e.g., 20 years) there may be differences in temperature trends. In addition, spatial

---

[1] Climate change in IPCC usage refers to any change in climate over time, whether due to natural variability or as a result of human activity. This usage differs from that in the Framework Convention on Climate Change, where climate change refers to a change of climate that is attributed directly or indirectly to human activity that alters the composition of the global atmosphere and that is in addition to natural climate variability observed over comparable time periods.

[2] In total 122 Co-ordinating Lead Authors and Lead Authors, 515 Contributing Authors, 21 Review Editors and 420 Expert Reviewers.

[3] Delegations of 99 IPCC member countries participated in the Eighth Session of Working Group I in Shanghai on 17 to 20 January 2001.

[4] The IPCC Second Assessment Report is referred to in this Summary for Policymakers as the SAR.

[5] Generally temperature trends are rounded to the nearest 0.05°C per unit time, the periods often being limited by data availability.

[6] In general, a 5% statistical significance level is used, and a 95% confidence level.

[7] In this Summary for Policymakers and in the Technical Summary, the following words have been used where appropriate to indicate judgmental estimates of confidence: *virtually certain* (greater than 99% chance that a result is true); *very likely* (90-99% chance); *likely* (66-90% chance); *medium likelihood* (33-66% chance); *unlikely* (10-33% chance); *very unlikely* (1-10% chance); *exceptionally unlikely* (less than 1% chance). The reader is referred to individual chapters for more details.

Variations of the Earth's surface temperature for:

(a) the past 140 years

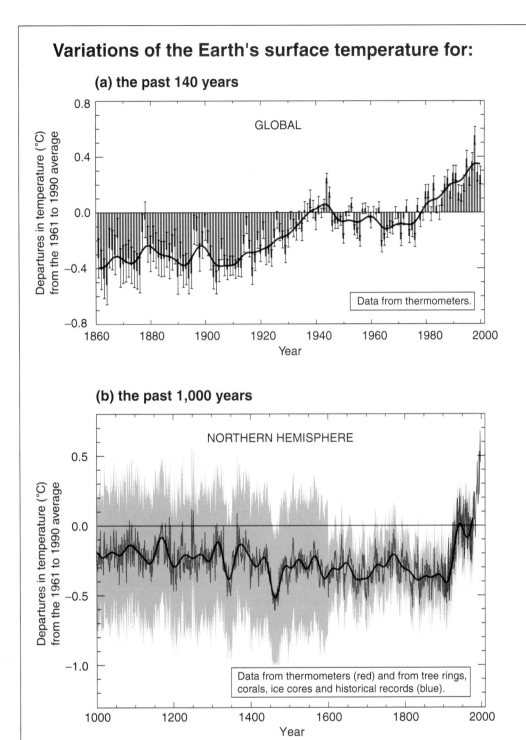

(b) the past 1,000 years

Figure 1: Variations of the Earth's surface temperature over the last 140 years and the last millennium. (a) The Earth's surface temperature is shown year by year (red bars) and approximately decade by decade (black line, a filtered annual curve suppressing fluctuations below near decadal time-scales). There are uncertainties in the annual data (thin black whisker bars represent the 95% confidence range) due to data gaps, random instrumental errors and uncertainties, uncertainties in bias corrections in the ocean surface temperature data and also in adjustments for urbanisation over the land. Over both the last 140 years and 100 years, the best estimate is that the global average surface temperature has increased by 0.6 ± 0.2°C. (b) Additionally, the year by year (blue curve) and 50 year average (black curve) variations of the average surface temperature of the Northern Hemisphere for the past 1000 years have been reconstructed from "proxy" data calibrated against thermometer data (see list of the main proxy data in the diagram). The 95% confidence range in the annual data is represented by the grey region. These uncertainties increase in more distant times and are always much larger than in the instrumental record due to the use of relatively sparse proxy data. Nevertheless the rate and duration of warming of the 20th century has been much greater than in any of the previous nine centuries. Similarly, it is likely[7] that the 1990s have been the warmest decade and 1998 the warmest year of the millennium. [Based upon (a) Chapter 2, Figure 2.7c and (b) Chapter 2, Figure 2.20]

sampling techniques can also explain some of the differences in trends, but these differences are not fully resolved.

Snow cover and ice extent have decreased.

- Satellite data show that there are very likely[7] to have been decreases of about 10% in the extent of snow cover since the late 1960s, and ground-based observations show that there is very likely[7] to have been a reduction of about two weeks in the annual duration of lake and river ice cover in the mid- and high latitudes of the Northern Hemisphere, over the 20th century.
- There has been a widespread retreat of mountain glaciers in non-polar regions during the 20th century.
- Northern Hemisphere spring and summer sea-ice extent has decreased by about 10 to 15% since the 1950s. It is likely[7] that there has been about a 40% decline in Arctic sea-ice thickness during late summer to early autumn in recent decades and a considerably slower decline in winter sea-ice thickness.

Global average sea level has risen and ocean heat content has increased.

- Tide gauge data show that global average sea level rose between 0.1 and 0.2 metres during the 20th century.
- Global ocean heat content has increased since the late 1950s, the period for which adequate observations of sub-surface ocean temperatures have been available.

Changes have also occurred in other important aspects of climate.

- It is very likely[7] that precipitation has increased by 0.5 to 1% per decade in the 20th century over most mid- and high latitudes of the Northern Hemisphere continents, and it is likely[7] that rainfall has increased by 0.2 to 0.3% per decade over the tropical ($10°$ N to $10°$ S) land areas. Increases in the tropics are not evident over the past few decades. It is also likely[7] that rainfall has decreased over much of the Northern Hemisphere sub-tropical ($10°$ N to $30°$ N) land areas during the 20th century by about 0.3% per decade. In contrast to the Northern Hemisphere, no comparable systematic changes have been detected in broad latitudinal averages over the Southern Hemisphere. There are insufficient data to establish trends in precipitation over the oceans.
- In the mid- and high latitudes of the Northern Hemisphere over the latter half of the 20th century, it is likely[7] that there has been a 2 to 4% increase in the frequency of heavy precipitation events. Increases in heavy precipitation events can arise from a number of causes, e.g., changes in atmospheric moisture, thunderstorm activity and large-scale storm activity.
- It is likely[7] that there has been a 2% increase in cloud cover over mid- to high latitude land areas during the 20th century.

In most areas the trends relate well to the observed decrease in daily temperature range.
- Since 1950 it is very likely[7] that there has been a reduction in the frequency of extreme low temperatures, with a smaller increase in the frequency of extreme high temperatures.
- Warm episodes of the El Niño-Southern Oscillation (ENSO) phenomenon (which consistently affects regional variations of precipitation and temperature over much of the tropics, sub-tropics and some mid-latitude areas) have been more frequent, persistent and intense since the mid-1970s, compared with the previous 100 years.
- Over the 20th century (1900 to 1995), there were relatively small increases in global land areas experiencing severe drought or severe wetness. In many regions, these changes are dominated by inter-decadal and multi-decadal climate variability, such as the shift in ENSO towards more warm events.
- In some regions, such as parts of Asia and Africa, the frequency and intensity of droughts have been observed to increase in recent decades.

Some important aspects of climate appear not to have changed.

- A few areas of the globe have not warmed in recent decades, mainly over some parts of the Southern Hemisphere oceans and parts of Antarctica.
- No significant trends of Antarctic sea-ice extent are apparent since 1978, the period of reliable satellite measurements.
- Changes globally in tropical and extra-tropical storm intensity and frequency are dominated by inter-decadal to multi-decadal variations, with no significant trends evident over the 20th century. Conflicting analyses make it difficult to draw definitive conclusions about changes in storm activity, especially in the extra-tropics.
- No systematic changes in the frequency of tornadoes, thunder days, or hail events are evident in the limited areas analysed.

Emissions of greenhouse gases and aerosols due to human activities continue to alter the atmosphere in ways that are expected to affect the climate.

Changes in climate occur as a result of both internal variability within the climate system and external factors (both natural and anthropogenic). The influence of external factors on climate can be broadly compared using the concept of radiative forcing[8]. A positive radiative forcing, such as that produced by increasing concentrations of greenhouse gases, tends to warm the surface. A negative radiative forcing, which can arise

---

[8] *Radiative forcing* is a measure of the influence a factor has in altering the balance of incoming and outgoing energy in the Earth-atmosphere system, and is an index of the importance of the factor as a potential climate change mechanism. It is expressed in Watts per square metre (Wm^{-2}).

from an increase in some types of aerosols (microscopic airborne particles) tends to cool the surface. Natural factors, such as changes in solar output or explosive volcanic activity, can also cause radiative forcing. Characterisation of these climate forcing agents and their changes over time (see Figure 2) is required to understand past climate changes in the context of natural variations and to project what climate changes could lie ahead. Figure 3 shows current estimates of the radiative forcing due to increased concentrations of atmospheric constituents and other mechanisms.

Concentrations of atmospheric greenhouse gases and their radiative forcing have continued to increase as a result of human activities.

- The atmospheric concentration of carbon dioxide (CO_2) has increased by 31% since 1750. The present CO_2 concentration has not been exceeded during the past 420,000 years and likely[7] not during the past 20 million years. The current rate of increase is unprecedented during at least the past 20,000 years.

Indicators of the human influence on the atmosphere during the Industrial Era

(a) Global atmospheric concentrations of three well mixed greenhouse gases

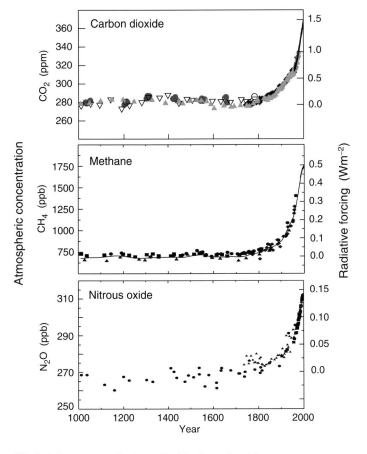

(b) Sulphate aerosols deposited in Greenland ice

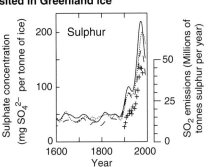

Figure 2: Long records of past changes in atmospheric composition provide the context for the influence of anthropogenic emissions. (a) shows changes in the atmospheric concentrations of carbon dioxide (CO_2), methane (CH_4), and nitrous oxide (N_2O) over the past 1000 years. The ice core and firn data for several sites in Antarctica and Greenland (shown by different symbols) are supplemented with the data from direct atmospheric samples over the past few decades (shown by the line for CO_2 and incorporated in the curve representing the global average of CH_4). The estimated positive radiative forcing of the climate system from these gases is indicated on the right-hand scale. Since these gases have atmospheric lifetimes of a decade or more, they are well mixed, and their concentrations reflect emissions from sources throughout the globe. All three records show effects of the large and increasing growth in anthropogenic emissions during the Industrial Era.

(b) illustrates the influence of industrial emissions on atmospheric sulphate concentrations, which produce negative radiative forcing. Shown is the time history of the concentrations of sulphate, not in the atmosphere but in ice cores in Greenland (shown by lines; from which the episodic effects of volcanic eruptions have been removed). Such data indicate the local deposition of sulphate aerosols at the site, reflecting sulphur dioxide (SO_2) emissions at mid-latitudes in the Northern Hemisphere. This record, albeit more regional than that of the globally-mixed greenhouse gases, demonstrates the large growth in anthropogenic SO_2 emissions during the Industrial Era. The pluses denote the relevant regional estimated SO_2 emissions (right-hand scale). [Based upon (a) Chapter 3, Figure 3.2b (CO_2); Chapter 4, Figure 4.1a and b (CH_4) and Chapter 4, Figure 4.2 (N_2O) and (b) Chapter 5, Figure 5.4a]

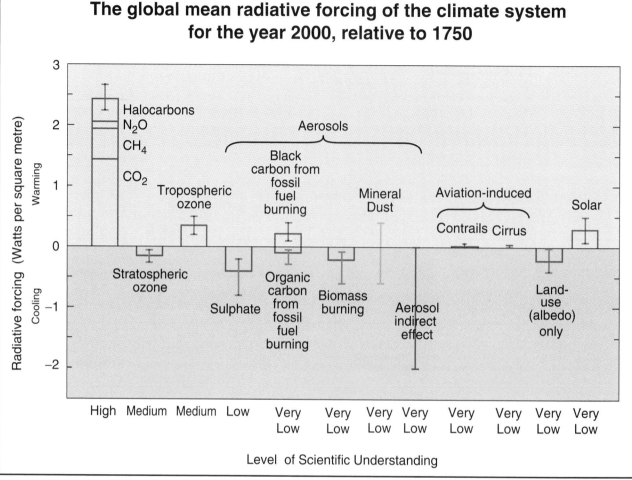

The global mean radiative forcing of the climate system for the year 2000, relative to 1750

Figure 3: Many external factors force climate change. These radiative forcings arise from changes in the atmospheric composition, alteration of surface reflectance by land use, and variation in the output of the sun. Except for solar variation, some form of human activity is linked to each. The rectangular bars represent estimates of the contributions of these forcings - some of which yield warming, and some cooling. Forcing due to episodic volcanic events, which lead to a negative forcing lasting only for a few years, is not shown. The indirect effect of aerosols shown is their effect on the size and number of cloud droplets. A second indirect effect of aerosols on clouds, namely their effect on cloud lifetime, which would also lead to a negative forcing, is not shown. Effects of aviation on greenhouse gases are included in the individual bars. The vertical line about the rectangular bars indicates a range of estimates, guided by the spread in the published values of the forcings and physical understanding. Some of the forcings possess a much greater degree of certainty than others. A vertical line without a rectangular bar denotes a forcing for which no best estimate can be given owing to large uncertainties. The overall level of scientific understanding for each forcing varies considerably, as noted. Some of the radiative forcing agents are well mixed over the globe, such as CO_2, thereby perturbing the global heat balance. Others represent perturbations with stronger regional signatures because of their spatial distribution, such as aerosols. For this and other reasons, a simple sum of the positive and negative bars cannot be expected to yield the net effect on the climate system. The simulations of this assessment report (for example, Figure 5) indicate that the estimated net effect of these perturbations is to have warmed the global climate since 1750. [Based upon Chapter 6, Figure 6.6]

- About three-quarters of the anthropogenic emissions of CO_2 to the atmosphere during the past 20 years is due to fossil fuel burning. The rest is predominantly due to land-use change, especially deforestation.
- Currently the ocean and the land together are taking up about half of the anthropogenic CO_2 emissions. On land, the uptake of anthropogenic CO_2 very likely[7] exceeded the release of CO_2 by deforestation during the 1990s.
- The rate of increase of atmospheric CO_2 concentration has been about 1.5 ppm[9] (0.4%) per year over the past two decades. During the 1990s the year to year increase varied

from 0.9 ppm (0.2%) to 2.8 ppm (0.8%). A large part of this variability is due to the effect of climate variability (e.g., El Niño events) on CO_2 uptake and release by land and oceans.
- The atmospheric concentration of methane (CH_4) has increased by 1060 ppb[9] (151%) since 1750 and continues

---

[9] ppm (parts per million) or ppb (parts per billion, 1 billion = 1,000 million) is the ratio of the number of greenhouse gas molecules to the total number of molecules of dry air. For example: 300 ppm means 300 molecules of a greenhouse gas per million molecules of dry air.

to increase. The present CH_4 concentration has not been exceeded during the past 420,000 years. The annual growth in CH_4 concentration slowed and became more variable in the 1990s, compared with the 1980s. Slightly more than half of current CH_4 emissions are anthropogenic (e.g., use of fossil fuels, cattle, rice agriculture and landfills). In addition, carbon monoxide (CO) emissions have recently been identified as a cause of increasing CH_4 concentration.

· The atmospheric concentration of nitrous oxide (N_2O) has increased by 46 ppb (17%) since 1750 and continues to increase. The present N_2O concentration has not been exceeded during at least the past thousand years. About a third of current N_2O emissions are anthropogenic (e.g., agricultural soils, cattle feed lots and chemical industry).

· Since 1995, the atmospheric concentrations of many of those halocarbon gases that are both ozone-depleting and greenhouse gases (e.g., $CFCl_3$ and CF_2Cl_2), are either increasing more slowly or decreasing, both in response to reduced emissions under the regulations of the Montreal Protocol and its Amendments. Their substitute compounds (e.g., CHF_2Cl and CF_3CH_2F) and some other synthetic compounds (e.g., perfluorocarbons (PFCs) and sulphur hexafluoride (SF_6)) are also greenhouse gases, and their concentrations are currently increasing.

· The radiative forcing due to increases of the well-mixed greenhouse gases from 1750 to 2000 is estimated to be 2.43 Wm^{-2}: 1.46 Wm^{-2} from CO_2; 0.48 Wm^{-2} from CH_4; 0.34 Wm^{-2} from the halocarbons; and 0.15 Wm^{-2} from N_2O. (See Figure 3, where the uncertainties are also illustrated.)

· The observed depletion of the stratospheric ozone (O_3) layer from 1979 to 2000 is estimated to have caused a negative radiative forcing (−0.15 Wm^{-2}). Assuming full compliance with current halocarbon regulations, the positive forcing of the halocarbons will be reduced as will the magnitude of the negative forcing from stratospheric ozone depletion as the ozone layer recovers over the 21st century.

· The total amount of O_3 in the troposphere is estimated to have increased by 36% since 1750, due primarily to anthropogenic emissions of several O_3-forming gases. This corresponds to a positive radiative forcing of 0.35 Wm^{-2}. O_3 forcing varies considerably by region and responds much more quickly to changes in emissions than the long-lived greenhouse gases, such as CO_2.

Anthropogenic aerosols are short-lived and mostly produce negative radiative forcing.

· The major sources of anthropogenic aerosols are fossil fuel and biomass burning. These sources are also linked to degradation of air quality and acid deposition.

· Since the SAR, significant progress has been achieved in better characterising the direct radiative roles of different types of aerosols. Direct radiative forcing is estimated to be −0.4 Wm^{-2} for sulphate, −0.2 Wm^{-2} for biomass burning

aerosols, −0.1 Wm^{-2} for fossil fuel organic carbon and +0.2 Wm^{-2} for fossil fuel black carbon aerosols. There is much less confidence in the ability to quantify the total aerosol direct effect, and its evolution over time, than that for the gases listed above. Aerosols also vary considerably by region and respond quickly to changes in emissions.

· In addition to their direct radiative forcing, aerosols have an indirect radiative forcing through their effects on clouds. There is now more evidence for this indirect effect, which is negative, although of very uncertain magnitude.

Natural factors have made small contributions to radiative forcing over the past century.

· The radiative forcing due to changes in solar irradiance for the period since 1750 is estimated to be about +0.3 Wm^{-2}, most of which occurred during the first half of the 20th century. Since the late 1970s, satellite instruments have observed small oscillations due to the 11-year solar cycle. Mechanisms for the amplification of solar effects on climate have been proposed, but currently lack a rigorous theoretical or observational basis.

· Stratospheric aerosols from explosive volcanic eruptions lead to negative forcin g, which lasts a few years. Several major eruptions occurred in the periods 1880 to 1920 and 1960 to 1991.

· The combined change in radiative forcing of the two major natural factors (solar variation and volcanic aerosols) is estimated to be negative for the past two, and possibly the past four, decades.

Confidence in the ability of models to project future climate has increased.

Complex physically-based climate models are required to provide detailed estimates of feedbacks and of regional features. Such models cannot yet simulate all aspects of climate (e.g., they still cannot account fully for the observed trend in the surface-troposphere temperature difference since 1979) and there are particular uncertainties associated with clouds and their interaction with radiation and aerosols. Nevertheless, confidence in the ability of these models to provide useful projections of future climate has improved due to their demonstrated performance on a range of space and time-scales.

· Understanding of climate processes and their incorporation in climate models have improved, including water vapour, sea-ice dynamics, and ocean heat transport.

· Some recent models produce satisfactory simulations of current climate without the need for non-physical adjustments of heat and water fluxes at the ocean-atmosphere interface used in earlier models.

· Simulations that include estimates of natural and anthropogenic forcing reproduce the observed large-scale changes in surface temperature over the 20th century (Figure 4). However, contributions from some additional processes and forcings may not have been included in the models. Nevertheless, the large-scale consistency between models and observations can be used to provide an independent check on projected warming rates over the next few decades under a given emissions scenario.

· Some aspects of model simulations of ENSO, monsoons and the North Atlantic Oscillation, as well as selected periods of past climate, have improved.

There is new and stronger evidence that most of the warming observed over the last 50 years is attributable to human activities.

The SAR concluded: "The balance of evidence suggests a discernible human influence on global climate". That report also noted that the anthropogenic signal was still emerging from the background of natural climate variability. Since the SAR, progress has been made in reducing uncertainty, particularly with respect to distinguishing and quantifying the magnitude of responses to different external influences. Although many of the sources of uncertainty identified in the SAR still remain to some degree, new evidence and improved understanding support an updated conclusion.

· There is a longer and more closely scrutinised temperature record and new model estimates of variability. The warming over the past 100 years is very unlikely[7] to be due to internal variability alone, as estimated by current models. Reconstructions of climate data for the past 1,000 years (Figure 1b) also indicate that this warming was unusual and is unlikely[7] to be entirely natural in origin.

· There are new estimates of the climate response to natural and anthropogenic forcing, and new detection techniques have been applied. Detection and attribution studies consistently find evidence for an anthropogenic signal in the climate record of the last 35 to 50 years.

· Simulations of the response to natural forcings alone (i.e., the response to variability in solar irradiance and volcanic eruptions) do not explain the warming in the second half of the 20th century (see for example Figure 4a). However, they indicate that natural forcings may have contributed to the observed warming in the first half of the 20th century.

· The warming over the last 50 years due to anthropogenic greenhouse gases can be identified despite uncertainties in forcing due to anthropogenic sulphate aerosol and natural factors (volcanoes and solar irradiance). The anthropogenic sulphate aerosol forcing, while uncertain, is negative over this period and therefore cannot explain the warming. Changes in natural forcing during most of this period are also estimated to be negative and are unlikely[7] to explain the warming.

· Detection and attribution studies comparing model simulated changes with the observed record can now take into account uncertainty in the magnitude of modelled response to external forcing, in particular that due to uncertainty in climate sensitivity.

· Most of these studies find that, over the last 50 years, the estimated rate and magnitude of warming due to increasing concentrations of greenhouse gases alone are comparable with, or larger than, the observed warming. Furthermore, most model estimates that take into account both greenhouse gases and sulphate aerosols are consistent with observations over this period.

· The best agreement between model simulations and observations over the last 140 years has been found when all the above anthropogenic and natural forcing factors are combined, as shown in Figure 4c. These results show that the forcings included are sufficient to explain the observed changes, but do not exclude the possibility that other forcings may also have contributed.

In the light of new evidence and taking into account the remaining uncertainties, most of the observed warming over the last 50 years is likely[7] to have been due to the increase in greenhouse gas concentrations.

Furthermore, it is very likely[7] that the 20th century warming has contributed significantly to the observed sea level rise, through thermal expansion of sea water and widespread loss of land ice. Within present uncertainties, observations and models are both consistent with a lack of significant acceleration of sea level rise during the 20th century.

Human influences will continue to change atmospheric composition throughout the 21st century.

Models have been used to make projections of atmospheric concentrations of greenhouse gases and aerosols, and hence of future climate, based upon emissions scenarios from the IPCC Special Report on Emission Scenarios (SRES) (Figure 5). These scenarios were developed to update the IS92 series, which were used in the SAR and are shown for comparison here in some cases.

Greenhouse gases

· Emissions of CO_2 due to fossil fuel burning are virtually certain[7] to be the dominant influence on the trends in atmospheric CO_2 concentration during the 21st century.

· As the CO_2 concentration of the atmosphere increases, ocean and land will take up a decreasing fraction of anthropogenic CO_2 emissions. The net effect of land and ocean climate feedbacks as indicated by models is to further increase projected atmospheric CO_2 concentrations, by reducing both the ocean and land uptake of CO_2.

Simulated annual global mean surface temperatures

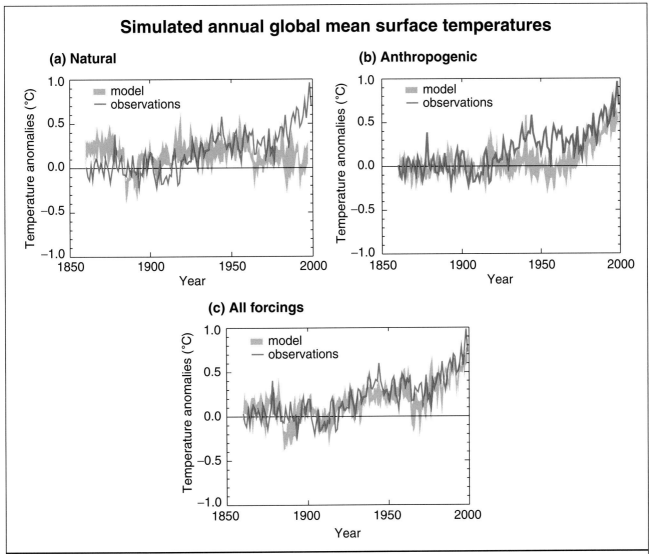

(a) Natural

(b) Anthropogenic

(c) All forcings

Figure 4: Simulating the Earth's temperature variations, and comparing the results to measured changes, can provide insight into the underlying causes of the major changes. A climate model can be used to simulate the temperature changes that occur both from natural and anthropogenic causes. The simulations represented by the band in (a) were done with only natural forcings: solar variation and volcanic activity. Those encompassed by the band in (b) were done with anthropogenic forcings: greenhouse gases and an estimate of sulphate aerosols, and those encompassed by the band in (c) were done with both natural and anthropogenic forcings included. From (b), it can be seen that inclusion of anthropogenic forcings provides a plausible explanation for a substantial part of the observed temperature changes over the past century, but the best match with observations is obtained in (c) when both natural and anthropogenic factors are included. These results show that the forcings included are sufficient to explain the observed changes, but do not exclude the possibility that other forcings may also have contributed. The bands of model results presented here are for four runs from the same model. Similar results to those in (b) are obtained with other models with anthropogenic forcing. [Based upon Chapter 12, Figure 12.7]

· By 2100, carbon cycle models project atmospheric CO_2 concentrations of 540 to 970 ppm for the illustrative SRES scenarios (90 to 250% above the concentration of 280 ppm in the year 1750), Figure 5b. These projections include the land and ocean climate feedbacks. Uncertainties, especially about the magnitude of the climate feedback from the terrestrial biosphere, cause a variation of about −10 to +30% around each scenario. The total range is 490 to 1260 ppm (75 to 350% above the 1750 concentration).

· Changing land use could influence atmospheric CO_2 concentration. Hypothetically, if all of the carbon released by historical land-use changes could be restored to the terrestrial biosphere over the course of the century (e.g., by reforestation), CO_2 concentration would be reduced by 40 to 70 ppm.

· Model calculations of the concentrations of the non-CO_2 greenhouse gases by 2100 vary considerably across the SRES illustrative scenarios, with CH_4 changing by −190 to +1,970 ppb (present concentration 1,760 ppb), N_2O changing by +38 to +144 ppb (present concentration 316 ppb), total tropospheric O_3 changing by −12 to +62%, and a wide range of changes in concentrations of HFCs, PFCs and SF_6, all

The global climate of the 21st century

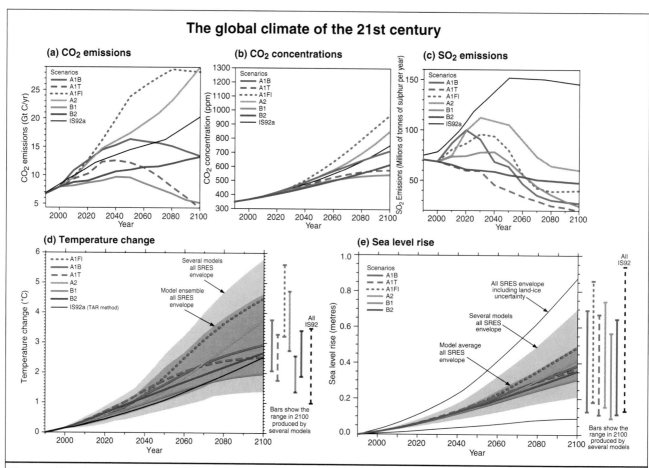

Figure 5: The global climate of the 21st century will depend on natural changes and the response of the climate system to human activities. Climate models project the response of many climate variables – such as increases in global surface temperature and sea level – to various scenarios of greenhouse gas and other human-related emissions. (a) shows the CO_2 emissions of the six illustrative SRES scenarios, which are summarised in the box on page 18, along with IS92a for comparison purposes with the SAR. (b) shows projected CO_2 concentrations. (c) shows anthropogenic SO_2 emissions. Emissions of other gases and other aerosols were included in the model but are not shown in the figure. (d) and (e) show the projected temperature and sea level responses, respectively. The "several models all SRES envelope" in (d) and (e) shows the temperature and sea level rise, respectively, for the simple model when tuned to a number of complex models with a range of climate sensitivities. All SRES envelopes refer to the full range of 35 SRES scenarios. The "model average all SRES envelope" shows the average from these models for the range of scenarios. Note that the warming and sea level rise from these emissions would continue well beyond 2100. Also note that this range does not allow for uncertainty relating to ice dynamical changes in the West Antarctic ice sheet, nor does it account for uncertainties in projecting non-sulphate aerosols and greenhouse gas concentrations. [Based upon (a) Chapter 3, Figure 3.12, (b) Chapter 3, Figure 3.12, (c) Chapter 5, Figure 5.13, (d) Chapter 9, Figure 9.14, (e) Chapter 11, Figure 11.12, Appendix II]

relative to the year 2000. In some scenarios, total tropospheric O_3 would become as important a radiative forcing agent as CH_4 and, over much of the Northern Hemisphere, would threaten the attainment of current air quality targets.

· Reductions in greenhouse gas emissions and the gases that control their concentration would be necessary to stabilise radiative forcing. For example, for the most important anthropogenic greenhouse gas, carbon cycle models indicate that stabilisation of atmospheric CO_2 concentrations at 450, 650 or 1,000 ppm would require global anthropogenic CO_2 emissions to drop below 1990 levels, within a few decades, about a century, or about two centuries, respectively, and continue to decrease steadily thereafter. Eventually CO_2

emissions would need to decline to a very small fraction of current emissions.

Aerosols

· The SRES scenarios include the possibility of either increases or decreases in anthropogenic aerosols (e.g., sulphate aerosols (Figure 5c), biomass aerosols, black and organic carbon aerosols) depending on the extent of fossil fuel use and policies to abate polluting emissions. In addition, natural aerosols (e.g., sea salt, dust and emissions leading to the production of sulphate and carbon aerosols) are projected to increase as a result of changes in climate.

Radiative forcing over the 21st century

· For the SRES illustrative scenarios, relative to the year 2000, the global mean radiative forcing due to greenhouse gases continues to increase through the 21st century, with the fraction due to CO_2 projected to increase from slightly more than half to about three quarters. The change in the direct plus indirect aerosol radiative forcing is projected to be smaller in magnitude than that of CO_2.

Global average temperature and sea level are projected to rise under all IPCC SRES scenarios.

In order to make projections of future climate, models incorporate past, as well as future emissions of greenhouse gases and aerosols. Hence, they include estimates of warming to date and the commitment to future warming from past emissions.

Temperature

· The globally averaged surface temperature is projected to increase by 1.4 to 5.8°C (Figure 5d) over the period 1990 to 2100. These results are for the full range of 35 SRES scenarios, based on a number of climate models[10, 11].
· Temperature increases are projected to be greater than those in the SAR, which were about 1.0 to 3.5°C based on the six IS92 scenarios. The higher projected temperatures and the wider range are due primarily to the lower projected sulphur dioxide emissions in the SRES scenarios relative to the IS92 scenarios.
· The projected rate of warming is much larger than the observed changes during the 20th century and is very likely[7] to be without precedent during at least the last 10,000 years, based on palaeoclimate data.
· By 2100, the range in the surface temperature response across the group of climate models run with a given scenario is comparable to the range obtained from a single model run with the different SRES scenarios.
· On timescales of a few decades, the current observed rate of warming can be used to constrain the projected response to a given emissions scenario despite uncertainty in climate

sensitivity. This approach suggests that anthropogenic warming is likely[7] to lie in the range of 0.1 to 0.2°C per decade over the next few decades under the IS92a scenario, similar to the corresponding range of projections of the simple model used in Figure 5d.
· Based on recent global model simulations, it is very likely[7] that nearly all land areas will warm more rapidly than the global average, particularly those at northern high latitudes in the cold season. Most notable of these is the warming in the northern regions of North America, and northern and central Asia, which exceeds global mean warming in each model by more than 40%. In contrast, the warming is less than the global mean change in south and southeast Asia in summer and in southern South America in winter.
· Recent trends for surface temperature to become more El Niño-like in the tropical Pacific, with the eastern tropical Pacific warming more than the western tropical Pacific, with a corresponding eastward shift of precipitation, are projected to continue in many models.

Precipitation

· Based on global model simulations and for a wide range of scenarios, global average water vapour concentration and precipitation are projected to increase during the 21st century. By the second half of the 21st century, it is likely[7] that precipitation will have increased over northern mid- to high latitudes and Antarctica in winter. At low latitudes there are both regional increases and decreases over land areas. Larger year to year variations in precipitation are very likely[7] over most areas where an increase in mean precipitation is projected.

Extreme Events

Table 1 depicts an assessment of confidence in observed changes in extremes of weather and climate during the latter half of the 20th century (left column) and in projected changes during the 21st century (right column)[a]. This assessment relies on observational and modelling studies, as well as the physical plausibility of future projections across all commonly-used scenarios and is based on expert judgement[7].

· For some other extreme phenomena, many of which may have important impacts on the environment and society, there is currently insufficient information to assess recent trends, and climate models currently lack the spatial detail required to make confident projections. For example, very small-scale phenomena, such as thunderstorms, tornadoes, hail and lightning, are not simulated in climate models.

---

[10] Complex physically based climate models are the main tool for projecting future climate change. In order to explore the full range of scenarios, these are complemented by simple climate models calibrated to yield an equivalent response in temperature and sea level to complex climate models. These projections are obtained using a simple climate model whose climate sensitivity and ocean heat uptake are calibrated to each of seven complex climate models. The climate sensitivity used in the simple model ranges from 1.7 to 4.2°C, which is comparable to the commonly accepted range of 1.5 to 4.5°C.

[11] This range does not include uncertainties in the modelling of radiative forcing, e.g. aerosol forcing uncertainties. A small carbon-cycle climate feedback is included.

---

[a] For more details see Chapter 2 (observations) and Chapter 9, 10 (projections).

Table 1: Estimates of confidence in observed and projected changes in extreme weather and climate events.[a]

Confidence in observed changes (latter half of the 20th century)	Changes in Phenomenon	Confidence in projected changes (during the 21st century)
Likely[7]	**Higher maximum temperatures and more hot days over nearly all land areas**	Very likely[7]
Very likely[7]	**Higher minimum temperatures, fewer cold days and frost days over nearly all land areas**	Very likely[7]
Very likely[7]	**Reduced diurnal temperatures range over most land areas**	Very likely[7]
Likely[7], over many areas	**Increase of heat index[12] over land areas**	Very likely[7], over most areas
Likely, over many Northern Hemisphere mid- to high latitude land areas	**More intense percipitation events[b]**	Very likely[7], over many areas
Likely[7], in a few areas	**Increased summer continental drying and assosciated risk of drought**	Likely[7], over most mid-latitude continental interiors (lack of consistent projections in other areas)
Not observed in the few analyses available	**Increase in tropical cyclone peak wind intensities[c]**	Likely[7], over some areas
Insufficient data for assassement	**Increase in tropical cyclone mean and peak precipitation intensities[c]**	Likely[7], over some areas

[a] For more details see Chapter 2 (observations) and Chapter 9, 10 (projections).
[b] For other areas, there are either insufficient data or conflicting analyses.
[c] Past and future changes in tropical cyclone location and frequency are uncertain.

El Niño

· Confidence in projections of changes in future frequency, amplitude, and spatial pattern of El Niño events in the tropical Pacific is tempered by some shortcomings in how well El Niño is simulated in complex models. Current projections show little change or a small increase in amplitude for El Niño events over the next 100 years.
· Even with little or no change in El Niño amplitude, global warming is likely[7] to lead to greater extremes of drying and heavy rainfall and increase the risk of droughts and floods that occur with El Niño events in many different regions.

Monsoons

· It is likely[7] that warming associated with increasing greenhouse gas concentrations will cause an increase of Asian summer monsoon precipitation variability. Changes in monsoon mean duration and strength depend on the details of the emission scenario. The confidence in such projections is also limited by how well the climate models simulate the detailed seasonal evolution of the monsoons.

[12] Heat index: A combination of temperature and humidity that measures effects on human comfort.

Thermohaline circulation

· Most models show weakening of the ocean thermohaline circulation which leads to a reduction of the heat transport into high latitudes of the Northern Hemisphere. However, even in models where the thermohaline circulation weakens, there is still a warming over Europe due to increased greenhouse gases. The current projections using climate models do not exhibit a complete shut-down of the thermohaline circulation by 2100. Beyond 2100, the thermohaline circulation could completely, and possibly irreversibly, shut-down in either hemisphere if the change in radiative forcing is large enough and applied long enough.

Snow and ice

· Northern Hemisphere snow cover and sea-ice extent are projected to decrease further.
· Glaciers and ice caps are projected to continue their widespread retreat during the 21st century.
· The Antarctic ice sheet is likely[7] to gain mass because of greater precipitation, while the Greenland ice sheet is likely[7] to lose mass because the increase in runoff will exceed the precipitation increase.

Concerns have been expressed about the stability of the West Antarctic ice sheet because it is grounded below sea level. However, loss of grounded ice leading to substantial sea level rise from this source is now widely agreed to be very unlikely[7] during the 21st century, although its dynamics are still inadequately understood, especially for projections on longer time-scales.

Sea level

Global mean sea level is projected to rise by 0.09 to 0.88 metres between 1990 and 2100, for the full range of SRES scenarios. This is due primarily to thermal expansion and loss of mass from glaciers and ice caps (Figure 5e). The range of sea level rise presented in the SAR was 0.13 to 0.94 metres based on the IS92 scenarios. Despite the higher temperature change projections in this assessment, the sea level projections are slightly lower, primarily due to the use of improved models, which give a smaller contribution from glaciers and ice sheets.

Anthropogenic climate change will persist for many centuries.

Emissions of long-lived greenhouse gases (i.e., CO_2, N_2O, PFCs, SF_6) have a lasting effect on atmospheric composition, radiative forcing and climate. For example, several centuries after CO_2 emissions occur, about a quarter of the increase in CO_2 concentration caused by these emissions is still present in the atmosphere.

After greenhouse gas concentrations have stabilised, global average surface temperatures would rise at a rate of only a few tenths of a degree per century rather than several degrees per century as projected for the 21st century without stabilisation. The lower the level at which concentrations are stabilised, the smaller the total temperature change.

Global mean surface temperature increases and rising sea level from thermal expansion of the ocean are projected to continue for hundreds of years after stabilisation of greenhouse gas concentrations (even at present levels), owing to the long timescales on which the deep ocean adjusts to climate change.

Ice sheets will continue to react to climate warming and contribute to sea level rise for thousands of years after climate has been stabilised. Climate models indicate that the local warming over Greenland is likely[7] to be one to three times the global average. Ice sheet models project that a local warming of larger than $3°C$, if sustained for millennia, would lead to virtually a complete melting of the Greenland ice sheet with a resulting sea level rise of about 7 metres. A local warming of $5.5°C$, if sustained for 1000 years, would be likely[7] to result in a contribution from Greenland of about 3 metres to sea level rise.

Current ice dynamic models suggest that the West Antarctic ice sheet could contribute up to 3 metres to sea level rise over the next 1000 years, but such results are strongly dependent on model assumptions regarding climate change scenarios, ice dynamics and other factors.

Further action is required to address remaining gaps in information and understanding.

Further research is required to improve the ability to detect, attribute and understand climate change, to reduce uncertainties and to project future climate changes. In particular, there is a need for additional systematic and sustained observations, modelling and process studies. A serious concern is the decline of observational networks. The following are high priority areas for action.

Systematic observations and reconstructions:
  - Reverse the decline of observational networks in many parts of the world.
  - Sustain and expand the observational foundation for climate studies by providing accurate, long-term, consistent data including implementation of a strategy for integrated global observations.
  - Enhance the development of reconstructions of past climate periods.
  - Improve the observations of the spatial distribution of greenhouse gases and aerosols.

Modelling and process studies:
  - Improve understanding of the mechanisms and factors leading to changes in radiative forcing.
  - Understand and characterise the important unresolved processes and feedbacks, both physical and biogeochemical, in the climate system.
  - Improve methods to quantify uncertainties of climate projections and scenarios, including long-term ensemble simulations using complex models.
  - Improve the integrated hierarchy of global and regional climate models with a focus on the simulation of climate variability, regional climate changes and extreme events.
  - Link more effectively models of the physical climate and the biogeochemical system, and in turn improve coupling with descriptions of human activities.

Cutting across these foci are crucial needs associated with strengthening international co-operation and co-ordination in order to better utilise scientific, computational and observational resources. This should also promote the free exchange of data among scientists. A special need is to increase the observational and research capacities in many regions, particularly in developing countries. Finally, as is the goal of this assessment, there is a continuing imperative to communicate research advances in terms that are relevant to decision making.

The Emissions Scenarios of the Special Report on Emissions Scenarios (SRES)

A1. The A1 storyline and scenario family describes a future world of very rapid economic growth, global population that peaks in mid-century and declines thereafter, and the rapid introduction of new and more efficient technologies. Major underlying themes are convergence among regions, capacity building and increased cultural and social interactions, with a substantial reduction in regional differences in per capita income. The A1 scenario family develops into three groups that describe alternative directions of technological change in the energy system. The three A1 groups are distinguished by their technological emphasis: fossil intensive (A1FI), non-fossil energy sources (A1T), or a balance across all sources (A1B) (where balanced is defined as not relying too heavily on one particular energy source, on the assumption that similar improvement rates apply to all energy supply and end use technologies).

A2. The A2 storyline and scenario family describes a very heterogeneous world. The underlying theme is self-reliance and preservation of local identities. Fertility patterns across regions converge very slowly, which results in continuously increasing population. Economic development is primarily regionally oriented and per capita economic growth and technological change more fragmented and slower than other storylines.

B1. The B1 storyline and scenario family describes a convergent world with the same global population, that peaks in mid-century and declines thereafter, as in the A1 storyline, but with rapid change in economic structures toward a service and information economy, with reductions in material intensity and the introduction of clean and resource-efficient technologies. The emphasis is on global solutions to economic, social and environmental sustainability, including improved equity, but without additional climate initiatives.

B2. The B2 storyline and scenario family describes a world in which the emphasis is on local solutions to economic, social and environmental sustainability. It is a world with continuously increasing global population, at a rate lower than A2, intermediate levels of economic development, and less rapid and more diverse technological change than in the B1 and A1 storylines. While the scenario is also oriented towards environmental protection and social equity, it focuses on local and regional levels.

An illustrative scenario was chosen for each of the six scenario groups A1B, A1FI, A1T, A2, B1 and B2. All should be considered equally sound.

The SRES scenarios do not include additional climate initiatives, which means that no scenarios are included that explicitly assume implementation of the United Nations Framework Convention on Climate Change or the emissions targets of the Kyoto Protocol.

Source Information: Summary for Policymakers

This appendix provides the cross-reference of the topics in the Summary for Policymakers (page and bullet point topic) to the sections of the chapters of the full report that contain expanded information about the topic.

An increasing body of observations gives a collective picture of a warming world and other changes in the climate system.

SPM Page   Cross-Reference: *SPM Topic* · **Chapter Section**
152   *The global average surface temperature has increased over the 20th century by about 0.6°C.*
     · Chapter 2.2.2 · Chapter 2.2.2 · Chapter 2.3
     · Chapter 2.2.2
152-154   *Temperatures have risen during the past four decades in the lowest 8 kilometres of the atmosphere.*
     · Chapter 2.2.3 and 2.2.4 · Chapter 2.2.3 and 2.2.4
     · Chapter 2.2.3, 2.2.4 and Chapter 12.3.2
154   *Snow cover and ice extent have decreased.*
     All three bullet points: Chapter 2.2.5 and 2.2.6

SPM Page   Cross-Reference: *SPM Topic* · **Chapter Section**
154   *Global average sea level has risen and ocean heat content has increased.*
     · Chapter 11.3.2 · Chapter 2.2.2 and Chapter 11.2.1
154   *Changes have also occurred in other important aspects of climate.*
     · Chapter 2.5.2 · Chapter 2.7.2 · Chapter 2.2.2 and 2.5.5 · Chapter 2.7.2 · Chapter 2.6.2 and 2.6.3
     · Chapter 2.7.3 · Chapter 2.7.3
154   *Some important aspects of climate appear not to have changed.*
     · Chapter 2.2.2 · Chapter 2.2.5 · Chapter 2.7.3
     · Chapter 2.7.3

Emissions of greenhouse gases and aerosols due to human activities continue to alter the atmosphere in ways that are expected to affect the climate system.

SPM Page   Cross-Reference: *SPM Topic* · **Chapter Section**
154-155   Chapeau: "Changes in climate occur ..."
     Chapter 1, Chapter 3.1, Chapter 4.1, Chapter 5.1,
     Chapter 6.1, 6.2, 6.9, 6.11 and 6.13

Confidence in the ability of models to project future climate has increased

There is new and stronger evidence that most of the warming observed over the last 50 years is attributable to human activities

Human influences will continue to change atmospheric composition throughout the 21st century.

Global average temperature and sea level are projected to rise under all IPCC SRES scenarios.

Anthropogenic climate change will persist for many centuries.

Further work is required to address remaining gaps in information and understanding.

SPM Page Cross-Reference: *SPM Topic* · **Chapter Section**
163   All bullet points: Chapter 14, Executive Summary

Climate Change 2001:
The Scientific Basis

Technical Summary

A Report Accepted by Working Group I of the IPCC but not Approved in Detail

"Acceptance" of IPCC Reports at a Session of the Working Group or Panel signifies that the material has not been subject to line by line discussion and agreement, but nevertheless presents a comprehensive, objective and balanced view of the subject matter.

Co-ordinating Lead Authors
D.L. Albritton (USA), L.G. Meira Filho (Brazil)

Lead Authors
U. Cubasch (Germany), X. Dai (China), Y. Ding (China), D.J. Griggs (UK), B. Hewitson (South Africa), J.T. Houghton (UK), I. Isaksen (Norway), T. Karl (USA), M. McFarland (USA), V.P. Meleshko (Russia), J.F.B. Mitchell (UK), M. Noguer (UK), B.S. Nyenzi (Tanzania), M. Oppenheimer (USA), J.E. Penner (USA), S. Pollonais (Trinidad and Tobago), T. Stocker (Switzerland), K.E. Trenberth (USA)

Contributing Authors
M.R. Allen (UK), A.P.M. Baede (Netherlands), J.A. Church (Australia), D.H. Ehhalt (Germany), C.K. Folland (UK), F. Giorgi (Italy), J.M. Gregory (UK), J.M. Haywood (UK), J.I. House (Germany), M. Hulme (UK), V.J. Jaramillo (Mexico), A. Jayaraman (India), C.A. Johnson (UK), S. Joussaume (France), D.J. Karoly (Australia), H. Kheshgi (USA), C. Le Quéré (France), L.J. Mata (Germany), B.J. McAvaney (Australia), L.O. Mearns (USA), G.A. Meehl (USA), B. Moore III (USA), R.K. Mugara (Zambia), M. Prather (USA), C. Prentice (Germany), V. Ramaswamy (USA), S.C.B. Raper (UK), M.J. Salinger (New Zealand), R. Scholes (S. Africa), S. Solomon (USA), R. Stouffer (USA), M-X. Wang (China), R.T. Watson (USA), K-S. Yap (Malaysia)

Review Editors
F. Joos (Switzerland), A. Ramirez-Rojas (Venezuela), J.M.R. Stone (Canada), J. Zillman (Australia)

A. Introduction

A.1. The IPCC and its Working Groups

The Intergovernmental Panel on Climate Change (IPCC) was established by the World Meteorological Organisation (WMO) and the United Nations Environment Programme (UNEP) in 1988. The aim was, and remains, to provide an assessment of the understanding of all aspects of climate change[1], including how human activities can cause such changes and can be impacted by them. It had become widely recognised that human-influenced emissions of greenhouse gases have the potential to alter the climate system (see Box 1), with possible deleterious or beneficial effects. It was also recognised that addressing such global issues required organisation on a global scale, including assessment of the understanding of the issue by the worldwide expert communities.

At its first session, the IPCC was organised into three Working Groups. The current remits of the Working Groups are for Working Group I to address the scientific aspects of the climate system and climate change, Working Group II to address the impacts of and adaptations to climate change, and Working Group III to address the options for the mitigation of climate change. The IPCC provided its first major assessment report in 1990 and its second major assessment report in 1996.

The IPCC reports are (i) up-to-date descriptions of the knowns and unknowns of the climate system and related factors, (ii) based on the knowledge of the international expert communities, (iii) produced by an open and peer-reviewed professional process, and (iv) based upon scientific publications whose findings are summarised in terms useful to decision makers. While the assessed information is policy relevant, the IPCC does not establish or advocate public policy.

The scope of the assessments of Working Group I includes observations of the current changes and trends in the climate system, a reconstruction of past changes and trends, an understanding of the processes involved in those changes, and the incorporation of this knowledge into models that can attribute the causes of changes and that can provide simulation of natural and human-induced future changes in the climate system.

A.2. The First and Second Assessment Reports of Working Group I

In the First Assessment Report in 1990, Working Group I broadly described the status of the understanding of the climate system and climate change that had been gained over the preceding decades of research. Several major points were emphasised. The greenhouse effect is a natural feature of the planet, and its fundamental physics is well understood. The atmospheric abundances of greenhouse gases were increasing, due largely to human activities. Continued future growth in greenhouse gas emissions was predicted to lead to significant increases in the average surface temperature of the planet, increases that would exceed the natural variation of the past several millennia and that could be reversed only slowly. The past century had, at that time, seen a surface warming of nearly 0.5°C, which was broadly consistent with that predicted by climate models for the greenhouse gas increases, but was also comparable to what was then known about natural variation. Lastly, it was pointed out that the current level of understanding at that time and the existing capabilities of climate models limited the prediction of changes in the climate of specific regions.

Based on the results of additional research and Special Reports produced in the interim, IPCC Working Group I assessed the new state of understanding in its Second Assessment Report (SAR[2]) in 1996. The report underscored that greenhouse gas abundances continued to increase in the atmosphere and that very substantial cuts in emissions would be required for stabilisation of greenhouse gas concentrations in the atmosphere (which is the ultimate goal of Article 2 of the Framework Convention on Climate Change). Further, the general increase in global temperature continued, with recent years being the warmest since at least 1860. The ability of climate models to simulate observed events and trends had improved, particularly with the inclusion of sulphate aerosols and stratospheric ozone as radiative forcing agents in climate models. Utilising this simulative capability to compare to the observed patterns of regional temperature changes, the report concluded that the ability to quantify the human influence on global climate was limited. The limitations arose because the expected signal was still emerging from the noise of natural variability and because of uncertainties in other key factors. Nevertheless, the report also concluded that "the balance of evidence suggests a discernible human influence on global climate". Lastly, based on a range of scenarios of future greenhouse gas abundances, a set of responses of the climate system was simulated.

---

[1] *Climate change* in IPCC usage refers to any change in climate over time, whether due to natural variability or as a result of human activity. This usage differs from that in the Framework Convention on Climate Change, where *climate change* refers to a change of climate that is attributed directly or indirectly to human activity that alters the composition of the global atmosphere and that is in addition to natural climate variability observed over comparable time periods. For a definition of scientific and technical terms: see the Glossary in Appendix I.

[2] The IPCC Second Assessment Report is referred to in this Technical Summary as the SAR.

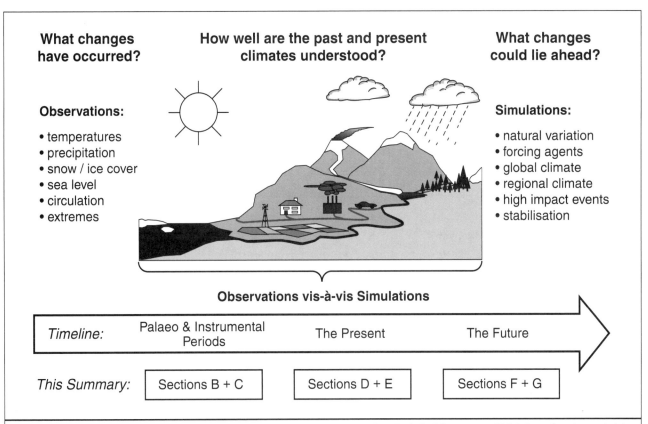

What changes have occurred?

How well are the past and present climates understood?

What changes could lie ahead?

Observations:

- temperatures
- precipitation
- snow / ice cover
- sea level
- circulation
- extremes

Simulations:

- natural variation
- forcing agents
- global climate
- regional climate
- high impact events
- stabilisation

Observations vis-à-vis Simulations

Timeline:	Palaeo & Instrumental Periods	The Present	The Future
This Summary:	Sections B + C	Sections D + E	Sections F + G

Figure 1: Key questions about the climate system and its relation to humankind. This Technical Summary, which is based on the underlying information in the chapters, is a status report on the answers, presented in the structure indicated.

A.3. The Third Assessment Report: This Technical Summary

The third major assessment report of IPCC Working Group I builds upon these past assessments and incorporates the results of the past five years of climate research. This Technical Summary is based on the underlying information of the chapters, which is cross-referenced in the Source Notes in the Appendix. This Summary aims to describe the major features (see Figure 1) of the understanding of the climate system and climate change at the outset of the 21st century. Specifically:

- What does the observational record show with regard to past climate changes, both globally and regionally and both on the average and in the extremes? (Section B)
- How quantitative is the understanding of the agents that cause climate to change, including both those that are natural (e.g., solar variation) and human-related (e.g., greenhouse gases) phenomena? (Section C)
- What is the current ability to simulate the responses of the climate system to these forcing agents? In particular, how well are key physical and biogeochemical processes described by present global climate models? (Section D)

- Based on today's observational data and today's climate predictive capabilities, what does the comparison show regarding a human influence on today's climate? (Section E)
- Further, using current predictive tools, what could the possible climate future be? Namely, for a wide spectrum of projections for several climate-forcing agents, what does current understanding project for global temperatures, regional patterns of precipitation, sea levels, and changes in extremes? (Section F)

Finally, what are the most urgent research activities that need to be addressed to improve our understanding of the climate system and to reduce our uncertainty regarding future climate change?

The Third Assessment Report of IPCC Working Group I is the product of hundreds of scientists from the developed and developing world who contributed to its preparation and review. What follows is a summary of their understanding of the climate system.

Box 1: What drives changes in climate?

The Earth absorbs radiation from the Sun, mainly at the surface. This energy is then redistributed by the atmospheric and oceanic circulations and radiated back to space at longer (infrared) wavelengths. For the annual mean and for the Earth as a whole, the incoming solar radiation energy is balanced approximately by the outgoing terrestrial radiation. Any factor that alters the radiation received from the Sun or lost to space, or that alters the redistribution of energy within the atmosphere and between the atmosphere, land, and ocean, can affect climate. A change in the net radiative energy available to the global Earth-atmosphere system is termed here, and in previous IPCC reports, a radiative forcing. Positive radiative forcings tend to warm the Earth's surface and lower atmosphere. Negative radiative forcings tend to cool them.

Increases in the concentrations of greenhouse gases will reduce the efficiency with which the Earth's surface radiates to space. More of the outgoing terrestrial radiation from the surface is absorbed by the atmosphere and re-emitted at higher altitudes and lower temperatures. This results in a positive radiative forcing that tends to warm the lower atmosphere and surface. Because less heat escapes to space, this is the enhanced greenhouse effect – an enhancement of an effect that has operated in the Earth's atmosphere for billions of years due to the presence of naturally occurring greenhouse gases: water vapour, carbon dioxide, ozone, methane and nitrous oxide. The amount of radiative forcing depends on the size of the increase in concentration of each greenhouse gas, the radiative properties of the gases involved, and the concentrations of other greenhouse gases already present in the atmosphere. Further, many greenhouse gases reside in the atmosphere for centuries after being emitted, thereby introducing a long-term commitment to positive radiative forcing.

Anthropogenic aerosols (microscopic airborne particles or droplets) in the troposphere, such as those derived from fossil fuel and biomass burning, can reflect solar radiation, which leads to a cooling tendency in the climate system. Because it can absorb solar radiation, black carbon (soot) aerosol tends to warm the climate system. In addition, changes in aerosol concentrations can alter cloud amount and cloud reflectivity through their effect on cloud properties and lifetimes. In most cases, tropospheric aerosols tend to produce a negative radiative forcing and a cooler climate. They have a much shorter lifetime (days to weeks) than most greenhouse gases (decades to centuries), and, as a result, their concentrations respond much more quickly to changes in emissions.

Volcanic activity can inject large amounts of sulphur-containing gases (primarily sulphur dioxide) into the stratosphere, which are transformed into sulphate aerosols. Individual eruptions can produce a large, but transitory, negative radiative forcing, tending to cool the Earth's surface and lower atmosphere over periods of a few years.

The Sun's output of energy varies by small amounts (0.1%) over an 11-year cycle and, in addition, variations over longer periods may occur. On time-scales of tens to thousands of years, slow variations in the Earth's orbit, which are well understood, have led to changes in the seasonal and latitudinal distribution of solar radiation. These changes have played an important part in controlling the variations of climate in the distant past, such as the glacial and inter-glacial cycles.

When radiative forcing changes, the climate system responds on various time-scales. The longest of these are due to the large heat capacity of the deep ocean and dynamic adjustment of the ice sheets. This means that the transient response to a change (either positive or negative) may last for thousands of years. Any changes in the radiative balance of the Earth, including those due to an increase in greenhouse gases or in aerosols, will alter the global hydrological cycle and atmospheric and oceanic circulation, thereby affecting weather patterns and regional temperatures and precipitation.

Any human-induced changes in climate will be embedded in a background of natural climatic variations that occur on a whole range of time- and space-scales. Climate variability can occur as a result of natural changes in the forcing of the climate system, for example variations in the strength of the incoming solar radiation and changes in the concentrations of aerosols arising from volcanic eruptions. Natural climate variations can also occur in the absence of a change in external forcing, as a result of complex interactions between components of the climate system, such as the coupling between the atmosphere and ocean. The El Niño-Southern Oscillation (ENSO) phenomenon is an example of such natural "internal" variability on interannual time-scales. To distinguish anthropogenic climate changes from natural variations, it is necessary to identify the anthropogenic "signal" against the background "noise" of natural climate variability.

B. The Observed Changes in the Climate System

Is the Earth's climate changing? The answer is unequivocally "Yes". A suite of observations supports this conclusion and provides insight about the rapidity of those changes. These data are also the bedrock upon which to construct the answer to the more difficult question: "Why is it changing?", which is addressed in later Sections.

This Section provides an updated summary of the observations that delineate how the climate system has changed in the past. Many of the variables of the climate system have been measured directly, i.e., the "instrumental record". For example, widespread direct measurements of surface temperature began around the middle of the 19th century. Near global observations of other surface "weather" variables, such as precipitation and winds, have been made for about a hundred years. Sea level measurements have been made for over 100 years in some places, but the network of tide gauges with long records provides only limited global coverage. Upper air observations have been made systematically only since the late 1940s. There are also long records of surface oceanic observations made from ships since the mid-19th century and by dedicated buoys since about the late 1970s. Sub-surface oceanic temperature measurements with near global coverage are now available from the late 1940s. Since the late 1970s, other data from Earth-observation satellites have been used to provide a wide range of global observations of various components of the climate system. In addition, a growing set of palaeoclimatic data, e.g., from trees, corals, sediments, and ice, are giving information about the Earth's climate of centuries and millennia before the present.

This Section places particular emphasis on current knowledge of past changes in key climate variables: temperature, precipitation and atmospheric moisture, snow cover, extent of land and sea ice, sea level, patterns in atmospheric and oceanic circulation, extreme weather and climate events, and overall features of the climate variability. The concluding part of this Section compares the observed trends in these various climate indicators to see if a collective picture emerges. The degree of this internal consistency is a critical factor in assessing the level of confidence in the current understanding of the climate system.

B.1. Observed Changes in Temperature

Temperatures in the instrumental record for land and oceans

The global average surface temperature has increased by 0.6 ± 0.2°C[3] since the late 19th century. It is very likely that the 1990s was the warmest decade and 1998 the warmest year in the instrumental record since 1861 (see Figure 2). The main cause of the increased estimate of global warming of 0.15°C since the SAR is related to the record warmth of the additional six years (1995 to 2000) of data. A secondary reason is related to improved methods of estimating change. The current, slightly larger uncertainty range (±0.2°C, 95% confidence interval) is also more objectively based. Further, the scientific basis for confidence in the estimates of the increase in global temperature since the late 19th century has been strengthened since the SAR. This is due to the improvements derived from several new studies. These include an independent test of the corrections used for time-dependent biases in the sea surface temperature data and new analyses of the effect of urban "heat island" influences on global land-temperature trends. As indicated in Figure 2, most of the increase in global temperature

---

[3] Generally, temperature trends are rounded to the nearest 0.05°C per unit of time, the periods often being limited by data availability.

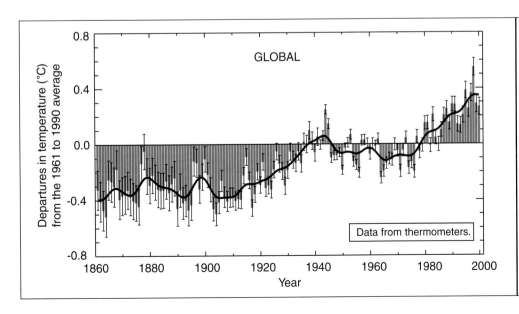

Figure 2: Combined annual land-surface air and sea surface temperature anomalies (°C) 1861 to 2000, relative to 1961 to 1990. Two standard error uncertainties are shown as bars on the annual number. [Based on Figure 2.7c]

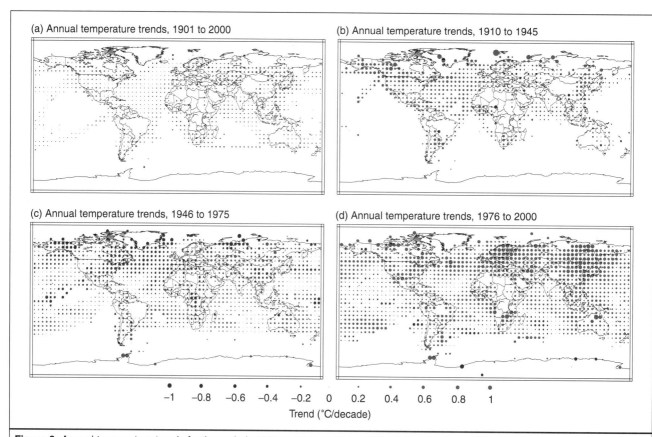

Figure 3: Annual temperature trends for the periods 1901 to 2000, 1910 to 1945, 1946 to 1975 and 1976 to 2000 respectively. Trends are represented by the area of the circle with red representing increases, blue representing decreases, and green little or no change. Trends were calculated from annually averaged gridded anomalies with the requirement that the calculation of annual anomalies include a minimum of 10 months of data. For the period 1901 to 2000, trends were calculated only for those grid boxes containing annual anomalies in at least 66 of the 100 years. The minimum number of years required for the shorter time periods (1910 to 1945, 1946 to 1975, and 1976 to 2000) was 24, 20, and 16 years respectively. [Based on Figure 2.9]

since the late 19th century has occurred in two distinct periods: 1910 to 1945 and since 1976. The rate of increase of temperature for both periods is about 0.15°C/decade. Recent warming has been greater over land compared to oceans; the increase in sea surface temperature over the period 1950 to 1993 is about half that of the mean land-surface air temperature. The high global temperature associated with the 1997 to 1998 El Niño event stands out as an extreme event, even taking into account the recent rate of warming.

The regional patterns of the warming that occurred in the early part of the 20th century were different than those that occurred in the latter part. Figure 3 shows the regional patterns of the warming that have occurred over the full 20th century, as well as for three component time periods. The most recent period of warming (1976 to 1999) has been almost global, but the largest increases in temperature have occurred over the mid- and high latitudes of the continents in the Northern Hemisphere. Year-round cooling is evident in the north-western North Atlantic and the central North Pacific Oceans, but the North Atlantic cooling trend has recently reversed. The recent regional patterns of temperature change have been shown to

be related, in part, to various phases of atmospheric-oceanic oscillations, such as the North Atlantic-Arctic Oscillation and possibly the Pacific Decadal Oscillation. Therefore, regional temperature trends over a few decades can be strongly influenced by regional variability in the climate system and can depart appreciably from a global average. The 1910 to 1945 warming was initially concentrated in the North Atlantic. By contrast, the period 1946 to 1975 showed significant cooling in the North Atlantic, as well as much of the Northern Hemisphere, and warming in much of the Southern Hemisphere.

New analyses indicate that global ocean heat content has increased significantly since the late 1950s. More than half of the increase in heat content has occurred in the upper 300 m of the ocean, equivalent to a rate of temperature increase in this layer of about 0.04°C/decade.

New analyses of daily maximum and minimum land-surface temperatures for 1950 to 1993 continue to show that this measure of diurnal temperature range is decreasing very widely, although not everywhere. On average, minimum temperatures are increasing at about twice the rate of maximum temperatures (0.2 versus 0.1°C/decade).

Temperatures above the surface layer from satellite and weather balloon records

Surface, balloon and satellite temperature measurements show that the troposphere and Earth's surface have warmed and that the stratosphere has cooled. Over the shorter time period for which there have been both satellite and weather balloon data (since 1979), the balloon and satellite records show significantly less lower-tropospheric warming than observed at the surface. Analyses of temperature trends since 1958 for the lowest 8 km of the atmosphere and at the surface are in good agreement, as shown in Figure 4a, with a warming of about 0.1°C per decade. However, since the beginning of the satellite record in 1979, the temperature data from both satellites and weather balloons show a warming in the global middle-to-lower troposphere at a rate of approximately 0.05 ± 0.10°C per decade. The global average surface temperature has increased significantly by 0.15 ± 0.05°C/decade. The difference in the warming rates is statistically significant. By contrast, during the period 1958 to 1978, surface temperature trends were near zero, while trends for the lowest 8 km of the atmosphere were near 0.2°C/decade. About half of the observed difference in warming since 1979 is likely[4] to be due to the combination of the differences in spatial coverage of the surface and tropospheric observations and the physical effects of the sequence of volcanic eruptions and a substantial El Niño (see Box 4 for a general description of ENSO) that occurred within this period. The remaining difference is very likely real and not an observing bias. It arises primarily due to differences in the rate of temperature change over the tropical and sub-tropical regions, which were faster in the lowest 8 km of the atmosphere before about 1979, but which have been slower since then. There are no significant differences in warming rates over mid-latitude continental regions in the Northern Hemisphere. In the upper troposphere, no significant global temperature trends have been detected since the early 1960s. In the stratosphere, as shown in Figure 4b, both satellites and balloons show substantial cooling, punctuated by sharp warming episodes of one to two years long that are due to volcanic eruptions.

Surface temperatures during the pre-instrumental period from the proxy record

It is likely that the rate and duration of the warming of the 20th century is larger than any other time during the last

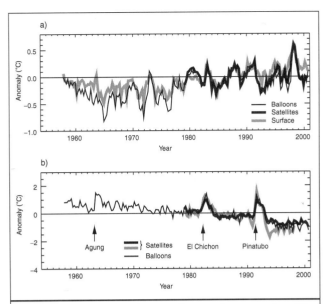

Figure 4: (a) Time-series of seasonal temperature anomalies of the troposphere based on balloons and satellites in addition to the surface. (b) Time-series of seasonal temperature anomalies of the lower stratosphere from balloons and satellites. [Based on Figure 2.12]

1,000 years. The 1990s are likely to have been the warmest decade of the millennium in the Northern Hemisphere, and 1998 is likely to have been the warmest year. There has been a considerable advance in understanding of temperature change that occurred over the last millennium, especially from the synthesis of individual temperature reconstructions. This new detailed temperature record for the Northern Hemisphere is shown in Figure 5. The data show a relatively warm period associated with the 11th to 14th centuries and a relatively cool period associated with the 15th to 19th centuries in the Northern Hemisphere. However, evidence does not support these "Medieval Warm Period" and "Little Ice Age" periods, respectively, as being globally synchronous. As Figure 5 indicates, the rate and duration of warming of the Northern Hemisphere in the 20th century appears to have been unprecedented during the millennium, and it cannot simply be considered as a recovery from the "Little Ice Age" of the 15th to 19th centuries. These analyses are complemented by sensitivity analysis of the spatial representativeness of available palaeoclimatic data, indicating that the warmth of the recent decade is outside the 95% confidence interval of temperature uncertainty, even during the warmest periods of the last millennium. Moreover, several different analyses have now been completed, each suggesting that the Northern Hemisphere temperatures of the past decade have been warmer than any other time in the past six to ten centuries. This is the time-span over which temperatures with annual resolution can be calculated using hemispheric-wide tree-ring, ice-cores, corals, and and other annually-resolved proxy data. Because less data are available, less is known about annual averages prior to 1,000 years before the present and for conditions prevailing in most of the Southern Hemisphere prior to 1861.

---

[4] In this Technical Summary and in the Summary for Policymakers, the following words have been used to indicate approximate judgmental estimates of confidence: *virtually certain* (greater than 99% chance that a result is true); *very likely* (90-99% chance); *likely* (66-90% chance); *medium likelihood* (33-66% chance); *unlikely* (10-33% chance); *very unlikely* (1-10% chance); *exceptionally unlikely* (less than 1% chance). The reader is referred to individual chapters for more details.

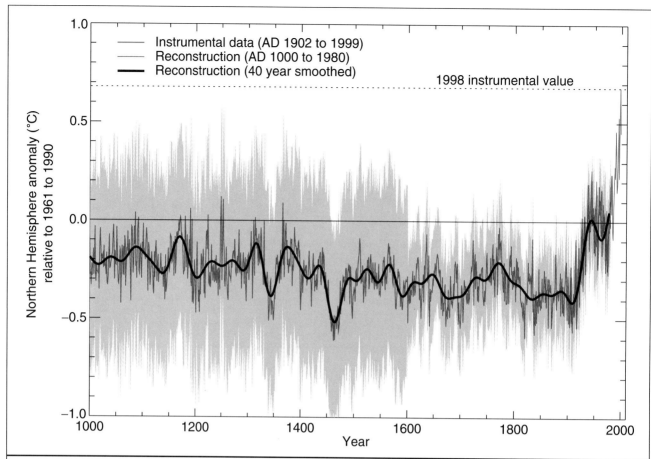

Figure 5: Millennial Northern Hemisphere (NH) temperature reconstruction (blue – tree rings, corals, ice cores, and historical records) and instrumental data (red) from AD 1000 to 1999. Smoother version of NH series (black), and two standard error limits (gray shaded) are shown. [Based on Figure 2.20]

It is likely that large rapid decadal temperature changes occurred during the last glacial and its deglaciation (between about 100,000 and 10,000 years ago), particularly in high latitudes of the Northern Hemisphere. In a few places during the deglaciation, local increases in temperature of 5 to 10°C are likely to have occurred over periods as short as a few decades. During the last 10,000 years, there is emerging evidence of significant rapid regional temperature changes, which are part of the natural variability of climate.

B.2. Observed Changes in Precipitation and Atmospheric Moisture

Since the time of the SAR, annual land precipitation has continued to increase in the middle and high latitudes of the Northern Hemisphere (very likely to be 0.5 to 1%/decade), except over Eastern Asia. Over the sub-tropics (10° N to 30° N), land-surface rainfall has decreased on average (likely to be about 0.3%/decade), although this has shown signs of recovery in recent years. Tropical land-surface precipitation measurements indicate that precipitation likely has increased by about 0.2 to 0.3%/ decade over the 20th century, but increases are not evident over the past few decades and the amount of tropical

land (versus ocean) area for the latitudes 10° N to 10° S is relatively small. Nonetheless, direct measurements of precipitation and model reanalyses of inferred precipitation indicate that rainfall has also increased over large parts of the tropical oceans. Where and when available, changes in annual streamflow often relate well to changes in total precipitation. The increases in precipitation over Northern Hemisphere mid- and high latitude land areas have a strong correlation to long-term increases in total cloud amount. In contrast to the Northern Hemisphere, no comparable systematic changes in precipitation have been detected in broad latitudinal averages over the Southern Hemisphere.

It is likely that total atmospheric water vapour has increased several per cent per decade over many regions of the Northern Hemisphere. Changes in water vapour over approximately the past 25 years have been analysed for selected regions using in situ surface observations, as well as lower-tropospheric measurements from satellites and weather balloons. A pattern of overall surface and lower-tropospheric water vapour increases over the past few decades is emerging from the most reliable data sets, although there are likely to be time-dependent biases in these data and regional variations in the trends.

Water vapour in the lower stratosphere is also likely to have increased by about 10% per decade since the beginning of the observational record (1980).

Changes in total cloud amounts over Northern Hemisphere mid- and high latitude continental regions indicate a likely increase in cloud cover of about 2% since the beginning of the 20th century, which has now been shown to be positively correlated with decreases in the diurnal temperature range. Similar changes have been shown over Australia, the only Southern Hemisphere continent where such an analysis has been completed. Changes in total cloud amount are uncertain both over sub-tropical and tropical land areas, as well as over the oceans.

B.3. Observed Changes in Snow Cover and Land- and Sea-Ice Extent

Decreasing snow cover and land-ice extent continue to be positively correlated with increasing land-surface temperatures. Satellite data show that there are very likely to have been decreases of about 10% in the extent of snow cover since the late 1960s. There is a highly significant correlation between increases in Northern Hemisphere land temperatures and the decreases. There is now ample evidence to support a major retreat of alpine and continental glaciers in response to 20th century warming. In a few maritime regions, increases in precipitation due to regional atmospheric circulation changes have overshadowed increases in temperature in the past two decades, and glaciers have re-advanced. Over the past 100 to 150 years, ground-based observations show that there is very likely to have been a reduction of about two weeks in the annual duration of lake and river ice in the mid- to high latitudes of the Northern Hemisphere.

Northern Hemisphere sea-ice amounts are decreasing, but no significant trends in Antarctic sea-ice extent are apparent. A retreat of sea-ice extent in the Arctic spring and summer of 10 to 15% since the 1950s is consistent with an increase in spring temperatures and, to a lesser extent, summer temperatures in the high latitudes. There is little indication of reduced Arctic sea-ice extent during winter when temperatures have increased in the surrounding region. By contrast, there is no readily apparent relationship between decadal changes of Antarctic temperatures and sea-ice extent since 1973. After an initial decrease in the mid-1970s, Antarctic sea-ice extent has remained stable, or even slightly increased.

New data indicate that there likely has been an approximately 40% decline in Arctic sea-ice thickness in late summer to early autumn between the period of 1958 to 1976 and the mid-1990s, and a substantially smaller decline in winter. The relatively short record length and incomplete sampling limit the interpretation of these data. Interannual variability and inter-decadal variability could be influencing these changes.

B.4. Observed Changes in Sea Level

Changes during the instrumental record

Based on tide gauge data, the rate of global mean sea level rise during the 20th century is in the range 1.0 to 2.0 mm/yr, with a central value of 1.5 mm/yr (the central value should not be interpreted as a best estimate). (See Box 2 for the factors that influence sea level.) As Figure 6 indicates, the longest instrumental records (two or three centuries at most) of local sea level come from tide gauges. Based on the very few long tide-gauge records, the average rate of sea level rise has been larger during the 20th century than during the 19th century. No significant acceleration in the rate of sea level rise during the 20th century has been detected. This is not inconsistent with model results due to the possibility of compensating factors and the limited data.

Changes during the pre-instrumental record

Since the last glacial maximum about 20,000 years ago, the sea level in locations far from present and former ice sheets has risen by over 120 m as a result of loss of mass from these ice sheets. Vertical land movements, both upward and downward, are still occurring in response to these large transfers of mass from ice sheets to oceans. The most rapid rise in global sea level was between 15,000 and 6,000 years ago, with an average rate of about 10 mm/yr. Based on geological data, eustatic sea level (i.e., corresponding to a change in ocean volume) may have risen at an average rate of 0.5 mm/yr over the past 6,000 years and at an average rate of 0.1 to 0.2 mm/yr over the last 3,000 years. This rate is about one tenth of that occurring during the 20th century. Over the past 3,000 to 5,000 years, oscillations in global sea level on time-scales of 100 to 1,000 years are unlikely to have exceeded 0.3 to 0.5 m.

B.5. Observed Changes in Atmospheric and Oceanic Circulation Patterns

The behaviour of ENSO (see Box 4 for a general description), has been unusual since the mid-1970s compared with the previous 100 years, with warm phase ENSO episodes being relatively more frequent, persistent, and intense than the opposite cool phase. This recent behaviour of ENSO is reflected in variations in precipitation and temperature over much of the global tropics and sub-tropics. The overall effect is likely to have been a small contribution to the increase in global temperatures during the last few decades. The Inter-decadal Pacific Oscillation and the Pacific Decadal Oscillation are associated with decadal to multidecadal climate variability over the Pacific basin. It is likely that these oscillations modulate ENSO-related climate variability.

Other important circulation features that affect the climate in large regions of the globe are being characterised. The North

Box 2: What causes sea level to change?

The level of the sea at the shoreline is determined by many factors in the global environment that operate on a great range of time-scales, from hours (tidal) to millions of years (ocean basin changes due to tectonics and sedimentation). On the time-scale of decades to centuries, some of the largest influences on the average levels of the sea are linked to climate and climate change processes.

Firstly, as ocean water warms, it expands. On the basis of observations of ocean temperatures and model results, thermal expansion is believed to be one of the major contributors to historical sea level changes. Further, thermal expansion is expected to contribute the largest component to sea level rise over the next hundred years. Deep ocean temperatures change only slowly; therefore, thermal expansion would continue for many centuries even if the atmospheric concentrations of greenhouse gases were to stabilise.

The amount of warming and the depth of water affected vary with location. In addition, warmer water expands more than colder water for a given change in temperature. The geographical distribution of sea level change results from the geographical variation of thermal expansion, changes in salinity, winds, and ocean circulation. The range of regional variation is substantial compared with the global average sea level rise.

Sea level also changes when the mass of water in the ocean increases or decreases. This occurs when ocean water is exchanged with the water stored on land. The major land store is the water frozen in glaciers or ice sheets. Indeed, the main reason for the lower sea level during the last glacial period was the amount of water stored in the large extension of the ice sheets on the continents of the Northern Hemisphere. After thermal expansion, the melting of mountain glaciers and ice caps is expected to make the largest contribution to the rise of sea level over the next hundred years. These glaciers and ice caps make up only a few per cent of the world's land-ice area, but they are more sensitive to climate change than the larger ice sheets in Greenland and Antarctica, because the ice sheets are in colder climates with low precipitation and low melting rates. Consequently, the large ice sheets are expected to make only a small net contribution to sea level change in the coming decades.

Sea level is also influenced by processes that are not explicitly related to climate change. Terrestrial water storage (and hence, sea level) can be altered by extraction of ground water, building of reservoirs, changes in surface runoff, and seepage into deep aquifers from reservoirs and irrigation. These factors may be offsetting a significant fraction of the expected acceleration in sea level rise from thermal expansion and glacial melting. In addition, coastal subsidence in river delta regions can also influence local sea level. Vertical land movements caused by natural geological processes, such as slow movements in the Earth's mantle and tectonic displacements of the crust, can have effects on local sea level that are comparable to climate-related impacts. Lastly, on seasonal, interannual, and decadal time-scales, sea level responds to changes in atmospheric and ocean dynamics, with the most striking example occurring during El Niño events.

Atlantic Oscillation (NAO) is linked to the strength of the westerlies over the Atlantic and extra-tropical Eurasia. During winter the NAO displays irregular oscillations on interannual to multi-decadal time-scales. Since the 1970s, the winter NAO has often been in a phase that contributes to stronger westerlies, which correlate with cold season warming over Eurasia. New evidence indicates that the NAO and changes in Arctic sea ice are likely to be closely coupled. The NAO is now believed to be part of a wider scale atmospheric Arctic Oscillation that affects much of the extratropical Northern Hemisphere. A similar Antarctic Oscillation has been in an enhanced positive phase during the last 15 years, with stronger westerlies over the Southern Oceans.

B.6. Observed Changes in Climate Variability and Extreme Weather and Climate Events

New analyses show that in regions where total precipitation has increased, it is very likely that there have been even more pronounced increases in heavy and extreme precipitation events. The converse is also true. In some regions, however, heavy and extreme events (i.e., defined to be within the upper or lower ten percentiles) have increased despite the fact that total precipitation has decreased or remained constant. This is attributed to a decrease in the frequency of precipitation events. Overall, it is likely that for many mid- and high latitude areas, primarily in the Northern Hemisphere, statistically significant increases have occurred in the proportion of total annual precipitation derived from heavy and extreme precipitation events; it is likely that there has been a 2 to 4% increase in the frequency of heavy precipitation events over the latter half of the 20th century. Over the 20th century (1900 to 1995), there were relatively small increases in global land areas experiencing severe drought or severe wetness. In some regions, such as parts of Asia and Africa, the frequency and intensity of drought have been observed to increase in recent decades. In many regions, these changes are dominated by inter-decadal and multi-decadal climate variability, such as the shift in ENSO towards more warm events. In many regions, inter-daily temperature variability has decreased, and increases in the daily

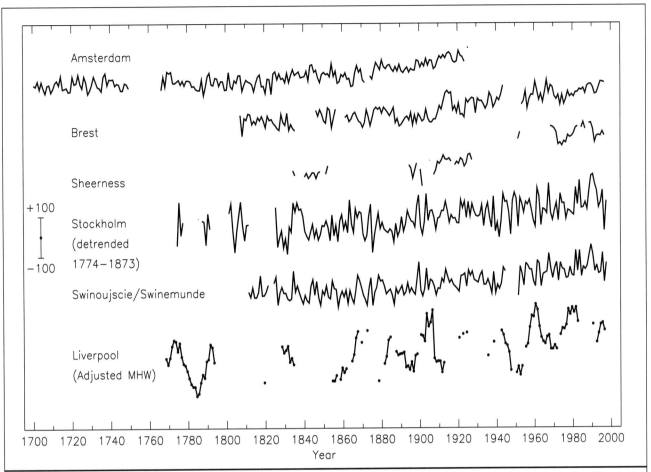

Figure 6: Time-series of relative sea level for the past 300 years from Northern Europe: Amsterdam, Netherlands; Brest, France; Sheerness, UK; Stockholm, Sweden (detrended over the period 1774 to 1873 to remove to first order the contribution of post-glacial rebound); Swinoujscie, Poland (formerly Swinemunde, Germany); and Liverpool, UK. Data for the latter are of "Adjusted Mean High Water" rather than Mean Sea Level and include a nodal (18.6 year) term. The scale bar indicates ±100 mm. [Based on Figure 11.7]

minimum temperature are lengthening the freeze-free period in most mid- and high latitude regions. Since 1950 it is very likely that there has been a significant reduction in the frequency of much-below-normal seasonal mean temperatures across much of the globe, but there has been a smaller increase in the frequency of much-above-normal seasonal temperatures.

There is no compelling evidence to indicate that the characteristics of tropical and extratropical storms have changed. Changes in tropical storm intensity and frequency are dominated by interdecadal to multidecadal variations, which may be substantial, e.g., in the tropical North Atlantic. Owing to incomplete data and limited and conflicting analyses, it is uncertain as to whether there have been any long-term and large-scale increases in the intensity and frequency of extra-tropical cyclones in the Northern Hemisphere. Regional increases have been identified in the North Pacific, parts of North America, and Europe over the past several decades. In the Southern Hemisphere, fewer analyses have been completed, but they suggest a decrease in extra-tropical cyclone activity since the 1970s. Recent analyses of changes in severe local weather (e.g., tornadoes, thunderstorm days, and hail) in a few selected regions do not provide compelling evidence to suggest long-term changes. In general, trends in severe weather events are notoriously difficult to detect because of their relatively rare occurrence and large spatial variability.

B.7. The Collective Picture: A Warming World and Other Changes in the Climate System

As summarised above, a suite of climate changes is now well-documented, particularly over the recent decades to century time period, with its growing set of direct measurements. Figure 7 illustrates these trends in temperature indicators (Figure 7a) and hydrological and storm-related indicators (Figure 7b), as well as also providing an indication of certainty about the changes.

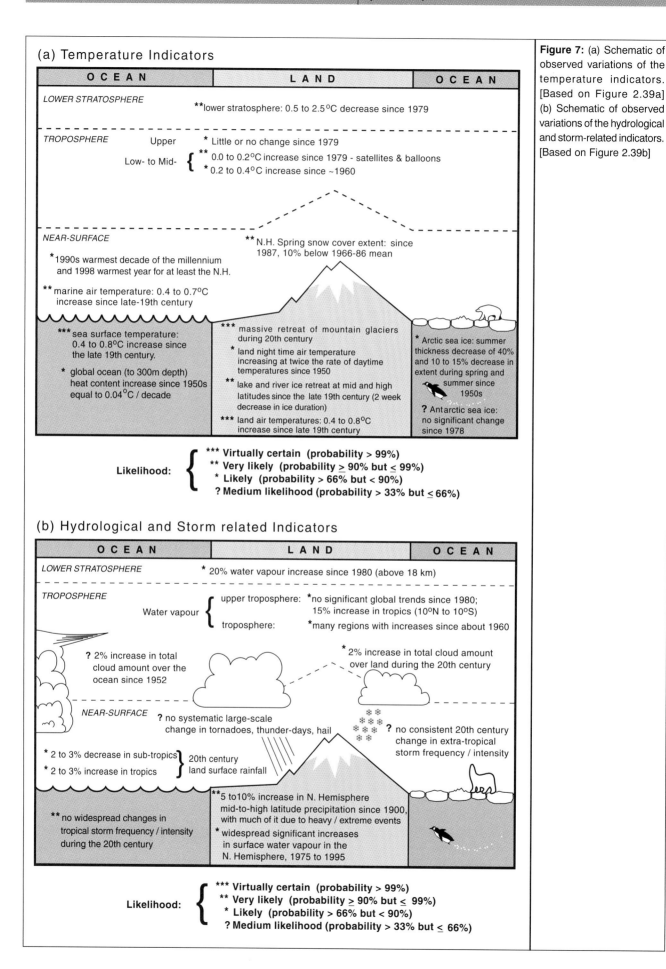

(a) Temperature Indicators

| OCEAN | LAND | OCEAN |

LOWER STRATOSPHERE

**lower stratosphere: 0.5 to 2.5°C decrease since 1979

TROPOSPHERE   Upper   * Little or no change since 1979

Low- to Mid- { ** 0.0 to 0.2°C increase since 1979 - satellites & balloons
* 0.2 to 0.4°C increase since ~1960

NEAR-SURFACE

** N.H. Spring snow cover extent: since 1987, 10% below 1966-86 mean

*1990s warmest decade of the millennium and 1998 warmest year for at least the N.H.

** marine air temperature: 0.4 to 0.7°C increase since late-19th century

*** sea surface temperature: 0.4 to 0.8°C increase since the late 19th century.

* global ocean (to 300m depth) heat content increase since 1950s equal to 0.04°C / decade

*** massive retreat of mountain glaciers during 20th century

* land night time air temperature increasing at twice the rate of daytime temperatures since 1950

** lake and river ice retreat at mid and high latitudes since the late 19th century (2 week decrease in ice duration)

*** land air temperatures: 0.4 to 0.8°C increase since late 19th century

* Arctic sea ice: summer thickness decrease of 40% and 10 to 15% decrease in extent during spring and summer since 1950s

? Antarctic sea ice: no significant change since 1978

Likelihood: {
*** **Virtually certain** (probability > 99%)
** **Very likely** (probability ≥ 90% but ≤ 99%)
* **Likely** (probability > 66% but < 90%)
? **Medium likelihood** (probability > 33% but ≤ 66%)
}

(b) Hydrological and Storm related Indicators

| OCEAN | LAND | OCEAN |

LOWER STRATOSPHERE   * 20% water vapour increase since 1980 (above 18 km)

TROPOSPHERE

Water vapour {
upper troposphere: *no significant global trends since 1980; 15% increase in tropics (10°N to 10°S)
troposphere: *many regions with increases since about 1960
}

? 2% increase in total cloud amount over the ocean since 1952

* 2% increase in total cloud amount over land during the 20th century

NEAR-SURFACE

? no systematic large-scale change in tornadoes, thunder-days, hail

? no consistent 20th century change in extra-tropical storm frequency / intensity

* 2 to 3% decrease in sub-tropics
* 2 to 3% increase in tropics
} 20th century land surface rainfall

** no widespread changes in tropical storm frequency / intensity during the 20th century

** 5 to10% increase in N. Hemisphere mid-to-high latitude precipitation since 1900, with much of it due to heavy / extreme events

* widespread significant increases in surface water vapour in the N. Hemisphere, 1975 to 1995

Likelihood: {
*** **Virtually certain** (probability > 99%)
** **Very likely** (probability ≥ 90% but ≤ 99%)
* **Likely** (probability > 66% but < 90%)
? **Medium likelihood** (probability > 33% but ≤ 66%)
}

Figure 7: (a) Schematic of observed variations of the temperature indicators. [Based on Figure 2.39a] (b) Schematic of observed variations of the hydrological and storm-related indicators. [Based on Figure 2.39b]

Taken together, these trends illustrate a collective picture of a warming world:

· Surface temperature measurements over the land and oceans (with two separate estimates over the latter) have been measured and adjusted independently. All data sets show quite similar upward trends globally, with two major warming periods globally: 1910 to 1945 and since 1976. There is an emerging tendency for global land-surface air temperatures to warm faster than the global ocean-surface temperatures.
· Weather balloon measurements show that lower-tropospheric temperatures have been increasing since 1958, though only slightly since 1979. Since 1979, satellite data are available and show similar trends to balloon data.
· The decrease in the continental diurnal temperature range coincides with increases in cloud amount, precipitation, and increases in total water vapour.
· The nearly worldwide decrease in mountain glacier extent and ice mass is consistent with worldwide surface temperature increases. A few recent exceptions in coastal regions are consistent with atmospheric circulation variations and related precipitation increases.
· The decreases in snow cover and the shortening seasons of lake and river ice relate well to increases in Northern Hemispheric land-surface temperatures.
· The systematic decrease in spring and summer sea-ice extent and thickness in the Arctic is consistent with increases in temperature over most of the adjacent land and ocean.
· Ocean heat content has increased, and global average sea level has risen.
· The increases in total tropospheric water vapour in the last 25 years are qualitatively consistent with increases in tropospheric temperatures and an enhanced hydrologic cycle, resulting in more extreme and heavier precipitation events in many areas with increasing precipitation, e.g., middle and high latitudes of the Northern Hemisphere.

Some important aspects of climate appear not to have changed.

· A few areas of the globe have not warmed in recent decades, mainly over some parts of the Southern Hemisphere oceans and parts of Antarctica.
· No significant trends in Antarctic sea-ice extent are apparent over the period of systematic satellite measurements (since 1978).
· Based on limited data, the observed variations in the intensity and frequency of tropical and extra-tropical cyclones and severe local storms show no clear trends in the last half of the 20th century, although multi-decadal fluctuations are sometimes apparent.

The variations and trends in the examined indicators imply that it is virtually certain that there has been a generally increasing trend in global surface temperature over the 20th century, although short-term and regional deviations from this trend occur.

C. The Forcing Agents That Cause Climate Change

In addition to the past variations and changes in the Earth's climate, observations have also documented the changes that have occurred in agents that can cause climate change. Most notable among these are increases in the atmospheric concentrations of greenhouse gases and aerosols (microscopic airborne particles or droplets) and variations in solar activity, both of which can alter the Earth's radiation budget and hence climate. These observational records of climate-forcing agents are part of the input needed to understand the past climate changes noted in the preceding Section and, very importantly, to predict what climate changes could lie ahead (see Section F).

Like the record of past climate changes, the data sets for forcing agents are of varying length and quality. Direct measurements of solar irradiance exist for only about two decades. The sustained direct monitoring of the atmospheric concentrations of carbon dioxide (CO_2) began about the middle of the 20th century and, in later years, for other long-lived, well-mixed gases such as methane. Palaeo-atmospheric data from ice cores reveal the concentration changes occurring in earlier millennia for some greenhouse gases. In contrast, the time-series measurements for the forcing agents that have relatively short residence times in the atmosphere (e.g., aerosols) are more recent and are far less complete, because they are harder to measure and are spatially heterogeneous. Current data sets show the human influence on atmospheric concentrations of both the long-lived greenhouse gases and short-lived forcing agents during the last part of the past millennium. Figure 8 illustrates the effects of the large growth over the Industrial Era in the anthropogenic emissions of greenhouse gases and sulphur dioxide, the latter being a precursor of aerosols.

A change in the energy available to the global Earth-atmosphere system due to changes in these forcing agents is termed radiative forcing (Wm^{-2}) of the climate system (see Box 1). Defined in this manner, radiative forcing of climate change constitutes an index of the relative global mean impacts on the surface-troposphere system due to different natural and anthropogenic causes. This Section updates the knowledge of the radiative forcing of climate change that has occurred from pre-industrial times to the present. Figure 9 shows the estimated radiative forcings from the beginning of the Industrial Era (1750) to 1999 for the quantifiable natural and anthropogenic forcing agents. Although not included in the figure due to their episodic nature, volcanic eruptions are the source of another important natural forcing. Summaries of the information about each forcing agent follow in the sub-sections below.

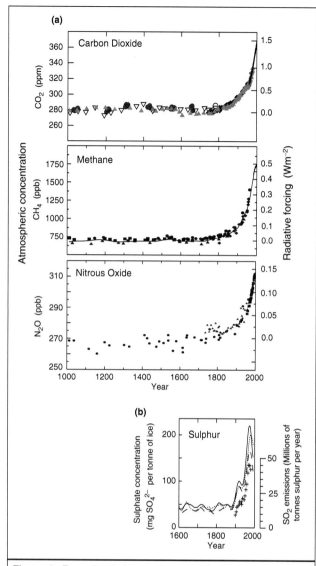

Figure 8: Records of changes in atmospheric composition. (a) Atmospheric concentrations of CO_2, CH_4 and N_2O over the past 1,000 years. Ice core and firn data for several sites in Antarctica and Greenland (shown by different symbols) are supplemented with the data from direct atmospheric samples over the past few decades (shown by the line for CO_2 and incorporated in the curve representing the global average of CH_4). The estimated radiative forcing from these gases is indicated on the right-hand scale. (b) Sulphate concentration in several Greenland ice cores with the episodic effects of volcanic eruptions removed (lines) and total SO_2 emissions from sources in the US and Europe (crosses). [Based on (a) Figure 3.2b (CO_2), Figure 4.1a and b (CH_4) and Figure 4.2 (N_2O) and (b) Figure 5.4a]

The forcing agents included in Figure 9 vary greatly in their form, magnitude and spatial distribution. Some of the greenhouse gases are emitted directly into the atmosphere; some are chemical products from other emissions. Some greenhouse gases have long atmospheric residence times and, as a result, are well-mixed throughout the atmosphere. Others are short-lived and have heterogeneous regional concentrations. Most of the gases originate from both natural and anthropogenic

sources. Lastly, as shown in Figure 9, the radiative forcings of individual agents can be positive (i.e., a tendency to warm the Earth's surface) or negative (i.e., a tendency to cool the Earth's surface).

C.1. Observed Changes in Globally Well-Mixed Greenhouse Gas Concentrations and Radiative Forcing

Over the millennium before the Industrial Era, the atmospheric concentrations of greenhouse gases remained relatively constant. Since then, however, the concentrations of many greenhouse gases have increased directly or indirectly because of human activities.

Table 1 provides examples of several greenhouse gases and summarises their 1750 and 1998 concentrations, their change during the 1990s, and their atmospheric lifetimes. The contribution of a species to radiative forcing of climate change depends on the molecular radiative properties of the gas, the size of the increase in atmospheric concentration, and the residence time of the species in the atmosphere, once emitted. *The latter – the atmospheric residence time of the greenhouse gas – is a highly policy relevant characteristic. Namely, emissions of a greenhouse gas that has a long atmospheric residence time is a quasi-irreversible commitment to sustained radiative forcing over decades, centuries, or millennia, before natural processes can remove the quantities emitted.*

Carbon dioxide (CO_2)

The atmospheric concentration of CO_2 has increased from 280 ppm[5] in 1750 to 367 ppm in 1999 (31%, Table 1). Today's CO_2 concentration has not been exceeded during the past 420,000 years and likely not during the past 20 million years. The rate of increase over the past century is unprecedented, at least during the past 20,000 years (Figure 10). The CO_2 isotopic composition and the observed decrease in Oxygen (O_2) demonstrates that the observed increase in CO_2 is predominately due to the oxidation of organic carbon by fossil-fuel combustion and deforestation. An expanding set of palaeo-atmospheric data from air trapped in ice over hundreds of millennia provide a context for the increase in CO_2 concentrations during the Industrial Era (Figure 10). Compared to the relatively stable CO_2 concentrations (280 ± 10 ppm) of the preceding several thousand years, the increase during the Industrial Era is dramatic. The average rate of increase since 1980 is 0.4%/yr. The increase is a consequence of CO_2 emissions. Most of the emissions during the past 20 years are due to fossil fuel burning,

---

[5] Atmospheric abundances of trace gases are reported here as the mole fraction (molar mixing ratio) of the gas relative to dry air (ppm = 10^{-6}, ppb = 10^{-9}, ppt = 10^{-12}). Atmospheric burden is reported as the total mass of the gas (e.g., Mt = Tg = 10^{12} g). The global carbon cycle is expressed in PgC = GtC.

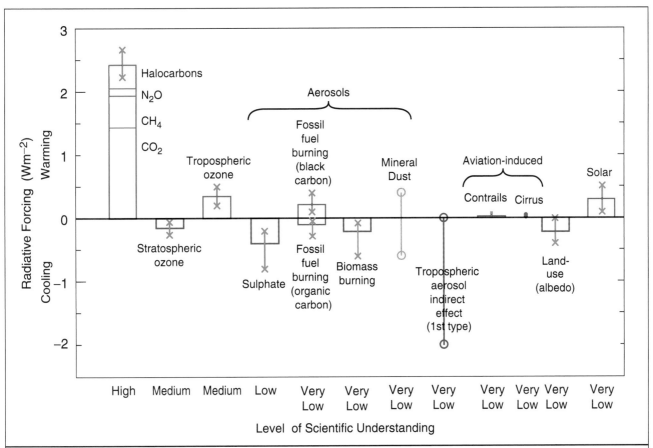

Figure 9: Global, annual-mean radiative forcings (Wm⁻²) due to a number of agents for the period from pre-industrial (1750) to present (late 1990s; about 2000) (numerical values are also listed in Table 6.11 of Chapter 6). For detailed explanations, see Chapter 6.13. The height of the rectangular bar denotes a central or best estimate value, while its absence denotes no best estimate is possible. The vertical line about the rectangular bar with "x" delimiters indicates an estimate of the uncertainty range, for the most part guided by the spread in the published values of the forcing. A vertical line without a rectangular bar and with "o" delimiters denotes a forcing for which no central estimate can be given owing to large uncertainties. The uncertainty range specified here has no statistical basis and therefore differs from the use of the term elsewhere in this document. A "level of scientific understanding" index is accorded to each forcing, with high, medium, low and very low levels, respectively. This represents the subjective judgement about the reliability of the forcing estimate, involving factors such as the assumptions necessary to evaluate the forcing, the degree of knowledge of the physical/chemical mechanisms determining the forcing, and the uncertainties surrounding the quantitative estimate of the forcing (see Table 6.12). The well-mixed greenhouse gases are grouped together into a single rectangular bar with the individual mean contributions due to CO_2, CH_4, N_2O and halocarbons shown (see Tables 6.1 and 6.11). Fossil fuel burning is separated into the "black carbon" and "organic carbon" components with its separate best estimate and range. The sign of the effects due to mineral dust is itself an uncertainty. The indirect forcing due to tropospheric aerosols is poorly understood. The same is true for the forcing due to aviation via its effects on contrails and cirrus clouds. Only the "first" type of indirect effect due to aerosols as applicable in the context of liquid clouds is considered here. The "second" type of effect is conceptually important, but there exists very little confidence in the simulated quantitative estimates. The forcing associated with stratospheric aerosols from volcanic eruptions is highly variable over the period and is not considered for this plot (however, see Figure 6.8). All the forcings shown have distinct spatial and seasonal features (Figure 6.7) such that the global, annual means appearing on this plot do not yield a complete picture of the radiative perturbation. They are only intended to give, in a relative sense, a first-order perspective on a global, annual mean scale and cannot be readily employed to obtain the climate response to the total natural and/or anthropogenic forcings. As in the SAR, it is emphasised that the positive and negative global mean forcings cannot be added up and viewed *a priori* as providing offsets in terms of the complete global climate impact. [Based on Figure 6.6]

the rest (10 to 30%) is predominantly due to land-use change, especially deforestation. As shown in Figure 9, CO_2 is the dominant human-influenced greenhouse gas, with a current radiative forcing of 1.46 Wm⁻², being 60% of the total from the changes in concentrations of all of the long-lived and globally mixed greenhouse gases.

Direct atmospheric measurements of CO_2 concentrations made over the past 40 years show that year to year fluctuations in the rate of increase of atmospheric CO_2 are large. In the 1990s, the annual rates of CO_2 increase in the atmosphere varied from 0.9 to 2.8 ppm/yr, equivalent to 1.9 to 6.0 PgC/yr. Such annual changes can be related statistically to short-term climate variability, which alters the rate at which atmospheric CO_2 is

Table 1: Example of greenhouse gases that are affected by human activities. [Based upon Chapter 3 and Table 4.1]

	CO_2 (Carbon Dioxide)	CH_4 (Methane)	N_2O (Nitrous Oxide)	CFC-11 (Chlorofluoro-carbon-11)	HCF-23 (Hydrofluoro-carbon-23)	CF_4 (Perfluoro-methane)
Pre-industrial concentration	about 280 ppm	about 700 ppb	about 270 ppb	zero	zero	40 ppt
Concentration in 1998	365 ppm	1745 ppb	314 ppb	268 ppt	14 ppt	80 ppt
Rate of concentration change[b]	1.5 ppm/yr[a]	7.0 ppb/yr[a]	0.8 ppb/yr	−1.4 ppt/yr	0.55 ppt/yr	1 ppt/yr
Atmospheric lifetime	5 to 200 yr[c]	12 yr[d]	114 yr[d]	45 yr	260 yr	>50.000 yr

[a] Rate has fluctuated between 0.9 ppm/yr and 2.8 ppm/yr for CO_2 and between 0 and 13 ppb/yr for CH_4 over the period 1990 to 1999.
[b] Rate is calculated over the period 1990 to 1990.
[c] No single lifetime can be defined for CO_2 because of different rates of uptake by different removal processes.
[d] This lifetime has been defined as an "adjustment time" that takes into account the indirect effect of the gas on its own residence time.

Table 2: Global CO_2 budgets (in PgC/yr) based on measurements of atmospheric CO_2 and O_2. Positive values are fluxes to the atmosphere; negative values represent uptake from the atmosphere. [Based upon Tables 3.1 and 3.3]

	SAR[a,b]	This report[a]	
	1980 to 1989	1980 to 1989	1990 to 1999
Atmospheric increase	3.3 ± 0.1	3.3 ± 0.1	3.2 ± 0.1
Emissions (fossil fuel, cement)[c]	5.5 ± 0.3	5.4 ± 0.3	6.3 ± 0.4
Ocean-atmosphere flux	−2.0 ± 0.5	−1.9 ± 0.6	−1.7 ± 0.5
Land-atmosphere flux[d]	−0.2 ± 0.6	−0.2 ± 0.7	−1.4 ± 0.7

[a] Note that the uncertainties cited in this table are ±1 standard error. The uncertainties cited in the SAR were ±1.6 standard error (i.e., approximately 90% confidence interval). Uncertainties cited from the SAR were adjusted to ±1 standard error. Error bars denote uncertainty, not interannual variability, which is substantially greater.
[b] Previous IPCC carbon budgets calculated ocean uptake from models and the land-atmosphere flux was inferred by difference.
[c] The fossil fuel emissions term for the 1980s has been revised slightly downward since the SAR.
[d] The land-atmosphere flux represents the balance of a positive term due to land-use change and a residual terrestrial sink. The two terms cannot be separated on the basis of current atmospheric measurements. Using independent analyses to estimate the land-use change component for 1980 to 1989, the residual terrestrial sink can be inferred as follows: Land-use change 1.7 PgC/yr (0.6 to 2.5); Residual terrestrial sink −1.9 PgC/yr (−3.8 to 0.3). Comparable data for the 1990s are not yet available.

taken up and released by the oceans and land. The highest rates of increase in atmospheric CO_2 have typically been in strong El Niño years (Box 4). These higher rates of increase can be plausibly explained by reduced terrestrial uptake (or terrestrial outgassing) of CO_2 during El Niño years, overwhelming the tendency of the ocean to take up more CO_2 than usual.

Partitioning of anthropogenic CO_2 between atmospheric increases and land and ocean uptake for the past two decades can now be calculated from atmospheric observations. Table 2 presents a global CO_2 budget for the 1980s (which proves to be similar to the one constructed with the help of ocean model results in the SAR) and for the 1990s. Measurements of the decrease in atmospheric oxygen (O_2) as well as the increase in CO_2 were used in the construction of these new budgets. Results from this approach are consistent with other analyses based on the isotopic composition of atmospheric CO_2 and with independent estimates based on measurements of CO_2 and $^{13}CO_2$ in seawater. The 1990s

budget is based on newly available measurements and updates the budget for 1989 to 1998 derived using SAR methodology for the IPCC Special Report on Land Use, Land-Use Change and Forestry (2000). The terrestrial biosphere as a whole has gained carbon during the 1980s and 1990s; i.e., the CO_2 released by land-use change (mainly tropical deforestation) was more than compensated by other terrestrial sinks, which are likely located in both the northern extra-tropics and in the tropics. There remain large uncertainties associated with estimating the CO_2 release due to land-use change (and, therefore, with the magnitude of the residual terrestrial sink).

Process-based modelling (terrestrial and ocean carbon models) has allowed preliminary quantification of mechanisms in the global carbon cycle. Terrestrial model results indicate that enhanced plant growth due to higher CO_2 (CO_2 fertilisation) and anthropogenic nitrogen deposition contribute significantly to CO_2 uptake, i.e., are potentially responsible for the residual

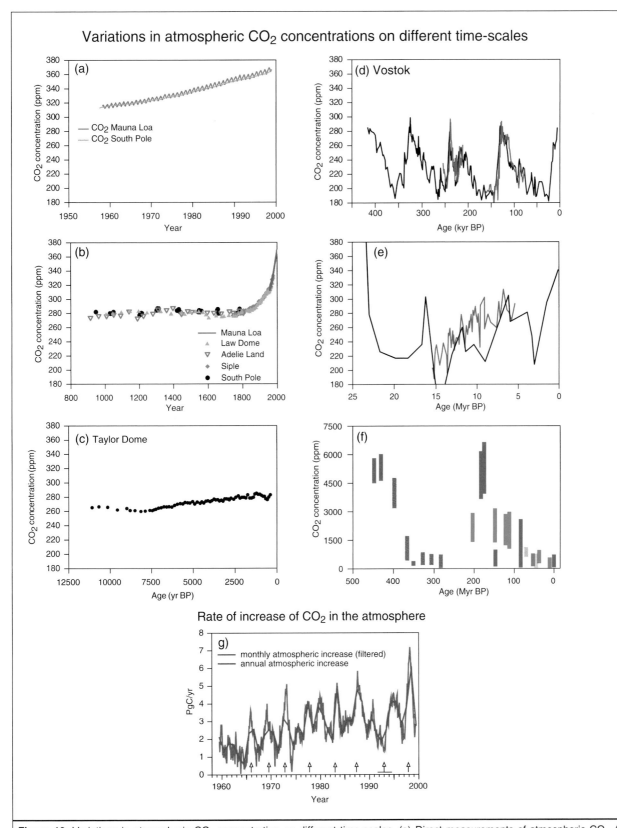

Figure 10: Variations in atmospheric CO_2 concentration on different time-scales. (a) Direct measurements of atmospheric CO_2. (b) CO_2 concentration in Antarctic ice cores for the past millenium. Recent atmospheric measurements (Mauna Loa) are shown for comparison. (c) CO_2 concentration in the Taylor Dome Antarctic ice core. (d) CO_2 concentration in the Vostok Antarctic ice core. (Different colours represent results from different studies.) (e to f) Geochemically inferred CO_2 concentrations. (Coloured bars and lines represent different published studies) (g) Annual atmospheric increases in CO_2. Monthly atmospheric increases have been filtered to remove the seasonal cycle. Vertical arrows denote El Niño events. A horizontal line defines the extended El Niño of 1991 to 1994. [Based on Figures 3.2 and 3.3]

terrestrial sink described above, along with other proposed mechanisms, such as changes in land-management practices. The modelled effects of climate change during the 1980s on the terrestrial sink are small and of uncertain sign.

Methane (CH_4)

Atmospheric methane (CH_4) concentrations have increased by about 150% (1,060 ppb) since 1750. The present CH_4 concentration has not been exceeded during the past 420,000 years. Methane (CH_4) is a greenhouse gas with both natural (e.g., wetlands) and human-influenced sources (e.g., agriculture, natural gas activities, and landfills). Slightly more than half of current CH_4 emissions are anthropogenic. It is removed from the atmosphere by chemical reactions. As Figure 11 shows, systematic, globally representative measurements of the concentration of CH_4 in the atmosphere have been made since 1983, and the record of atmospheric concentrations has been extended to earlier times from air extracted from ice cores and firn layers. The current direct radiative forcing of 0.48 Wm^{-2} from CH_4 is 20% of the total from all of the long-lived and globally mixed greenhouse gases (see Figure 9).

The atmospheric abundance of CH_4 continues to increase, from about 1,610 ppb in 1983 to 1,745 ppb in 1998, but the observed annual increase has declined during this period. The increase was highly variable during the 1990s; it was near zero in 1992 and as large as 13 ppb during 1998. There is no clear quantitative explanation for this variability. Since the SAR, quantification of certain anthropogenic sources of CH_4, such as that from rice production, has improved.

The rate of increase in atmospheric CH_4 is due to a small imbalance between poorly characterised sources and sinks, which makes the prediction of future concentrations problematic. Although the major contributors to the global CH_4 budget likely have been identified, most of them are quite uncertain quantitatively because of the difficulty in assessing emission rates of highly variable biospheric sources. The limitations of poorly quantified and characterised CH_4 source strengths inhibit the prediction of future CH_4 atmospheric concentrations (and hence its contribution to radiative forcing) for any given anthropogenic emission scenario, particularly since both natural emissions and the removal of CH_4 can be influenced substantially by climate change.

Nitrous oxide (N_2O)

The atmospheric concentration of nitrous oxide (N_2O) has steadily increased during the Industrial Era and is now 16% (46 ppb) larger than in 1750. The present N_2O concentration has not been exceeded during at least the past thousand years. Nitrous oxide is another greenhouse gas with both natural and anthropogenic sources, and it is removed from the atmosphere by chemical reactions. Atmospheric concentrations of N_2O

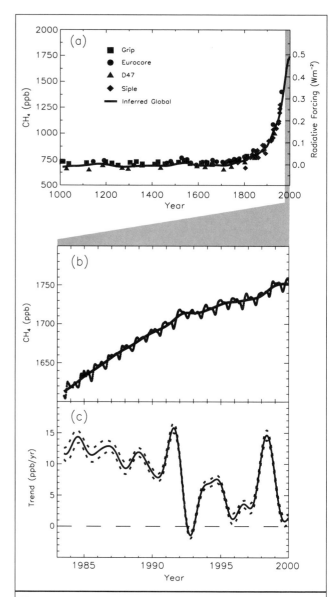

Figure 11: (a) Change in CH_4 abundance (mole fraction, in ppb = 10^{-9}) determined from ice cores, firn, and whole air samples plotted for the last 1,000 years. Radiative forcing, approximated by a linear scale since the pre-industrial era, is plotted on the right axis. (b) Globally averaged CH_4 (monthly varying) and deseasonalised CH_4 (smooth line) abundance plotted for 1983 to 1999. (c) Instantaneous annual growth rate (ppb/yr) in global atmospheric CH_4 abundance from 1983 through 1999 calculated as the derivative of the deseasonalised trend curve above. Uncertainties (dotted lines) are ±1 standard deviation. [Based on Figure 4.1]

continue to increase at a rate of 0.25%/yr (1980 to 1998). Significant interannual variations in the upward trend of N_2O concentrations are observed, e.g., a 50% reduction in annual growth rate from 1991 to 1993. Suggested causes are several-fold: a decrease in use of nitrogen-based fertiliser, lower biogenic emissions, and larger stratospheric losses due to volcanic-induced circulation changes. Since 1993, the growth of N_2O concentrations has returned to rates closer to those observed during the 1980s. While this observed multi-year variance has provided some potential insight into what processes control the behaviour of atmospheric N₂O, the multi-year trends of this greenhouse gas remain largely unexplained.

The global budget of nitrous oxide is in better balance than in the SAR, but uncertainties in the emissions from individual sources are still quite large. Natural sources of N_2O are estimated to be approximately 10 TgN/yr (1990), with soils being about 65% of the sources and oceans about 30%. New, higher estimates of the emissions from anthropogenic sources (agriculture, biomass burning, industrial activities, and livestock management) of approximately 7 TgN/yr have brought the source/sink estimates closer in balance, compared with the SAR. However, the predictive understanding associated with this significant, long-lived greenhouse gas has not improved significantly since the last assessment. The radiative forcing is estimated at 0.15 Wm⁻², which is 6% of the total from all of the long-lived and globally mixed greenhouse gases (see Figure 9).

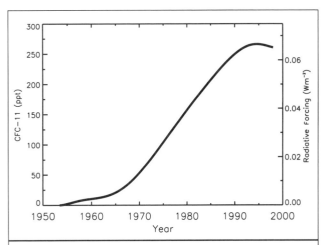

Figure 12: Global mean CFC-11 (CFCl₃) tropospheric abundance (ppt) from 1950 to 1998 based on smoothed measurements and emission models. CFC-11's radiative forcing is shown on the right axis. [Based on Figure 4.6]

Halocarbons and related compounds

The atmospheric concentrations of many of those gases that are both ozone-depleting and greenhouse gases are either decreasing (CFC-11, CFC-113, CH₃CCl₃ and CCl₄) or increasing more slowly (CFC-12) in response to reduced emissions under the regulations of the Montreal Protocol and its Amendments. Many of these halocarbons are also radiatively effective, long-lived greenhouse gases. Halocarbons are carbon compounds that contain fluorine, chlorine, bromine or iodine. For most of these compounds, human activities are the sole source. Halocarbons that contain chlorine (e.g., chlorofluorocarbons – CFCs) and bromine (e.g., halons) cause depletion of the stratospheric ozone layer and are controlled under the Montreal Protocol. The combined tropospheric abundance of ozone-depleting gases peaked in 1994 and is slowly declining. The atmospheric abundances of some of the major greenhouse halocarbons have peaked, as shown for CFC-11 in Figure 12. The concentrations of CFCs and chlorocarbons in the troposphere are consistent with reported emissions. Halocarbons contribute a radiative forcing of 0.34 Wm⁻², which is 14% of the radiative forcing from all of the globally mixed greenhouse gases (Figure 9).

The observed atmospheric concentrations of the substitutes for the CFCs are increasing, and some of these compounds are greenhouse gases. The abundances of the hydrochlorofluoro-carbons (HCFCs) and hydrofluorocarbons (HFCs) are increasing as a result of continuation of earlier uses and of their use as substitutes for the CFCs. For example, the concentration of HFC-23 has increased by more than a factor of three between 1978 and 1995. Because current concentrations are relatively low, the present contribution of HFCs to radiative forcing is relatively small. The present contribution of HCFCs to radiative forcing is also relatively small, and future emissions of these gases are limited by the Montreal Protocol.

The perfluorocarbons (PFCs, e.g., CF₄ and C₂F₆) and sulphur hexafluoride (SF₆) have anthropogenic sources, have extremely long atmospheric residence times, and are strong absorbers of infrared radiation. Therefore, these compounds, even with relatively small emissions, have the potential to influence climate far into the future. Perfluoromethane (CF₄) resides in the atmosphere for at least 50,000 years. It has a natural background; however, current anthropogenic emissions exceed natural ones by a factor of 1,000 or more and are responsible for the observed increase. Sulphur hexafluoride (SF₆) is 22,200 times more effective a greenhouse gas than CO₂ on a per-kg basis. The current atmospheric concentrations are very small (4.2 ppt), but have a significant growth rate (0.24 ppt/yr). There is good agreement between the observed atmospheric growth rate of SF₆ and the emissions based on revised sales and storage data.

C.2. Observed Changes in Other Radiatively Important Gases

Atmospheric ozone (O₃)

Ozone (O₃) is an important greenhouse gas present in both the stratosphere and troposphere. The role of ozone in the atmospheric radiation budget is strongly dependent on the altitude at which changes in ozone concentrations occur. The changes in ozone concentrations are also spatially variable. Further, ozone is not a directly emitted species, but rather it is formed in the atmosphere from photochemical processes involving both natural and human-influenced precursor species. Once formed, the residence time of ozone in the atmosphere is relatively short, varying from weeks to months. As a result, estimation of ozone's radiative role is more complex and much less certain than for the above long-lived and globally well-mixed greenhouse gases.

The observed losses of stratospheric ozone layer over the past two decades have caused a negative forcing of 0.15 ± 0.1 Wm⁻² (i.e., a tendency toward cooling) of the surface troposphere system. It was reported in Climate Change 1992: The Supplementary Report to the IPCC Scientific Assessment, that depletion of the ozone layer by anthropogenic halocarbons introduces a negative radiative forcing. The estimate shown in Figure 9 is slightly larger in magnitude than that given in the SAR, owing to the ozone depletion that has continued over the past five years, and it is more certain as a result of an increased number of modelling studies. Studies with General Circulation Models indicate that, despite the inhomogeneity in ozone loss (i.e., lower stratosphere at high latitudes), such a negative forcing does relate to a surface temperature decrease in proportion to the magnitude of the negative forcing. Therefore, this negative forcing over the past two decades has offset some of the positive forcing that is occurring from the long-lived and globally well-mixed greenhouse gases (Figure 9). A major source of uncertainty in the estimation of the negative forcing is due to incomplete knowledge of ozone depletion near the tropopause. Model calculations indicate that increased penetration of ultraviolet radiation to the troposphere, as a result of stratospheric ozone depletion, leads to enhanced removal rates of gases like CH_4, thus amplifying the negative forcing due to ozone depletion. As the ozone layer recovers in future decades because of the effects of the Montreal Protocol, relative to the present, future radiative forcing associated with stratospheric ozone is projected to become positive.

The global average radiative forcing due to increases in tropospheric ozone since pre-industrial times is estimated to have enhanced the anthropogenic greenhouse gas forcing by 0.35 ± 0.2 Wm⁻². This makes tropospheric ozone the third most important greenhouse gas after CO_2 and CH_4. Ozone is formed by photochemical reactions and its future change will be determined by, among other things, emissions of CH_4 and pollutants (as noted below). Ozone concentrations respond relatively quickly to changes in the emissions of pollutants. On the basis of limited observations and several modelling studies, tropospheric ozone is estimated to have increased by about 35% since the Pre-industrial Era, with some regions experiencing larger and some with smaller increases. There have been few observed increases in ozone concentrations in the global troposphere since the mid-1980s at most of the few remote locations where it is regularly measured. The lack of observed increase over North America and Europe is related to the lack of a sustained increase in ozone-precursor emissions from those continents. However, some Asian stations indicate a possible rise in tropospheric ozone, which could be related to the increase in East Asian emissions. As a result of more modelling studies than before, there is now an increased confidence in the estimates of tropospheric ozone forcing. The confidence, however, is still much less than that for the well-mixed greenhouse gases, but more so than that for aerosol forcing. Uncertainties arise because of limited information on pre-industrial ozone

distributions and limited information to evaluate modelled global trends in the modern era (i.e., post-1960).

Gases with only indirect radiative influences

Several chemically reactive gases, including reactive nitrogen species (NO_x), carbon monoxide (CO), and the volatile organic compounds (VOCs), control, in part, the oxidising capacity of the troposphere, as well as the abundance of ozone. These pollutants act as indirect greenhouse gases through their influence not only on ozone, but also on the lifetimes of CH_4 and other greenhouse gases. The emissions of NO_x and CO are dominated by human activities.

Carbon monoxide is identified as an important indirect greenhouse gas. Model calculations indicate that emission of 100 Mt of CO is equivalent in terms of greenhouse gas perturbations to the emission of about 5 Mt of CH_4. The abundance of CO in the Northern Hemisphere is about twice that in the Southern Hemisphere and has increased in the second half of the 20th century along with industrialisation and population.

The reactive nitrogen species NO and NO_2, (whose sum is denoted NO_x), are key compounds in the chemistry of the troposphere, but their overall radiative impact remains difficult to quantify. The importance of NO_x in the radiation budget is because increases in NO_x concentrations perturb several greenhouse gases; for example, decreases in methane and the HFCs and increases in tropospheric ozone. Deposition of the reaction products of NO_x fertilises the biosphere, thereby decreasing atmospheric CO_2. While difficult to quantify, increases in NO_x that are projected to the year 2100 would cause significant changes in greenhouse gases.

C.3. Observed and Modelled Changes in Aerosols

Aerosols (very small airborne particles and droplets) are known to influence significantly the radiative budget of the Earth/atmosphere. Aerosol radiative effects occur in two distinct ways: (i) the direct effect, whereby aerosols themselves scatter and absorb solar and thermal infrared radiation, and (ii) the indirect effect, whereby aerosols modify the microphysical and hence the radiative properties and amount of clouds. Aerosols are produced by a variety of processes, both natural (including dust storms and volcanic activity) and anthropogenic (including fossil fuel and biomass burning). The atmospheric concentrations of tropospheric aerosols are thought to have increased over recent years due to increased anthropogenic emissions of particles and their precursor gases, hence giving rise to radiative forcing. Most aerosols are found in the lower troposphere (below a few kilometres), but the radiative effect of many aerosols is sensitive to the vertical distribution. Aerosols undergo chemical and physical changes while in the atmosphere, notably within clouds, and are removed largely and relatively rapidly by precipitation (typically within a week). Because of this short residence time and the inhomogeneity of sources, aerosols are distributed inhomogeneously in the troposphere, with maxima near the sources. The radiative forcing due to aerosols depends not only on these spatial distributions, but also on the size, shape, and chemical composition of the particles and various aspects (e.g., cloud formation) of the hydrological cycle as well. As a result of all of these factors, obtaining accurate estimates of this forcing has been very challenging, from both the observational and theoretical standpoints.

*Nevertheless, substantial progress has been achieved in better defining the **direct effect** of a wider set of different aerosols.* The SAR considered the direct effects of only three anthropogenic aerosol species: sulphate aerosols, biomass-burning aerosols, and fossil fuel black carbon (or soot). Observations have now shown the importance of organic materials in both fossil fuel carbon aerosols and biomass-burning carbon aerosols. Since the SAR, the inclusion of estimates for the abundance of fossil fuel organic carbon aerosols has led to an increase in the predicted total optical depth (and consequent negative forcing) associated with industrial aerosols. Advances in observations and in aerosol and radiative models have allowed quantitative estimates of these separate components, as well as an estimate for the range of radiative forcing associated with mineral dust, as shown in Figure 9. Direct radiative forcing is estimated to be -0.4 Wm^{-2} for sulphate, -0.2 Wm^{-2} for biomass-burning aerosols, -0.1 Wm^{-2} for fossil fuel organic carbon, and $+0.2$ Wm^{-2} for fossil fuel black carbon aerosols. Uncertainties remain relatively large, however. These arise from difficulties in determining the concentration and radiative characteristics of atmospheric aerosols and the fraction of the aerosols that are of anthropogenic origin, particularly the knowledge of the sources of carbonaceous aerosols. This leads to considerable differences (i.e., factor of two to three range) in the burden and substantial differences in the vertical distribution (factor of ten). Anthropogenic dust aerosol is also poorly quantified. Satellite observations, combined with model calculations, are enabling the identification of the spatial signature of the total aerosol radiative effect in clear skies; however, the quantitative amount is still uncertain.

*Estimates of the **indirect radiative** forcing by anthropogenic aerosols remain problematic, although observational evidence points to a negative aerosol-induced indirect forcing in warm clouds.* Two different approaches exist for estimating the indirect effect of aerosols: empirical methods and mechanistic methods. The former have been applied to estimate the effects of industrial aerosols, while the latter have been applied to estimate the effects of sulphate, fossil fuel carbonaceous aerosols, and biomass aerosols. In addition, models for the indirect effect have been used to estimate the effects of the initial change in droplet size and concentrations (a first indirect effect), as well as the effects of the subsequent change in

precipitation efficiency (a second indirect effect). The studies represented in Figure 9 provide an expert judgement for the range of the first of these; the range is now slightly wider than in the SAR; the radiative perturbation associated with the second indirect effect is of the same sign and could be of similar magnitude compared to the first effect.

The indirect radiative effect of aerosols is now understood to also encompass effects on ice and mixed-phase clouds, but the magnitude of any such indirect effect is not known, although it is likely to be positive. It is not possible to estimate the number of anthropogenic ice nuclei at the present time. Except at cold temperatures (below $-45°$C) where homogeneous nucleation is expected to dominate, the mechanisms of ice formation in these clouds are not yet known.

C.4. Observed Changes in Other Anthropogenic Forcing Agents

Land-use (albedo) change

Changes in land use, deforestation being the major factor, appear to have produced a negative radiative forcing of -0.2 ± 0.2 Wm^{-2} (Figure 8). The largest effect is estimated to be at the high latitudes. This is because deforestation has caused snow-covered forests with relatively low albedo to be replaced with open, snow-covered areas with higher albedo. The estimate given above is based on simulations in which pre-industrial vegetation is replaced by current land-use patterns. However, the level of understanding is very low for this forcing, and there have been far fewer investigations of this forcing compared to investigations of other factors considered in this report.

C.5. Observed and Modelled Changes in Solar and Volcanic Activity

Radiative forcing of the climate system due to solar irradiance change is estimated to be 0.3 ± 0.2 Wm^{-2} for the period 1750 to the present (Figure 8), and most of the change is estimated to have occurred during the first half of the 20th century. The fundamental source of all energy in the Earth's climate system is radiation from the Sun. Therefore, variation in solar output is a radiative forcing agent. The absolute value of the spectrally integrated total solar irradiance (TSI) incident on the Earth is not known to better than about 4 Wm^{-2}, but satellite observations since the late 1970s show relative variations over the past two solar 11-year activity cycles of about 0.1%, which is equivalent to a variation in radiative forcing of about 0.2 Wm^{-2}. Prior to these satellite observations, reliable direct measurements of solar irradiance are not available. Variations over longer periods may have been larger, but the techniques used to reconstruct historical values of TSI from proxy observations (e.g., sunspots) have not been adequately verified. Solar variation varies more substantially in the ultraviolet region, and studies with climate models suggest that inclusion of spectrally resolved solar irradiance variations and solar-induced stratospheric ozone changes may improve the realism of model simulations of the impact of solar variability on climate. Other mechanisms for the amplification of solar effects on climate have been proposed, but do not have a rigorous theoretical or observational basis.

Stratospheric aerosols from explosive volcanic eruptions lead to negative forcing that lasts a few years. Several explosive eruptions occurred in the periods 1880 to 1920 and 1960 to 1991, and no explosive eruptions since 1991. Enhanced stratospheric aerosol content due to volcanic eruptions, together with the small solar irradiance variations, result in a net negative natural radiative forcing over the past two, and possibly even the past four, decades.

C.6. Global Warming Potentials

Radiative forcings and Global Warming Potentials (GWPs) are presented in Table 3 for an expanded set of gases. GWPs are a measure of the relative radiative effect of a given substance compared to CO_2, integrated over a chosen time horizon. New categories of gases in Table 3 include fluorinated organic molecules, many of which are ethers that are proposed as halocarbon substitutes. Some of the GWPs have larger uncertainties than that of others, particularly for those gases where detailed laboratory data on lifetimes are not yet available. The direct GWPs have been calculated relative to CO_2 using an improved calculation of the CO_2 radiative forcing, the SAR response function for a CO_2 pulse, and new values for the radiative forcing and lifetimes for a number of halocarbons. Indirect GWPs, resulting from indirect radiative forcing effects, are also estimated for some new gases, including carbon monoxide. The direct GWPs for those species whose lifetimes are well characterised are estimated to be accurate within ±35%, but the indirect GWPs are less certain.

D. The Simulation of the Climate System and its Changes

The preceding two Sections reported on the climate from the distant past to the present day through the observations of climate variables and the forcing agents that cause climate to change. This Section bridges to the climate of the future by describing the only tool that provides quantitative estimates of future climate changes, namely, numerical models. The basic understanding of the energy balance of the Earth system means that quite simple models can provide a broad quantitative estimate of some globally averaged variables, but more accurate estimates of feedbacks and of regional detail can only come from more elaborate climate models. The complexity of the processes in the climate system prevents the use of extrapolation of past trends or statistical and other purely empirical techniques for projections. Climate models can be used to simulate the climate responses to different input scenarios of future forcing agents

(Section F). Similarly, projection of the fate of emitted CO_2 (i.e., the relative sequestration into the various reservoirs) and other greenhouse gases requires an understanding of the biogeochemical processes involved and incorporating these into a numerical carbon cycle model.

A climate model is a simplified mathematical representation of the Earth's climate system (see Box 3). The degree to which the model can simulate the responses of the climate system hinges to a very large degree on the level of understanding of the physical, geophysical, chemical and biological processes that govern the climate system. Since the SAR, researchers have made substantial improvements in the simulation of the Earth's climate system with models. First, the current understanding of some of the most important processes that govern the climate system and how well they are represented in present climate models are summarised here. Then, this Section presents an assessment of the overall ability of present models to make useful projections of future climate.

D.1. Climate Processes and Feedbacks

Processes in the climate system determine the natural variability of the climate system and its response to perturbations, such as the increase in the atmospheric concentrations of greenhouse gases. Many basic climate processes of importance are well-known and modelled exceedingly well. Feedback processes amplify (a positive feedback) or reduce (a negative feedback) changes in response to an initial perturbation and hence are very important for accurate simulation of the evolution of climate.

Water vapour

A major feedback accounting for the large warming predicted by climate models in response to an increase in CO_2 is the increase in atmospheric water vapour. An increase in the temperature of the atmosphere increases its water-holding capacity; however, since most of the atmosphere is undersaturated, this does not automatically mean that water vapour, itself, must increase. Within the boundary layer (roughly the lowest 1 to 2 km of the atmosphere), water vapour increases with increasing temperature. In the free troposphere above the boundary layer, where the water vapour greenhouse effect is most important, the situation is harder to quantify. Water vapour feedback, as derived from current models, approximately doubles the warming from what it would be for fixed water vapour. Since the SAR, major improvements have occurred in the treatment of water vapour in models, although detrainment of moisture from clouds remains quite uncertain and discrepancies exist between model water vapour distributions and those observed. Models are capable of simulating the moist and very dry regions observed in the tropics and sub-tropics and how they evolve with the seasons and from year to year. While reassuring, this does not provide a check of the feedbacks, although the balance of evidence favours a positive clear-sky water vapour

Table 3: Direct Global Warming Potentials (GWPs) relative to carbon dioxide (for gases for which the lifetimes have been adequately characterised). GWPs are an index for estimating relative global warming contribution due to atmospheric emission of a kg of a particular greenhouse gas compared to emission of a kg of carbon dioxide. GWPs calculated for different time horizons show the effects of atmospheric lifetimes of the different gases. [Based upon Table 6.7]

Gas		Lifetime (years)	Global Warming Potential (Time Horizon in years)		
			20 yrs	*100 yrs*	*500 yrs*
Carbon dioxide	CO_2		1	1	1
Methane[a]	CH_4	12.0^b	62	23	7
Nitrous oxide	N_2O	114^b	275	296	156
Hydrofluorocarbons					
HFC-23	CHF_3	260	9400	12000	10000
HFC-32	CH_2F_2	5.0	1800	550	170
HFC-41	CH_3F	2.6	330	97	30
HFC-125	CHF_2CF_3	29	5900	3400	1100
HFC-134	CHF_2CHF_2	9.6	3200	1100	330
HFC-134a	CH_2FCF_3	13.8	3300	1300	400
HFC-143	CHF_2CH_2F	3.4	1100	330	100
HFC-143a	CF_3CH_3	52	5500	4300	1600
HFC-152	CH_2FCH_2F	0.5	140	43	13
HFC-152a	CH_3CHF_2	1.4	410	120	37
HFC-161	CH_3CH_2F	0.3	40	12	4
HFC-227ea	CF_3CHFCF_3	33	5600	3500	1100
HFC-236cb	$CH_2FCF_2CF_3$	13.2	3300	1300	390
HFC-236ea	CHF_2CHFCF_3	10	3600	1200	390
HFC-236fa	$CF_3CH_2CF_3$	220	7500	9400	7100
HFC-245ca	$CH_2FCF_2CHF_2$	5.9	2100	640	200
HFC-245fa	$CHF_2CH_2CF_3$	7.2	3000	950	300
HFC-365mfc	$CF_3CH_2CF_2CH_3$	9.9	2600	890	280
HFC-43-10mee	$CF_3CHFCHFCF_2CF_3$	15	3700	1500	470
Fully fluorinated species					
SF_6		3200	15100	22200	32400
CF_4		50000	3900	5700	8900
C_2F_6		10000	8000	11900	18000
C_3F_8		2600	5900	8600	12400
C_4F_{10}		2600	5900	8600	12400
$c\text{-}C_4F_8$		3200	6800	10000	14500
C_5F_{12}		4100	6000	8900	13200
C_6F_{14}		3200	6100	9000	13200
Ethers and Halogenated Ethers					
CH_3OCH_3		0.015	1	1	<<1
HFE-125	CF_3OCHF_2	150	12900	14900	9200
HFE-134	CHF_2OCHF_2	26.2	10500	6100	2000
HFE-143a	CH_3OCF_3	4.4	2500	750	230
HCFE-235da2	$CF_3CHClOCHF_2$	2.6	1100	340	110
HFE-245fa2	$CF_3CH_2OCHF_2$	4.4	1900	570	180
HFE-254cb2	$CHF_2CF_2OCH_3$	0.22	99	30	9
HFE-7100	$C_4F_9OCH_3$	5.0	1300	390	120
HFE-7200	$C_4F_9OC_2H_5$	0.77	190	55	17
H-Galden 1040x	$CHF_2OCF_2OC_2F_4OCHF_2$	6.3	5900	1800	560
HG-10	$CHF_2OCF_2OCHF_2$	12.1	7500	2700	850
HG-01	$CHF_2OCF_2CF_2OCHF_2$	6.2	4700	1500	450

[a] The methane GWPs include an indirect contribution from stratospheric H_2O and O_3 production.
[b] The values for methane and nitrous oxide are adjustment times, which incorporate the indirect effects of emission of each gas on its own lifetime.

Box 3. Climate Models: How are they built and how are they applied?

Comprehensive climate models are based on physical laws represented by mathematical equations that are solved using a three-dimensional grid over the globe. For climate simulation, the major components of the climate system must be represented in sub-models (atmosphere, ocean, land surface, cryosphere and biosphere), along with the processes that go on within and between them. Most results in this report are derived from the results of models, which include some representation of all these components. Global climate models in which the atmosphere and ocean components have been coupled together are also known as Atmosphere-Ocean General Circulation Models (AOGCMs). In the atmospheric module, for example, equations are solved that describe the large-scale evolution of momentum, heat and moisture. Similar equations are solved for the ocean. Currently, the resolution of the atmospheric part of a typical model is about 250 km in the horizontal and about 1 km in the vertical above the boundary layer. The resolution of a typical ocean model is about 200 to 400 m in the vertical, with a horizontal resolution of about 125 to 250 km. Equations are typically solved for every half hour of a model integration. Many physical processes, such as those related to clouds or ocean convection, take place on much smaller spatial scales than the model grid and therefore cannot be modelled and resolved explicitly. Their average effects are approximately included in a simple way by taking advantage of physically based relationships with the larger-scale variables. This technique is known as parametrization.

In order to make quantitative projections of future climate change, it is necessary to use climate models that simulate all the important processes governing the future evolution of the climate. Climate models have developed over the past few decades as computing power has increased. During that time, models of the main components, atmosphere, land, ocean and sea ice have been developed separately and then gradually integrated. This coupling of the various components is a difficult process. Most recently, sulphur cycle components have been incorporated to represent the emissions of sulphur and how they are oxidised to form aerosol particles. Currently in progress, in a few models, is the coupling of the land carbon cycle and the ocean carbon cycle. The atmospheric chemistry component currently is modelled outside the main climate model. The ultimate aim is, of course, to model as much as possible of the whole of the Earth's climate system so that all the components can interact and, thus, the predictions of climate change will continuously take into account the effect of feedbacks among components. The Figure above shows the past, present and possible future evolution of climate models.

Some models offset errors and surface flux imbalances through "flux adjustments", which are empirically determined systematic adjustments at the atmosphere-ocean interface held fixed in time in order to bring the simulated climate closer to the observed state. A strategy has been designed for carrying out climate experiments that removes much of the effects of some model errors on results. What is often done is that first a "control" climate simulation is run with the model. Then, the climate change experiment simulation is run, for example, with increased CO_2 in the model atmosphere. Finally, the difference is taken to provide an estimate of the change in climate due to the perturbation. The differencing technique removes most of the effects of any artificial adjustments in the model, as well as systematic errors that are common to both runs. However, a comparison of different model results makes it apparent that the nature of some errors still influences the outcome.

Many aspects of the Earth's climate system are chaotic – its evolution is sensitive to small perturbations in initial conditions. This sensitivity limits predictability of the detailed evolution of weather to about two weeks. However, predictability of climate is not so limited because of the systematic influences on the atmosphere of the more slowly varying components of the climate system. Nevertheless, to be able to make reliable forecasts in the presence of both initial condition and model uncertainty, it is desirable to repeat the prediction many times

feedback of the magnitude comparable to that found in simulations.

Clouds

As has been the case since the first IPCC Assessment Report in 1990, probably the greatest uncertainty in future projections of climate arises from clouds and their interactions with radiation. Clouds can both absorb and reflect solar radiation (thereby cooling the surface) and absorb and emit long wave radiation (thereby warming the surface). The competition between these effects depends on cloud height, thickness and radiative properties. The radiative properties and evolution of clouds depend on the distribution of atmospheric water vapour, water drops, ice particles, atmospheric aerosols and cloud thickness. The physical basis of cloud parametrizations is greatly improved in models through inclusion of bulk representation of cloud microphysical properties in a cloud water budget equation, although considerable uncertainty remains. Clouds represent a significant source of potential error in climate simulations. The possibility that models underestimate systematically solar absorption in clouds remains a controversial

The Development of Climate models, Past, Present and Future

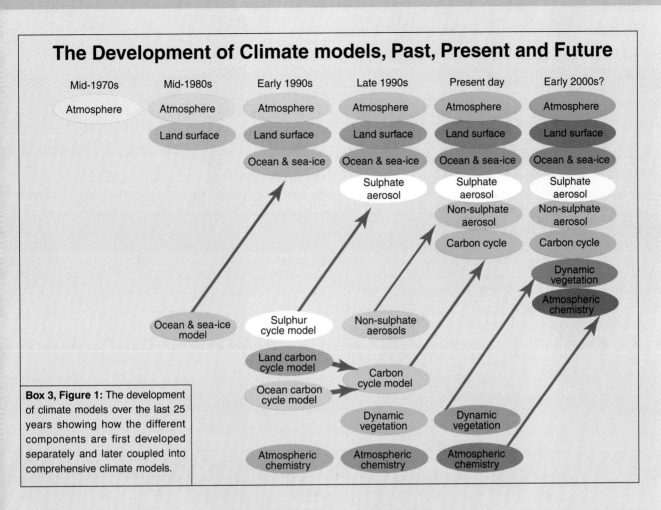

Box 3, Figure 1: The development of climate models over the last 25 years showing how the different components are first developed separately and later coupled into comprehensive climate models.

from different perturbed initial states and using different global models. These ensembles are the basis of probability forecasts of the climate state.

Comprehensive AOGCMs are very complex and take large computer resources to run. To explore different scenarios of emissions of greenhouse gases and the effects of assumptions or approximations in parameters in the model more thoroughly, simpler models are also widely used. The simplifications may include coarser resolution and simplified dynamics and physical processes. Together, simple, intermediate, and comprehensive models form a "hierarchy of climate models", all of which are necessary to explore choices made in parametrizations and assess the robustness of climate changes.

matter. The sign of the net cloud feedback is still a matter of uncertainty, and the various models exhibit a large spread. Further uncertainties arise from precipitation processes and the difficulty in correctly simulating the diurnal cycle and precipitation amounts and frequencies.

Stratosphere

There has been a growing appreciation of the importance of the stratosphere in the climate system because of changes in its structure and recognition of the vital role of both radiative and dynamical processes. The vertical profile of temperature change in the atmosphere, including the stratosphere, is an important indicator in detection and attribution studies. Most of the observed decreases in lower-stratospheric temperatures have been due to ozone decreases, of which the Antarctic "ozone hole" is a part, rather than increased CO_2 concentrations. Waves generated in the troposphere can propagate into the stratosphere where they are absorbed. As a result, stratospheric changes alter where and how these waves are absorbed, and the effects can extend downward into the troposphere. Changes in solar irradiance, mainly in the ultraviolet (UV), lead to photochemically-induced ozone changes and, hence, alter the stratospheric heating rates, which can alter the tropospheric circulation. Limitations in resolution and relatively poor representation of some stratospheric processes adds uncertainty to model results.

Ocean

Major improvements have taken place in modelling ocean processes, in particular heat transport. These improvements, in conjunction with an increase in resolution, have been important in reducing the need for flux adjustment in models and in producing realistic simulations of natural large-scale circulation patterns and improvements in simulating El Niño (see Box 4). Ocean currents carry heat from the tropics to higher latitudes. The ocean exchanges heat, water (through evaporation and precipitation) and CO_2 with the atmosphere. Because of its huge mass and high heat capacity, the ocean slows climate change and influences the time-scales of variability in the ocean-atmosphere system. Considerable progress has been made in the understanding of ocean processes relevant for climate change. Increases in resolution, as well as improved representation (parametrization) of important sub-grid scale processes (e.g., mesoscale eddies), have increased the realism of simulations. Major uncertainties still exist with the representation of small-scale processes, such as overflows (flow through narrow channels, e.g., between Greenland and Iceland), western boundary currents (i.e., large-scale narrow currents along coastlines), convection and mixing. Boundary currents in climate simulations are weaker and wider than in nature, although the consequences of this for climate are not clear.

Cryosphere

The representation of sea-ice processes continues to improve, with several climate models now incorporating physically based treatments of ice dynamics. The representation of land-ice processes in global climate models remains rudimentary. The cryosphere consists of those regions of Earth that are seasonally or perennially covered by snow and ice. Sea ice is important because it reflects more incoming solar radiation than the sea surface (i.e., it has a higher albedo), and it insulates the sea from heat loss during the winter. Therefore, reduction of sea ice gives a positive feedback on climate warming at high latitudes. Furthermore, because sea ice contains less salt than sea water, when sea ice is formed the salt content (salinity) and density of the surface layer of the ocean is increased. This promotes an exchange of water with deeper layers of the ocean, affecting ocean circulation. The formation of icebergs and the melting of ice shelves returns fresh water from the land to the ocean, so that changes in the rates of these processes could affect ocean circulation by changing the surface salinity. Snow has a higher albedo than the land surface; hence, reductions in snow cover lead to a similar positive albedo feedback, although weaker than for sea ice. Increasingly complex snow schemes

and sub-grid scale variability in ice cover and thickness, which can significantly influence albedo and atmosphere-ocean exchanges, are being introduced in some climate models.

Land surface

Research with models containing the latest representations of the land surface indicates that the direct effects of increased CO_2 on the physiology of plants could lead to a relative reduction in evapotranspiration over the tropical continents, with associated regional warming and drying over that predicted for conventional greenhouse warming effects. Land surface changes provide important feedbacks as anthropogenic climate changes (e.g., increased temperature, changes in precipitation, changes in net radiative heating, and the direct effects of CO_2) will influence the state of the land surface (e.g., soil moisture, albedo, roughness and vegetation). Exchanges of energy, momentum, water, heat and carbon between the land surface and the atmosphere can be defined in models as functions of the type and density of the local vegetation and the depth and physical properties of the soil, all based on land-surface data bases that have been improved using satellite observations. Recent advances in the understanding of vegetation photosynthesis and water use have been used to couple the terrestrial energy, water and carbon cycles within a new generation of land surface parametrizations, which have been tested against field observations and implemented in a few GCMs, with demonstrable improvements in the simulation of land-atmosphere fluxes. However, significant problems remain to be solved in the areas of soil moisture processes, runoff prediction, land-use change and the treatment of snow and sub-grid scale heterogeneity.

Changes in land-surface cover can affect global climate in several ways. Large-scale deforestation in the humid tropics (e.g., South America, Africa, and Southeast Asia) has been identified as the most important ongoing land-surface process, because it reduces evaporation and increases surface temperature. These effects are qualitatively reproduced by most models. However, large uncertainties still persist on the quantitative impact of large-scale deforestation on the hydrological cycle, particularly over Amazonia.

Carbon cycle

Recent improvements in process-based terrestrial and ocean carbon cycle models and their evaluation against observations have given more confidence in their use for future scenario studies. CO_2 naturally cycles rapidly among the atmosphere, oceans and land. However, the removal of the CO_2 perturbation added by human activities from the atmosphere takes far longer. This is because of processes that limit the rate at which ocean and terrestrial carbon stocks can increase. Anthropogenic CO_2 is taken up by the ocean because of its high solubility (caused by the nature of carbonate chemistry), but the rate of uptake is limited by the finite speed of vertical mixing. Anthropogenic CO_2 is taken up by terrestrial ecosystems through several possible mechanisms, for example, land management, CO_2 fertilisation (the enhancement of plant growth as a result of increased atmospheric CO_2 concentration) and increasing anthropogenic inputs of nitrogen. This uptake is limited by the relatively small fraction of plant carbon that can enter long-term storage (wood and humus). The fraction of emitted CO_2 that can be taken up by the oceans and land is expected to decline with increasing CO_2 concentrations. Process-based models of the ocean and land carbon cycles (including representations of physical, chemical and biological processes) have been developed and evaluated against measurements pertinent to the natural carbon cycle. Such models have also been set up to mimic the human perturbation of the carbon cycle and have been able to generate time-series of ocean and land carbon uptake that are broadly consistent with observed global trends. There are still substantial differences among models, especially in how they treat the physical ocean circulation and in regional responses of terrestrial ecosystem processes to climate. Nevertheless, current models consistently indicate that when the effects of climate change are considered, CO_2 uptake by oceans and land becomes smaller.

D.2. The Coupled Systems

As noted in Section D.1, many feedbacks operate within the individual components of the climate system (atmosphere, ocean, cryosphere and land surface). However, many important processes and feedbacks occur through the coupling of the climate system components. Their representation is important to the prediction of large-scale responses.

Modes of natural variability

There is an increasing realisation that natural circulation patterns, such as ENSO and NAO, play a fundamental role in global climate and its interannual and longer-term variability. The strongest natural fluctuation of climate on interannual time-scales is the ENSO phenomenon (see Box 4). It is an inherently coupled atmosphere-ocean mode with its core activity in the tropical Pacific, but with important regional climate impacts throughout the world. Global climate models are only now beginning to exhibit variability in the tropical Pacific that resembles ENSO, mainly through increased meridional resolution at the equator. Patterns of sea surface temperature and atmospheric circulation similar to those occurring during ENSO on interannual time-scales also occur on decadal and longer time-scales.

The North Atlantic Oscillation (NAO) is the dominant pattern of northern wintertime atmospheric circulation variability and is increasingly being simulated realistically. The NAO is closely related to the Arctic Oscillation (AO), which has an additional annular component around the Arctic. There is strong evidence that the NAO arises mainly from internal atmospheric processes involving the entire troposphere-stratosphere system. Fluctuations in Atlantic Sea Surface Temperatures (SSTs) are related to the strength of the NAO, and a modest two-way interaction between the NAO and the Atlantic Ocean, leading to decadal variability, is emerging as important in projecting climate change.

Climate change may manifest itself both as shifting means, as well as changing preference of specific climate regimes, as evidenced by the observed trend toward positive values for the last 30 years in the NAO index and the climate "shift" in the tropical Pacific about 1976. While coupled models simulate features of observed natural climate variability, such as the NAO and ENSO, which suggests that many of the relevant processes are included in the models, further progress is needed to depict these natural modes accurately. Moreover, because ENSO and NAO are key determinants of regional climate change and can possibly result in abrupt and counter intuitive changes, there has been an increase in uncertainty in those aspects of climate change that critically depend on regional changes.

The thermohaline circulation (THC)

The thermohaline circulation (THC) is responsible for the major part of the meridional heat transport in the Atlantic Ocean. The THC is a global-scale overturning in the ocean driven by density differences arising from temperature and salinity effects. In the Atlantic, heat is transported by warm surface waters flowing northward and cold saline waters from the North Atlantic returning at depth. Reorganisations in the Atlantic THC can be triggered by perturbations in the surface buoyancy, which is influenced by precipitation, evaporation, continental runoff, sea-ice formation, and the exchange of heat, processes that could all change with consequences for regional and global climate. Interactions between the atmosphere and the ocean are also likely to be of considerable importance on decadal and longer time-scales, where the THC is involved. The interplay between the large-scale atmospheric forcing, with warming and evaporation in low latitudes and cooling and increased precipitation at high latitudes, forms the basis of a potential instability of the present Atlantic THC. ENSO may also influence the Atlantic THC by altering the fresh water balance of the tropical Atlantic, therefore providing a coupling between low and high latitudes. Uncertainties in the representation of small-scale flows over sills and through narrow straits and of ocean convection limit the ability of models to simulate situations involving substantial changes in the THC. The less saline North Pacific means that a deep THC does not occur in the Pacific.

Non-linear events and rapid climate change

The possibility for rapid and irreversible changes in the climate system exists, but there is a large degree of uncertainty about

Box 4: The El Niño-Southern Oscillation (ENSO)

The strongest natural fluctuation of climate on interannual time-scales is the El Niño-Southern Oscillation (ENSO) phenomenon. The term "El Niño" originally applied to an annual weak warm ocean current that ran southwards along the coast of Peru about Christmas-time and only subsequently became associated with the unusually large warmings. The coastal warming, however, is often associated with a much more extensive anomalous ocean warming to the International Dateline, and it is this Pacific basinwide phenomenon that forms the link with the anomalous global climate patterns. The atmospheric component tied to "El Niño" is termed the "Southern Oscillation". Scientists often call this phenomenon, where the atmosphere and ocean collaborate together, ENSO (El Niño-Southern Oscillation).

ENSO is a natural phenomenon, and there is good evidence from cores of coral and glacial ice in the Andes that it has been going on for millennia. The ocean and atmospheric conditions in the tropical Pacific are seldom average, but instead fluctuate somewhat irregularly between El Niño events and the opposite "La Niña" phase, consisting of a basinwide cooling of the tropical Pacific, with a preferred period of about three to six years. The most intense phase of each event usually lasts about a year.

A distinctive pattern of sea surface temperatures in the Pacific Ocean sets the stage for ENSO events. Key features are the "warm pool" in the tropical western Pacific, where the warmest ocean waters in the world reside, much colder waters in the eastern Pacific, and a cold tongue along the equator that is most pronounced about October and weakest in March. The atmospheric easterly trade winds in the tropics pile up the warm waters in the west, producing an upward slope of sea level along the equator of 0.60 m from east to west. The winds drive the surface ocean currents, which determine where the surface waters flow and diverge. Thus, cooler nutrient-rich waters upwell from below along the equator and western coasts of the Americas, favouring development of phytoplankton, zooplankton, and hence fish. Because convection and thunderstorms preferentially occur over warmer waters, the pattern of sea surface temperatures determines the distribution of rainfall in the tropics, and this in turn determines the atmospheric heating patterns through the release of latent heat. The heating drives the large-scale monsoonal-type circulations in the tropics, and consequently determines the winds. This strong coupling between the atmosphere and ocean in the tropics gives rise to the El Niño phenomenon.

During El Niño, the warm waters from the western tropical Pacific migrate eastward as the trade winds weaken, shifting the pattern of tropical rainstorms, further weakening the trade winds, and thus reinforcing the changes in sea temperatures. Sea level drops in the west, but rises in the east by as much as 0.25 m, as warm waters surge eastward along the equator. However, the changes in atmospheric circulation are not confined to the tropics, but extend globally and influence the jet streams and storm tracks in mid-latitudes. Approximately reverse patterns occur during the opposite La Niña phase of the phenomenon.

Changes associated with ENSO produce large variations in weather and climate around the world from year to year. These often have a profound impact on humanity and society because of associated droughts, floods, heat waves and other changes that can severely disrupt agriculture, fisheries, the environment, health, energy demand, air quality and also change the risks of fire. ENSO also plays a prominent role in modulating exchanges of CO_2 with the atmosphere. The normal upwelling of cold nutrient-rich and CO_2-rich waters in the tropical Pacific is suppressed during El Niño.

the mechanisms involved and hence also about the likelihood or time-scales of such transitions. The climate system involves many processes and feedbacks that interact in complex non-linear ways. This interaction can give rise to thresholds in the climate system that can be crossed if the system is perturbed sufficiently. There is evidence from polar ice cores suggesting that atmospheric regimes can change within a few years and that large-scale hemispheric changes can evolve as fast as a few decades. For example, the possibility of a threshold for a rapid transition of the Atlantic THC to a collapsed state has been demonstrated with a hierarchy of models. It is not yet clear what this threshold is and how likely it is that human activity would lead to it being exceeded (see Section F.6). Atmospheric circulation can be characterised by different preferred patterns; e.g., arising from ENSO and the NAO/AO, and changes in their phase can occur rapidly. Basic theory and models suggest that climate change may be first expressed in changes in the frequency of occurrence of these patterns. Changes in vegetation, through either direct anthropogenic deforestation or those caused by global warming, could occur rapidly and could induce further climate change. It is supposed that the rapid creation of the Sahara about 5,500 years ago represents an example of such a non-linear change in land cover.

D.3. Regionalisation Techniques

Regional climate information was only addressed to a limited degree in the SAR. Techniques used to enhance regional detail have been substantially improved since the SAR and have become more widely applied. They fall into three categories: high and variable resolution AOGCMs; regional (or nested limited area) climate models (RCMs); and empirical/statistical and statistical/dynamical methods. The techniques exhibit different strengths and weaknesses and their use at the continental scale strongly depends on the needs of specific applications.

Coarse resolution AOGCMs simulate atmospheric general circulation features well in general. At the regional scale, the models display area-average biases that are highly variable from region to region and among models, with sub-continental area averaged seasonal temperature biases typically ±4°C and precipitation biases between −40 and +80%. These represent an important improvement compared to AOGCMs evaluated in the SAR.

The development of high resolution/variable resolution Atmospheric General Circulation Models (AGCMs) since the SAR generally shows that the dynamics and large-scale flow in the models improves as resolution increases. In some cases, however, systematic errors are worsened compared to coarser resolution models, although only very few results have been documented.

High resolution RCMs have matured considerably since the SAR. Regional models consistently improve the spatial detail of simulated climate compared to AGCMs. RCMs driven by observed boundary conditions evidence area-averaged temperature biases (regional scales of 10^5 to 10^6 km^2) generally below 2°C, while precipitation biases are below 50%. Regionalisation work indicates at finer scales that the changes can be substantially different in magnitude or sign from the large area-average results. A relatively large spread exists among models, although attribution of the cause of these differences is unclear.

D.4. Overall Assessment of Abilities

Coupled models have evolved and improved significantly since the SAR. In general, they provide credible simulations of climate, at least down to sub-continental scales and over temporal scales from seasonal to decadal. Coupled models, as a class, are considered to be suitable tools to provide useful projections of future climates. These models cannot yet simulate all aspects of climate (e.g., they still cannot account fully for the observed trend in the surface-troposphere temperature differences since 1979). Clouds and humidity also remain sources of significant uncertainty, but there have been incremental improvements in simulations of these quantities. No single model can be considered "best", and it is important to utilise results from a range of carefully evaluated coupled models to explore effects of different formulations. The rationale for increased confidence in models arises from model performance in the following areas.

Flux adjustment

The overall confidence in model projections is increased by the improved performance of several models that do not use flux adjustment. These models now maintain stable, multi-century simulations of surface climate that are considered to be of sufficient quality to allow their use for climate change projections. The changes whereby many models can now run without flux adjustment have come from improvements in both the atmospheric and oceanic components. In the model atmosphere, improvements in convection, the boundary layer, clouds, and surface latent heat fluxes are most notable. In the model ocean, the improvements are in resolution, boundary layer mixing, and in the representation of eddies. The results from climate change studies with flux adjusted and non-flux adjusted models are broadly in agreement; nonetheless, the development of stable non-flux adjusted models increases confidence in their ability to simulate future climates.

Climate of the 20th century

Confidence in the ability of models to project future climates is increased by the ability of several models to reproduce warming trends in the 20th century surface air temperature when driven by increased greenhouse gases and sulphate aerosols. This is illustrated in Figure 13. However, only idealized scenarios of sulphate aerosols have been used and contributions from some additional processes and forcings may not have been included in the models. Some modelling studies suggest that inclusion of additional forcings like solar variability and volcanic aerosols may improve some aspects of the simulated climate variability of the 20th century.

Extreme events

Analysis of and confidence in extreme events simulated within climate models are still emerging, particularly for storm tracks and storm frequency. "Tropical-cyclone-like" vortices are being simulated in climate models, although enough uncertainty remains over their interpretation to warrant caution in projections of tropical cyclone changes. However, in general, the analysis of extreme events in both observations (see Section B.6) and coupled models is underdeveloped.

Interannual variability

The performance of coupled models in simulating ENSO has improved; however, its variability is displaced westward and its strength is generally underestimated. When suitably initialised with surface wind and sub-surface ocean data, some coupled models have had a degree of success in predicting ENSO events.

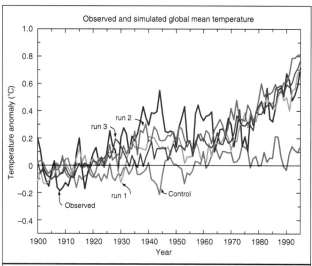

Figure 13: Observed and modelled global annual mean temperature anomalies (°C) relative to the average of the observations over the period 1900 to 1930. The control and three independent simulations with the same greenhouse gas plus aerosol forcing and slightly different initial conditions are shown from an AOGCM. The three greenhouse gas plus aerosol simulations are labeled 'run 1', 'run 2', and 'run 3' respectively. [Based on Figure 8.15]

Model intercomparisons

The growth in systematic intercomparisons of models provides the core evidence for the growing capabilities of climate models. For example, the Coupled Model Intercomparison Project (CMIP) is enabling a more comprehensive and systematic evaluation and intercomparison of coupled models run in a standardised configuration and responding to standardised forcing. Some degree of quantification of improvements in coupled model performance has now been demonstrated. The Palaeoclimate Model Intercomparison Project (PMIP) provides intercomparisons of models for the mid-Holocene (6,000 years before present) and for the Last Glacial Maximum (21,000 years before present). The ability of these models to simulate some aspects of palaeoclimates, compared to a range of palaeoclimate proxy data, gives confidence in models (at least the atmospheric component) over a range of difference forcings.

E. The Identification of a Human Influence on Climate Change

Sections B and C characterised the observed past changes in climate and in forcing agents, respectively. Section D examined the capabilities of climate models to predict the response of the climate system to such changes in forcing. This Section uses that information to examine the question of whether a human influence on climate change to date can be identified.

This is an important point to address. The SAR concluded that "the balance of evidence suggests that there is a discernible human influence on global climate". It noted that the detection and attribution of anthropogenic climate change signals will be accomplished through a gradual accumulation of evidence. The SAR also noted uncertainties in a number of factors, including internal variability and the magnitude and patterns of forcing and response, which prevented them from drawing a stronger conclusion.

E.1. The Meaning of Detection and Attribution

Detection is the process of demonstrating that an observed change is significantly different (in a statistical sense) than can be explained by natural variability. Attribution is the process of establishing cause and effect with some defined level of confidence, including the assessment of competing hypotheses. The response to anthropogenic changes in climate forcing occurs against a backdrop of natural internal and externally forced climate variability. Internal climate variability, i.e., climate variability not forced by external agents, occurs on all time-scales from weeks to centuries and even millennia. Slow climate components, such as the ocean, have particularly important roles on decadal and century time-scales because they integrate weather variability. Thus, the climate is capable of producing long time-scale variations of considerable magnitude without external influences. Externally forced climate variations (signals) may be due to changes in natural forcing factors, such as solar radiation or volcanic aerosols, or to changes in anthropogenic forcing factors, such as increasing concentrations of greenhouse gases or aerosols. The presence of this natural climate variability means that the detection and attribution of anthropogenic climate change is a statistical "signal to noise" problem. *Detection* studies demonstrate whether or not an observed change is highly unusual in a statistical sense, but this does not necessarily imply that we understand its causes. The *attribution* of climate change to anthropogenic causes involves statistical analysis and the careful assessment of multiple lines of evidence to demonstrate, within a pre-specified margin of error, that the observed changes are:

· unlikely to be due entirely to internal variability;
· consistent with the estimated responses to the given combination of anthropogenic and natural forcing; and
· not consistent with alternative, physically plausible explanations of recent climate change that exclude important elements of the given combination of forcings.

E.2. A Longer and More Closely Scrutinised Observational Record

Three of the last five years (1995, 1997 and 1998) were the warmest globally in the instrumental record. The impact of observational sampling errors has been estimated for the global and hemispheric mean temperature record. There is also a better understanding of the errors and uncertainties in the satellite-based (Microwave Sounding Unit, MSU) temperature record. Discrepancies between MSU and radiosonde data have largely been resolved, although the observed trend in the difference between the surface and lower tropospheric temperatures cannot fully be accounted for (see Section B). New reconstructions of temperature over the last 1,000 years indicate that the temperature changes over the last hundred years are unlikely to be entirely natural in origin, even taking into account the large uncertainties in palaeo-reconstructions (see Section B).

E.3. New Model Estimates of Internal Variability

The warming over the past 100 years is very unlikely to be due to internal variability alone, as estimated by current models. The instrumental record is short and covers the period of human influence and palaeo-records include natural forced variations, such as those due to variations in solar irradiance and in the frequency of major volcanic eruptions. These limitations leave few alternatives to using long "control" simulations with coupled models for the estimation of internal climate variability. Since the SAR, more models have been used to estimate the magnitude of internal climate variability, a representative sample of which is given in Figure 14. As can be seen, there is a wide range of global scale internal variability in these models. Estimates of the longer time-scale variability relevant to detection and attribution studies is uncertain, but, on interannual and decadal time-scales, some models show similar or larger variability than observed, even though models do not include variance from external sources. Conclusions on detection of an anthropogenic signal are insensitive to the model used to estimate internal variability, and recent changes cannot be accounted for as pure internal variability, even if the amplitude of simulated internal variations is increased by a factor of two or perhaps more. Most recent detection and attribution studies find no evidence that model-estimated internal variability at the surface is inconsistent with the residual variability that remains in the observations after removal of the estimated anthropogenic signals on the large spatial and long time-scales used in detection and attribution studies. Note, however, the ability to detect inconsistencies is limited. As Figure 14 indicates, no model control simulation shows a trend in surface air temperature as large as the observed trend over the last 1,000 years.

E.4. New Estimates of Responses to Natural Forcing

Assessments based on physical principles and model simulations indicate that natural forcing alone is unlikely to explain the recent observed global warming or the observed changes in vertical temperature structure of the atmosphere. Fully coupled ocean-atmosphere models have used reconstructions of solar and volcanic forcings over the last one to three

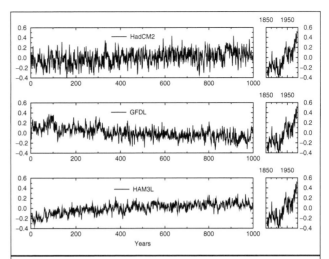

Figure 14: Global mean surface air temperature anomalies from 1,000 year control simulations with three different climate models, – Hadley, Geophysical Fluid Dynamics Laboratory and Hamburg, compared to the recent instrumental record. No model control simulation shows a trend in surface air temperature as large as the observed trend. If internal variability is correct in these models, the recent warming is likely not due to variability produced within the climate system alone. [Based on Figure 12.1]

centuries to estimate the contribution of natural forcing to climate variability and change. Although the reconstruction of natural forcings is uncertain, including their effects produces an increase in variance at longer (multi-decadal) time-scales. This brings the low-frequency variability closer to that deduced from palaeo-reconstructions. It is likely that the net natural forcing (i.e., solar plus volcanic) has been negative over the past two decades, and possibly even the past four decades. Statistical assessments confirm that simulated natural variability, both internal and naturally forced, is unlikely to explain the warming in the latter half of the 20th century (see Figure 15). However, there is evidence for a detectable volcanic influence on climate and evidence that suggests a detectable solar influence, especially in the early part of the 20th century. Even if the models underestimate the magnitude of the response to solar or volcanic forcing, the spatial and temporal patterns are such that these effects alone cannot explain the observed temperature changes over the 20th century.

E.5. Sensitivity to Estimates of Climate Change Signals

There is a wide range of evidence of qualitative consistencies between observed climate changes and model responses to anthropogenic forcing. Models and observations show increasing global temperature, increasing land-ocean temperature contrast, diminishing sea-ice extent, glacial retreat, and increases in precipitation at high latitudes in the Northern Hemisphere. Some qualitative inconsistencies remain, including the fact that models predict a faster rate of warming in the mid- to upper

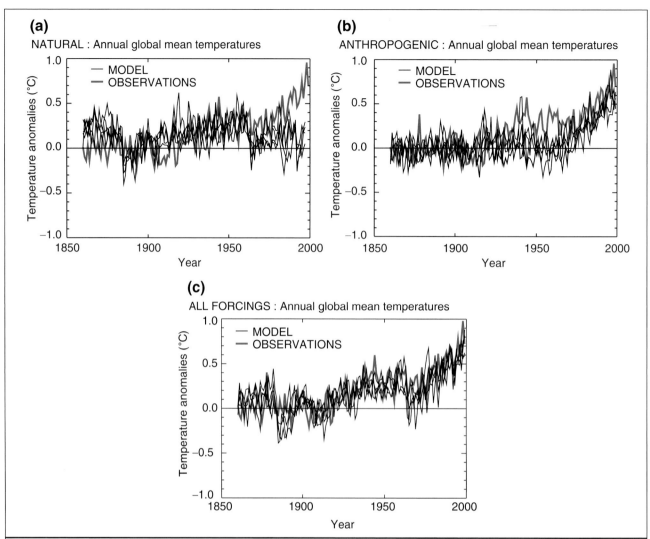

(a)

NATURAL : Annual global mean temperatures

(b)

ANTHROPOGENIC : Annual global mean temperatures

(c)

ALL FORCINGS : Annual global mean temperatures

Figure 15: Global mean surface temperature anomalies relative to the 1880 to 1920 mean from the instrumental record compared with ensembles of four simulations with a coupled ocean-atmosphere climate model forced (a) with solar and volcanic forcing only, (b) with anthropogenic forcing including well mixed greenhouse gases, changes in stratospheric and tropospheric ozone and the direct and indirect effects of sulphate aerosols, and (c) with all forcings, both natural and anthropogenic. The thick line shows the instrumental data while the thin lines show the individual model simulations in the ensemble of four members. Note that the data are annual mean values. The model data are only sampled at the locations where there are observations. The changes in sulphate aerosol are calculated interactively, and changes in tropospheric ozone were calculated offline using a chemical transport model. Changes in cloud brightness (the first indirect effect of sulphate aerosols) were calculated by an offline simulation and included in the model. The changes in stratospheric ozone were based on observations. The volcanic and solar forcing were based on published combinations of measured and proxy data. The net anthropogenic forcing at 1990 was 1.0 Wm^{-2} including a net cooling of 1.0 Wm^{-2} due to sulphate aerosols. The net natural forcing for 1990 relative to 1860 was 0.5 Wm^{-2}, and for 1992 was a net cooling of 2.0 Wm^{-2} due to Mount Pinatubo. Other models forced with anthropogenic forcing give similar results to those shown in (b). [Based on Figure 12.7]

troposphere than is observed in either satellite or radiosonde tropospheric temperature records.

All simulations with greenhouse gases and sulphate aerosols that have been used in detection studies have found that a significant anthropogenic contribution is required to account for surface and tropospheric trends over at least the last 30 years. Since the SAR, more simulations with increases in greenhouse gases and some representation of aerosol effects have become available. Several studies have included an explicit representation of greenhouse gases (as opposed to an equivalent

increase in CO_2). Some have also included tropospheric ozone changes, an interactive sulphur cycle, an explicit radiative treatment of the scattering of sulphate aerosols, and improved estimates of the changes in stratospheric ozone. Overall, while detection of the climate response to these other anthropogenic factors is often ambiguous, detection of the influence of greenhouse gases on the surface temperature changes over the past 50 years is robust. In some cases, ensembles of simulations have been run to reduce noise in the estimates of the time-dependent response. Some studies have evaluated seasonal variation of the response. Uncertainties in the estimated climate change signals have made it difficult to attribute the observed climate change to one specific combination of anthropogenic and natural influences, but all studies have found a significant anthropogenic contribution is required to account for surface and tropospheric trends over at least the last thirty years.

E.6. A Wider Range of Detection Techniques

Temperature

Evidence of a human influence on climate is obtained over a substantially wider range of detection techniques. A major advance since the SAR is the increase in the range of techniques used and the evaluation of the degree to which the results are independent of the assumptions made in applying those techniques. There have been studies using pattern correlations, optimal detection studies using one or more fixed patterns and time-varying patterns, and a number of other techniques. The increase in the number of studies, breadth of techniques, increased rigour in the assessment of the role of anthropogenic forcing in climate, and the robustness of results to the assumptions made using those techniques, has increased the confidence in these aspects of detection and attribution.

Results are sensitive to the range of temporal and spatial scales that are considered. Several decades of data are necessary to separate forced signals from internal variability. Idealised studies have demonstrated that surface temperature changes are detectable only on scales in the order of 5,000 km. Such studies show that the level of agreement found between simulations and observations in pattern correlation studies is close to what one would expect in theory.

Most attribution studies find that, over the last 50 years, the estimated rate and magnitude of global warming due to increasing concentrations of greenhouse gases alone are comparable with or larger than the observed warming. Attribution studies address the question of "whether the magnitude of the simulated response to a particular forcing agent is consistent with observations". The use of multi-signal techniques has enabled studies that discriminate between the effects of different factors on climate. The inclusion of the time dependence of signals has helped to distinguish between natural and anthropogenic forcings. As more response patterns are included, the problem of degeneracy (different combinations of patterns yielding near identical fits to the observations) inevitably arises. Nevertheless, even with all the major responses that have been included in the analysis, a distinct greenhouse gas signal remains detectable. Furthermore, most model estimates that take into account both greenhouse gases and sulphate aerosols are consistent with observations over this period. The best agreement between model simulations and observations over the last 140 years is found when both anthropogenic and natural factors are included (see Figure 15). These results show that the forcings included are sufficient to explain the observed

changes, but do not exclude the possibility that other forcings have also contributed. Overall, the magnitude of the temperature response to increasing concentrations of greenhouse gases is found to be consistent with observations on the scales considered (see Figure 16), but there remain discrepies between modelled and observed response to other natural and anthropogenic factors.

Uncertainties in other forcings that have been included do not prevent identification of the effect of anthropogenic greenhouse gases over the last 50 years. The sulphate forcing, while uncertain, is negative over this period. Changes in natural forcing during most of this period are also estimated to be negative. Detection of the influence of anthropogenic greenhouse gases therefore cannot be eliminated either by the uncertainty in sulphate aerosol forcing or because natural forcing has not been included in all model simulations. Studies that distinguish the separate responses to greenhouse gas, sulphate aerosol and natural forcing produce uncertain estimates of the amplitude of the sulphate aerosol and natural signals, but almost all studies are nevertheless able to detect the presence of the anthropogenic greenhouse gas signal in the recent climate record.

The detection and attribution methods used should not be sensitive to errors in the amplitude of the global mean response to individual forcings. In the signal-estimation methods used in this report, the amplitude of the signal is estimated from the observations and not the amplitude of the simulated response. Hence the estimates are independent of those factors determining the simulated amplitude of the response, such as the climate sensitivity of the model used. In addition, if the signal due to a given forcing is estimated individually, the amplitude is largely independent of the magnitude of the forcing used to derive the response. Uncertainty in the amplitude of the solar and indirect sulphate aerosol forcing should not affect the magnitude of the estimated signal.

Sea level

It is very likely that the 20th century warming has contributed significantly to the observed sea level rise, through thermal expansion of sea water and widespread loss of land ice. Within present uncertainties, observations and models are both consistent with a lack of significant acceleration of sea level rise during the 20th century.

E.7. Remaining Uncertainties in Detection and Attribution

Some progress has been made in reducing uncertainty, though many of the sources of uncertainty identified in the SAR still exist. These include:
· *Discrepancies between the vertical profile of temperature change in the troposphere seen in observations and models.* These have been reduced as more realistic forcing histories have been used in models, although not fully resolved.

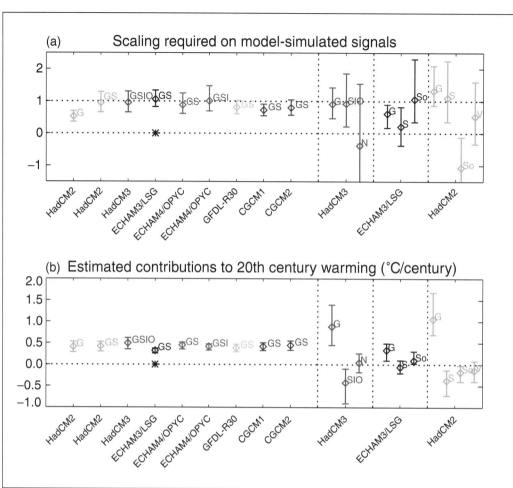

Figure 16: (a) Estimates of the "scaling factors" by which the amplitude of several model-simulated signals must be multiplied to reproduce the corresponding changes in the observed record. The vertical bars indicate the 5 to 95% uncertainty range due to internal variability. A range encompassing unity implies that this combination of forcing amplitude and model-simulated response is consistent with the corresponding observed change, while a range encompassing zero implies that this model-simulated signal is not detectable. Signals are defined as the ensemble mean response to external forcing expressed in large-scale (>5,000 km) near-surface temperatures over the 1946 to 1996 period relative to the 1896 to 1996 mean. The first entry (G) shows the scaling factor and 5 to 95% confidence interval obtained with the assumption that the observations consist only of a response to greenhouse gases plus internal variability. The range is significantly less than one (consistent with results from other models), meaning that models forced with greenhouse gases alone significantly over predict the observed warming signal. The next eight entries show scaling factors for model-simulated responses to greenhouse and sulphate forcing (GS), with two cases including indirect sulphate and tropospheric ozone forcing, one of these also including stratospheric ozone depletion (GSI and GSIO, respectively). All but one (CGCM1) of these ranges is consistent with unity. Hence there is little evidence that models are systematically over- or under-predicting the amplitude of the observed response under the assumption that model-simulated GS signals and internal variability are an adequate representation (i.e., that natural forcing has had little net impact on this diagnostic). Observed residual variability is consistent with this assumption in all but one case (ECHAM3, indicated by the asterisk). One is obliged to make this assumption to include models for which only a simulation of the anthropogenic response is available, but uncertainty estimates in these single signal cases are incomplete since they do not account for uncertainty in the naturally forced response. These ranges indicate, however, the high level of confidence with which internal variability, as simulated by these various models, can be rejected as an explanation of recent near-surface temperature change. A more complete uncertainty analysis is provided by the next three entries, which show corresponding scaling factors on individual greenhouse (G), sulphate (S), solar-plus-volcanic (N), solar-only (So) and volcanic-only (V) signals for those cases in which the relevant simulations have been performed. In these cases, multiple factors are estimated simultaneously to account for uncertainty in the amplitude of the naturally forced response. The uncertainties increase but the greenhouse signal remains consistently detectable. In one case (ECHAM3) the model appears to be overestimating the greenhouse response (scaling range in the G signal inconsistent with unity), but this result is sensitive to which component of the control is used to define the detection space. It is also not known how it would respond to the inclusion of a volcanic signal. In cases where both solar and volcanic forcing is included (HadCM2 and HadCM3), G and S signals remain detectable and consistent with unity independent of whether natural signals are estimated jointly or separately (allowing for different errors in S and V responses). (b) Estimated contributions to global mean warming over the 20th century, based on the results shown in (a), with 5 to 95% confidence intervals. Although the estimates vary depending on which model's signal and what forcing is assumed, and are less certain if more than one signal is estimated, all show a significant contribution from anthropogenic climate change to 20th century warming. [Based on Figure 12.12]

Also, the difference between observed surface and lower-tropospheric trends over the last two decades cannot be fully reproduced by model simulations.

· *Large uncertainties in estimates of internal climate variability from models and observations.* Although as noted above, these are unlikely (bordering on very unlikely) to be large enough to nullify the claim that a detectable climate change has taken place.

· *Considerable uncertainty in the reconstructions of solar and volcanic forcing which are based on proxy or limited observational data for all but the last two decades.* Detection of the influence of greenhouse gases on climate appears to be robust to possible amplification of the solar forcing by ozone-solar or solar-cloud interactions, provided these do not alter the pattern or time-dependence of the response to solar forcing. Amplification of the solar signal by these processes, which are not yet included in models, remains speculative.

· *Large uncertainties in anthropogenic forcing are associated with the effects of aerosols.* The effects of some anthropogenic factors, including organic carbon, black carbon, biomass aerosols, and changes in land use, have not been included in detection and attribution studies. Estimates of the size and geographic pattern of the effects of these forcings vary considerably, although individually their global effects are estimated to be relatively small.

· Large differences in the response of different models to the same forcing. These differences, which are often greater than the difference in response in the same model with and without aerosol effects, highlight the large uncertainties in climate change prediction and the need to quantify uncertainty and reduce it through better observational data sets and model improvement.

E.8. Synopsis

In the light of new evidence and taking into account the remaining uncertainties, most of the observed warming over the last 50 years is likely to have been due to the increase in greenhouse gas concentrations.

F. The Projections of the Earth's Future Climate

The tools of climate models are used with future scenarios of forcing agents (e.g., greenhouse gases and aerosols) as input to make a suite of projected future climate changes that illustrates the possibilities that could lie ahead. Section F.1 provides a description of the future scenarios of forcing agents given in the IPCC Special Report on Emission Scenarios (SRES) on which, wherever possible, the future changes presented in this section are based. Sections F.2 to F.9 present the resulting projections of changes to the future climate. Finally, Section F.10 presents the results of future projections based on scenarios of a future where greenhouse gas concentrations are stabilised.

F.1. The IPCC Special Report on Emissions Scenarios (SRES)

In 1996, the IPCC began the development of a new set of emissions scenarios, effectively to update and replace the well-known IS92 scenarios. The approved new set of scenarios is described in the IPCC Special Report on Emission Scenarios (SRES). Four different narrative storylines were developed to describe consistently the relationships between the forces driving emissions and their evolution and to add context for the scenario quantification. The resulting set of 40 scenarios (35 of which contain data on the full range of gases required to force climate models) cover a wide range of the main demographic, economic and technological driving forces of future greenhouse gas and sulphur emissions. Each scenario represents a specific quantification of one of the four storylines. All the scenarios based on the same storyline constitute a scenario "family" (See Box 5, which briefly describes the main characteristics of the four SRES storylines and scenario families). The SRES scenarios do not include additional climate initiatives, which means that no scenarios are included that explicitly assume implementation of the United Nations Framework Convention on Climate Change or the emissions targets of the Kyoto Protocol. However, greenhouse gas emissions are directly affected by non-climate change policies designed for a wide range of other purposes (e.g., air quality). Furthermore, government policies can, to varying degrees, influence the greenhouse gas emission drivers, such as demographic change, social and economic development, technological change, resource use, and pollution management. This influence is broadly reflected in the storylines and resulting scenarios.

Since the SRES was not approved until 15 March 2000, it was too late for the modelling community to incorporate the final approved scenarios in their models and have the results available in time for this Third Assessment Report. However, draft scenarios were released to climate modellers earlier to facilitate their input to the Third Assessment Report, in accordance with a decision of the IPCC Bureau in 1998. At that time, one marker scenario was chosen from each of four of the scenario groups based directly on the storylines (A1B, A2, B1, and B2). The choice of the markers was based on which of the initial quantifications best reflected the storyline and features of specific models. Marker scenarios are no more or less likely than any other scenarios, but are considered illustrative of a particular storyline. Scenarios were also selected later to illustrate the other two scenario groups (A1FI and A1T) within the A1 family, which specifically explore alternative technology developments, holding the other driving forces constant. Hence there is an illustrative scenario for each of the six scenario groups, and all are equally plausible. Since the latter two illustrative scenarios were selected at a late stage

Box 5: The Emissions Scenarios of the Special Report on Emissions Scenarios (SRES)

A1. The A1 storyline and scenario family describes a future world of very rapid economic growth, global population that peaks in mid-century and declines thereafter, and the rapid introduction of new and more efficient technologies. Major underlying themes are convergence among regions, capacity building and increased cultural and social interactions, with a substantial reduction in regional differences in per capita income. The A1 scenario family develops into three groups that describe alternative directions of technological change in the energy system. The three A1 groups are distinguished by their technological emphasis: fossil intensive (A1FI), non-fossil energy sources (A1T), or a balance across all sources (A1B) (where balanced is defined as not relying too heavily on one particular energy source, on the assumption that similar improvement rates apply to all energy supply and end use technologies).

A2. The A2 storyline and scenario family describes a very heterogeneous world. The underlying theme is self-reliance and preservation of local identities. Fertility patterns across regions converge very slowly, which results in continuously increasing population. Economic development is primarily regionally oriented and per capita economic growth and technological change more fragmented and slower than other storylines.

B1. The B1 storyline and scenario family describes a convergent world with the same global population, that peaks in mid-century and declines thereafter, as in the A1 storyline, but with rapid change in economic structures toward a service and information economy, with reductions in material intensity and the introduction of clean and resource-efficient technologies. The emphasis is on global solutions to economic, social and environmental sustainability, including improved equity, but without additional climate initiatives.

B2. The B2 storyline and scenario family describes a world in which the emphasis is on local solutions to economic, social and environmental sustainability. It is a world with continuously increasing global population, at a rate lower than A2, intermediate levels of economic development, and less rapid and more diverse technological change than in the B1 and A1 storylines. While the scenario is also oriented towards environmental protection and social equity, it focuses on local and regional levels.

---

in the process, the AOGCM modelling results presented in this report only use two of the four draft marker scenarios. At present, only scenarios A2 and B2 have been integrated by more than one AOGCM. The AOGCM results have been augmented by results from simple climate models that cover all six illustrative scenarios. The IS92a scenario is also presented in a number of cases to provide direct comparison with the results presented in the SAR.

The final four marker scenarios contained in the SRES differ in minor ways from the draft scenarios used for the AOGCM experiments described in this report. In order to ascertain the likely effect of differences in the draft and final SRES scenarios, each of the four draft and final marker scenarios were studied using a simple climate model. For three of the four marker scenarios (A1B, A2, and B2) temperature change from the draft and marker scenarios are very similar. The primary difference is a change to the standardised values for 1990 to 2000, which is common to all these scenarios. This results in a higher forcing early in the period. There are further small differences in net forcing, but these decrease until, by 2100, differences in temperature change in the two versions of these scenarios are in the range 1 to 2%. For the B1 scenario, however, temperature change is significantly lower in the final version, leading to a difference in the temperature change in 2100 of almost 20%, as a result of generally lower emissions across the full range of greenhouse gases.

Anthropogenic emissions of the three main greenhouse gases, CO_2, CH_4 and N_2O, together with anthropogenic sulphur dioxide emissions, are shown for the six illustrative SRES scenarios in Figure 17. It is evident that these scenarios encompass a wide range of emissions. For comparison, emissions are also shown for IS92a. Particularly noteworthy are the much lower future sulphur dioxide emissions for the six SRES scenarios, compared to the IS92 scenarios, due to structural changes in the energy system as well as concerns about local and regional air pollution.

F.2. Projections of Future Changes in Greenhouse Gases and Aerosols

Models indicate that the illustrative SRES scenarios lead to very different CO_2 concentration trajectories (see Figure 18). By 2100, carbon cycle models project atmospheric CO_2 concentrations of 540 to 970 ppm for the illustrative SRES scenarios (90 to 250% above the concentration of 280 ppm in 1750). The net effect of land and ocean climate feedbacks as indicated by models is to further increase projected atmospheric CO_2 concentrations by reducing both the ocean and land uptake of CO_2. These projections include the land and ocean climate feedbacks. Uncertainties, especially about the magnitude of the climate feedback from the terrestrial biosphere, cause a variation of about -10 to $+30\%$ around each scenario. The total range is 490 to 1260 ppm (75 to 350% above the 1750 concentration).

Measures to enhance carbon storage in terrestrial ecosystems could influence atmospheric CO_2 concentration, but the upper bound for reduction of CO_2 concentration by such means is 40 to 70 ppm. If all the carbon released by historic land-use changes could be restored to the terrestrial biosphere over the course of the century (e.g., by reforestation), CO_2 concentration would be reduced by 40 to 70 ppm. Thus, fossil fuel CO_2 emissions are virtually certain to remain the dominant control over trends in atmospheric CO_2 concentration during this century.

Model calculations of the abundances of the primary non-CO_2 greenhouse gases by the year 2100 vary considerably across the six illustrative SRES scenarios. In general A1B, A1T and B1 have the smallest increases, and A1FI and A2, the largest. The CH_4 changes from 1998 to 2100 range from -190 to $+1970$ ppb (-11 to $+112\%$), and N_2O increases from $+38$ to $+144$ ppb ($+12$ to $+46\%$) (see Figures 17b and c). The HFCs (134a, 143a, and 125) reach abundances of a few hundred to a thousand ppt from negligible levels today. The

PFC CF_4 is projected to increase to 200 to 400 ppt, and SF_6 is projected to increase to 35 to 65 ppt.

For the six illustrative SRES emissions scenarios, projected emissions of indirect greenhouse gases (NO$_x$, CO, VOC), together with changes in CH_4, are projected to change the global mean abundance of the tropospheric hydroxyl radical (OH), by -20% to $+6\%$ over the next century. Because of the importance of OH in tropospheric chemistry, comparable, but opposite sign, changes occur in the atmospheric lifetimes of the greenhouse gases CH_4 and HFCs. This impact depends in large part on the magnitude of and the balance between NO$_x$ and CO emissions. Changes in tropospheric O_3 of -12 to $+62\%$ are calculated from 2000 until 2100. The largest increase predicted for the 21st century is for scenarios A1FI and A2 and would be more than twice as large as that experienced since the Pre-industrial Era. These O_3 increases are attributable to the concurrent and large increases in anthropogenic NO$_x$ and CH_4 emissions.

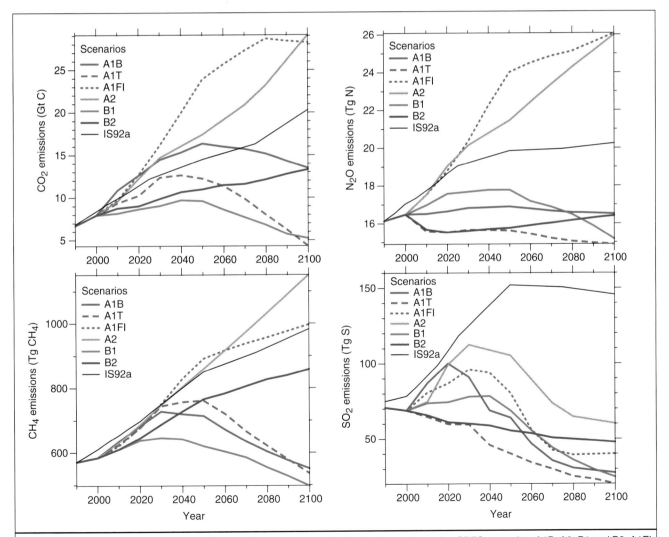

Figure 17: Anthropogenic emissions of CO_2, CH_4, N_2O and sulphur dioxide for the six illustrative SRES scenarios, A1B, A2, B1 and B2, A1FI and A1T. For comparison the IS92a scenario is also shown. [Based on IPCC Special Report on Emissions Scenarios.]

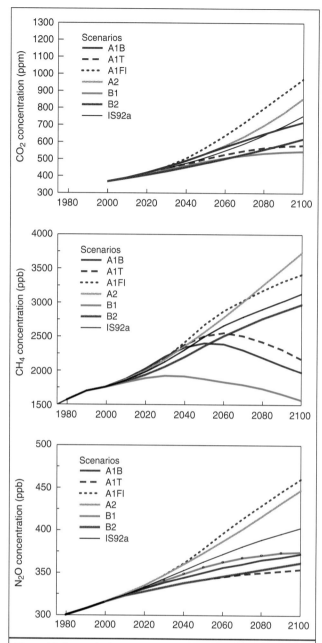

Figure 18: Atmospheric concentrations of CO_2, CH_4 and N_2O resulting from the six SRES scenarios and from the IS92a scenario computed with current methodology. [Based on Figures 3.12 and 4.14]

productivity. This problem reaches across continental boundaries and couples emissions of NO_x on a hemispheric scale.

Except for sulphate and black carbon, models show an approximately linear dependence of the abundance of aerosols on emissions. The processes that determine the removal rate for black carbon differ substantially between the models, leading to major uncertainty in the future projections of black carbon. Emissions of natural aerosols such as sea salt, dust, and gas phase precursors of aerosols such as terpenes, sulphur dioxide (SO_2), and dimethyl sulphide oxidation may increase as a result of changes in climate and atmospheric chemistry.

The six illustrative SRES scenarios cover nearly the full range of forcing that results from the full set of SRES scenarios. Estimated total historical anthropogenic radiative forcing from 1765 to 1990 followed by forcing resulting from the six SRES scenarios are shown in Figure 19. The forcing from the full range of 35 SRES scenarios is shown on the figure as a shaded envelope, since the forcings resulting from individual scenarios cross with time. The direct forcing from biomass-burning aerosols is scaled with deforestation rates. The SRES scenarios include the possibility of either increases or decreases in anthropogenic aerosols (e.g., sulphate aerosols, biomass aerosols, and black and organic carbon aerosols), depending on the extent of fossil fuel use and policies to abate polluting emissions. The SRES scenarios do not include emissions estimates for non-sulphate aerosols. Two methods for projecting these emissions were considered in this report: the first scales the emissions of fossil fuel and biomass aerosols with CO while the second scales the emissions with SO_2 and deforestation. Only the second method was used for climate projections. For comparison, radiative forcing is also shown for the IS92a scenario. It is evident that the range for the new SRES scenarios is shifted higher compared to the IS92 scenarios. This is mainly due to the reduced future SO_2 emissions of the SRES scenarios compared to the IS92 scenarios, but also to the slightly larger cumulative carbon emissions featured in some SRES scenarios.

In almost all SRES scenarios, the radiative forcing due to CO_2, CH_4, N_2O and tropospheric O_3 continue to increase, with the fraction of the total radiative forcing due to CO_2 projected to increase from slightly more than half to about three-quarters of the total. The radiative forcing due to O_3-depleting gases decreases due to the introduction of emission controls aimed at curbing stratospheric ozone depletion. The direct aerosol (sulphate and black and organic carbon components taken together) radiative forcing (evaluated relative to present day, 2000) varies in sign for the different scenarios. The direct plus indirect aerosol effects are projected to be smaller in magnitude than that of CO_2. No estimates are made for the spatial aspects of the future forcings. The indirect effect of aerosols on clouds is included in simple climate model calculations and scaled non-linearly with SO_2 emissions, assuming a present day value of -0.8 Wm^{-2}, as in the SAR.

The large growth in emissions of greenhouse gases and other pollutants as projected in some of the six illustrative SRES scenarios for the 21st century will degrade the global environment in ways beyond climate change. Changes projected in the SRES A2 and A1FI scenarios would degrade air quality over much of the globe by increasing background levels of tropospheric O_3. In northern mid-latitudes during summer, the zonal average of O_3 increases near the surface are about 30 ppb or more, raising background levels to about 80 ppb, threatening the attainment of current air quality standards over most metropolitan and even rural regions and compromising crop and forest

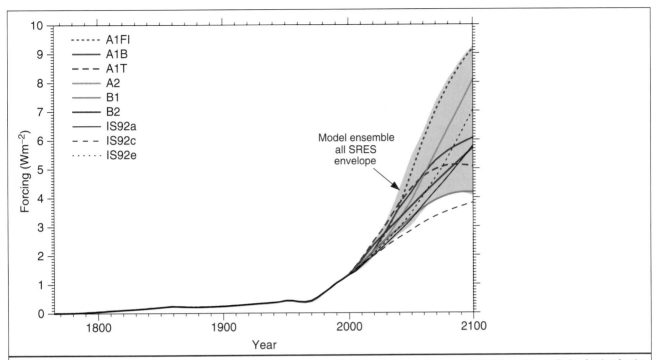

Figure 19: Simple model results: estimated historical anthropogenic radiative forcing up to the year 2000 followed by radiative forcing for the six illustrative SRES scenarios. The shading shows the envelope of forcing that encompasses the full set of thirty five SRES scenarios. The method of calculation closely follows that explained in the chapters. The values are based on the radiative forcing for a doubling of CO_2 from seven AOGCMs. The IS92a, IS92c, and IS92e forcing is also shown following the same method of calculation. [Based on Figure 9.13a]

F.3. Projections of Future Changes in Temperature

AOGCM results

Climate sensitivity is likely to be in the range of 1.5 to 4.5 °C. This estimate is unchanged from the first IPCC Assessment Report in 1990 and the SAR. The climate sensitivity is the equilibrium response of global surface temperature to a doubling of equivalent CO_2 concentration. The range of estimates arises from uncertainties in the climate models and their internal feedbacks, particularly those related to clouds and related processes. Used for the first time in this IPCC report is the Transient Climate Response (TCR). The TCR is defined as the globally averaged surface air temperature change, at the time of doubling of CO_2, in a 1%/yr CO_2-increase experiment. This rate of CO_2 increase is assumed to represent the radiative forcing from all greenhouse gases. The TCR combines elements of model sensitivity and factors that affect response (e.g., ocean heat uptake). The range of the TCR for current AOGCMs is 1.1 to 3.1 °C.

Including the direct effect of sulphate aerosols reduces global mean mid-21st century warming. The surface temperature response pattern for a given model, with and without sulphate aerosols, is more similar than the pattern between two models using the same forcing.

Models project changes in several broad-scale climate variables. As the radiative forcing of the climate system changes, the land warms faster and more than the ocean, and there is greater relative warming at high latitudes. Models project a smaller surface air temperature increase in the North Atlantic and circumpolar southern ocean regions relative to the global mean. There is projected to be a decrease in diurnal temperature range in many areas, with night-time lows increasing more than daytime highs. A number of models show a general decrease of daily variability of surface air temperature in winter and increased daily variability in summer in the Northern Hemisphere land areas. As the climate warms, the Northern Hemisphere snow cover and sea-ice extent are projected to decrease. Many of these changes are consistent with recent observational trends, as noted in Section B.

Multi-model ensembles of AOGCM simulations for a range of scenarios are being used to quantify the mean climate change and uncertainty based on the range of model results. For the end of the 21st century (2071 to 2100), the mean change in global average surface air temperature, relative to the period 1961 to 1990, is 3.0° C (with a range of 1.3 to 4.5° C) for the A2 draft marker scenario and 2.2° C (with a range of 0.9 to 3.4° C) for the B2 draft marker scenario. The B2 scenario produces a smaller warming that is consistent with its lower rate of increased CO_2 concentration.

On time-scales of a few decades, the current observed rate of warming can be used to constrain the projected response to a given emissions scenario despite uncertainty in climate sensitivity. Analysis of simple models and intercomparisons of AOGCM responses to idealised forcing scenarios suggest that, for most scenarios over the coming decades, errors in large-scale temperature projections are likely to increase in proportion to the magnitude of the overall response. The estimated size of and uncertainty in current observed warming rates attributable to human influence thus provides a relatively model-independent estimate of uncertainty in multi-decade projections under most scenarios. To be consistent with recent observations, anthropogenic warming is likely to lie in the range 0.1 to 0.2°C/ decade over the next few decades under the IS92a scenario. This is similar to the range of responses to this scenario based on the seven versions of the simple model used in Figure 22.

Most of the features of the geographical response in the SRES scenario experiments are similar for different scenarios (see Figure 20) and are similar to those for idealised 1% CO_2-increase integrations. The biggest difference between the 1% CO_2-increase experiments, which have no sulphate aerosol, and the SRES experiments is the regional moderating of the warming over industrialised areas, in the SRES experiments, where the negative forcing from sulphate aerosols is greatest. This regional effect was noted in the SAR for only two models, but this has now been shown to be a consistent response across the greater number of more recent models.

It is very likely that nearly all land areas will warm more rapidly than the global average, particularly those at northern high latitudes in the cold season. Results (see Figure 21) from recent AOGCM simulations forced with SRES A2 and B2 emissions scenarios indicate that in winter the warming for all high-latitude northern regions exceeds the global mean warming in each model by more than 40% (1.3 to 6.3°C for the range of models and scenarios considered). In summer, warming is in excess of 40% above the global mean change in central and northern Asia. Only in south Asia and southern South America in June/July/August, and Southeast Asia for both seasons, do the models consistently show warming less than the global average.

Simple climate model results

Due to computational expense, AOGCMs can only be run for a limited number of scenarios. A simple model can be calibrated to represent globally averaged AOGCM responses and run for a much larger number of scenarios.

The globally averaged surface temperature is projected to increase by 1.4 to 5.8°C (Figure 22(a)) over the period 1990 to 2100. These results are for the full range of 35 SRES scenarios, based on a number of climate models.[6,7] Temperature increases are projected to be greater than those in the SAR, which were about 1.0 to 3.5°C based on six IS92 scenarios. The higher projected temperatures and the wider range are due primarily to the lower projected SO_2 emissions in the SRES scenarios relative to the IS92 scenarios. The projected rate of warming is much larger than the observed changes during the 20th century and is very likely to be without precedent during at least the last 10,000 years, based on palaeoclimate data.

The relative ranking of the SRES scenarios in terms of global mean temperature changes with time. In particular, for scenarios with higher fossil fuel use (hence, higher carbon dioxide emissions, e.g., A2), the SO_2 emissions are also higher. In the near term (to around 2050), the cooling effect of higher sulphur dioxide emissions significantly reduces the warming caused by increased emissions of greenhouse gases in scenarios such as A2. The opposite effect is seen for scenarios B1 and B2, which have lower fossil fuel emissions as well as lower SO_2 emissions, and lead to a larger near-term warming. In the longer term, however, the level of emissions of long-lived greenhouse gases such as CO_2 and N_2O become the dominant determinants of the resulting climate changes.

By 2100, differences in emissions in the SRES scenarios and different climate model responses contribute similar uncertainty to the range of global temperature change. Further uncertainties arise due to uncertainties in the radiative forcing. The largest forcing uncertainty is that due to the sulphate aerosols.

F.4. Projections of Future Changes in Precipitation

Globally averaged water vapour, evaporation and precipitation are projected to increase. At the regional scale both increases and decreases in precipitation are seen. Results (see Figure 23) from recent AOGCM simulations forced with SRES A2 and B2 emissions scenarios indicate that it is likely for precipitation to increase in both summer and winter over high-latitude regions. In winter, increases are also seen over northern mid-latitudes, tropical Africa and Antarctica, and in summer in southern and eastern Asia. Australia, central America, and southern Africa show consistent decreases in winter rainfall.

Based on patterns emerging from a limited number of studies with current AOGCMs, older GCMs, and regionalisation studies, there is a strong correlation between precipitation interannual variability and mean precipitation. Future increases in mean precipitation will likely lead to increases in variability. Conversely, precipitation variability will likely decrease only in areas of reduced mean precipitation.

---

[6] Complex physically based climate models are the main tool for projecting future climate change. In order to explore the range of scenarios, these are complemented by simple climate models calibrated to yield an equivalent response in temperature and sea level to complex climate models. These projections are obtained using a simple climate model whose climate sensitivity and ocean heat uptake are calibrated to each of 7 complex climate models. The climate sensitivity used in the simple model ranges from 1.7 to 4.2°C, which is comparable to the commonly accepted range of 1.5 to 4.5°C.

[7] This range does not include uncertainties in the modelling of radiative forcing, e.g. aerosol forcing uncertainties. A small carbon cycle climate feedback is included.

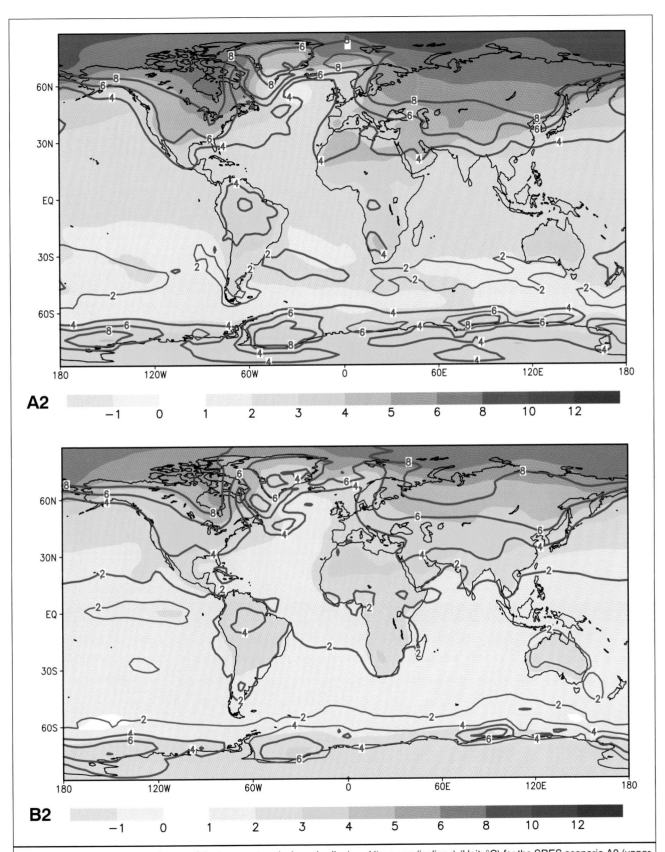

Figure 20: The annual mean change of the temperature (colour shading) and its range (isolines) (Unit: °C) for the SRES scenario A2 (upper panel) and the SRES scenario B2 (lower panel). Both SRES scenarios show the period 2071 to 2100 relative to the period 1961 to 1990 and were performed by OAGCMs. [Based on Figures 9.10d and 9.10e]

F.5. Projections of Future Changes in Extreme Events

It is only recently that changes in extremes of weather and climate observed to date have been compared to changes projected by models (Table 4). More hot days and heat waves are very likely over nearly all land areas. These increases are projected to be largest mainly in areas where soil moisture decreases occur. Increases in daily minimum temperature are projected to occur over nearly all land areas and are generally larger where snow and ice retreat. Frost days and cold waves are very likely to become fewer. The changes in surface air temperature and surface absolute humidity are projected to result in increases in the heat index (which is a measure of the combined effects of temperature and moisture). The increases in surface air temperature are also projected to result in an increase in the "cooling degree days" (which is a measure of the amount of cooling required on a given day once the temperature exceeds a given threshold) and a decrease in "heating degree days". Precipitation extremes are projected to increase more than the mean and the intensity of precipitation events are projected to increase. The frequency of extreme precipitation events is projected to increase almost everywhere. There is projected to be a general drying of the mid-continental areas during summer. This is ascribed to a combination of increased temperature and potential evaporation that is not balanced by increases of precipitation. There is little agreement yet among models concerning future changes in mid-latitude storm intensity, frequency, and variability. There is little consistent evidence that shows changes in the projected frequency of tropical cyclones and areas of formation. However, some measures of intensities show projected increases, and some theoretical and modelling studies suggest that the upper limit of these intensities could increase. Mean and peak precipitation intensities from tropical cyclones are likely to increase appreciably.

For some other extreme phenomena, many of which may have important impacts on the environment and society, there is currently insufficient information to assess recent trends, and confidence in models and understanding is inadequate to make firm projections. In particular, very small-scale phenomena such as thunderstorms, tornadoes, hail, and lightning are not simulated in global models. Insufficient analysis has occurred of how extra-tropical cyclones may change.

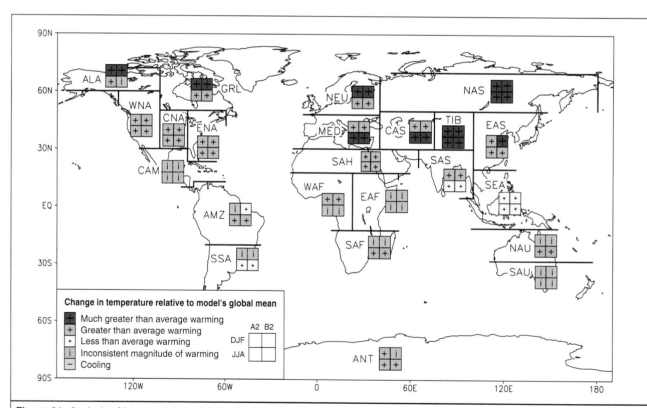

Figure 21: Analysis of inter-model consistency in regional relative warming (warming relative to each model's global average warming). Regions are classified as showing either agreement on warming in excess of 40% above the global average ('Much greater than average warming'), agreement on warming greater than the global average ('Greater than average warming'), agreement on warming less than the global average ('Less than average warming'), or disagreement amongst models on the magnitude of regional relative warming ('Inconsistent magnitude of warming'). There is also a category for agreement on cooling (which never occurs). A consistent result from at least seven of the nine models is deemed necessary for agreement. The global annual average warming of the models used span 1.2 to 4.5°C for A2 and 0.9 to 3.4°C for B2, and therefore a regional 40% amplification represents warming ranges of 1.7 to 6.3°C for A2 and 1.3 to 4.7°C for B2. [Based on Chapter 10, Box 1 , Figure 1]

F.6. Projections of Future Changes in Thermohaline Circulation

Most models show weakening of the Northern Hemisphere Thermohaline Circulation (THC), which contributes to a reduction of the surface warming in the northern North Atlantic. Even in models where the THC weakens, there is still a warming over Europe due to increased greenhouse gases. In experiments where the atmospheric greenhouse gas concentration is stabilised at twice its present day value, the North Atlantic THC is

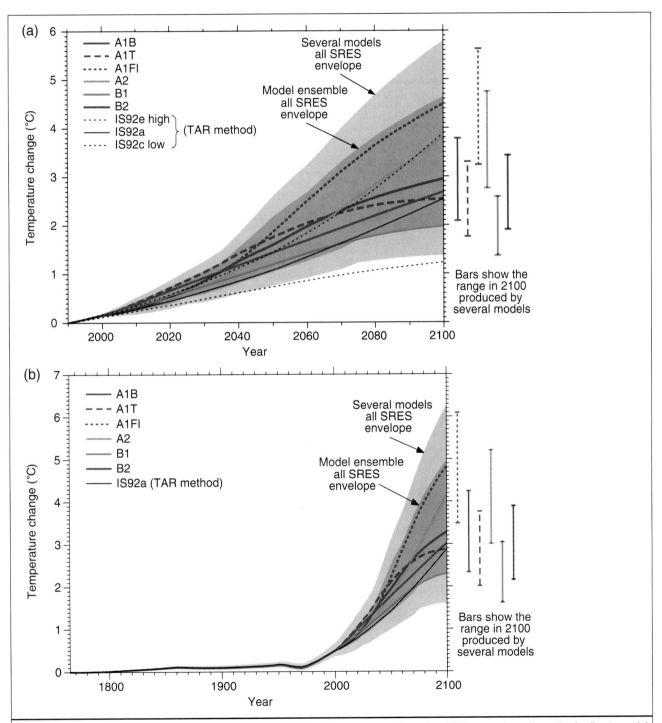

Figure 22: Simple model results: (a) global mean temperature projections for the six illustrative SRES scenarios using a simple climate model tuned to a number of complex models with a range of climate sensitivities. Also for comparison, following the same method, results are shown for IS92a. The darker shading represents the envelope of the full set of thirty-five SRES scenarios using the average of the model results (mean climate sensitivity is 2.8°C). The lighter shading is the envelope based on all seven model projections (with climate sensitivity in the range 1.7 to 4.2°C). The bars show, for each of the six illustrative SRES scenarios, the range of simple model results in 2100 for the seven AOGCM model tunings. (b) Same as (a) but results using estimated historical anthropogenic forcing are also used. [Based on Figures 9.14 and 9.13b]

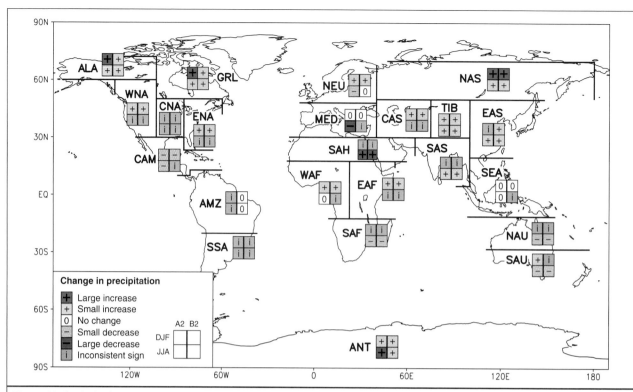

Figure 23: Analysis of inter-model consistency in regional precipitation change. Regions are classified as showing either agreement on increase with an average change of greater than 20% ('Large increase'), agreement on increase with an average change between 5 and 20% ('Small increase'), agreement on a change between −5 and +5% or agreement with an average change between −5 and 5% ('No change'), agreement on decrease with an average change between −5 and −20% ('Small decrease'), agreement on decrease with an average change of less than −20% ('Large decrease'), or disagreement ('Inconsistent sign'). A consistent result from at least seven of the nine models is deemed necessary for agreement. [Based on Chapter 10, Box 1, Figure 2]

projected to recover from initial weakening within one to several centuries. The THC could collapse entirely in either hemisphere if the rate of change in radiative forcing is large enough and applied long enough. Models indicate that a decrease of the THC reduces its resilience to perturbations, i.e., a once reduced THC appears to be less stable and a shut-down can become more likely. However, it is too early to say with confidence whether an irreversible collapse in the THC is likely or not, or at what threshold it might occur and what the climate implications could be. None of the current projections with coupled models exhibits a complete shut-down of the THC by 2100. Although the North Atlantic THC weakens in most models, the relative roles of surface heat and fresh water fluxes vary from model to model. Wind stress changes appear to play only a minor role in the transient response.

F.7. Projections of Future Changes in Modes of Natural Variability

Many models show a mean El Niño-like response in the tropical Pacific, with the central and eastern equatorial Pacific sea surface temperatures projected to warm more than the western equatorial Pacific and with a corresponding mean eastward shift of precipitation. Although many models show an El Niño-like

change of the mean state of tropical Pacific sea surface temperatures, the cause is uncertain. It has been related to changes in the cloud radiative forcing and/or evaporative damping of the east-west sea surface temperature gradient in some models. Confidence in projections of changes in future frequency, amplitude, and spatial pattern of El Niño events in the tropical Pacific is tempered by some shortcomings in how well El Niño is simulated in complex models. Current projections show little change or a small increase in amplitude for El Niño events over the next 100 years. However, even with little or no change in El Niño amplitude, global warming is likely to lead to greater extremes of drying and heavy rainfall and increase the risk of droughts and floods that occur with El Niño events in many regions. It also is likely that warming associated with increasing greenhouse gas concentrations will cause an increase of Asian summer monsoon precipitation variability. Changes in monsoon mean duration and strength depend on the details of the emission scenario. The confidence in such projections is limited by how well the climate models simulate the detailed seasonal evolution of the monsoons. There is no clear agreement on changes in frequency or structure of naturally occurring modes of variability, such as the North Atlantic Oscillation, i.e., the magnitude and character of the changes vary across the models.

Table 4: Estimates of confidence in observed and projected changes in extreme weather and climate events. The table depicts an assessment of confidence in observed changes in extremes of weather and climate during the latter half of the 20th century (left column) and in projected changes during the 21st century (right column)[a]. This assessment relies on observational and modelling studies, as well as physical plausibility of future projections across all commonly used scenarios and is based on expert judgement (see Footnote 4). [Based upon Table 9.6]

Confidence in observed changes (latter half of the 20th century)	Changes in Phenomenon	Confidence in projected changes (during the 21st century)
Likely	**Higher maximum temperatures and more hot days over nearly all land areas**	Very likely
Very likely	**Higher minimum temperatures, fewer cold days and frost days over nearly all land areas**	Very likely
Very likely	**Reduced diurnal temperatures range over most land areas**	Very likely
Likely, over many areas	**Increase of heat index[8] over land areas**	Very likely, over most areas
Likely, over many Northern Hemisphere mid- to high latitude land areas	**More intense percipitation events[b]**	Very likely, over many areas
Likely, in a few areas	**Increased summer continental drying and associated risk of drought**	Likely, over most mid-latitude continental interiors (lack of consistent projections in other areas)
Not observed in the few analyses available	**Increase in tropical cyclone peak wind intensities[c]**	Likely, over some areas
Insufficient data for assassement	**Increase in tropical cyclone mean and peak precipitation intensities[c]**	Likely, over some areas

[a] For more details see Chapter 2 (observations) and Chapter 9, 10 (projections).
[b] For other areas, there are either insufficient data or conflicting analyses.
[c] Past and future changes in tropical cyclone location and frequency are uncertain.

F.8. Projections of Future Changes in Land Ice (Glaciers, Ice Caps and Ice Sheets), Sea Ice and Snow Cover

Glaciers and ice caps will continue their widespread retreat during the 21st century and Northern Hemisphere snow cover and sea ice are projected to decrease further. Methods have been developed recently for estimating glacier melt from seasonally and geographically dependent patterns of surface air temperature change, that are obtained from AOGCM experiments. Modelling studies suggest that the evolution of glacial mass is controlled principally by temperature changes, rather than precipitation changes, on the global average.

The Antarctic ice sheet is likely to gain mass because of greater precipitation, while the Greenland ice sheet is likely to lose mass because the increase in runoff will exceed the precipitation increase. The West Antarctic Ice Sheet (WAIS) has attracted special attention because it contains enough ice to raise sea level

by 6 m and because of suggestions that instabilities associated with its being grounded below sea level may result in rapid ice discharge when the surrounding ice shelves are weakened. However, loss of grounded ice leading to substantial sea level rise from this source is now widely agreed to be very unlikely during the 21st century, although its dynamics are still inadequately understood, especially for projections on longer time-scales.

F.9. Projections of Future Changes in Sea Level

Projections of global average sea level rise from 1990 to 2100, using a range of AOGCMs following the IS92a scenario (including the direct effect of sulphate aerosol emissions), lie in the range 0.11 to 0.77 m. This range reflects the systematic uncertainty of modelling. The main contributions to this sea level rise are:
- a thermal expansion of 0.11 to 0.43 m, accelerating through the 21st century;
- a glacier contribution of 0.01 to 0.23 m;
- a Greenland contribution of −0.02 to 0.09 m; and
- an Antarctic contribution of −0.17 to +0.02 m.

---

[8] Heat index: A combination of temperature and humidity that measures effects on human comfort.

Also included in the computation of the total change are smaller contributions from thawing of permafrost, deposition of sediment, and the ongoing contributions from ice sheets as a result of climate change since the Last Glacial Maximum. To establish the range of sea level rise resulting from the choice of different SRES scenarios, results for thermal expansion and land-ice change from simple models tuned to several AOGCMs are used (as in Section F.3 for temperature).

For the full set of SRES scenarios, a sea level rise of 0.09 to 0.88 m is projected for 1990 to 2100 (see Figure 24), primarily from thermal expansion and loss of mass from glaciers and ice caps. The central value is 0.48 m, which corresponds to an average rate of about two to four times the rate over the 20th century. The range of sea level rise presented in the SAR was 0.13 to 0.94 m based on the IS92 scenarios. Despite higher temperature change projections in this assessment, the sea level projections are slightly lower, primarily due to the use of

improved models which give a smaller contribution from glaciers and ice sheets. If terrestrial storage continues at its current rates, the projections could be changed by −0.21 to 0.11 m. For an average of the AOGCMs, the SRES scenarios give results that differ by 0.02 m or less for the first half of the 21st century. By 2100, they vary over a range amounting to about 50% of the central value. Beyond the 21st century, sea level rise depends strongly on the emissions scenario.

Models agree on the qualitative conclusion that the range of regional variation in sea level change is substantial compared to global average sea level rise. However, confidence in the regional distribution of sea level change from AOGCMs is low because there is little similarity between models, although nearly all models project greater than average rise in the Arctic Ocean and less than average rise in the Southern Ocean. Further, land movements, both isostatic and tectonic, will continue through the 21st century at rates that are unaffected

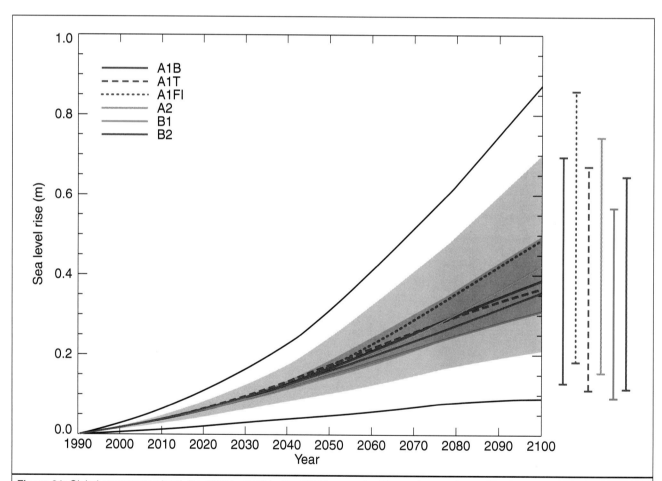

Figure 24: Global average sea level rise 1990 to 2100 for the SRES scenarios. Thermal expansion and land ice changes were calculated using a simple climate model calibrated separately for each of seven AOGCMs, and contributions from changes in permafrost, the effect of sediment deposition and the long-term adjustment of the ice sheets to past climate change were added. Each of the six lines appearing in the key is the average of AOGCMs for one of the six illustrative scenarios. The region in dark shading shows the range of the average of AOGCMs for all thirty five SRES scenarios. The region in light shading shows the range of all AOGCMs for all thirty five scenarios. The region delimited by the outermost lines shows the range of all AOGCMs and scenarios including uncertainty in land-ice changes, permafrost changes and sediment deposition. Note that this range does not allow for uncertainty relating to ice-dynamic changes in the West Antarctic ice sheet. [Based on Figure 11.12]

by climate change. It can be expected that by 2100, many regions currently experiencing relative sea level fall will instead have a rising relative sea level. Lastly, extreme high water levels will occur with increasing frequency as a result of mean sea level rise. Their frequency may be further increased if storms become more frequent or severe as a result of climate change.

F.10. Projections of Future Changes in Response to CO_2 Concentration Stabilisation Profiles

Greenhouse gases and aerosols

All of the stabilisation profiles studied require CO_2 emissions to eventually drop well below current levels. Anthropogenic CO_2 emission rates that arrive at stable CO_2 concentration levels from 450 to 1,000 ppm were deduced from the prescribed CO_2 profiles (Figure 25a). The results (Figure 25b) are not substantially different from those presented in the SAR; however, the range is larger, mainly due to the range of future terrestrial carbon uptake caused by different assumptions in the models. Stabilisation at 450, 650 or 1,000 ppm would require global anthropogenic emissions to drop below 1990 levels within a few decades, about a century, or about two centuries, respectively, and continue to steadily decrease thereafter. Although there is sufficient uptake capacity in the ocean to incorporate 70 to 80% of foreseeable anthropogenic CO_2 emissions to the atmosphere, this process takes centuries due to the rate of ocean mixing. As a result, even several centuries after emissions occurred, about a quarter of the increase in concentration caused by these emissions is still present in the atmosphere. To maintain constant CO_2 concentration beyond 2300 requires emissions to drop to match the rate of carbon sinks at that time. Natural land and ocean sinks with the capacity to persist for hundreds or thousands of years are small (<0.2 PgC/yr).

Temperature

Global mean temperature continues to increase for hundreds of years at a rate of a few tenths of a degree per century after concentrations of CO_2 have been stabilised, due to long time-scales in the ocean. The temperature implications of CO_2 concentration profiles leading to stabilisation from 450 ppm to 1,000 ppm were studied using a simple climate model tuned to seven AOGCMs with a mean climate sensitivity of 2.8°C. For all the pathways leading to stabilisation, the climate system shows considerable warming during the 21st century and beyond (see Figure 26). The lower the level at which concentrations stabilise, the smaller the total temperature change.

Sea level

If greenhouse gas concentrations were stabilised (even at present levels), sea level would nonetheless continue to rise for hundreds of years. After 500 years, sea level rise from thermal

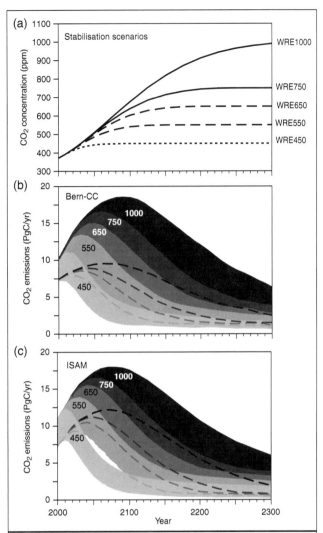

Figure 25: Projected CO_2 emissions permitting stabilisation of atmospheric CO_2 concentrations at different final values. Panel (a) shows the assumed trajectories of CO_2 concentration (WRE scenarios) and panels (b) and (c) show the implied CO_2 emissions, as projected with two fast carbon cycle models, Bern-CC and ISAM. The model ranges for ISAM were obtained by tuning the model to approximate the range of responses to CO_2 and climate from model intercomparisons. This approach yields a lower bound on uncertainties in the carbon cycle response. The model ranges for Bern-CC were obtained by combining different bounding assumptions about the behaviour of the CO_2 fertilisation effect, the response of heterotrophic respiration to temperature and the turnover time of the ocean, thus approaching an upper bound on uncertainties in the carbon cycle response. For each model, the upper and lower bounds are indicated by the top and bottom of the shaded area. Alternatively, the lower bound (where hidden) is indicated by a hatched line. [Based on Figure 3.13]

expansion may have reached only half of its eventual level, which models suggest may lie within a range of 0.5 to 2.0 m and 1 to 4 m for CO_2 levels of twice and four times pre-industrial, respectively. The long time-scale is characteristic of the weak diffusion and slow circulation processes that transport heat into the deep ocean.

The loss of a substantial fraction of the total glacier mass is likely. Areas that are currently marginally glaciated are most likely to become ice-free.

Ice sheets will continue to react to climatic change during the next several thousand years, even if the climate is stabilised. Together, the present Antarctic and Greenland ice sheets contain enough water to raise sea level by almost 70 m if they were to melt, so that only a small fractional change in their volume would have a significant effect.

Models project that a local annual average warming of larger than 3°C, sustained for millennia, would lead to virtually a complete melting of the Greenland ice sheet with a resulting sea level rise of about 7 m. Projected temperatures over Greenland are generally greater than globally averaged temperatures by a factor of 1.2 to 3.1 for the range of models used in Chapter 11. For a warming over Greenland of 5.5° C, consistent with mid-range stabilisation scenarios (see Figure 26), the Greenland ice sheet is likely to contribute about 3 m in 1,000 years. For a warming of 8° C, the contribution is about 6 m, the ice sheet being largely eliminated. For smaller warmings, the decay of the ice sheet would be substantially slower (see Figure 27).

Current ice dynamic models project that the West Antarctic ice sheet (WAIS) will contribute no more than 3 mm/yr to sea level rise over the next thousand years, even if significant changes were to occur in the ice shelves. Such results are strongly dependent on model assumptions regarding climate change scenarios, ice dynamics and other factors. Apart from the possibility of an internal ice dynamic instability, surface melting will affect the long-term viability of the Antarctic ice sheet. For warmings of more than 10°C, simple runoff models predict that a zone of net mass loss would develop on the ice sheet surface. Irreversible disintegration of the WAIS would result because the WAIS cannot retreat to higher ground once its margins are subjected to surface melting and begin to recede. Such a disintegration would take at least a few millennia. Thresholds for total disintegration of the East Antarctic ice sheet by surface melting involve warmings above 20°C, a situation that has not occurred for at least 15 million years and which is far more than predicted by any scenario of climate change currently under consideration.

G. Advancing Understanding

The previous sections have contained descriptions of the current state of knowledge of the climate of the past and present, the current understanding of the forcing agents and processes in the climate system and how well they can be represented in climate models. Given the knowledge possessed today, the best assessment was given whether climate change can be detected and whether that change can be attributed to human influence. With the best tools available today, projections

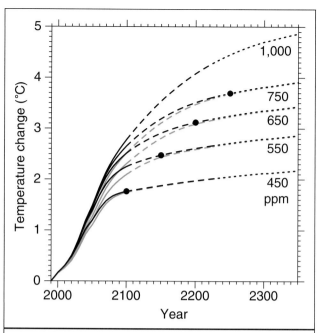

Figure 26: Simple model results: Projected global mean temperature changes when the concentration of CO_2 is stabilised following the WRE profiles (see Chapter 9 Section 9.3.3). For comparison, results based on the S profiles in the SAR are also shown in green (S1000 not available). The results are the average produced by a simple climate model tuned to seven AOGCMs. The baseline scenario is scenario A1B, this is specified only to 2100. After 2100, the emissions of gases other than CO_2 are assumed to remain constant at their A1B 2100 values. The projections are labelled according to the level of CO_2 stabilisation. The broken lines after 2100 indicate increased uncertainty in the simple climate model results beyond 2100. The black dots indicate the time of CO_2 stabilisation. The stabilisation year for the WRE1000 profile is 2375. [Based on Figure 9.16]

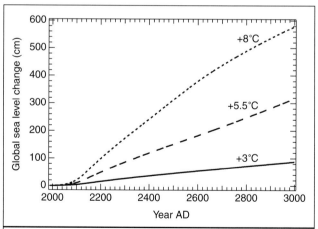

Figure 27: Response of the Greenland ice sheet to three climatic warming scenarios during the third millennium expressed in equivalent changes of global sea level. The curve labels refer to the mean temperature rise over Greenland by 3000 AD as predicted by two-dimensional climate and ocean model forced by greenhouse gas concentration rises until 2130 AD and kept constant after that. Note that projected temperatures over Greenland are generally greater globally averaged temperatures by a factor of 1.2 to 3.1 for the models used in Chapter 11. [Based on Figure 11.16]

were made of how the climate could change in the future for different scenarios of emissions of greenhouse gases.

This Section looks into the future in a different way. Uncertainties are present in each step of the chain from emissions of greenhouse gases and aerosols, through to the impacts that they have on the climate system and society (see Figure 28). Many factors continue to limit the ability to detect, attribute, and understand current climate change and to project what future climate changes may be. Further work is needed in nine broad areas.

G.1. Data

Arrest the decline of observational networks in many parts of the world. Unless networks are significantly improved, it may be difficult or impossible to detect climate change in many areas of the globe.

Expand the observational foundation for climate studies to provide accurate, long-term data with expanded temporal and spatial coverage. Given the complexity of the climate system and the inherent multi-decadal time-scale, there is a need for long-term consistent data to support climate and environmental change investigations and projections. Data from the present and recent past, climate-relevant data for the last few centuries, and for the last several millennia are all needed. There is a particular shortage of data in polar regions and data for the quantitative assessment of extremes on the global scale.

G.2. Climate Processes and Modelling

Estimate better future emissions and concentrations of greenhouse gases and aerosols. It is particularly important that improvements are realised in deriving concentrations from emissions of gases and particularly aerosols, in addressing biogeochemical sequestration and cycling, and specifically, in determining the spatial-temporal distribution of CO_2 sources and sinks, currently and in the future.

Understand and characterise more completely dominant processes (e.g., ocean mixing) and feedbacks (e.g., from clouds and sea ice) in the atmosphere, biota, land and ocean surfaces, and deep oceans. These sub-systems, phenomena, and processes are important and merit increased attention to improve prognostic capabilities generally. The interplay of observation and models will be the key for progress. The rapid forcing of a non-linear system has a high prospect of producing surprises.

Address more completely patterns of long-term climate variability. This topic arises both in model calculations and in the climate system. In simulations, the issue of climate drift within model calculations needs to be clarified better in part because it compounds the difficulty of distinguishing signal and noise. With respect to the long-term natural variability in

the climate system *per se*, it is important to understand this variability and to expand the emerging capability of predicting patterns of organised variability such as ENSO.

Explore more fully the probabilistic character of future climate states by developing multiple ensembles of model calculations. The climate system is a coupled non-linear chaotic system, and therefore the long-term prediction of future exact climate states is not possible. Rather the focus must be upon the prediction of the probability distribution of the system's future possible states by the generation of ensembles of model solutions.

Improve the integrated hierarchy of global and regional climate models with emphasis on improving the simulation of regional impacts and extreme weather events. This will require improvements in the understanding of the coupling between the major atmospheric, oceanic, and terrestrial systems, and extensive diagnostic modelling and observational studies that evaluate and improve simulative performance. A particularly important issue is the adequacy of data needed to attack the question of changes in extreme events.

G.3. Human Aspects

Link more formally physical climate-biogeochemical models with models of the human system and thereby provide the basis for expanded exploration of possible cause-effect-cause patterns linking human and non-human components of the Earth system. At present, human influences generally are treated only through emission scenarios that provide external forcings to the climate system. In future more comprehensive models are required in which human activities need to begin to interact with the dynamics of physical, chemical, and biological sub-systems through a diverse set of contributing activities, feedbacks and responses.

G.4. International Framework

Accelerate internationally progress in understanding climate change by strengthening the international framework that is needed to co-ordinate national and institutional efforts so that research, computational, and observational resources may be used to the greatest overall advantage. Elements of this framework exist in the international programmes supported by the International Council of Scientific Unions (ICSU), the World Meteorological Organization (WMO), the United Nations Environment Programme (UNEP), and the United Nations Education, Scientific and Cultural Organisation (UNESCO). There is a corresponding need for strengthening the co-operation within the international research community, building research capacity in many regions and, as is the goal of this assessment, effectively describing research advances in terms that are relevant to decision making.

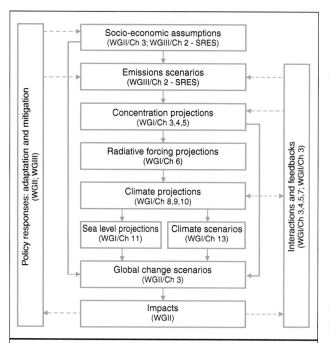

Figure 28: The cascade of uncertainties in projections to be considered in developing climate and related scenarios for climate change impact, adaptation, and mitigation assessment. [Based on Figure 13.2]

Source Information: Technical Summary

This Appendix provides the cross-reference of the topics in the Technical Summary (page and section) to the sections of the chapters that contain expanded information about the topic.

Section A: Introduction

TS Page	Technical Summary Section and Topic – Chapter Section
168	*A.1. The IPCC and its Working Groups* Introduction to the Intergovernmental Panel on Climate Change (from the IPCC Secretariat, Geneva) or the IPCC web page at http://www.ipcc.ch
168	*A.2. The First and Second Assessment Reports of Working Group I* IPCC, 1990a: Climate Change: The IPCC Scientific Assessment. J.T. Houghton, G.J. Jenkins and J.J. Ephraums (eds.), Cambridge University Press, Cambridge, United Kingdom, 365 pp. IPCC, 1992: Climate Change 1992: The Supplementary Report to the IPCC Scientific Assessment. J.T. Houghton, B.A. Callander and S.K. Varney (eds.), Cambridge University Press, Cambridge, United Kingdom, 198 pp. IPCC, 1994: Climate Change 1994: Radiative Forcing of Climate Change and an Evaluation of the IPCC IS92 Emission Scenarios. J.T. Houghton, L.G. Meira Filho, J. Bruce, Hoesung Lee, B.A. Callander, E. Haites, N. Harris and K. Maskell (eds.), Cambridge University Press, Cambridge, United Kingdom, 339 pp. IPCC, 1996a: Climate Change 1995: The Science of Climate Change. Contribution of Working Group I to the Second Assessment Report of the Intergovernmental Panel on Climate Change [Houghton, J.T., L.G. Meira Filho, B.A. Callander, N Harris, A. Kattenberg, and K. Maskell (eds.)]. Cambridge University Press, Cambridge, United Kingdom and New York, NY, USA, 572 pp.
169	*A.3. The Third Assessment Report: This Technical Summary* Background to these questions is in Chapter 1. Box 1: What drives changes in climate? – Chapter 1.

Section B: The Observed Changes in the Climate System

TS Page	Technical Summary Section and Topic – Chapter Section
171-174	*B.1. Observed Changes in Temperature* Temperatures in the instrumental record for land and oceans - Chapter 2.2.2 and 2.3. Temperatures above the surface layer from satellite

Section C: The Forcing Agents That Cause Climate Change

Section D: The Simulation of the Climate System and Its Changes

Climate Change 2001:
Impacts, Adaptation, and Vulnerability

Working Group II Summaries

Summary for Policymakers
A Report of Working Group II of the Intergovernmental Panel on Climate Change

Technical Summary of the Working Group II Report
A Report accepted by Working Group II of the Intergovernmental Panel on Climate Change but not approved in detail

Part of the Working Group II contribution to the Third Assessment Report
of the Intergovernmental Panel on Climate Change

Contents

Climate Change 2001:
Impacts, Adaptation, and Vulnerability

Summary for Policymakers

A Report of Working Group II of the Intergovernmental Panel on Climate Change

This summary, approved in detail at the Sixth Session of IPCC Working Group II (Geneva, Switzerland, 13-16 February 2001), represents the formally agreed statement of the IPCC concerning the sensitivity, adaptive capacity, and vulnerability of natural and human systems to climate change, and the potential consequences of climate change.

Based on a draft prepared by:

Q.K. Ahmad, Oleg Anisimov, Nigel Arnell, Sandra Brown, Ian Burton, Max Campos, Osvaldo Canziani, Timothy Carter, Stewart J. Cohen, Paul Desanker, William Easterling, B. Blair Fitzharris, Donald Forbes, Habiba Gitay, Andrew Githeko, Patrick Gonzalez, Duane Gubler, Sujata Gupta, Andrew Haines, Hideo Harasawa, Jarle Inge Holten, Bubu Pateh Jallow, Roger Jones, Zbigniew Kundzewicz, Murari Lal, Emilio Lebre La Rovere, Neil Leary, Rik Leemans, Chunzhen Liu, Chris Magadza, Martin Manning, Luis Jose Mata, James McCarthy, Roger McLean, Anthony McMichael, Kathleen Miller, Evan Mills, M. Monirul Qader Mirza, Daniel Murdiyarso, Leonard Nurse, Camille Parmesan, Martin Parry, Jonathan Patz, Michel Petit, Olga Pilifosova, Barrie Pittock, Jeff Price, Terry Root, Cynthia Rosenzweig, Jose Sarukhan, John Schellnhuber, Stephen Schneider, Robert Scholes, Michael Scott, Graham Sem, Barry Smit, Joel Smith, Brent Sohngen, Alla Tsyban, Jean-Pascal van Ypersele, Pier Vellinga, Richard Warrick, Tom Wilbanks, Alistair Woodward, David Wratt, and many reviewers

1. Introduction

The sensitivity, adaptive capacity, and vulnerability of natural and human systems to climate change, and the potential consequences of climate change, are assessed in the report of Working Group II of the Intergovernmental Panel on Climate Change (IPCC), *Climate Change 2001: Impacts, Adaptation, and Vulnerability*.[1] This report builds upon the past assessment reports of the IPCC, reexamining key conclusions of the earlier assessments and incorporating results from more recent research.[2, 3]

Observed changes in climate, their causes, and potential future changes are assessed in the report of Working Group I of the IPCC, *Climate Change 2001: The Scientific Basis*. The Working Group I report concludes, *inter alia*, that the globally averaged surface temperatures have increased by $0.6 \pm 0.2°$C over the 20th century; and that, for the range of scenarios developed in the IPCC *Special Report on Emission Scenarios* (SRES), the globally averaged surface air temperature is projected by models to warm 1.4 to $5.8°$C by 2100 relative to 1990, and globally averaged sea level is projected by models to rise 0.09 to 0.88 m by 2100. These projections indicate that the warming would vary by region, and be accompanied by increases and decreases in precipitation. In addition, there would be changes in the variability of climate, and changes in the frequency and intensity of some extreme climate phenomena. These general features of climate change act on natural and human systems and they set the context for the Working Group II assessment. The available literature has not yet investigated climate change impacts, adaptation, and vulnerability associated with the upper end of the projected range of warming.

This Summary for Policymakers, which was approved by IPCC member governments in Geneva in February 2001, describes the current state of understanding of the impacts, adaptation, and vulnerability to climate change and their uncertainties. Further details can be found in the underlying report.[4] Section 2 of the Summary presents a number of general findings that emerge from integration of information across the full report. Each of these findings addresses a different dimension of climate change impacts, adaptation, and vulnerability, and no one dimension is paramount. Section 3 presents findings regarding individual natural and human systems, and Section 4 highlights some of the issues of concern for different regions of the world. Section 5 identifies priority research areas to further advance understanding of the potential consequences of and adaptation to climate change.

2. Emergent Findings

2.1. Recent Regional Climate Changes, particularly Temperature Increases, have Already Affected Many Physical and Biological Systems

Available observational evidence indicates that regional changes in climate, particularly increases in temperature, have already affected a diverse set of physical and biological systems in many parts of the world. Examples of observed changes include shrinkage of glaciers, thawing of permafrost, later freezing and earlier break-up of ice on rivers and lakes, lengthening of mid- to high-latitude growing seasons, poleward and altitudinal shifts of plant and animal ranges, declines of some plant and animal populations, and earlier flowering of trees, emergence of insects, and egg-laying in birds (see Figure SPM-1). Associations between changes in regional temperatures and observed changes in physical and biological systems have been documented in many aquatic, terrestrial, and marine environments. [2.1, 4.3, 4.4, 5.7, and 7.1]

The studies mentioned above and illustrated in Figure SPM-1 were drawn from a literature survey, which identified long-term studies, typically 20 years or more, of changes in biological and physical systems that could be correlated with regional changes in temperature.[5] In most cases where changes in biological and physical systems were detected, the direction of change was that expected on the basis of known mechanisms. The probability that the observed changes in the expected direction (with no reference to magnitude) could occur by chance alone is negligible. In many parts of the world, precipitation-related impacts may be important. At present, there is a lack of systematic concurrent climatic and biophysical data of sufficient length (2 or more decades) that are considered necessary for assessment of precipitation impacts.

---

[1] *Climate change* in IPCC usage refers to any change in climate over time, whether due to natural variability or as a result of human activity. This usage differs from that in the Framework Convention on Climate Change, where *climate change* refers to a change of climate that is attributed directly or indirectly to human activity that alters the composition of the global atmosphere and that is in addition to natural climate variability observed over comparable time periods. Attribution of climate change to natural forcing and human activities has been addressed by Working Group I.

[2] The report has been written by 183 Coordinating Lead Authors and Lead Authors, and 243 Contributing Authors. It was reviewed by 440 government and expert reviewers, and 33 Review Editors oversaw the review process.

[3] Delegations from 100 IPCC member countries participated in the Sixth Session of Working Group II in Geneva on 13-16 February 2001.

[4] A more comprehensive summary of the report is provided in the Technical Summary, and relevant sections of that volume are referenced in brackets at the end of paragraphs of the Summary for Policymakers for readers who need more information.

[5] There are 44 regional studies of over 400 plants and animals, which varied in length from about 20 to 50 years, mainly from North America, Europe, and the southern polar region. There are 16 regional studies covering about 100 physical processes over most regions of the world, which varied in length from about 20 to 150 years. See Section 7.1 of the Technical Summary for more detail.

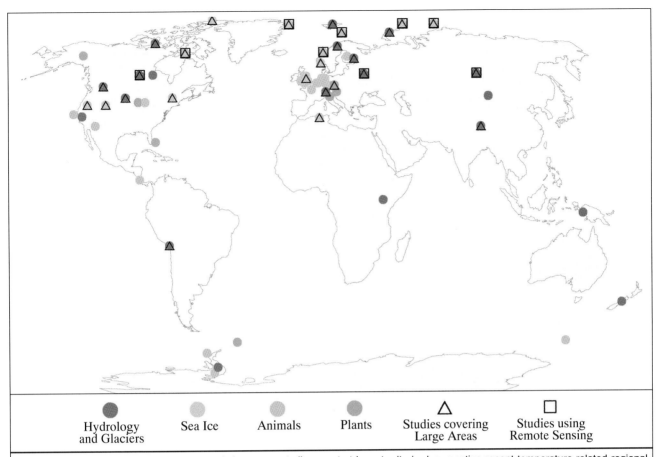

Figure SPM-1: Locations at which systematic long-term studies meet stringent criteria documenting recent temperature-related regional climate change impacts on physical and biological systems. Hydrology, glacial retreat, and sea-ice data represent decadal to century trends. Terrestrial and marine ecosystem data represent trends of at least 2 decades. Remote-sensing studies cover large areas. Data are for single or multiple impacts that are consistent with known mechanisms of physical/biological system responses to observed regional temperature-related changes. For reported impacts spanning large areas, a representative location on the map was selected.

Factors such as land-use change and pollution also act on these physical and biological systems, making it difficult to attribute changes to particular causes in some specific cases. However, taken together, the observed changes in these systems are consistent in direction and coherent across diverse localities and/or regions (see Figure SPM-1) with the expected effects of regional changes in temperature. Thus, from the collective evidence, there is *high confidence*[6] that recent regional changes in temperature have had discernible impacts on many physical and biological systems.

---

[6] In this Summary for Policymakers, the following words have been used where appropriate to indicate judgmental estimates of confidence (based upon the collective judgment of the authors using the observational evidence, modeling results, and theory that they have examined): *very high* (95% or greater), *high* (67-95%), *medium* (33-67%), *low* (5-33%), and *very low* (5% or less). In other instances, a qualitative scale to gauge the level of scientific understanding is used: *well established*, *established-but-incomplete*, *competing explanations*, and *speculative*. The approaches used to assess confidence levels and the level of scientific understanding, and the definitions of these terms, are presented in Section 1.4 of the Technical Summary. Each time these terms are used in the Summary for Policymakers, they are footnoted and in *italics*.

2.2. There are Preliminary Indications that Some Human Systems have been Affected by Recent Increases in Floods and Droughts

There is emerging evidence that some social and economic systems have been affected by the recent increasing frequency of floods and droughts in some areas. However, such systems are also affected by changes in socioeconomic factors such as demographic shifts and land-use changes. The relative impact of climatic and socioeconomic factors are generally difficult to quantify. [4.6 and 7.1]

2.3. Natural Systems are Vulnerable to Climate Change, and Some will be Irreversibly Damaged

Natural systems can be especially vulnerable to climate change because of limited adaptive capacity (see Box SPM-1), and some of these systems may undergo significant and irreversible damage. Natural systems at risk include glaciers, coral reefs and atolls, mangroves, boreal and tropical forests, polar and alpine ecosystems, prairie wetlands, and remnant native grasslands. While some species may increase in abundance or

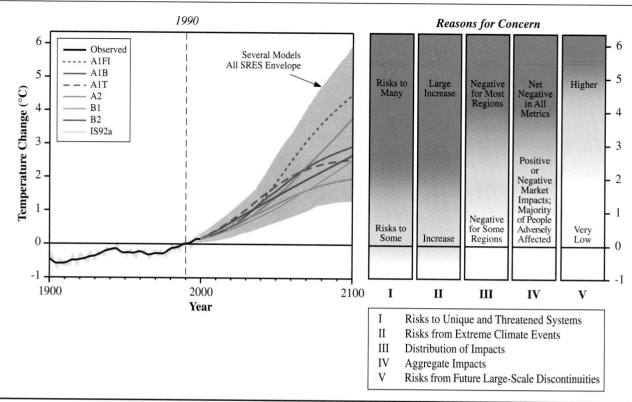

	Risks to Unique and Threatened Systems
I	Risks to Unique and Threatened Systems
II	Risks from Extreme Climate Events
III	Distribution of Impacts
IV	Aggregate Impacts
V	Risks from Future Large-Scale Discontinuities

Figure SPM-2: Reasons for concern about projected climate change impacts. The risks of adverse impacts from climate change increase with the magnitude of climate change. The left part of the figure displays the observed temperature increase relative to 1990 and the range of projected temperature increase after 1990 as estimated by Working Group I of the IPCC for scenarios from the *Special Report on Emissions Scenarios*. The right panel displays conceptualizations of five reasons for concern regarding climate change risks evolving through 2100. White indicates neutral or small negative or positive impacts or risks, yellow indicates negative impacts for some systems or low risks, and red means negative impacts or risks that are more widespread and/or greater in magnitude. The assessment of impacts or risks takes into account only the magnitude of change and not the rate of change. Global mean annual temperature change is used in the figure as a proxy for the magnitude of climate change, but projected impacts will be a function of, among other factors, the magnitude and rate of global and regional changes in mean climate, climate variability and extreme climate phenomena, social and economic conditions, and adaptation.

range, climate change will increase existing risks of extinction of some more vulnerable species and loss of biodiversity. It is *well-established*[6] that the geographical extent of the damage or loss, and the number of systems affected, will increase with the magnitude and rate of climate change (see Figure SPM-2). [4.3 and 7.2.1]

2.4. Many Human Systems are Sensitive to Climate Change, and Some are Vulnerable

Human systems that are sensitive to climate change include mainly water resources; agriculture (especially food security) and forestry; coastal zones and marine systems (fisheries); human settlements, energy, and industry; insurance and other financial services; and human health. The vulnerability of these systems varies with geographic location, time, and social, economic, and environmental conditions. [4.1, 4.2, 4.3, 4.4, 4.5, 4.6, and 4.7]

Projected adverse impacts based on models and other studies include:

- A general reduction in potential crop yields in most tropical and sub-tropical regions for most projected increases in temperature [4.2]
- A general reduction, with some variation, in potential crop yields in most regions in mid-latitudes for increases in annual-average temperature of more than a few °C [4.2]
- Decreased water availability for populations in many water-scarce regions, particularly in the sub-tropics [4.1]
- An increase in the number of people exposed to vector-borne (e.g., malaria) and water-borne diseases (e.g., cholera), and an increase in heat stress mortality [4.7]
- A widespread increase in the risk of flooding for many human settlements (tens of millions of inhabitants in settlements studied) from both increased heavy precipitation events and sea-level rise [4.5]
- Increased energy demand for space cooling due to higher summer temperatures. [4.5]

Box SPM-1. Climate Change Sensitivity, Adaptive Capacity, and Vulnerability

Sensitivity is the degree to which a system is affected, either adversely or beneficially, by climate-related stimuli. Climate-related stimuli encompass all the elements of climate change, including mean climate characteristics, climate variability, and the frequency and magnitude of extremes. The effect may be direct (e.g., a change in crop yield in response to a change in the mean, range, or variability of temperature) or indirect (e.g., damages caused by an increase in the frequency of coastal flooding due to sea-level rise).

Adaptive capacity is the ability of a system to adjust to climate change (including climate variability and extremes) to moderate potential damages, to take advantage of opportunities, or to cope with the consequences.

Vulnerability is the degree to which a system is susceptible to, or unable to cope with, adverse effects of climate change, including climate variability and extremes. Vulnerability is a function of the character, magnitude, and rate of climate change and variation to which a system is exposed, its sensitivity, and its adaptive capacity.

Projected beneficial impacts based on models and other studies include:

- Increased potential crop yields in some regions at mid-latitudes for increases in temperature of less than a few °C [4.2]
- A potential increase in global timber supply from appropriately managed forests [4.3]
- Increased water availability for populations in some water-scarce regions—for example, in parts of southeast Asia [4.1]
- Reduced winter mortality in mid- and high-latitudes [4.7]
- Reduced energy demand for space heating due to higher winter temperatures. [4.5]

2.5. Projected Changes in Climate Extremes could have Major Consequences

The vulnerability of human societies and natural systems to climate extremes is demonstrated by the damage, hardship, and death caused by events such as droughts, floods, heat waves, avalanches, and windstorms. While there are uncertainties attached to estimates of such changes, some extreme events are projected to increase in frequency and/or severity during the 21st century due to changes in the mean and/or variability of climate, so it can be expected that the severity of their impacts will also increase in concert with global warming (see Figure SPM-2). Conversely, the frequency and magnitude of extreme low temperature events, such as cold spells, is projected to decrease in the future, with both positive and negative impacts. The impacts of future changes in climate extremes are expected

to fall disproportionately on the poor. Some representative examples of impacts of these projected changes in climate variability and climate extremes are presented in Table SPM-1. [3.5, 4.6, 6, and 7.2.4]

2.6. The Potential for Large-Scale and Possibly Irreversible Impacts Poses Risks that have yet to be Reliably Quantified

Projected climate changes[7] during the 21st century have the potential to lead to future large-scale and possibly irreversible changes in Earth systems resulting in impacts at continental and global scales. These possibilities are very climate scenario-dependent and a full range of plausible scenarios has not yet been evaluated. Examples include significant slowing of the ocean circulation that transports warm water to the North Atlantic, large reductions in the Greenland and West Antarctic Ice Sheets, accelerated global warming due to carbon cycle feedbacks in the terrestrial biosphere, and releases of terrestrial carbon from permafrost regions and methane from hydrates in coastal sediments. The likelihood of many of these changes in Earth systems is not well-known, but is probably very low; however, their likelihood is expected to increase with the rate, magnitude, and duration of climate change (see Figure SPM-2). [3.5, 5.7, and 7.2.5]

If these changes in Earth systems were to occur, their impacts would be widespread and sustained. For example, significant slowing of the oceanic thermohaline circulation would impact deep-water oxygen levels and carbon uptake by oceans and marine ecosystems, and would reduce warming over parts of Europe. Disintegration of the West Antarctic Ice Sheet or melting of the Greenland Ice Sheet could raise global sea level up to 3 m each over the next 1,000 years[8], submerge many islands, and inundate extensive coastal areas. Depending on the rate of ice loss, the rate and magnitude of sea-level rise could greatly exceed the capacity of human and natural systems to adapt without substantial impacts. Releases of terrestrial carbon from permafrost regions and methane from hydrates in coastal sediments, induced by warming, would further increase greenhouse gas concentrations in the atmosphere and amplify climate change. [3.5, 5.7, and 7.2.5]

2.7. Adaptation is a Necessary Strategy at All Scales to Complement Climate Change Mitigation Efforts

Adaptation has the potential to reduce adverse impacts of climate change and to enhance beneficial impacts, but will incur costs

---

[7] Details of projected climate changes, illustrated in Figure SPM-2, are provided in the Working Group I Summary for Policymakers.

[8] Details of projected contributions to sea-level rise from the West Anarctic Ice Sheet and Greenland Ice Sheet are provided in the Working Group I Summary for Policymakers.

Table SPM-1: Examples of impacts resulting from projected changes in extreme climate events.

Projected Changes during the 21st Century in Extreme Climate Phenomena and their Likelihood[a]	Representative Examples of Projected Impacts[b] (all high confidence of occurrence in some areas[c])
Simple Extremes	
Higher maximum temperatures; more hot days and heat waves[d] over nearly all land areas (*very likely*[a])	· Increased incidence of death and serious illness in older age groups and urban poor [4.7] · Increased heat stress in livestock and wildlife [4.2 and 4.3] · Shift in tourist destinations [Table TS-4 and 5.8] · Increased risk of damage to a number of crops [4.2] · Increased electric cooling demand and reduced energy supply reliability [Table TS-4 and 4.5]
Higher (increasing) minimum temperatures; fewer cold days, frost days, and cold waves[d] over nearly all land areas (*very likely*[a])	· Decreased cold-related human morbidity and mortality [4.7] · Decreased risk of damage to a number of crops, and increased risk to others [4.2] · Extended range and activity of some pest and disease vectors [4.2 and 4.3] · Reduced heating energy demand [4.5]
More intense precipitation events (*very likely*[a] over many areas)	· Increased flood, landslide, avalanche, and mudslide damage [4.5] · Increased soil erosion [5.2.4] · Increased flood runoff could increase recharge of some floodplain aquifers [4.1] · Increased pressure on government and private flood insurance systems and disaster relief [Table TS-4 and 4.6]
Complex Extremes	
Increased summer drying over most mid-latitude continental interiors and associated risk of drought (*likely*[a])	· Decreased crop yields [4.2] · Increased damage to building foundations caused by ground shrinkage [Table TS-4] · Decreased water resource quantity and quality [4.1 and 4.5] · Increased risk of forest fire [5.4.2]
Increase in tropical cyclone peak wind intensities, mean and peak precipitation intensities (*likely*[a] over some areas)[e]	· Increased risks to human life, risk of infectious disease epidemics, and many other risks [4.7] · Increased coastal erosion and damage to coastal buildings and infrastructure [4.5 and 7.2.4] · Increased damage to coastal ecosystems such as coral reefs and mangroves [4.4]
Intensified droughts and floods associated with El Niño events in many different regions (*likely*[a]) (see also under droughts and intense precipitation events)	· Decreased agricultural and rangeland productivity in drought- and flood-prone regions [4.3] · Decreased hydro-power potential in drought-prone regions [5.1.1 and Figure TS-7]
Increased Asian summer monsoon precipitation variability (*likely*[a])	· Increased flood and drought magnitude and damages in temperate and tropical Asia [5.2.4]
Increased intensity of mid-latitude storms (little agreement between current models)[d]	· Increased risks to human life and health [4.7] · Increased property and infrastructure losses [Table TS-4] · Increased damage to coastal ecosystems [4.4]

[a] Likelihood refers to judgmental estimates of confidence used by TAR WGI: *very likely* (90-99% chance); *likely* (66-90% chance). Unless otherwise stated, information on climate phenomena is taken from the Summary for Policymakers, TAR WGI.

[b] These impacts can be lessened by appropriate response measures.

[c] High confidence refers to probabilities between 67 and 95% as described in Footnote 6.

[d] Information from TAR WGI, Technical Summary, Section F.5.

[e] Changes in regional distribution of tropical cyclones are possible but have not been established.

and will not prevent all damages. Extremes, variability, and rates of change are all key features in addressing vulnerability and adaptation to climate change, not simply changes in average climate conditions. Human and natural systems will to some degree adapt autonomously to climate change. Planned adaptation can supplement autonomous adaptation, though options and incentives are greater for adaptation of human systems than for adaptation to protect natural systems.

Adaptation is a necessary strategy at all scales to complement climate change mitigation efforts. [6]

Experience with adaptation to climate variability and extremes can be drawn upon to develop appropriate strategies for adapting to anticipated climate change. Adaptation to current climate variability and extremes often produces benefits as well as forming a basis for coping with future climate change. However, experience also demonstrates that there are constraints to achieving the full measure of potential adaptation. In addition, maladaptation, such as promoting development in risk-prone locations, can occur due to decisions based on short-term considerations, neglect of known climatic variability, imperfect foresight, insufficient information, and over-reliance on insurance mechanisms. [6]

2.8. Those with the Least Resources have the Least Capacity to Adapt and are the Most Vulnerable

The ability of human systems to adapt to and cope with climate change depends on such factors as wealth, technology, education, information, skills, infrastructure, access to resources, and management capabilities. There is potential for developed and developing countries to enhance and/or acquire adaptive capabilities. Populations and communities are highly variable in their endowments with these attributes, and the developing countries, particularly the least developed countries, are generally poorest in this regard. As a result, they have lesser capacity to adapt and are more vulnerable to climate change damages, just as they are more vulnerable to other stresses. This condition is most extreme among the poorest people. [6.1; see also 5.1.7, 5.2.7, 5.3.5, 5.4.6, 5.6.1, 5.6.2, 5.7, and 5.8.1 for regional-scale information]

Benefits and costs of climate change effects have been estimated in monetary units and aggregated to national, regional, and global scales. These estimates generally exclude the effects of changes in climate variability and extremes, do not account for the effects of different rates of change, and only partially account for impacts on goods and services that are not traded in markets. These omissions are likely to result in underestimates of economic losses and overestimates of economic gains. Estimates of aggregate impacts are controversial because they treat gains for some as canceling out losses for others and because the weights that are used to aggregate across individuals are necessarily subjective. [7.2.2 and 7.2.3]

Notwithstanding the limitations expressed above, based on a few published estimates, increases in global mean temperature[9] would produce net economic losses in many developing countries

for all magnitudes of warming studied (*low confidence*[6]), and losses would be greater in magnitude the higher the level of warming (*medium confidence*[6]). In contrast, an increase in global mean temperature of up to a few °C would produce a mixture of economic gains and losses in developed countries (*low confidence*[6]), with economic losses for larger temperature increases (*medium confidence*[6]). The projected distribution of economic impacts is such that it would increase the disparity in well-being between developed countries and developing countries, with disparity growing for higher projected temperature increases (*medium confidence*[6]). The more damaging impacts estimated for developing countries reflects, in part, their lesser adaptive capacity relative to developed countries. [7.2.3]

Further, when aggregated to a global scale, world gross domestic product (GDP) would change by ± a few percent for global mean temperature increases of up to a few °C (*low confidence*[6]), and increasing net losses would result for larger increases in temperature (*medium confidence*[6]) (see Figure SPM-2). More people are projected to be harmed than benefited by climate change, even for global mean temperature increases of less than a few °C (*low confidence*[6]). These results are sensitive to assumptions about changes in regional climate, level of development, adaptive capacity, rate of change, the valuation of impacts, and the methods used for aggregating monetary losses and gains, including the choice of discount rate. [7.2.2]

The effects of climate change are expected to be greatest in developing countries in terms of loss of life and relative effects on investment and the economy. For example, the relative percentage damages to GDP from climate extremes have been substantially greater in developing countries than in developed countries. [4.6]

2.9. Adaptation, Sustainable Development, and Enhancement of Equity can be Mutually Reinforcing

Many communities and regions that are vulnerable to climate change are also under pressure from forces such as population growth, resource depletion, and poverty. Policies that lessen pressures on resources, improve management of environmental risks, and increase the welfare of the poorest members of society can simultaneously advance sustainable development and equity, enhance adaptive capacity, and reduce vulnerability to climate and other stresses. Inclusion of climatic risks in the design and implementation of national and international development initiatives can promote equity and development that is more sustainable and that reduces vulnerability to climate change. [6.2]

---

[9] Global mean temperature change is used as an indicator of the magnitude of climate change. Scenario-dependent exposures taken into account in these studies include regionally differentiated changes in temperature, precipitation, and other climatic variables.

3. Effects on and Vulnerability of Natural and Human Systems

3.1. Hydrology and Water Resources

The effect of climate change on streamflow and groundwater recharge varies regionally and between climate scenarios, largely following projected changes in precipitation. A consistent projection across most climate change scenarios is for increases in annual mean streamflow in high latitudes and southeast Asia, and decreases in central Asia, the area around the Mediterranean, southern Africa, and Australia (*medium confidence*[6]) (see Figure SPM-3); the amount of change, however, varies between scenarios. For other areas, including mid-latitudes, there is no strong consistency in projections of streamflow, partly because of differences in projected rainfall and partly because of differences in projected evaporation, which can offset rainfall increases. The retreat of most glaciers is projected to accelerate, and many small glaciers may disappear (*high confidence*[6]). In general, the projected changes in average annual runoff are less robust than impacts based solely on temperature change because precipitation changes vary more between scenarios. At the catchment scale, the effect of a given change in climate varies with physical properties and vegetation of catchments, and may be in addition to land-cover changes. [4.1]

Approximately 1.7 billion people, one-third of the world's population, presently live in countries that are water-stressed (defined as using more than 20% of their renewable water supply, a commonly used indicator of water stress). This number is projected to increase to around 5 billion by 2025, depending on the rate of population growth. The projected climate change could further decrease the streamflow and groundwater recharge in many of these water-stressed countries—for example in central Asia, southern Africa, and countries around the Mediterranean Sea—but may increase it in some others. [4.1; see also 5.1.1, 5.2.3, 5.3.1, 5.4.1, 5.5.1, 5.6.2, and 5.8.4 for regional-scale information]

Demand for water is generally increasing due to population growth and economic development, but is falling in some countries because of increased efficiency of use. Climate change is unlikely to have a big effect on municipal and industrial water demands in general, but may substantially affect irrigation withdrawals, which depend on how increases in evaporation are offset or exaggerated by changes in precipitation. Higher temperatures, hence higher crop evaporative demand, mean that the general tendency would be towards an increase in irrigation demands. [4.1]

Flood magnitude and frequency could increase in many regions as a consequence of increased frequency of heavy precipitation events, which can increase runoff in most areas as well as groundwater recharge in some floodplains. Land-use change could exacerbate such events. Streamflow during seasonal low flow periods would decrease in many areas due to greater evaporation; changes in precipitation may exacerbate or offset the effects of increased evaporation. The projected climate change would degrade water quality through higher water temperatures and increased pollutant load from runoff and overflows of waste facilities. Quality would be degraded further where flows decrease, but increases in flows may mitigate to a certain extent some degradations in water quality by increasing dilution. Where snowfall is currently an important component of the water balance, a greater proportion of winter precipitation may fall as rain, and this can result in a more intense peak streamflow which in addition would move from spring to winter. [4.1]

The greatest vulnerabilities are likely to be in unmanaged water systems and systems that are currently stressed or poorly and unsustainably managed due to policies that discourage efficient water use and protection of water quality, inadequate watershed management, failure to manage variable water supply and demand, or lack of sound professional guidance. In unmanaged systems there are few or no structures in place to buffer the effects of hydrologic variability on water quality and supply. In unsustainably managed systems, water and land uses can add stresses that heighten vulnerability to climate change. [4.1]

Water resource management techniques, particularly those of integrated water resource management, can be applied to adapt to hydrologic effects of climate change, and to additional uncertainty, so as to lessen vulnerabilities. Currently, supply-side approaches (e.g., increasing flood defenses, building weirs, utilizing water storage areas, including natural systems, improving infrastructure for water collection and distribution) are more widely used than demand-side approaches (which alter the exposure to stress); the latter is the focus of increasing attention. However, the capacity to implement effective management responses is unevenly distributed around the world and is low in many transition and developing countries. [4.1]

3.2. Agriculture and Food Security

Based on experimental research, crop yield responses to climate change vary widely, depending upon species and cultivar; soil properties; pests and pathogens; the direct effects of carbon dioxide (CO_2) on plants; and interactions between CO_2, air temperature, water stress, mineral nutrition, air quality, and adaptive responses. Even though increased CO_2 concentration can stimulate crop growth and yield, that benefit may not always overcome the adverse effects of excessive heat and drought (*medium confidence*[6]). These advances, along with advances in research on agricultural adaptation, have been incorporated since the Second Assessment Report (SAR) into models used to assess the effects of climate change on crop yields, food supply, farm incomes, and prices. [4.2]

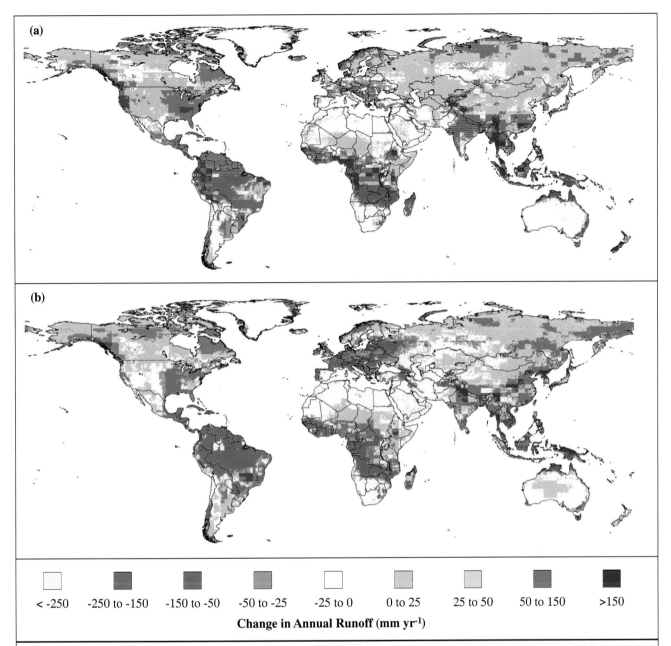

Figure SPM-3: Projected changes in average annual water runoff by 2050, relative to average runoff for 1961-1990, largely follow projected changes in precipitation. Changes in runoff are calculated with a hydrologic model using as inputs climate projections from two versions of the Hadley Centre atmosphere-ocean general circulation model (AOGCM) for a scenario of 1% per annum increase in effective carbon dioxide concentration in the atmosphere: (a) HadCM2 ensemble mean and (b) HadCM3. Projected increases in runoff in high latitudes and southeast Asia, and decreases in central Asia, the area around the Mediterranean, southern Africa, and Australia are broadly consistent across the Hadley Centre experiments, and with the precipitation projections of other AOGCM experiments. For other areas of the world, changes in precipitation and runoff are scenario- and model-dependent.

Costs will be involved in coping with climate-induced yield losses and adaptation of livestock production systems. These agronomic and husbandry adaptation options could include, for example, adjustments to planting dates, fertilization rates, irrigation applications, cultivar traits, and selection of animal species. [4.2]

When autonomous agronomic adaptation is included, crop modeling assessments indicate, with *medium* to *low confidence*[6], that climate change will lead to generally positive responses at less than a few °C warming and generally negative responses for more than a few °C in mid-latitude crop yields. Similar assessments indicate that yields of some crops in tropical locations would decrease generally with even minimal increases in temperature, because such crops are near their maximum temperature tolerance and dryland/rainfed agriculture predominates. Where there is also a large decrease in rainfall, tropical crop yields would be even more adversely affected. With autonomous agronomic adaptation, crop yields in the tropics tend to be less adversely affected by climate change than without adaptation, but they still tend to remain below levels estimated with current climate. [4.2]

Most global and regional economic studies not incorporating climate change indicate that the downward trend in global real commodity prices in the 20th century is likely to continue into the 21st, although confidence in these predictions decreases farther into the future. Economic modeling assessments indicate that impacts of climate change on agricultural production and prices are estimated to result in small percentage changes in global income (*low confidence*[6]), with larger increases in more developed regions and smaller increases or declines in developing regions. Improved confidence in this finding depends on further research into the sensitivity of economic modeling assessments to their base assumptions. [4.2 and Box 5-5]

Most studies indicate that global mean annual temperature increases of a few °C or greater would prompt food prices to increase due to a slowing in the expansion of global food supply relative to growth in global food demand (*established, but incomplete*[6]). At lesser amounts of warming than a few °C, economic models do not clearly distinguish the climate change signal from other sources of change based on those studies included in this assessment. Some recent aggregated studies have estimated economic impacts on vulnerable populations such as smallholder producers and poor urban consumers. These studies find that climate change would lower incomes of the vulnerable populations and increase the absolute number of people at risk of hunger, though this is uncertain and requires further research. It is established, though incompletely, that climate change, mainly through increased extremes and temporal/spatial shifts, will worsen food security in Africa. [4.2]

3.3. Terrestrial and Freshwater Ecosystems

Vegetation modeling studies continue to show the potential for significant disruption of ecosystems under climate change (*high confidence*[6]). Migration of ecosystems or biomes as discrete units is unlikely to occur; instead at a given site, species composition and dominance will change. The results of these changes will lag behind the changes in climate by years to decades to centuries (*high confidence*[6]). [4.3]

Distributions, population sizes, population density, and behavior of wildlife have been, and will continue to be, affected directly by changes in global or regional climate and indirectly through changes in vegetation. Climate change will lead to poleward movement of the boundaries of freshwater fish distributions along with loss of habitat for cold- and cool-water fishes and gain in habitat for warm-water fishes (*high confidence*[6]). Many species and populations are already at high risk, and are expected to be placed at greater risk by the synergy between climate change rendering portions of current habitat unsuitable for many species, and land-use change fragmenting habitats and raising obstacles to species migration. Without

appropriate management, these pressures will cause some species currently classified as "critically endangered" to become extinct and the majority of those labeled "endangered or vulnerable" to become rarer, and thereby closer to extinction, in the 21st century (*high confidence*[6]). [4.3]

Possible adaptation methods to reduce risks to species could include: 1) establishment of refuges, parks, and reserves with corridors to allow migration of species, and 2) use of captive breeding and translocation. However, these options may have limitations due to costs. [4.3]

Terrestrial ecosystems appear to be storing increasing amounts of carbon. At the time of the SAR, this was largely attributed to increasing plant productivity because of the interaction between elevated CO_2 concentration, increasing temperatures, and soil moisture changes. Recent results confirm that productivity gains are occurring but suggest that they are smaller under field conditions than indicated by plant-pot experiments (*medium confidence*[6]). Hence, the terrestrial uptake may be due more to change in uses and management of land than to the direct effects of elevated CO_2 and climate. The degree to which terrestrial ecosystems continue to be net sinks for carbon is uncertain due to the complex interactions between the factors mentioned above (e.g., arctic terrestrial ecosystems and wetlands may act as both sources and sinks) (*medium confidence*[6]). [4.3]

Contrary to the SAR, global timber market studies that include adaptations through land and product management, even without forestry projects that increase the capture and storage of carbon, suggest that a small amount of climate change would increase global timber supply and enhance existing market trends towards rising market share in developing countries (*medium confidence*[6]). Consumers may benefit from lower timber prices while producers may gain or lose depending on regional changes in timber productivity and potential dieback effects. [4.3]

3.4. Coastal Zones and Marine Ecosystems

Large-scale impacts of climate change on oceans are expected to include increases in sea surface temperature and mean global sea level, decreases in sea-ice cover, and changes in salinity, wave conditions, and ocean circulation. The oceans are an integral and responsive component of the climate system with important physical and biogeochemical feedbacks to climate. Many marine ecosystems are sensitive to climate change. Climate trends and variability as reflected in multiyear climate-ocean regimes (e.g., Pacific Decadal Oscillation) and switches from one regime to another are now recognized to strongly affect fish abundance and population dynamics, with significant impacts on fish-dependent human societies. [4.4]

Many coastal areas will experience increased levels of flooding, accelerated erosion, loss of wetlands and mangroves, and seawater intrusion into freshwater sources as a result of climate change. The extent and severity of storm impacts, including storm-surge floods and shore erosion, will increase as a result of climate change including sea-level rise. High-latitude coasts will experience added impacts related to higher wave energy and permafrost degradation. Changes in relative sea level will vary locally due to uplift and subsidence caused by other factors. [4.4]

Impacts on highly diverse and productive coastal ecosystems such as coral reefs, atolls and reef islands, salt marshes, and mangrove forests will depend upon the rate of sea-level rise relative to growth rates and sediment supply, space for and obstacles to horizontal migration, changes in the climate-ocean environment such as sea surface temperatures and storminess, and pressures from human activities in coastal zones. Episodes of coral bleaching over the past 20 years have been associated with several causes, including increased ocean temperatures. Future sea surface warming would increase stress on coral reefs and result in increased frequency of marine diseases (*high confidence*[6]). [4.4]

Assessments of adaptation strategies for coastal zones have shifted emphasis away from hard protection structures of shorelines (e.g., seawalls, groins) toward soft protection measures (e.g., beach nourishment), managed retreat, and enhanced resilience of biophysical and socioeconomic systems in coastal regions. Adaptation options for coastal and marine management are most effective when incorporated with policies in other areas, such as disaster mitigation plans and land-use plans. [4.4]

3.5. Human Health

The impacts of short-term weather events on human health have been further elucidated since the SAR, particularly in relation to periods of thermal stress, the modulation of air pollution impacts, the impacts of storms and floods, and the influences of seasonal and interannual climatic variability on infectious diseases. There has been increased understanding of the determinants of population vulnerability to adverse health impacts and the possibilities for adaptive responses. [4.7]

Many vector-, food-, and water-borne infectious diseases are known to be sensitive to changes in climatic conditions. From results of most predictive model studies, there is *medium* to *high confidence*[6] that, under climate change scenarios, there would be a net increase in the geographic range of potential transmission of malaria and dengue—two vector-borne infections each of which currently impinge on 40-50% of the world population.[10] Within their present ranges, these and many other infectious diseases would tend to increase in incidence and seasonality—although regional decreases would occur in some infectious diseases. In all cases, however, actual disease occurrence is strongly influenced by local environmental conditions, socioeconomic circumstances, and public health infrastructure. [4.7]

Projected climate change will be accompanied by an increase in heat waves, often exacerbated by increased humidity and urban air pollution, which would cause an increase in heat-related deaths and illness episodes. The evidence indicates that the impact would be greatest in urban populations, affecting particularly the elderly, sick, and those without access to air-conditioning (*high confidence*[6]). Limited evidence indicates that in some temperate countries reduced winter deaths would outnumber increased summer deaths (*medium confidence*[6]); yet, published research has been largely confined to populations in developed countries, thus precluding a generalized comparison of changes in summer and winter mortality. [3.5 and 4.7]

Extensive experience makes clear that any increase in flooding will increase the risk of drowning, diarrhoeal and respiratory diseases, and, in developing countries, hunger and malnutrition (*high confidence*[6]). If cyclones were to increase regionally, devastating impacts would often occur, particularly in densely settled populations with inadequate resources. A reduction in crop yields and food production because of climate change in some regions, particularly in the tropics, will predispose food-insecure populations to malnutrition, leading to impaired child development and decreased adult activity. Socioeconomic disruptions could occur in some regions, impairing both livelihoods and health. [3.5, 4.1, 4.2, 4.5, and 4.7]

For each anticipated adverse health impact there is a range of social, institutional, technological, and behavioral adaptation options to lessen that impact. Adaptations could, for example, encompass strengthening of the public health infrastructure, health-oriented management of the environment (including air and water quality, food safety, urban and housing design, and surface water management), and the provision of appropriate medical care facilities. Overall, the adverse health impacts of climate change will be greatest in vulnerable lower income populations, predominantly within tropical/subtropical countries. Adaptive policies would, in general, reduce these impacts. [4.7]

3.6. Human Settlements, Energy, and Industry

A growing and increasingly quantitative literature shows that human settlements are affected by climate change in one of three major ways:

1) The economic sectors that support the settlement are affected because of changes in resource productivity or changes in market demand for the goods and services produced there. [4.5]

---

[10] Eight studies have modeled the effects of climate change on these diseases, five on malaria and three on dengue. Seven use a biological or process-based approach, and one uses an empirical, statistical approach.

2) Some aspects of physical infrastructure (including energy transmission and distribution systems), buildings, urban services (including transportation systems), and specific industries (such as agroindustry, tourism, and construction) may be directly affected. [4.5]

3) Populations may be directly affected through extreme weather, changes in health status, or migration. The problems are somewhat different in the largest (<1 million) and mid- to small-sized population centers. [4.5]

The most widespread direct risk to human settlements from climate change is flooding and landslides, driven by projected increases in rainfall intensity and, in coastal areas, sea-level rise. Riverine and coastal settlements are particularly at risk (*high confidence*[6]), but urban flooding could be a problem anywhere that storm drains, water supply, and waste management systems have inadequate capacity. In such areas, squatter and other informal urban settlements with high population density, poor shelter, little or no access to resources such as safe water and public health services, and low adaptive capacity are highly vulnerable. Human settlements currently experience other significant environmental problems which could be exacerbated under higher temperature/increased precipitation regimes, including water and energy resources and infrastructure, waste treatment, and transportation [4.5]

Rapid urbanization in low-lying coastal areas of both the developing and developed world is greatly increasing population densities and the value of human-made assets exposed to coastal climatic extremes such as tropical cyclones. Model-based projections of the mean annual number of people who would be flooded by coastal storm surges increase several fold (by 75 to 200 million people depending on adaptive responses) for mid-range scenarios of a 40-cm sea-level rise by the 2080s relative to scenarios with no sea-level rise. Potential damages to infrastructure in coastal areas from sea-level rise have been projected to be tens of billions US$ for individual countries— for example, Egypt, Poland, and Vietnam. [4.5]

Settlements with little economic diversification and where a high percentage of incomes derive from climate-sensitive primary resource industries (agriculture, forestry, and fisheries) are more vulnerable than more diversified settlements (*high confidence*[6]). In developed areas of the Arctic, and where the permafrost is ice-rich, special attention will be required to mitigate the detrimental impacts of thawing, such as severe damage to buildings and transport infrastructure (*very high confidence*[6]). Industrial, transportation, and commercial infrastructure is generally vulnerable to the same hazards as settlement infrastructure. Energy demand is expected to increase for space cooling and decrease for space heating, but the net effect is scenario- and location-dependent. Some energy production and distribution systems may experience adverse impacts that would reduce supplies or system reliability while other energy systems may benefit. [4.5 and 5.7]

Possible adaptation options involve the planning of settlements and their infrastructure, placement of industrial facilities, and making similar long-lived decisions in a manner to reduce the adverse effects of events that are of low (but increasing) probability and high (and perhaps rising) consequences. [4.5]

3.7. Insurance and Other Financial Services

The costs of ordinary and extreme weather events have increased rapidly in recent decades. Global economic losses from catastrophic events increased 10.3-fold from 3.9 billion US$ yr^{-1} in the 1950s to 40 billion US$ yr^{-1} in the 1990s (all in 1999US$, unadjusted for purchasing power parity), with approximately one-quarter of the losses occurring in developing countries. The insured portion of these losses rose from a negligible level to 9.2 billion US$ yr^{-1} during the same period. Total costs are a factor of two larger when losses from smaller, non-catastrophic weather-related events are included. As a measure of increasing insurance industry vulnerability, the ratio of global property/casual insurance premiums to weather related losses fell by a factor of three between 1985 and 1999. [4.6]

The costs of weather events have risen rapidly despite significant and increasing efforts at fortifying infrastructure and enhancing disaster preparedness. Part of the observed upward trend in disaster losses over the past 50 years is linked to socioeconomic factors, such as population growth, increased wealth, and urbanization in vulnerable areas, and part is linked to climatic factors such as the observed changes in precipitation and flooding events. Precise attribution is complex and there are differences in the balance of these two causes by region and type of event. [4.6]

Climate change and anticipated changes in weather-related events perceived to be linked to climate change would increase actuarial uncertainty in risk assessment (*high confidence*[6]). Such developments would place upward pressure on insurance premiums and/or could lead to certain risks being reclassified as uninsurable with subsequent withdrawal of coverage. Such changes would trigger increased insurance costs, slow the expansion of financial services into developing countries, reduce the availability of insurance for spreading risk, and increase the demand for government-funded compensation following natural disasters. In the event of such changes, the relative roles of public and private entities in providing insurance and risk management resources can be expected to change. [4.6]

The financial services sector as a whole is expected to be able to cope with the impacts of climate change, although the historic record demonstrates that low-probability high-impact events or multiple closely spaced events severely affect parts of the sector, especially if adaptive capacity happens to be simultaneously depleted by non-climate factors (e.g., adverse financial market conditions). The property/casualty insurance and reinsurance segments and small specialized or undiversified companies have exhibited greater sensitivity, including reduced profitability and bankruptcy triggered by weather-related events. [4.6]

Adaptation to climate change presents complex challenges, but also opportunities, to the sector. Regulatory involvement in pricing, tax treatment of reserves, and the (in)ability of firms to withdraw from at-risk markets are examples of factors that influence the resilience of the sector. Public- and private-sector actors also support adaptation by promoting disaster preparedness, loss-prevention programs, building codes, and improved land-use planning. However, in some cases, public insurance and relief programs have inadvertently fostered complacency and maladaptation by inducing development in at-risk areas such as U.S. flood plains and coastal zones. [4.6]

The effects of climate change are expected to be greatest in the developing world, especially in countries reliant on primary production as a major source of income. Some countries experience impacts on their GDP as a consequence of natural disasters, with damages as high as half of GDP in one case. Equity issues and development constraints would arise if weather-related risks become uninsurable, prices increase, or availability becomes limited. Conversely, more extensive access to insurance and more widespread introduction of micro-financing schemes and development banking would increase the ability of developing countries to adapt to climate change. [4.6]

4. Vulnerability Varies across Regions

The vulnerability of human populations and natural systems to climate change differs substantially across regions and across populations within regions. Regional differences in baseline climate and expected climate change give rise to different exposures to climate stimuli across regions. The natural and social systems of different regions have varied characteristics, resources, and institutions, and are subject to varied pressures that give rise to differences in sensitivity and adaptive capacity. From these differences emerge different key concerns for each of the major regions of the world. Even within regions however, impacts, adaptive capacity, and vulnerability will vary. [5]

In light of the above, all regions are likely to experience some adverse effects of climate change. Table SPM-2 presents in a highly summarized fashion some of the key concerns for the different regions. Some regions are particularly vulnerable because of their physical exposure to climate change hazards and/or their limited adaptive capacity. Most less-developed regions are especially vulnerable because a larger share of their economies are in climate-sensitive sectors and their adaptive capacity is low due to low levels of human, financial, and natural resources, as well as limited institutional and technological capability. For example, small island states and low-lying coastal areas are particularly vulnerable to increases in sea level and storms, and most of them have limited capabilities for adaptation. Climate change impacts in polar regions are expected to be large and rapid, including reduction in sea-ice extent and thickness and degradation of permafrost. Adverse changes in seasonal river flows, floods and droughts, food security, fisheries, health effects, and loss of biodiversity are among the major regional vulnerabilities and concerns of Africa, Latin America, and Asia where adaptation opportunities are generally low. Even in regions with higher adaptive capacity, such as North America and Australia and New Zealand, there are vulnerable communities, such as indigenous peoples, and the possibility of adaptation of ecosystems is very limited. In Europe, vulnerability is significantly greater in the south and in the Arctic than elsewhere in the region. [5]

5. Improving Assessments of Impacts, Vulnerabilities, and Adaptation

Advances have been made since previous IPCC assessments in the detection of change in biotic and physical systems, and steps have been taken to improve the understanding of adaptive capacity, vulnerability to climate extremes, and other critical impact-related issues. These advances indicate a need for initiatives to begin designing adaptation strategies and building adaptive capacities. Further research is required, however, to strengthen future assessments and to reduce uncertainties in order to assure that sufficient information is available for policymaking about responses to possible consequences of climate change, including research in and by developing countries. [8]

The following are high priorities for narrowing gaps between current knowledge and policymaking needs:

- Quantitative assessment of the sensitivity, adaptive capacity, and vulnerability of natural and human systems to climate change, with particular emphasis on changes in the range of climatic variation and the frequency and severity of extreme climate events
- Assessment of possible thresholds at which strongly discontinuous responses to projected climate change and other stimuli would be triggered
- Understanding dynamic responses of ecosystems to multiple stresses, including climate change, at global, regional, and finer scales
- Development of approaches to adaptation responses, estimation of the effectiveness and costs of adaptation options, and identification of differences in opportunities for and obstacles to adaptation in different regions, nations, and populations
- Assessment of potential impacts of the full range of projected climate changes, particularly for non-market goods and services, in multiple metrics and with consistent treatment of uncertainties, including but not limited to numbers of people affected, land area affected, numbers of

Table SPM-2: Regional adaptive capacity, vulnerability, and key concerns.[a,b]

Region	Adaptive Capacity, Vulnerability, and Key Concerns
Africa	· Adaptive capacity of human systems in Africa is low due to lack of economic resources and technology, and vulnerability high as a result of heavy reliance on rain-fed agriculture, frequent droughts and floods, and poverty. [5.1.7] · Grain yields are projected to decrease for many scenarios, diminishing food security, particularly in small food-importing countries (*medium to high confidence*[6]). [5.1.2] · Major rivers of Africa are highly sensitive to climate variation; average runoff and water availability would decrease in Mediterranean and southern countries of Africa (*medium confidence*[6]). [5.1.1] · Extension of ranges of infectious disease vectors would adversely affect human health in Africa (*medium confidence*[6]). [5.1.4] · Desertification would be exacerbated by reductions in average annual rainfall, runoff, and soil moisture, especially in southern, North, and West Africa (*medium confidence*[6]). [5.1.6] · Increases in droughts, floods, and other extreme events would add to stresses on water resources, food security, human health, and infrastructures, and would constrain development in Africa (*high confidence*[6]). [5.1] · Significant extinctions of plant and animal species are projected and would impact rural livelihoods, tourism, and genetic resources (*medium confidence*[6]). [5.1.3] · Coastal settlements in, for example, the Gulf of Guinea, Senegal, Gambia, Egypt, and along the East-Southern African coast would be adversely impacted by sea-level rise through inundation and coastal erosion (*high confidence*[6]). [5.1.5]
Asia	· Adaptive capacity of human systems is low and vulnerability is high in the developing countries of Asia; the developed countries of Asia are more able to adapt and less vulnerable. [5.2.7] · Extreme events have increased in temperate and tropical Asia, including floods, droughts, forest fires, and tropical cyclones (*high confidence*[6]). [5.2.4] · Decreases in agricultural productivity and aquaculture due to thermal and water stress, sea-level rise, floods and droughts, and tropical cyclones would diminish food security in many countries of arid, tropical, and temperate Asia; agriculture would expand and increase in productivity in northern areas (*medium confidence*[6]). [5.2.1] · Runoff and water availability may decrease in arid and semi-arid Asia but increase in northern Asia (*medium confidence*[6]). [5.2.3] · Human health would be threatened by possible increased exposure to vector-borne infectious diseases and heat stress in parts of Asia (*medium confidence*[6]). [5.2.6] · Sea-level rise and an increase in the intensity of tropical cyclones would displace tens of millions of people in low-lying coastal areas of temperate and tropical Asia; increased intensity of rainfall would increase flood risks in temperate and tropical Asia (*high confidence*[6]). [5.2.5 and Table TS-8] · Climate change would increase energy demand, decrease tourism attraction, and influence transportation in some regions of Asia (*medium confidence*[6]). [5.2.4 and 5.2.7] · Climate change would exacerbate threats to biodiversity due to land-use and land-cover change and population pressure in Asia (*medium confidence*[6]). Sea-level rise would put ecological security at risk, including mangroves and coral reefs (*high confidence*[6]). [5.2.2] · Poleward movement of the southern boundary of the permafrost zones of Asia would result in a change of thermokarst and thermal erosion with negative impacts on social infrastructure and industries (*medium confidence*[6]). [5.2.2]
Australia and New Zealand	· Adaptive capacity of human systems is generally high, but there are groups in Australia and New Zealand, such as indigenous peoples in some regions, with low capacity to adapt and consequently high vulnerability. [5.3 and 5.3.5] · The net impact on some temperate crops of climate and CO_2 changes may initially be beneficial, but this balance is expected to become negative for some areas and crops with further climate change (*medium confidence*[6]). [5.3.3] · Water is likely to be a key issue (*high confidence*[6]) due to projected drying trends over much of the region and change to a more El Niño-like average state. [5.3 and 5.3.1] · Increases in the intensity of heavy rains and tropical cyclones (*medium confidence*[6]), and region-specific changes in the frequency of tropical cyclones, would alter the risks to life, property, and ecosystems from flooding, storm surges, and wind damage. [5.3.4] · Some species with restricted climatic niches and which are unable to migrate due to fragmentation of the landscape, soil differences, or topography could become endangered or extinct (*high confidence*[6]). Australian ecosystems that are particularly vulnerable to climate change include coral reefs, arid and semi-arid habitats in southwest and inland Australia, and Australian alpine systems. Freshwater wetlands in coastal zones in both Australia and New Zealand are vulnerable, and some New Zealand ecosystems are vulnerable to accelerated invasion by weeds. [5.3.2]

Table SPM-2 (continued)

Region	*Adaptive Capacity, Vulnerability, and Key Concerns*
Europe	· Adaptive capacity is generally high in Europe for human systems; southern Europe and the European Arctic are more vulnerable than other parts of Europe. [5.4 and 5.4.6] · Summer runoff, water availability, and soil moisture are likely to decrease in southern Europe, and would widen the difference between the north and drought-prone south; increases are likely in winter in the north and south (*high confidence*[6]). [5.4.1] · Half of alpine glaciers and large permafrost areas could disappear by end of the 21st century (*medium confidence*[6]). [5.4.1] · River flood hazard will increase across much of Europe (*medium to high confidence*[6]); in coastal areas, the risk of flooding, erosion, and wetland loss will increase substantially with implications for human settlement, industry, tourism, agriculture, and coastal natural habitats. [5.4.1 and 5.4.4] · There will be some broadly positive effects on agriculture in northern Europe (*medium confidence*[6]); productivity will decrease in southern and eastern Europe (*medium confidence*[6]). [5.4.3] · Upward and northward shift of biotic zones will take place. Loss of important habitats (wetlands, tundra, isolated habitats) would threaten some species (*high confidence*[6]). [5.4.2] · Higher temperatures and heat waves may change traditional summer tourist destinations, and less reliable snow conditions may impact adversely on winter tourism (*medium confidence*[6]). [5.4.4]
Latin America	· Adaptive capacity of human systems in Latin America is low, particularly with respect to extreme climate events, and vulnerability is high. [5.5] · Loss and retreat of glaciers would adversely impact runoff and water supply in areas where glacier melt is an important water source (*high confidence*[6]). [5.5.1] · Floods and droughts would become more frequent with floods increasing sediment loads and degrade water quality in some areas (*high confidence*[6]). [5.5] · Increases in intensity of tropical cyclones would alter the risks to life, property, and ecosystems from heavy rain, flooding, storm surges, and wind damages (*high confidence*[6]). [5.5] · Yields of important crops are projected to decrease in many locations in Latin America, even when the effects of CO_2 are taken into account; subsistence farming in some regions of Latin America could be threatened (*high confidence*[6]). [5.5.4] · The geographical distribution of vector-borne infectious diseases would expand poleward and to higher elevations, and exposures to diseases such as malaria, dengue fever, and cholera will increase (*medium confidence*[6]). [5.5.5] · Coastal human settlements, productive activities, infrastructure, and mangrove ecosystems would be negatively affected by sea-level rise (*medium confidence*[6]). [5.5.3] · The rate of biodiversity loss would increase (*high confidence*[6]). [5.5.2]
North America	· Adaptive capacity of human systems is generally high and vulnerability low in North America, but some communities (e.g., indigenous peoples and those dependent on climate-sensitive resources) are more vulnerable; social, economic, and demographic trends are changing vulnerabilities in subregions. [5.6 and 5.6.1] · Some crops would benefit from modest warming accompanied by increasing CO_2, but effects would vary among crops and regions (*high confidence*[6]), including declines due to drought in some areas of Canada's Prairies and the U.S. Great Plains, potential increased food production in areas of Canada north of current production areas, and increased warm-temperate mixed forest production (*medium confidence*[6]). However, benefits for crops would decline at an increasing rate and possibly become a net loss with further warming (*medium confidence*[6]). [5.6.4] · Snowmelt-dominated watersheds in western North America will experience earlier spring peak flows (*high confidence*[6]), reductions in summer flows (*medium confidence*[6]), and reduced lake levels and outflows for the Great Lakes-St. Lawrence under most scenarios (*medium confidence*[6]); adaptive responses would offset some, but not all, of the impacts on water users and on aquatic ecosystems (*medium confidence*[6]). [5.6.2] · Unique natural ecosystems such as prairie wetlands, alpine tundra, and cold-water ecosystems will be at risk and effective adaptation is unlikely (*medium confidence*[6]). [5.6.5] · Sea-level rise would result in enhanced coastal erosion, coastal flooding, loss of coastal wetlands, and increased risk from storm surges, particularly in Florida and much of the U.S. Atlantic coast (*high confidence*[6]). [5.6.1] · Weather-related insured losses and public sector disaster relief payments in North America have been increasing; insurance sector planning has not yet systematically included climate change information, so there is potential for surprise (*high confidence*[6]). [5.6.1] · Vector-borne diseases—including malaria, dengue fever, and Lyme disease—may expand their ranges in North America; exacerbated air quality and heat stress morbidity and mortality would occur (*medium confidence*[6]); socioeconomic factors and public health measures would play a large role in determining the incidence and extent of health effects. [5.6.6]

Table SPM-2 (continued)	
Region	**Adaptive Capacity, Vulnerability, and Key Concerns**
Polar	· Natural systems in polar regions are highly vulnerable to climate change and current ecosystems have low adaptive capacity; technologically developed communities are likely to adapt readily to climate change, but some indigenous communities, in which traditional lifestyles are followed, have little capacity and few options for adaptation. [5.7] · Climate change in polar regions is expected to be among the largest and most rapid of any region on the Earth, and will cause major physical, ecological, sociological, and economic impacts, especially in the Arctic, Antarctic Peninsula, and Southern Ocean (*high confidence*[6]). [5.7] · Changes in climate that have already taken place are manifested in the decrease in extent and thickness of Arctic sea ice, permafrost thawing, coastal erosion, changes in ice sheets and ice shelves, and altered distribution and abundance of species in polar regions (*high confidence*[6]). [5.7] · Some polar ecosystems may adapt through eventual replacement by migration of species and changing species composition, and possibly by eventual increases in overall productivity; ice edge systems that provide habitat for some species would be threatened (*medium confidence*[6]). [5.7] · Polar regions contain important drivers of climate change. Once triggered, they may continue for centuries, long after greenhouse gas concentrations are stabilized, and cause irreversible impacts on ice sheets, global ocean circulation, and sea-level rise (*medium confidence*[6]). [5.7]
Small Island States	· Adaptive capacity of human systems is generally low in small island states, and vulnerability high; small island states are likely to be among the countries most seriously impacted by climate change. [5.8] · The projected sea-level rise of 5 mm yr^{-1} for the next 100 years would cause enhanced coastal erosion, loss of land and property, dislocation of people, increased risk from storm surges, reduced resilience of coastal ecosystems, saltwater intrusion into freshwater resources, and high resource costs to respond to and adapt to these changes (*high confidence*[6]). [5.8.2 and 5.8.5] · Islands with very limited water supplies are highly vulnerable to the impacts of climate change on the water balance (*high confidence*[6]). [5.8.4] · Coral reefs would be negatively affected by bleaching and by reduced calcification rates due to higher CO_2 levels (*medium confidence*[6]); mangrove, sea grass bed, and other coastal ecosystems and the associated biodiversity would be adversely affected by rising temperatures and accelerated sea-level rise (*medium confidence*[6]). [4.4 and 5.8.3] · Declines in coastal ecosystems would negatively impact reef fish and threaten reef fisheries, those who earn their livelihoods from reef fisheries, and those who rely on the fisheries as a significant food source (*medium confidence*[6]). [4.4 and 5.8.4] · Limited arable land and soil salinization makes agriculture of small island states, both for domestic food production and cash crop exports, highly vulnerable to climate change (*high confidence*[6]). [5.8.4] · Tourism, an important source of income and foreign exchange for many islands, would face severe disruption from climate change and sea-level rise (*high confidence*[6]). [5.8.5]

[a] Because the available studies have not employed a common set of climate scenarios and methods, and because of uncertainties regarding the sensitivities and adaptability of natural and social systems, the assessment of regional vulnerabilities is necessarily qualitative.

[b] The regions listed in Table SPM-2 are graphically depicted in Figure TS-2 of the Technical Summary.

species at risk, monetary value of impact, and implications in these regards of different stabilization levels and other policy scenarios

· Improving tools for integrated assessment, including risk assessment, to investigate interactions between components of natural and human systems and the consequences of different policy decisions

· Assessment of opportunities to include scientific information on impacts, vulnerability, and adaptation in decisionmaking processes, risk management, and sustainable development initiatives

· Improvement of systems and methods for long-term monitoring and understanding the consequences of climate change and other stresses on human and natural systems.

Cutting across these foci are special needs associated with strengthening international cooperation and coordination for regional assessment of impacts, vulnerability, and adaptation, including capacity-building and training for monitoring, assessment, and data gathering, especially in and for developing countries (particularly in relation to the items identified above).

Climate Change 2001:
Impacts, Adaptation, and Vulnerability

Technical Summary

A Report of Working Group II of the Intergovernmental Panel on Climate Change

This summary was accepted but not approved in detail at the Sixth Session of IPCC Working Group II (Geneva, Switzerland, 13-16 February 2001). "Acceptance" of IPCC reports at a session of the Working Group or Panel signifies that the material has not been subject to line-by-line discussion and agreement, but nevertheless presents a comprehensive, objective, and balanced view of the subject matter.

Lead Authors
K.S. White (USA), Q.K. Ahmad (Bangladesh), O. Anisimov (Russia), N. Arnell (UK), S. Brown (USA), M. Campos (Costa Rica), T. Carter (Finland), Chunzhen Liu (China), S. Cohen (Canada), P. Desanker (Malawi), D.J. Dokken (USA), W. Easterling (USA), B. Fitzharris (New Zealand), H. Gitay (Australia), A. Githeko (Kenya), S. Gupta (India), H. Harasawa (Japan), B.P. Jallow (The Gambia), Z.W. Kundzewicz (Poland), E.L. La Rovere (Brazil), M. Lal (India), N. Leary (USA), C. Magadza (Zimbabwe), L.J. Mata (Venezuela), R. McLean (Australia), A. McMichael (UK), K. Miller (USA), E. Mills (USA), M.Q. Mirza (Bangladesh), D. Murdiyarso (Indonesia), L.A. Nurse (Barbados), C. Parmesan (USA), M.L. Parry (UK), O. Pilifosova (Kazakhstan), B. Pittock (Australia), J. Price (USA), T. Root (USA), C. Rosenzweig (USA), J. Sarukhan (Mexico), H.-J. Schellnhuber (Germany), S. Schneider (USA), M.J. Scott (USA), G. Sem (Papua New Guinea), B. Smit (Canada), J.B. Smith (USA), A. Tsyban (Russia), P. Vellinga (The Netherlands), R. Warrick (New Zealand), D. Wratt (New Zealand)

Review Editors
M. Manning (New Zealand) and C. Nobre (Brazil)

1. Scope and Approach of the Assessment

1.1. Mandate of the Assessment

The Intergovernmental Panel on Climate Change (IPCC) was established by the World Meteorological Organization (WMO) and United Nations Environment Programme (UNEP) in 1988 to assess scientific, technical, and socioeconomic information that is relevant in understanding human-induced climate change, its potential impacts, and options for mitigation and adaptation. The IPCC currently is organized into three Working Groups: Working Group I (WGI) addresses observed and projected changes in climate; Working Group II (WGII) addresses vulnerability, impacts, and adaptation related to climate change; and Working Group III (WGIII) addresses options for mitigation of climate change.

This volume—*Climate Change 2001: Impacts, Adaptation, and Vulnerability*—is the WGII contribution to the IPCC's Third Assessment Report (TAR) on scientific, technical, environmental, economic, and social issues associated with the climate system and climate change.[1] WGII's mandate for the TAR is to assess the vulnerability of ecological systems, socioeconomic sectors, and human health to climate change as well as potential impacts of climate change, positive and negative, on these systems. This assessment also examines the feasibility of adaptation to enhance the positive effects of climate change and ameliorate negative effects. This new assessment builds on previous IPCC assessments, reexamining key findings of earlier assessments and emphasizing new information and implications from more recent studies.

1.2. What is Potentially at Stake?

Human activities—primarily burning of fossil fuels and changes in land cover—are modifying the concentration of atmospheric constituents or properties of the surface that absorb or scatter radiant energy. The WGI contribution to the TAR—*Climate Change 2001: The Scientific Basis*—found, "In the light of new evidence and taking into account the remaining uncertainties, most of the observed warming over the last 50 years is likely to have been due to the increase in greenhouse gas concentrations." Future changes in climate are expected to include additional warming, changes in precipitation patterns and amounts, sea-level rise, and changes in the frequency and intensity of some extreme events.

[1] *Climate change* in IPCC usage refers to any change in climate over time, whether due to natural variability or as a result of human activity. This usage differs from the definition in Article 1 of the United Nations Framework Convention on Climate Change, where *climate change* refers to a change of climate which is attributed directly or indirectly to human activity that alters the composition of the global atmosphere and which is in addition to natural climate variability observed over comparable time periods.

The stakes associated with projected changes in climate are high. Numerous Earth systems that sustain human societies are sensitive to climate and will be impacted by changes in climate (very high confidence). Impacts can be expected in ocean circulation; sea level; the water cycle; carbon and nutrient cycles; air quality; the productivity and structure of natural ecosystems; the productivity of agricultural, grazing, and timber lands; and the geographic distribution, behavior, abundance, and survival of plant and animal species, including vectors and hosts of human disease. Changes in these systems in response to climate change, as well as direct effects of climate change on humans, would affect human welfare, positively and negatively. Human welfare would be impacted through changes in supplies of and demands for water, food, energy, and other tangible goods that are derived from these systems; changes in opportunities for nonconsumptive uses of the environment for recreation and tourism; changes in non-use values of the environment such as cultural and preservation values; changes in incomes; changes in loss of property and lives from extreme climate phenomena; and changes in human health. Climate change impacts will affect the prospects for sustainable development in different parts of the world and may further widen existing inequalities. Impacts will vary in distribution across people, places, and times (very high confidence), raising important questions about equity.

Although the stakes are demonstrably high, the risks associated with climate change are less easily established. Risks are a function of the probability and magnitude of different types of impacts. The WGII report assesses advances in the state of

---

Box TS-1. Climate Change Sensitivity, Adaptive Capacity, and Vulnerability

Sensitivity is the degree to which a system is affected, either adversely or beneficially, by climate-related stimuli. Climate-related stimuli encompass all the elements of climate change, including mean climate characteristics, climate variability, and the frequency and magnitude of extremes. The effect may be direct (e.g., a change in crop yield in response to a change in the mean, range or variability of temperature) or indirect (e.g., damages caused by an increase in the frequency of coastal flooding due to sea-level rise).

Adaptive capacity is the ability of a system to adjust to climate change, including climate variability and extremes, to moderate potential damages, to take advantage of opportunities, or to cope with the consequences.

Vulnerability is the degree to which a system is susceptible to, or unable to cope with, adverse effects of climate change, including climate variability and extremes. Vulnerability is a function of the character, magnitude and rate of climate change and variation to which a system is exposed, its sensitivity, and its adaptive capacity.

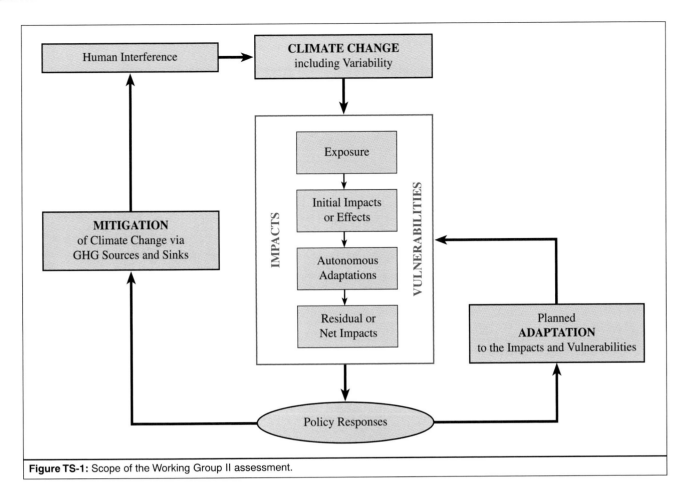

Figure TS-1: Scope of the Working Group II assessment.

knowledge regarding impacts of climate stimuli to which systems may be exposed, the sensitivity of exposed systems to changes in climate stimuli, their adaptive capacity to alleviate or cope with adverse impacts or enhance beneficial ones, and their vulnerability to adverse impacts (see Box TS-1). Possible impacts include impacts that threaten substantial and irreversible damage to or loss of some systems within the next century; modest impacts to which systems may readily adapt; and impacts that would be beneficial for some systems.

Figure TS-1 illustrates the scope of the WGII assessment and its relation to other parts of the climate change system. Human activities that change the climate expose natural and human systems to an altered set of stresses or stimuli. Systems that are sensitive to these stimuli are affected or impacted by the changes, which can trigger autonomous, or expected, adaptations. These autonomous adaptations will reshape the residual or net impacts of climate change. Policy responses in reaction to impacts already perceived or in anticipation of potential future impacts can take the form of planned adaptations to lessen adverse effects or enhance beneficial ones. Policy responses also can take the form of actions to mitigate climate change through greenhouse gas (GHG) emission reductions and enhancement of sinks. The WGII assessment focuses on the central box of Figure TS-1—exposure, impacts, and vulnerabilities—and the adaptation policy loop.

1.3. Approach of the Assessment

The assessment process involves evaluation and synthesis of available information to advance understanding of climate change impacts, adaptation, and vulnerability. The information comes predominantly from peer-reviewed published literature. Evidence also is drawn from published, non-peer-reviewed literature and unpublished sources, but only after evaluation of its quality and validity by the authors of this report.

WGII's assessment has been conducted by an international group of experts nominated by governments and scientific bodies and selected by the WGII Bureau of the IPCC for their scientific and technical expertise and to achieve broad geographical balance. These experts come from academia, governments, industry, and scientific and environmental organizations. They participate without compensation from the IPCC, donating substantial time to support the work of the IPCC.

This assessment is structured to examine climate change impacts, adaptations, and vulnerabilities of systems and regions and to provide a global synthesis of cross-system and cross-regional issues. To the extent feasible, given the available literature, climate change is examined in the context of sustainable development and equity. The first section sets the stage for the assessment by discussing the context of climate

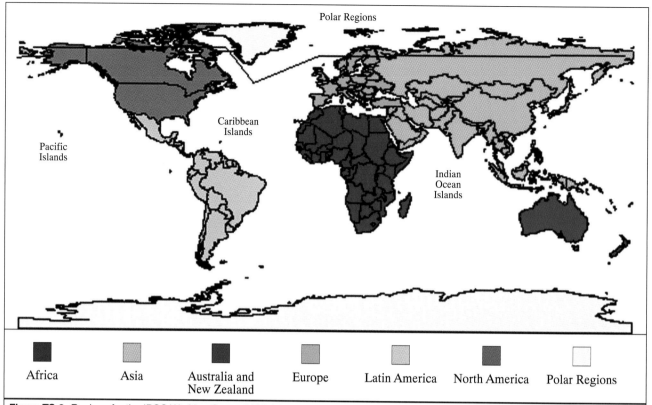

Figure TS-2: Regions for the IPCC Working Group II Third Assessment Report. Note that regions in which small island states are located include the Pacific, Indian, and Atlantic Oceans, and the Caribbean and Mediterranean Seas. The boundary between Europe and Asia runs along the eastern Ural Mountains, River Ural, and Caspian Sea. For the polar regions, the Arctic consists of the area north of the Arctic Circle, including Greenland; the Antarctic consists of the Antarctic continent, together with the Southern Ocean south of ~58°S.

change, methods and tools, and scenarios. Individual chapters assess vulnerabilities of water systems, terrestrial ecosystems (including agriculture and forestry), ocean and coastal systems, human settlements (including energy and industrial sectors), insurance and other financial services, and human health. A chapter is devoted to each of eight major regions of the world: Africa, Asia, Australia and New Zealand, Europe, Latin America, North America, polar regions, and small island states. These regions are shown in Figure TS-2. All of the regions are highly heterogeneous, and climate change impacts, adaptive capacity, and vulnerability will vary in important ways within each of the regions. The final section of the report synthesizes adaptation capacity and its potential to alleviate adverse impacts, enhance beneficial effects, and increase sustainable development and equity and reviews information that is relevant for interpretation of Article 2 of the United Nations Framework Convention on Climate Change (UNFCCC) and key provisions of international agreements to address climate change. The report also contains a Summary for Policymakers, which provides a brief synthesis of the conclusions of the report that have particular relevance to those who have responsibility for making climate change response decisions. This Technical Summary provides a more comprehensive summary of the assessment; it references sections of the underlying report in brackets at the end of the paragraphs for readers who would like more information on a particular topic. [1.1]

1.4. Treatment of Uncertainties

Since the SAR, greater emphasis has been placed on developing methods for characterizing and communicating uncertainties. Two approaches to evaluate uncertainties are applied in the WGII assessment. A quantitative approach is adopted to assess confidence levels in instances for which present understanding of relevant processes, system behavior, observations, model simulations, and estimates is sufficient to support broad agreement among authors of the report about Bayesian probabilities associated with selected findings. A more qualitative approach is used to assess and report the quality or level of scientific understanding that supports a conclusion (see Box TS-2). These approaches, and the rationale for them, are explained in more detail in *Third Assessment Report: Cross-Cutting Issues Guidance Papers* (http://www.gispri.or.jp), supporting material prepared by the IPCC to increase the use of consistent terms and concepts within and across the Working Group volumes of the TAR. [1.1, 2.6].

2. Methods and Tools of the Assessment

Assessment of climate change impacts, adaptations, and vulnerability draws on a wide range of physical, biological, and social science disciplines and consequently employs an

Box TS-2. Confidence Levels and State of Knowledge

Quantitative Assessment of Confidence Levels

In applying the *quantitative* approach, authors of the report assign a confidence level that represents the degree of belief among the authors in the validity of a conclusion, based on their collective expert judgment of observational evidence, modeling results, and theory that they have examined. Five confidence levels are used. In the tables of the Technical Summary, symbols are substituted for words:

Very High (*****)   95% or greater
High (****)   67-95%
Medium(***)   33-67%
Low (**)   5-33%
Very Low (*)   5% or less

Qualitative Assessment of the State of Knowledge

In applying the *qualitative* approach, authors of the report evaluate the level of scientific understanding in support of a conclusion, based on the amount of supporting evidence and the level of agreement among experts about the interpretation of the evidence. Four qualitative classifications are employed:

· *Well-established:* Models incorporate known processes, observations are consistent with models, or multiple lines of evidence support the finding.
· *Established but incomplete:* Models incorporate most known processes, although some parameterizations may not be well tested; observations are somewhat consistent but incomplete; current empirical estimates are well founded, but the possibility of changes in governing processes over time is considerable; or only one or a few lines of evidence support the finding.
· *Competing explanations:* Different model representations account for different aspects of observations or evidence or incorporate different aspects of key processes, leading to competing explanations.
· *Speculative:* Conceptually plausible ideas that are not adequately represented in the literature or that contain many difficult-to-reduce uncertainties. [Box 1-1]

enormous variety of methods and tools. Since the SAR, such methods have improved detection of climate change in biotic and physical systems and produced new substantive findings. In addition, cautious steps have been taken since the SAR to expand the "tool-box" to address more effectively the human dimensions of climate as both causes and consequences of change and to deal more directly with cross-sectoral issues concerning vulnerability, adaptation, and decisionmaking. In particular, a greater number of studies have begun to apply methods and tools for costing and valuing effects, treating uncertainties, integrating effects across sectors and regions, and applying decision analytic frameworks for evaluating adaptive capacity. Overall, these modest methodological developments are encouraging analyses that will build a more solid foundation for understanding how decisions regarding adaptation to future climate change might be taken. [2.8]

2.1. Detecting Responses to Climate Change using Indicator Species or Systems

Since the SAR, methods have been developed and applied to the detection of present impacts of 20th century climate change on abiotic and biotic systems. Assessment of impacts on human and natural systems that already have occurred as a result of recent climate change is an important complement to model projections of future impacts. Such detection is impeded by multiple, often inter-correlated, nonclimatic forces that concurrently affect those systems. Attempts to overcome this problem have involved the use of indicator species (e.g., butterflies, penguins, frogs, and sea anemones) to detect responses to climate change and to infer more general impacts of climate change on natural systems (e.g., in native meadows, coastal Antarctica, tropical cloud forest, and the Pacific rocky intertidal, respectively). An important component of this detection process is the search for systematic patterns of change across many studies that are consistent with expectations, based on observed or predicted changes in climate. Confidence in attribution of these observed changes to climate change increases as studies are replicated across diverse systems and geographic regions. Even though studies now number in the hundreds, some regions and systems remain underrepresented. [2.2]

To investigate possible links between observed changes in regional climate and biological or physical processes in ecosystems, the author team gathered more than 2,500 articles on climate and one of the following entities: animals, plants, glaciers, sea ice, and ice on lakes or streams. To determine if these entities have been influenced by changing climate, only studies meeting at least two of the following criteria were included:

· A trait of these entities (e.g., range boundary, melting date) shows a change over time.
· The trait is correlated with changes in local temperature.
· Local temperature changed over time.

At least two of these three criteria had to exhibit a statistically significant correlation. Only temperature was considered because it is well established in the literature how it influences the entities examined and because temperature trends are more globally homogeneous than other locally varying climatic factors, such as precipitation changes. Selected studies must also have examined at least 10 years of data; more than 90% had a time span of more than 20 years.

These stringent criteria reduced the number of studies used in the analysis to 44 animal and plant studies that cover more than 600 species. Of these species, about 90% (more than 550) show changes in traits over time. Of these 550+ species, about 80% (more than 450) show change in a direction expected given scientific understanding of known mechanisms that relate temperature to each of the species traits. The probability that more than 450 species of 550+ would show changes in the directions expected by random chance is negligible.

Sixteen studies examining glaciers, sea ice, snow cover extent/ snow melt, or ice on lakes or streams included more than 150 sites. Of these 150+ sites, 67% (100+) show changes in traits over time. Of these 100+ sites, about 99% (99+) exhibited trends in a direction expected, given scientific understanding of known mechanisms that relate temperatures to physical processes that govern change in that trait. The probability that 99+ of 100+ sites would show changes in the directions expected by chance alone is negligible. [5.2, 5.4, 19.2]

2.2. Anticipating the Effects of Future Climate Change

Since the SAR, improvements in methods and tools for studying impacts of future changes in climate have included greater emphasis on the use of process-oriented models, transient climate change scenarios, refined socioeconomic baselines, and higher resolution spatial and temporal scales. Country studies and regional assessments in every continent have tested models and tools in a variety of contexts. First-order impact models have been linked to global systems models. Adaptation has been included in many assessments, often for the first time.

Methodological gaps remain concerning scales, data, validation, and integration of adaptation and the human dimensions of climate change. Procedures for assessing regional and local vulnerability and long-term adaptation strategies require high-resolution assessments, methodologies to link scales, and dynamic modeling that uses corresponding and new data sets. Validation at different scales often is lacking. Regional integration across sectors is required to place vulnerability in the context of local and regional development. Methods and tools to assess vulnerability to extreme events have improved but are constrained by low confidence in climate change scenarios and the sensitivity of impact models to major climatic anomalies. Understanding and integrating higher order economic effects and other human dimensions of global change are required. Adaptation models and vulnerability indices to prioritize adaptation options are at early stages of development in many fields. Methods to enable stakeholder participation in assessments need improvement. [2.3]

2.3. Integrated Assessment

Integrated assessment is an interdisciplinary process that combines, interprets, and communicates knowledge from diverse scientific disciplines from the natural and social sciences to investigate and understand causal relationships within and between complicated systems. Methodological approaches employed in such assessments include computer-aided modeling, scenario analyses, simulation gaming and participatory integrated assessment, and qualitative assessments that are based on existing experience and expertise. Since the SAR, significant progress has been made in developing and applying such approaches to integrated assessment, globally and regionally.

However, progress to date, particularly with regard to integrated modeling, has focused largely on mitigation issues at the global or regional scale and only secondarily on issues of impacts, vulnerability, and adaptation. Greater emphasis on the development of methods for assessing vulnerability is required, especially at national and subnational scales where impacts of climate change are felt and responses are implemented. Methods designed to include adaptation and adaptive capacity explicitly in specific applications must be developed. [2.4]

2.4. Costing and Valuation

Methods of economic costing and valuation rely on the notion of opportunity cost of resources used, degraded, or saved. Opportunity cost depends on whether the market is competitive or monopolistic and on whether any externalities are internalized. It also depends on the rate at which the future is discounted, which can vary across countries, over time, and over generations. The impact of uncertainty also can be valued if the probabilities of different possible outcomes are known. Public and nonmarket goods and services can be valued through willingness to pay for them or willingness to accept compensation for lack of them. Impacts on different groups, societies, nations, and species must be assessed. Comparison of alternative distributions of welfare across individuals and groups within a country can be justified if they are made according to internally consistent norms. Comparisons across nations with different societal, ethical, and governmental structures cannot yet be made meaningfully.

Since the SAR, no new fundamental developments in costing and valuation methodology have taken place. Many new applications of existing methods to a widening range of climate change issues have demonstrated, however, the strengths and limitations of some of these methods. Research efforts are required to strengthen methods for multi-objective assessments. Multi-objective assessments are increasingly preferred, but the means by which their underlying metrics might more accurately reflect diverse social, political, economic, and cultural contexts must be developed. In addition, methods for integrating across these multiple metrics are still missing from the methodological repertoire. [2.5]

2.5. Decision Analytic Frameworks

Policymakers who are responsible for devising and implementing adaptive policies should be able to rely on results from one or more of a diverse set of decision analytical frameworks. Commonly used methods include cost-benefit and cost-effectiveness analysis, various types of decision analysis (including multi-objective studies), and participatory techniques such as policy exercises.

Very few cases in which policymakers have used decision analytical frameworks in evaluating adaptation options have been reported. Among the large number of assessments of climate change impacts reviewed in the TAR, only a small fraction include comprehensive and quantitative estimates of adaptation options and their costs, benefits, and uncertainty characteristics. This information is necessary for meaningful applications of any decision analytical method to issues of adaptation. Greater use of such methods in support of adaptation decisions is needed to establish their efficacy and to identify directions for necessary research in the context of vulnerability and adaptation to climate change. [2.7]

3. Scenarios of Future Change

3.1. Scenarios and their Role

A scenario is a coherent, internally consistent, and plausible description of a possible future state of the world. Scenarios are commonly required in climate change impact, adaptation, and vulnerability assessments to provide alternative views of future conditions considered likely to influence a given system or activity. A distinction is made between climate scenarios, which describe the forcing factor of focal interest to the IPCC, and nonclimatic scenarios, which provide the socioeconomic and environmental context within which climate forcing operates. Most assessments of the impacts of future climate change are based on results from impact models that rely on quantitative climate and nonclimatic scenarios as inputs. [3.1.1, Box 3-1]

3.2. Socioeconomic, Land-Use, and Environmental Scenarios

Nonclimatic scenarios describing future socioeconomic, land-use, and environmental changes are important for characterizing the sensitivity of systems to climate change, their vulnerability, and the capacity for adaptation. Such scenarios only recently have been widely adopted in impact assessments alongside climate scenarios.

Socioeconomic scenarios. Socioeconomic scenarios have been used more extensively for projecting GHG emissions than for assessing climate vulnerability and adaptive capacity. Most socioeconomic scenarios identify several different topics or domains, such as population or economic activity, as well as background factors such as the structure of governance, social values, and patterns of technological change. Scenarios make it possible to establish baseline socioeconomic vulnerability, pre-climate change; determine climate change impacts; and assess post-adaptation vulnerability. [3.2]

Land-use and land-cover change scenarios. Land-use change and land-cover change (LUC-LCC) involve several processes that are central to the estimation of climate change and its impacts. First, LUC-LCC influences carbon fluxes and GHG emissions, which directly alter atmospheric composition and radiative forcing properties. Second, LUC-LCC modifies land-surface characteristics and, indirectly, climatic processes. Third, land-cover modification and conversion may alter the properties of ecosystems and their vulnerability to climate change. Finally, several options and strategies for mitigating GHG emissions involve land cover and changed land-use practices. A great diversity of LUC-LCC scenarios have been constructed. Most of these scenarios do not address climate change issues explicitly, however; they focus on other issues—for example, food security and carbon cycling. Large improvements have been made since the SAR in defining current and historic land-use and land-cover patterns, as well as in estimating future scenarios. Integrated assessment models currently are the most appropriate tools for developing LUC-LCC scenarios. [3.3.1, 3.3.2]

Environmental scenarios. Environmental scenarios refer to changes in environmental factors other than climate that will occur in the future regardless of climate change. Because these factors could have important roles in modifying the impacts of future climate change, scenarios are required to portray possible future environmental conditions such as atmospheric composition [e.g., carbon dioxide (CO_2), tropospheric ozone, acidifying compounds, and ultraviolet-B (UV-B) radiation]; water availability, use, and quality; and marine pollution. Apart from the direct effects of CO_2 enrichment, changes in other environmental factors rarely have been considered alongside climate changes in past impact assessments, although their use is increasing with the emergence of integrated assessment methods. [3.4.1]

3.3. Sea-Level Rise Scenarios

Sea-level rise scenarios are required to evaluate a diverse range of threats to human settlements, natural ecosystems, and landscape in the coastal zone. Relative sea-level scenarios (i.e., sea-level rise with reference to movements of the local land surface) are of most interest for impact and adaptation assessments. Tide gauge and wave-height records of 50 years or more are required, along with information on severe weather and coastal processes, to establish baseline levels or trends. Recent techniques of satellite altimetry and geodetic leveling have enhanced and standardized baseline determinations of relative sea level over large areas of the globe. [3.6.2]

Table TS-1: The SRES scenarios and their implications for atmospheric composition, climate, and sea level. Values of population, GDP, and per capita income ratio (a measure of regional equity) are those applied in integrated assessment models used to estimate emissions (based on Tables 3-2 and 3-9).

Date	Global Population (billions)[a]	Global GDP (10^{12} US$ yr^{-1})[b]	Per Capita Income Ratio[c]	Ground Level O_3 Concentration (ppm)[d]	CO_2 Concentration (ppm)[e]	Global Temperature Change (°C)[f]	Global Sea-Level Rise (cm)[g]
1990	5.3	21	16.1	—	354	0	0
2000	6.1-6.2	25-28	12.3-14.2	40	367	0.2	2
2050	8.4-11.3	59-187	2.4-8.2	~60	463-623	0.8-2.6	5-32
2100	7.0-15.1	197-550	1.4-6.3	>70	478-1099	1.4-5.8	9-88

[a] Values for 2000 show range across the six illustrative SRES emissions scenarios; values for 2050 and 2100 show range across all 40 SRES scenarios.

[b] See footnote a; gross domestic product (trillion 1990 US$ yr^{-1}).

[c] See footnote a; ratio of developed countries and economies-in-transition (Annex I) to developing countries (non-Annex I).

[d] Model estimates for industrialized continents of northern hemisphere assuming emissions for 2000, 2060, and 2100 from the A1F and A2 illustrative SRES emissions scenarios at high end of the SRES range (Chapter 4, TAR WGI).

[e] Observed 1999 value (Chapter 3, WGI TAR); values for 1990, 2050, and 2100 are from simple model runs across the range of 35 fully quantified SRES emissions scenarios and accounting for uncertainties in carbon cycle feedbacks related to climate sensitivity (data from S.C.B. Raper, Chapter 9, WGI TAR). Note that the ranges for 2050 and 2100 differ from those presented by TAR WGI (Appendix II), which were ranges across the six illustrative SRES emissions scenarios from simulations using two different carbon cycle models.

[f] Change in global mean annual temperature relative to 1990 averaged across simple climate model runs emulating results of seven AOGCMs with an average climate sensitivity of 2.8°C for the range of 35 fully quantified SRES emissions scenarios (Chapter 9, WGI TAR).

[g] Based on global mean temperature changes but also accounting for uncertainties in model parameters for land ice, permafrost, and sediment deposition (Chapter 11, WGI TAR).

Although some components of future sea-level rise can be modeled regionally by using coupled ocean-atmosphere models, the most common method of obtaining scenarios is to apply global mean estimates from simple models. Changes in the occurrence of extreme events such as storm surges and wave setup, which can lead to major coastal impacts, sometimes are investigated by superimposing historically observed events onto a rising mean sea level. More recently, some studies have begun to express future sea-level rise in probabilistic terms, enabling rising levels to be evaluated in terms of the risk of exceeding a critical threshold of impact. [3.6.3, 3.6.4, 3.6.5, 3.6.6]

3.4. Climate Scenarios

Three main types of climate scenarios have been employed in impact assessments: incremental scenarios, analog scenarios, and climate model-based scenarios. Incremental scenarios are simple adjustments of the baseline climate according to anticipated future changes that can offer a valuable aid for testing system sensitivity to climate. However, because they involve arbitrary adjustments, they may not be realistic meteorologically. Analogs of a changed climate from the past record or from other regions may be difficult to identify and are seldom applied, although they sometimes can provide useful insights into impacts of climate conditions outside the present-day range. [3.5.2]

The most common scenarios use outputs from general circulation models (GCMs) and usually are constructed by adjusting a

baseline climate (typically based on regional observations of climate over a reference period such as 1961-1990) by the absolute or proportional change between the simulated present and future climates. Most recent impact studies have constructed scenarios on the basis of transient GCM outputs, although some still apply earlier equilibrium results. The great majority of scenarios represent changes in mean climate; some recent scenarios, however, also have incorporated changes in variability and extreme weather events, which can lead to important impacts for some systems. Regional detail is obtained from the coarse-scale outputs of GCMs by using three main methods: simple interpolation, statistical downscaling, and high-resolution dynamical modeling. The simple method, which reproduces the GCM pattern of change, is the most widely applied in scenario development. In contrast, the statistical and modeling approaches can produce local climate changes that are different from large-scale GCM estimates. More research is needed to evaluate the value added to impact studies of such regionalization exercises. One reason for this caution is the large uncertainty of GCM projections, which requires further quantification through model intercomparisons, new model simulations, and pattern scaling methods. [3.5.2, 3.5.4, 3.5.5]

3.5. Scenarios of the 21st Century

In 2000, the IPCC completed a *Special Report on Emissions Scenarios* (SRES) to replace the earlier set of six IS92 scenarios developed for the IPCC in 1992. These newer scenarios consider the period 1990 to 2100 and include a range of socioeconomic assumptions (e.g., global population and gross domestic product). Their implications for other aspects of global change also have been calculated; some of these implications are summarized for 2050 and 2100 in Table TS-1. For example, mean ground-level ozone concentrations in July over the industrialized continents of the northern hemisphere are projected to rise from about 40 ppb in 2000 to more than 70 ppb in 2100 under the highest illustrative SRES emissions scenarios; by comparison, the clean-air standard is below 80 ppb. Peak levels of ozone in local smog events could be many times higher. Estimates of CO_2 concentration range from 478 ppm to1099 ppm by 2100, given the range of SRES emissions and uncertainties about the carbon cycle (Table TS-1). This range of implied radiative forcing gives rise to an estimated global warming from 1990 to 2100 of 1.4-5.8°C, assuming a range of climate sensitivities. This range is higher than the 0.7-3.5°C of the SAR because of higher levels of radiative forcing in the SRES scenarios than in the IS92a-f scenarios—primarily as a result of lower sulfate aerosol emissions, especially after 2050. The equivalent range of estimates of global sea-level rise (for this range of global temperature change in combination with a range of ice melt sensitivities) to 2100 is 9-88 cm (compared to 15-95 cm in the SAR). [3.2.4.1, 3.4.4, 3.8.1, 3.8.2]

In terms of *mean changes in regional climate*, results from GCMs that have been run assuming the new SRES emissions scenarios display many similarities with previous runs. The WGI contribution to the TAR concludes that rates of warming are expected to be greater than the global average over most land areas and will be most pronounced at high latitudes in winter. As warming proceeds, northern hemisphere snow cover and sea-ice extent will be reduced. Models indicate warming below the global average in the north Atlantic and circumpolar southern ocean regions, as well as in southern and southeast Asia and southern South America in June-August. Globally, there will be increases in average water vapor and precipitation. Regionally, December-February precipitation is expected to increase over the northern extratropics, Antarctica, and tropical Africa. Models also agree on a decrease in precipitation over Central America and little change in southeast Asia. Precipitation in June-August is estimated to increase in high northern latitudes, Antarctica, and south Asia; it is expected to change little in southeast Asia and to decrease in Central America, Australia, southern Africa, and the Mediterranean region.

Changes in the frequency and intensity of extreme climate events also can be expected. Based on the conclusions of the WGI report and the likelihood scale employed therein, under GHG forcing to 2100, it is very likely that daytime maximum and minimum temperatures will increase, accompanied by an increased frequency of hot days (see Table TS-2). It also is very likely that heat waves will become more frequent, and the number of cold waves and frost days (in applicable regions) will decline. Increases in high-intensity precipitation events are likely at many locations; Asian summer monsoon precipitation

variability also is likely to increase. The frequency of summer drought will increase in many interior continental locations, and droughts—as well as floods—associated with El Niño events are likely to intensify. Peak wind intensity and mean and peak precipitation intensities of tropical cyclones are likely to increase. The direction of changes in the average intensity of mid-latitude storms cannot be determined with current climate models. [Table 3-10]

3.6. How can We Improve Scenarios and their Use?

Some features of scenario development and application that are now well established and tested include continued development of global and regional databases for defining baseline conditions, widespread use of incremental scenarios to explore system sensitivity prior to application of model-based scenarios, improved availability and wider application of estimates of long-term mean global changes on the basis of projections produced by specialized international organizations or the use of simple models, and a growing volume of accessible information that enables construction of regional scenarios for some aspects of global change. [3.9.1]

There also are numerous shortcomings of current scenario development, many of which are being actively investigated. These investigations include efforts to properly represent socioeconomic, land-use, and environmental changes in scenarios; to obtain scenarios at higher resolution (in time and space); and to incorporate changes in variability as well as mean conditions in scenarios. Increasing attention is required on construction of scenarios that address policy-related issues such as stabilization of GHG concentrations or adaptation, as well as improving the representation of uncertainties in projections, possibly within a risk assessment framework. [3.9.2]

4. Natural and Human Systems

Natural and human systems are expected to be exposed to climatic variations such as changes in the average, range, and variability of temperature and precipitation, as well as the frequency and severity of weather events. Systems also would be exposed to indirect effects from climate change such as sea-level rise, soil moisture changes, changes in land and water condition, changes in the frequency of fire and pest infestation, and changes in the distribution of infectious disease vectors and hosts. The sensitivity of a system to these exposures depends on system characteristics and includes the potential for adverse and beneficial effects. The potential for a system to sustain adverse impacts is moderated by adaptive capacity. The capacity to adapt human management of systems is determined by access to resources, information and technology, the skill and knowledge to use them, and the stability and effectiveness of cultural, economic, social, and governance institutions that facilitate or constrain how human systems respond.

Table TS-2: Examples of impacts resulting from projected changes in extreme climate events.

Projected Changes during the 21st Century in Extreme Climate Phenomena and their Likelihood[a]	Representative Examples of Projected Impacts[b] (all high confidence of occurrence in some areas[c])
Simple Extremes	
Higher maximum temperatures; more hot days and heat waves[d] over nearly all land areas (*very likely*[a])	· Increased incidence of death and serious illness in older age groups and urban poor · Increased heat stress in livestock and wildlife · Shift in tourist destinations · Increased risk of damage to a number of crops · Increased electric cooling demand and reduced energy supply reliability
Higher (increasing) minimum temperatures; fewer cold days, frost days, and cold waves[d] over nearly all land areas (*very likely*[a])	· Decreased cold-related human morbidity and mortality · Decreased risk of damage to a number of crops, and increased risk to others · Extended range and activity of some pest and disease vectors · Reduced heating energy demand
More intense precipitation events (*very likely*[a] over many areas)	· Increased flood, landslide, avalanche, and mudslide damage · Increased soil erosion · Increased flood runoff could increase recharge of some floodplain aquifers · Increased pressure on government and private flood insurance systems and disaster relief
Complex Extremes	
Increased summer drying over most mid-latitude continental interiors and associated risk of drought (*likely*[a])	· Decreased crop yields · Increased damage to building foundations caused by ground shrinkage · Decreased water resource quantity and quality · Increased risk of forest fire
Increase in tropical cyclone peak wind intensities, mean and peak precipitation intensities (*likely*[a] over some areas)[e]	· Increased risks to human life, risk of infectious disease epidemics, and many other risks · Increased coastal erosion and damage to coastal buildings and infrastructure · Increased damage to coastal ecosystems such as coral reefs and mangroves
Intensified droughts and floods associated with El Niño events in many different regions (*likely*[a]) (see also under droughts and intense precipitation events)	· Decreased agricultural and rangeland productivity in drought- and flood-prone regions · Decreased hydro-power potential in drought-prone regions
Increased Asian summer monsoon precipitation variability (*likely*[a])	· Increased flood and drought magnitude and damages in temperate and tropical Asia
Increased intensity of mid-latitude storms (little agreement between current models)[d]	· Increased risks to human life and health · Increased property and infrastructure losses · Increased damage to coastal ecosystems

[a] Likelihood refers to judgmental estimates of confidence used by TAR WGI: *very likely* (90-99% chance); *likely* (66-90% chance). Unless otherwise stated, information on climate phenomena is taken from the Summary for Policymakers, TAR WGI.
[b] These impacts can be lessened by appropriate response measures.
[c] High confidence refers to probabilities between 67 and 95% as described in Footnote 6 of TAR WGII, Summary for Policymakers.
[d] Information from TAR WGI, Technical Summary, Section F.5.
[e] Changes in regional distribution of tropical cyclones are possible but have not been established.

4.1. Water Resources

There are apparent trends in streamflow volumes—increases and decreases—in many regions. However, confidence that these trends are a result of climate change is low because of factors such as the variability of hydrological behavior over time, the brevity of instrumental records, and the response of river flows to stimuli other than climate change. *In contrast, there is high confidence that observations of widespread accelerated glacier retreat and shifts in the timing of streamflow from spring toward winter in many areas are associated with observed increases in temperature.* High confidence in these findings exists because these changes are driven by rising temperature and are unaffected by factors that influence streamflow

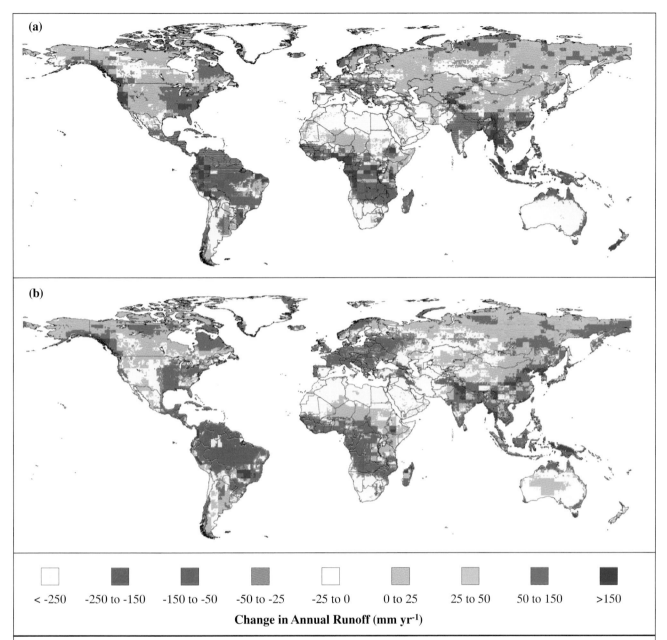

Figure TS-3: The pattern of changes in runoff largely follows the pattern of simulated changes in precipitation, which varies between climate models. The modeled increases in runoff shown in both maps [(a) HadCM2 ensemble mean and (b) HadCM3; see Section 4.3.6.2 of Chapter 4 for discussion of models and scenarios used] for high latitudes and southeast Asia, and decreases in central Asia, the area around the Mediterranean, southern Africa, and Australia are broadly consistent—in terms of direction of change—across most climate models. In other parts of the world, changes in precipitation and runoff vary between climate change scenarios.

volumes. Glacier retreat will continue, and many small glaciers may disappear (high confidence). The rate of retreat will depend on the rate of temperature rise. [4.3.6.1, 4.3.11]

The effect of climate change on streamflow and groundwater recharge varies regionally and among scenarios, largely following projected changes in precipitation. In some parts of the world, the direction of change is consistent between scenarios, although the magnitude is not. In other parts of the world, the direction of change is uncertain. Possible streamflow changes under two climate change scenarios are shown in

Figure TS-3. Confidence in the projected direction and magnitude of change in streamflow and groundwater recharge is largely dependent on confidence in the projected changes in precipitation. The mapped increase in streamflow in high latitudes and southeast Asia and the decrease in streamflow in central Asia, the area around the Mediterranean, and southern Africa are broadly consistent across climate models. Changes in other areas vary between climate models. [4.3.5, 4.3.6.2]

Peak streamflow will move from spring to winter in many areas where snowfall currently is an important component of the

water balance (high confidence). Higher temperatures mean that a greater proportion of winter precipitation falls as rain rather than snow and therefore is not stored on the land surface until it melts in spring. In particularly cold areas, an increase in temperature would still mean that winter precipitation falls as snow, so there would be little change in streamflow timing in these regions. The greatest changes therefore are likely to be in "marginal" zones—including central and eastern Europe and the southern Rocky Mountain chain—where a small temperature rise reduces snowfall substantially. [4.3.6.2]

Water quality generally would be degraded by higher water temperatures (high confidence). The effect of temperature on water quality would be modified by changes in flow volume, which may either exacerbate or lessen the effect of temperature, depending on the direction of change in flow volume. Other things being equal, increasing water temperature alters the rate of operation of biogeochemical processes (some degrading, some cleaning) and, most important, lowers the dissolved oxygen concentration of water. In rivers this effect may be offset to an extent by increased streamflow—which would dilute chemical concentrations further—or enhanced by lower streamflow, which would increase concentrations. In lakes, changes in mixing may offset or exaggerate the effects of increased temperature. [4.3.10]

Flood magnitude and frequency are likely to increase in most regions, and low flows are likely to decrease in many regions. The general direction of change in extreme flows and flow variability is broadly consistent among climate change scenarios, although confidence in the potential magnitude of change in any catchment is low. The general increase in flood magnitude and frequency is a consequence of a projected general increase in the frequency of heavy precipitation events, although the effect of a given change in precipitation depends on catchment characteristics. Changes in low flows are a function of changes in precipitation and evaporation. Evaporation generally is projected to increase, which may lead to lower low flows even where precipitation increases or shows little change. [4.3.8, 4.3.9]

Approximately 1.7 billion people, one-third of the world's population, presently live in countries that are water-stressed (i.e., using more than 20% of their renewable water supply—a commonly used indicator of water stress). This number is projected to increase to about 5 billion by 2025, depending on the rate of population growth. Projected climate change could further decrease streamflow and groundwater recharge in many of these water-stressed countries—for example, in central Asia, southern Africa, and countries around the Mediterranean Sea—but may increase it in some others.

Demand for water generally is increasing, as a result of population growth and economic development, but is falling in some countries. Climate change may decrease water availability in some water-stressed regions and increase it in others.

Climate change is unlikely to have a large effect on municipal and industrial demands but may substantially affect irrigation withdrawals. In the municipal and industrial sectors, it is likely that nonclimatic drivers will continue to have very substantial effects on demand for water. Irrigation withdrawals, however, are more climatically determined, but whether they increase or decrease in a given area depends on the change in precipitation: Higher temperatures, hence crop evaporative demand, would mean that the general tendency would be toward an increase in irrigation demands. [4.4.2, 4.4.3, 4.5.2]

The impact of climate change on water resources depends not only on changes in the volume, timing, and quality of streamflow and recharge but also on system characteristics, changing pressures on the system, how management of the system evolves, and what adaptations to climate change are implemented. Nonclimatic changes may have a greater impact on water resources than climate change. Water resources systems are evolving continually to meet changing management challenges. Many of the increased pressures will increase vulnerability to climate change, but many management changes will reduce vulnerability. Unmanaged systems are likely to be most vulnerable to climate change. By definition, these systems have no management structures in place to buffer the effects of hydrological variability. [4.5.2]

Climate change challenges existing water resources management practices by adding uncertainty. Integrated water resources management will enhance the potential for adaptation to change. The historic basis for designing and operating infrastructure no longer holds with climate change because it cannot be assumed that the future hydrological regime will be the same as that of the past. The key challenge, therefore, is incorporating uncertainty into water resources planning and management. Integrated water resources management is an increasingly used means of reconciling different and changing water uses and demands, and it appears to offer greater flexibility than conventional water resources management. Improved ability to forecast streamflow weeks or months ahead also would significantly enhance water management and its ability to cope with a changing hydrological variability. [4.6]

Adaptive capacity (specifically, the ability to implement integrated water resources management), however, is very unevenly distributed across the world. In practice, it may be very difficult to change water management practices in a country where, for example, management institutions and market-like processes are not well developed. The challenge, therefore, is to develop ways to introduce integrated water management practices into specific institutional settings—which is necessary even in the absence of climate change to improve the effectiveness of water management. [4.6.4]

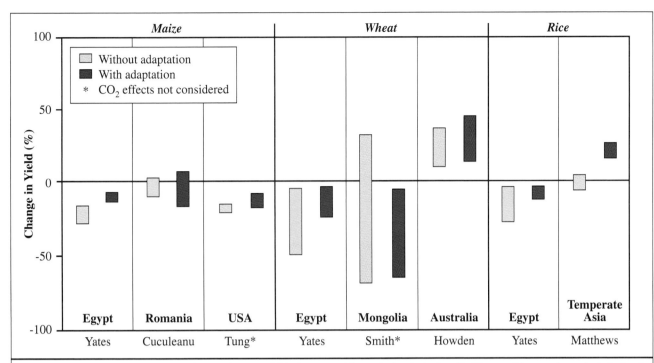

Figure TS-4: Ranges of percentage changes in crop yields (expressed in vertical extent of vertical bars only) spanning selected climate change scenarios—with and without agronomic adaptation—from paired studies listed in Table 5-4. Each pair of ranges is differentiated by geographic location and crop. Pairs of vertical bars represent the range of percentage changes with and without adaptation. Endpoints of each range represent collective high and low percentage change values derived from all climate scenarios used in the study. The horizontal extent of the bars is not meaningful. On the x-axis, the last name of the lead author is listed as it appears in Table 5-4; full source information is provided in the WGII TAR Chapter 5 reference list.

4.2. Agriculture and Food Security

The response of crop yields to climate change varies widely, depending on the species, cultivar, soil conditions, treatment of CO_2 direct effects, and other locational factors. It is established with medium confidence that a few degrees of projected warming will lead to general increases in temperate crop yields, with some regional variation (Table 5-4). At larger amounts of projected warming, most temperate crop yield responses become generally negative. Autonomous agronomic adaptation ameliorates temperate crop yield loss and improves gain in most cases (Figure TS-4). In the tropics, where some crops are near their maximum temperature tolerance and where dryland agriculture predominates, yields would decrease generally with even minimal changes in temperature; where there is a large decrease in rainfall, crop yields would be even more adversely affected (medium confidence). With autonomous agronomic adaptation, it is established with medium confidence that crop yields in the tropics tend to be less adversely affected by climate change than without adaptation, but they still tend to remain below baseline levels. Extreme events also will affect crop yields. Higher minimum temperatures will be beneficial to some crops, especially in temperate regions, and detrimental to other crops, especially in low latitudes (high confidence). Higher maximum temperatures will be generally detrimental to numerous crops (high confidence). [5.3.3]

Important advances in research since the SAR on the direct effects of CO_2 on crops suggest that beneficial effects may be greater under certain stressful conditions, including warmer temperatures and drought. Although these effects are well established for a few crops under experimental conditions, knowledge of them is incomplete for suboptimal conditions of actual farms. Research on agricultural adaptation to climate change also has made important advances. Inexpensive, farm-level (autonomous) agronomic adaptations such as altering of planting dates and cultivar selections have been simulated in crop models extensively. More expensive, directed adaptations—such as changing land-use allocations and developing and using irrigation infrastructure—have been examined in a small but growing number of linked crop-economic models, integrated assessment models, and econometric models.

Degradation of soil and water resources is one of the major future challenges for global agriculture. It is established with high confidence that those processes are likely to be intensified by adverse changes in temperature and precipitation. Land use and management have been shown to have a greater impact on soil conditions than the indirect effect of climate change; thus, adaptation has the potential to significantly mitigate these impacts. A critical research need is to assess whether resource degradation will significantly increase the risks faced by vulnerable agricultural and rural populations [5.3.2, 5.3.4, 5.3.6].

In the absence of climate change, most global and regional studies project declining real prices for agricultural commodities. Confidence in these projections declines farther into the future. *The impacts of climate change on agriculture are estimated to result in small percentage changes in global income, with positive changes in more developed regions and smaller or negative changes in developing regions (low to medium confidence).* The effectiveness of adaptation (agronomic and economic) in ameliorating the impacts of climate change will vary regionally and depend a great deal on regional resource endowments, including stable and effective institutions. [5.3.1, 5.3.5]

Most studies indicate that mean annual temperature increases of 2.5° C or greater would prompt food prices to increase (low confidence) as a result of slowing in the expansion of global food capacity relative to growth in global food demand. At lesser amounts of warming than 2.5° C, global impact assessment models cannot distinguish the climate change signal from other sources of change. Some recent aggregated studies have estimated economic impacts on vulnerable populations such as smallholder producers and poor urban consumers. These studies indicate that climate change will lower the incomes of vulnerable populations and increase the absolute number of people at risk of hunger (low confidence). [5.3.5, 5.3.6]

Without autonomous adaptation, increases in extreme events are likely to increase heat stress-related livestock deaths, although winter warming may reduce neonatal deaths at temperate latitudes (established but incomplete). Strategies to adapt livestock to physiological stresses of warming are considered effective; however, adaptation research is hindered by the lack of experimentation and simulation. [5.3.3]

Confidence in specific numerical estimates of climate change impacts on production, income, and prices obtained from large, aggregated, integrated assessment models is considered to be low because there are several remaining uncertainties. The models are highly sensitive to some parameters that have been subjected to sensitivity analysis, yet sensitivity to a large number of other parameters has not been reported. Other uncertainties include the magnitude and persistence of effects of rising atmospheric CO_2 on crop yield under realistic farming conditions; potential changes in crop and animal pest losses; spatial variability in crop responses to climate change; and the effects of changes in climate variability and extreme events on crops and livestock. [Box 5-3]

4.3. Terrestrial and Freshwater Ecosystems

Ecosystems are subject to many pressures, such as land-use changes, deposition of nutrients and pollutants, harvesting, grazing by livestock, introduction of exotic species, and natural climate variability. Climate change constitutes an additional pressure that could change or endanger these systems. The impact of climate change on these systems will be influenced by land and water management adaptation and interactions with other pressures. Adaptive capacity is greater for more intensively managed lands and waters and in production of marketed goods (e.g., timber production in plantations) than in less intensively managed lands and nonmarket values of those lands and waters. [5.1, 5.2]

Populations of many species already are threatened and are expected to be placed at greater risk by the synergy between the stresses of changing climate, rendering portions of current habitat unsuitable, and land-use change that fragments habitats. Without adaptation, some species that currently are classified as "critically endangered" will become extinct, and the majority of those labeled "endangered or vulnerable" will become much rarer in the 21st century (high confidence). This may have the greatest impact on the lowest income human societies, which rely on wildlife for subsistence living. In addition, there is high confidence that loss or reduction of species would impact the services provided by wildlife through roles within an ecosystem (e.g., pollination, natural pest control), recreation (e.g., sport hunting, wildlife viewing), and cultural and religious practices of indigenous people. Possible adaptation methods to reduce risks to species could include establishment of refuges, parks, and reserves with corridors to allow migration of species, as well as use of captive breeding and translocation. However, these options may have limitations of cost. [5.4]

There are now substantial observational and experimental studies demonstrating the link between change in regional climate and biological or physical processes in ecosystems. These include a lengthening of vegetative growing season by 1.2 to 3.6 days per decade in the high northern latitudes (one factor leading to community composition changes); warming of lakes and rivers as a result of shortening duration of ice cover; upward range shifts in alpine herbs; and increased mortality and range contraction of wildlife as a result of heat stress. Others include changes in population sizes, body sizes, and migration times (see TS 2.1 and 7.1, Figure TS-11, and Table TS-16 for additional information). [5.2.1]

Vegetation distribution models since the SAR suggest that mass ecosystem or biome movement is most unlikely to occur because of different climatic tolerance of the species involved, different migration abilities, and the effects of invading species. Species composition and dominance will change, resulting in ecosystem types that may be quite different from those we see today. These changes will lag the changes in climate by years to decades to centuries (high confidence). The effects of changes in disturbances such as fire, blowdown, or pest attacks on vegetation have not been included in these studies. [5.2]

Recent modeling studies continue to show potential for significant disruption of ecosystems under climate change (high confidence). Further development of simple correlative models that were available at the time of the SAR point to areas where ecosystem disruption and the potential for ecosystem migration are high. Observational data and newer dynamic vegetation models linked to transient climate models are refining the projections. However, the precise outcomes depend on processes that are too subtle to be fully captured by current models. [5.2]

Increasing CO_2 concentration would increase net primary productivity (plant growth, litterfall, and mortality) in most systems, whereas increasing temperature may have positive or negative effects (high confidence). Experiments on tree species grown under elevated CO_2 over several years show continued and consistent stimulation of photosynthesis and little evidence of long-term loss of sensitivity to CO_2. However, changes in net ecosystem productivity (which includes plant growth, litterfall, mortality, litter decomposition, and soil carbon dynamics) and net biome productivity (which includes those effects plus the effects of fire or other disturbances) are less likely to be positive and may be generally negative. Research reported since the SAR confirms the view that the largest and earliest impacts induced by climate change are likely to occur in boreal forests, through changes in weather-related disturbance regimes and nutrient cycling. [5.6.1.1, 5.6.3.1]

Terrestrial ecosystems appear to be storing increasing amounts of carbon. At the time of the SAR, this was attributed largely to increasing plant productivity because of the interaction among elevated CO_2 concentration, increasing temperatures, and soil moisture changes. Recent results confirm that productivity gains are occurring but suggest that they are smaller under field conditions than plant-pot experiments indicate (*medium confidence*). Hence, the terrestrial uptake may be caused more by change in uses and management of land than by the direct effects of elevated CO_2 and climate. The degree to which terrestrial ecosystems continue to be net sinks for carbon is uncertain because of the complex interactions between the aforementioned factors (e.g., arctic terrestrial ecosystems and wetlands may act as sources and sinks) (*medium confidence*).

In arid or semi-arid areas (e.g., rangelands, dry forests/woodlands) where climate change is likely to decrease available soil moisture, productivity is expected to decrease. Increased CO_2 concentrations may counteract some of these losses. However, many of these areas are affected by El Niño/La Niña, other climatic extremes, and disturbances such as fire. Changes in the frequencies of these events and disturbances could lead to loss of productivity thus potential land degradation, potential loss of stored carbon, or decrease in the rate of carbon uptake (medium confidence). [5.5]

Some wetlands will be replaced by forests or heathlands, and those overlying permafrost are likely to be disrupted as a result of thawing of permafrost (high confidence). The initial net effect of warming on carbon stores in high-latitude ecosystems is likely to be negative because decomposition initially may respond more rapidly than production. In these systems, changes in albedo and energy absorption during winter are likely to act as a positive feedback to regional warming as a result of earlier melting of snow and, over decades to centuries, poleward movement of the treeline. [5.8, 5.9]

Most wetland processes are dependent on catchment-level hydrology; thus, adaptations for projected climate change may be practically impossible. Arctic and subarctic ombrotrophic bog communities on permafrost, as well as more southern depressional wetlands with small catchment areas, are likely to be most vulnerable to climate change. The increasing speed of peatland conversion and drainage in southeast Asia is likely to place these areas at a greatly increased risk of fires and affect the viability of tropical wetlands. [5.8]

Opportunities for adapting to expected changes in high-latitude and alpine ecosystems are limited because these systems will respond most strongly to globally induced changes in climate. Careful management of wildlife resources could minimize climatic impacts on indigenous peoples. Many high-latitude regions depend strongly on one or a few resources, such as timber, oil, reindeer, or wages from fighting fires. Economic diversification would reduce the impacts of large changes in the availability or economic value of particular goods and services. High levels of endemism in many alpine floras and their inability to migrate upward means that these species are very vulnerable. [5.9]

Contrary to the SAR, global timber market studies that include adaptations through land and product management suggest that climate change would increase global timber supply (medium confidence). At the regional and global scales, the extent and nature of adaptation will depend primarily on wood and non-wood product prices, the relative value of substitutes, the cost of management, and technology. On specific sites, changes in forest growth and productivity will constrain—and could limit—choices regarding adaptation strategies (high confidence). In markets, prices will mediate adaptation through land and product management. Adaptation in managed forests will include salvaging dead and dying timber, replanting new species that are better suited to the new climate, planting genetically modified species, and intensifying or decreasing management. Consumers will benefit from lower timber prices; producers may gain or lose, depending on regional changes in timber productivity and potential dieback effects. [5.6]

Climate change will lead to poleward movement of the southern and northern boundaries of fish distributions, loss of habitat for cold- and coolwater fish, and gain in habitat for warmwater fish (high confidence). As a class of ecosystems, inland waters are vulnerable to climatic change and other pressures owing to their small size and position downstream from many human activities (high confidence). The most vulnerable elements include reduction and loss of lake and river ice (very high confidence), loss of habitat for coldwater fish (very high confidence), increases in extinctions and invasions of exotics (high confidence), and potential exacerbation of existing pollution problems such as eutrophication, toxics, acid rain, and UV-B radiation (medium confidence). [5.7]

4.4. Coastal Zones and Marine Ecosystems

Global climate change will result in increases in sea-surface temperature (SST) and sea level; decreases in sea-ice cover; and changes in salinity, wave climate, and ocean circulation. Some of these changes already are taking place. Changes in oceans are expected to have important feedback effects on global climate and on the climate of the immediate coastal area (see TAR WGI). They also would have profound impacts on the biological production of oceans, including fish production. For instance, changes in global water circulation and vertical mixing will affect the distribution of biogenic elements and the efficiency of CO_2 uptake by the ocean; changes in upwelling rates would have major impacts on coastal fish production and coastal climates. [6.3]

If warm events associated with El Niños increase in frequency, plankton biomass and fish larvae abundance would decline and adversely impact fish, marine mammals, seabirds, and ocean biodiversity (high confidence). In addition to El Niño-Southern Oscillation (ENSO) variability, the persistence of multi-year climate-ocean regimes and switches from one regime to another have been recognized since the SAR. Changes in recruitment patterns of fish populations have been linked to such switches. Fluctuations in fish abundance are increasingly regarded as biological responses to medium-term climate fluctuations in addition to overfishing and other anthropogenic factors. Similarly, survival of marine mammals and seabirds also is affected by interannual and longer term variability in several oceanographic and atmospheric properties and processes, especially in high latitudes. [6.3.4]

Growing recognition of the role of the climate-ocean system in the management of fish stocks is leading to new adaptive strategies that are based on the determination of acceptable removable percentages of fish and stock resilience. Another consequence of the recognition of climate-related changes in the distribution of marine fish populations suggests that the sustainability of many nations' fisheries will depend on adaptations that increase flexibility in bilateral and multilateral fishing agreements, coupled with international stock assessments and management plans. Creating sustainable fisheries also depends on understanding synergies between climate-related impacts on fisheries and factors such as harvest pressure and habitat conditions. [6.3.4, 6.6.4]

Adaptation by expansion of marine aquaculture may partly compensate for potential reductions in ocean fish catch. Marine aquaculture production has more than doubled since 1990, and in 1997 represented approximately 30% of total commercial fish and shellfish production for human consumption. However, future aquaculture productivity may be limited by ocean stocks of herring, anchovies, and other species that are used to provide fishmeal and fish oils to feed cultured species, which may be negatively impacted by climate change. Decreases in dissolved oxygen levels associated with increased seawater temperatures and enrichment of organic matter creates conditions for the spread of diseases in wild and aquaculture fisheries, as well as outbreaks of algal blooms in coastal areas. Pollution and habitat destruction that can accompany aquaculture also may place limits on its expansion and on the survival success of wild stocks. [6.3.5]

Many coastal areas already are experiencing increased levels of sea flooding, accelerated coastal erosion, and seawater intrusion into freshwater sources; these processes will be exacerbated by climate change and sea-level rise. Sea-level rise in particular has contributed to erosion of sandy and gravel beaches and barriers; loss of coastal dunes and wetlands; and drainage problems in many low-lying, mid-latitude coastal areas. Highly diverse and productive coastal ecosystems, coastal settlements, and island states will continue to be exposed to pressures whose impacts are expected to be largely negative and potentially disastrous in some instances. [6.4]

Low-latitude tropical and subtropical coastlines, particularly in areas where there is significant human population pressure, are highly susceptible to climate change impacts. These impacts will exacerbate many present-day problems. For instance, human activities have increased land subsidence in many deltaic regions by increasing subsurface water withdrawals, draining wetland soils, and reducing or cutting off riverine sediment loads. Problems of inundation, salinization of potable groundwater, and coastal erosion will all be accelerated with global sea-level rise superimposed on local submergence. Especially at risk are large delta regions of Asia and small islands whose vulnerability was recognized more than a decade ago and continues to increase. [6.4.3, 6.5.3]

High-latitude (polar) coastlines also are susceptible to climate warming impacts, although these impacts have been less studied. Except on rock-dominated or rapidly emerging coasts, a combination of accelerated sea-level rise, more energetic wave climate with reduced sea-ice cover, and increased ground temperatures that promote thaw of permafrost and ground ice (with consequent volume loss in coastal landforms) will have severe impacts on settlements and infrastructure and will result in rapid coastal retreat. [6.4.6]

Coastal ecosystems such as coral reefs and atolls, salt marshes and mangrove forests, and submergered aquatic vegetation will be impacted by sea-level rise, warming SSTs, and any changes in storm frequency and intensity. Impacts of sea-level rise on mangroves and salt marshes will depend on the rate of rise relative to vertical accretion and space for horizontal migration, which can be limited by human development in coastal areas. Healthy coral reefs are likely to be able to keep up with sea-level rise, but this is less certain for reefs degraded by coral bleaching, UV-B radiation, pollution, and other stresses. Episodes of coral bleaching over the past 20 years have been associated with several causes, including increased ocean temperatures. Future sea-surface warming would increase stress on coral reefs and result in increased frequency of marine diseases (high confidence). Changes in ocean chemistry resulting from higher CO_2 levels may have a negative impact on coral reef development and health, which would have a detrimental effect on coastal fisheries and on social and economic uses of reef resources. [6.4.4, 6.4.5]

Few studies have examined potential changes in prevailing ocean wave heights and directions and storm waves and surges as a consequence of climate change. Such changes can be expected to have serious impacts on natural and human-modified coasts because they will be superimposed on a higher sea level than at present.

Vulnerabilities have been documented for a variety of coastal settings, initially by using a common methodology developed in the early 1990s. These and subsequent studies have confirmed the spatial and temporal variability of coastal vulnerability at national and regional levels. Within the common methodology, three coastal adaptation strategies have been identified: protect, accommodate, and retreat. Since the SAR, adaptation strategies for coastal zones have shifted in emphasis away from hard protection structures (e.g., seawalls, groins) toward soft protection measures (e.g., beach nourishment), managed retreat, and enhanced resilience of biophysical and socioeconomic systems, including the use of flood insurance to spread financial risk. [6.6.1, 6.6.2]

Integrated assessments of coastal zones and marine ecosystems and better understanding of their interaction with human development and multi-year climate variability could lead to improvements in sustainable development and management. Adaptation options for coastal and marine management are most effective when they are incorporated with policies in other areas, such as disaster mitigation plans and land-use plans.

4.5. Human Settlements, Energy, and Industry

Human settlements are integrators of many of the climate impacts initially felt in other sectors and differ from each other in geographic location, size, economic circumstances, and political and institutional capacity. As a consequence, it is difficult to make blanket statements concerning the importance of climate or climate change that will not have numerous exceptions. However, classifying human settlements by considering pathways by which climate may affect them, size or other obvious physical considerations, and adaptive capacities (wealth, education of the populace, technological and institutional capacity) helps to explain some of the differences in expected impacts. [7.2]

Human settlements are affected by climate in one of three major ways:

1) Economic sectors that support the settlement are affected because of changes in productive capacity (e.g., in agriculture or fisheries) or changes in market demand for goods and services produced there (including demand from people living nearby and from tourism). The importance of this impact depends in part on whether the settlement is rural—which generally means that it is dependent on one or two resource-based industries—or urban, in which case there usually (but not always) is a broader array of alternative resources. It also depends on the adaptive capacity of the settlement. [7.1]

2) Some aspects of physical infrastructure (including energy transmission and distribution systems), buildings, urban services (including transportation systems), and specific industries (such as agroindustry, tourism, and construction) may be directly affected. For example, buildings and infrastructure in deltaic areas may be affected by coastal and river flooding; urban energy demand may increase or decrease as a result of changed balances in space heating and space cooling; and coastal and mountain tourism may be affected by changed seasonal temperature and precipitation patterns and sea-level rise. Concentration of population and infrastructure in urban areas can mean higher numbers of persons and higher value of physical capital at risk, although there also are many economies of scale and proximity in ensuring well-managed infrastructure and service provision. When these factors are combined with other prevention measures, risks can be reduced considerably. However, some larger urban centers in Africa, Asia, Latin America, and the Caribbean, as well as smaller settlements (including villages and small urban centers), often have less wealth, political power, and institutional capacity to reduce risks in this way. [7.1]

3) Population may be directly affected through extreme weather, changes in health status, or migration. Extreme weather episodes may lead to changes in deaths, injuries, or illness. For example, health status may improve as a result of reduced cold stress or deteriorate as a result of increased heat stress and disease. Population movements caused by climate changes may affect the size and characteristics of settlement populations, which in turn changes the demand for urban services. The problems are somewhat different in the largest population centers (e.g., those of more than 1 million population) and mid-sized to small-sized regional centers. The former are more likely to be destinations for migrants from rural areas and smaller settlements and cross-border areas, but larger settlements

Table TS-3: Impacts of climate change on human settlements, by impact type and settlement type (impact mechanism). [a,b]

Type of Settlement, Importance Rating, and Reference

Impact Type	Resource-Dependent (Effects on Resources)				Coastal-Riverine-Steeplands (Effects on Buildings and Infrastructure)				Urban 1+ M (Effects on Populations)		Urban <1 M (Effects on Populations)		Confidence [c]
	Urban, High Capacity	Urban, Low Capacity	Rural, High Capacity	Rural, Low Capacity	Urban, High Capacity	Urban, Low Capacity	Rural, High Capacity	Rural, Low Capacity	High Capacity	Low Capacity	High Capacity	Low Capacity	
Flooding, landslides	L-M	M-H	L-M	M-H	L-M	M-H	M-H	M-H	M	M-H	M	M-H	****
Tropical cyclone	L-M	M-H	L-M	M-H	L-M	M-H	M	M-H	L-M	M	L	L-M	***
Water quality	L-M	M	L-M	M-H	L-M	M-H	L-M	M-H	L-M	M-H	L-M	M-H	***
Sea-level rise	L-M	M-H	L-M	M-H	M	M-H	M	M-H	L	L-M	L	L-M	**** (** for resource-dependent)
Heat/cold waves	L-M	M-H	L-M	M-H	L-M	L-M	L-M	L	L-M	M-H	L-M	M-H	*** (**** for urban)
Water shortage	L	L-M	M	M-H	L	L-M	L-M	M-H	L	M	L-M	M	*** (** for urban)
Fires	L-M	L-M	L-M	M-H	L-M	L-M	L-M	L-M	L-M	L-M	L-M	M	* (*** for urban)
Hail, windstorm	L-M	L-M	L-M	M-H	L-M	L-M	L-M	M	L-M	L-M	L-M	L-M	**
Agriculture/forestry/fisheries productivity	L-M	L-M	L-M	M-H	L	L	L	L	L	L-M	L-M	M	***
Air pollution	L-M	L-M	L	L	-	-	-	-	L-M	M-H	L-M	M-H	***
Permafrost melting	L	L	L-M	L-M	L	L	L	L	-	-	L-M	L-M	****
Heat islands	L	L	-	-	L	L	-	-	M	L-M	L-M	L-M	***

a Values in cells in the table were assigned by authors on the basis of direct evidence in the literature or inference from impacts shown in other cells. Typeface indicates source of rating: Boldface indicates direct evidence or study; italic indicates direct inference from similar impacts; regular typeface indicates logical conclusion from settlement type, but cannot be directly corroborated from a study or inferred from similar impacts.

b Impacts ratings: Low (L) = impacts are barely discernible or easily overcome; moderate (M) = impacts are clearly noticeable, although not disruptive, and may require significant expense or difficulty in adapting; high (H) = impacts are clearly disruptive and may not be overcome or adaptation is so costly that it is disruptive (impacts generally based on $2xCO_2$ scenarios or studies describing impact of current weather events, but have been placed in context of the IPCC transient scenarios for mid-to late 21st century). Note that "Urban 1+ M" and "Urban <1 M" refer to populations above and below 1 million, respectively.

c See Section 1.4 of Technical Summary for key to confidence-level rankings.

generally have much greater command over national resource. Thus, smaller settlements actually may be more vulnerable. Informal settlements surrounding large and medium-size cities in the developing world remain a cause for concern because they exhibit several current health and environmental hazards that could be exacerbated by global warming and have limited command over resources. [7.1]

Table TS-3 classifies several types of climate-caused environmental changes discussed in the climate and human settlement literatures. The table features three general types of settlements, each based on one of the three major mechanisms by which climate affects settlements. The impacts correspond to the mechanism of the effect. Thus, a given settlement may be affected positively by effects of climate change on its resource base (e.g., more agricultural production) and negatively by effects on its infrastructure (e.g., more frequent flooding of its water works and overload of its electrical system). Different types of settlements may experience these effects in different relative intensities (e.g., noncoastal settlements do not directly experience impacts through sea-level rise); the impacts are ranked from overall highest to lowest importance. Most settlement effects literature is based on $2xCO_2$ scenarios or studies describing the impact of current weather events (analogs) but has been placed in context of the IPCC transient scenarios. [7.1]

Climate change has the potential to create local and regional conditions that involve water deficits and surpluses, sometimes seasonally in the same geographic locations. *The most widespread serious potential impacts are flooding, landslides, mudslides, and avalanches driven by projected increases in rainfall intensity and sea-level rise.* A growing literature suggests that a very wide variety of settlements in nearly every climate zone may be affected (established but incomplete). Riverine and coastal settlements are believed to be particularly at risk, but urban flooding could be a problem anywhere storm drains, water supply, and waste management systems are not designed with enough capacity or sophistication (including conventional hardening and more advanced system design) to avoid being overwhelmed. The next most serious threats are tropical cyclones (hurricanes or typhoons), which may increase in peak intensity in a warmer world. Tropical cyclones combine the effects of heavy rainfall, high winds, and storm surge in coastal areas and can be disruptive far inland, but they are not as universal in location as floods and landslides. Tens of millions of people live in the settlements potentially flooded. For example, estimates of the mean annual number of people who would be flooded by coastal storm surges increase several-fold (by 75 million to 200 million people, depending on adaptive responses) for mid-range scenarios of a 40-cm sea-level rise by the 2080s relative to scenarios with no sea-level rise. Potential damages to infrastructure in coastal areas from sea-level rise have been estimated to be tens of billions of dollars for individual countries such as Egypt, Poland, and Vietnam. In the middle of Table TS-3 are effects such as heat or cold waves, which

can be disruptive to the resource base (e.g., agriculture), human health, and demand for heating and cooling energy. Environmental impacts such as reduced air and water quality also are included. Windstorms, water shortages, and fire also are expected to be moderately important in many regions. At the lower end are effects such as permafrost melting and heat island effects—which, although important locally, may not apply to as wide a variety of settlements or hold less importance once adaptation is taken into account. [7.2, 7.3]

Global warming is expected to result in increases in energy demand for spacing cooling and in decreased energy use for space heating. Increases in heat waves add to cooling energy demand, and decreases in cold waves reduce heating energy demand. The projected net effect on annual energy consumption is scenario- and location-specific. Adapting human settlements, energy systems, and industry to climate change provides challenges for the design and operation of settlements (in some cases) during more severe weather and opportunities to take advantage (in other cases) of more benign weather. For instance, transmission systems of electric systems are known to be adversely affected by extreme events such as tropical cyclones, tornadoes, and ice storms. The existence of local capacity to limit environmental hazards or their health consequences in any settlement generally implies local capacity to adapt to climate change, unless adaptation implies particularly expensive infrastructure investment. Adaptation to warmer climate will require local tuning of settlements to a changing environment, not just warmer temperatures. Urban experts are unanimous that successful environmental adaptation cannot occur without locally based, technically and institutionally competent, and politically supported leadership that have good access to national-level resources. [7.2, 7.3, 7.4, 7.5]

Possible adaptation options involve planning of settlements and their infrastructure, placement of industrial facilities, and making similar long-lived decisions to reduce the adverse effects of events that are of low (but increasing) probability and high (and perhaps rising) consequences. Many specific conventional and advanced techniques can contribute to better environmental planning and management, including market-based tools for pollution control, demand management and waste reduction, mixed-use zoning and transport planning (with appropriate provision for pedestrians and cyclists), environmental impact assessments, capacity studies, strategic environmental plans, environmental audit procedures, and state-of-the-environment reports. Many cities have used a combination of these strategies in developing "Local Agenda 21s." Many Local Agenda 21s deal with a list of urban problems that could closely interact with climate change in the future. [7.2, 7.5]

4.6. Insurance and Other Financial Services

The financial services sector—broadly defined as private and public institutions that offer insurance and disaster relief, banking, and asset management services—is a unique indicator of potential socioeconomic impacts of climate change because it is sensitive to climate change and it integrates effects on other sectors. The sector is a key agent of adaptation (e.g., through support of building codes and, to a limited extent, land-use planning), and financial services represent risk-spreading mechanisms through which the costs of weather-related events are distributed among other sectors and throughout society. However, insurance, whether provided by public or private entities, also can encourage complacency and maladaptation by fostering development in at-risk areas such as U.S. floodplains or coastal zones. The effects of climate change on the financial services sector are likely to manifest primarily through changes in spatial distribution, frequencies, and intensities of extreme weather events (Table TS-4). [8.1, 8.2, 15.2.7]

The costs of extreme weather events have exhibited a rapid upward trend in recent decades. Yearly global economic losses from large events increased from US$3.9 billion yr^{-1} in the 1950s to US$40 billion yr^{-1} in the 1990s (all 1999 US$, uncorrected for purchasing power parity). Approximately one-quarter of the losses occurred in developing countries. The insured portion of these losses rose from a negligible level to US$9.2 billion annually during the same period. Including events of all sizes doubles these loss totals (see Figure TS-5). The costs of weather events have risen rapidly, despite

significant and increasing efforts at fortifying infrastructure and enhancing disaster preparedness. These efforts dampen to an unknown degree the observed rise in loss costs, although the literature attempting to separate natural from human driving forces has not quantified this effect. As a measure of increasing insurance industry vulnerability, the ratio of global property/casualty insurance premiums to weather-related losses—an important indicator of adaptive capacity—fell by a factor of three between 1985 and 1999. [8.3]

Part of the observed upward trend in historical disaster losses is linked to socioeconomic factors—such as population growth, increased wealth, and urbanization in vulnerable areas—and part is linked to climatic factors such as observed changes in precipitation, flooding, and drought events. Precise attribution is complex, and there are differences in the balance of these two causes by region and by type of event. Many of the observed trends in weather-related losses are consistent with what would be expected under climate change. Notably, the growth rate in human-induced and non-weather-related losses has been far lower than that of weather-related events. [8.2.2]

Recent history has shown that weather-related losses can stress insurance companies to the point of impaired profitability, consumer price increases, withdrawal of coverage, and elevated demand for publicly funded compensation and relief. Increased uncertainty will increase the vulnerability of the

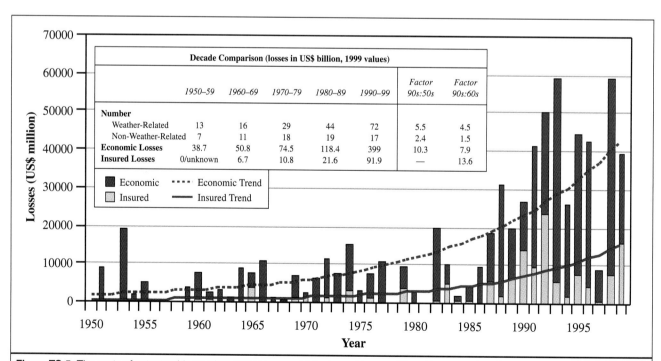

Figure TS-5: The costs of catastrophic weather events have exhibited a rapid upward trend in recent decades. Yearly economic losses from large events increased 10.3-fold from US$4 billion yr^{-1} in the 1950s to US$40 billion yr^{-1} in the 1990s (all in 1999 US$). The insured portion of these losses rose from a negligible level to US$9.2 billion annually during the same period, and the ratio of premiums to catastrophe losses fell by two-thirds. Notably, costs are larger by a factor of 2 when losses from ordinary, noncatastrophic weather-related events are included. The numbers generally include "captive" self-insurers but not the less-formal types of self-insurance.

Table TS-4: Extreme climate-related phenomena and their effects on the insurance industry: observed changes and projected changes during 21st century (after Table 3-10; see also Table 8-1).

Changes in Extreme Climate Phenomena	Observed Changes / *Likelihood*	Projected Changes	Type of Event Relevant to Insurance Sector	Relevant Time Scale	Sensitive Sectors/ Activities	Sensitive Insurance Branches
Temperature Extremes						
Higher maximum temperatures, more hot days and heat waves[b] over nearly all land areas	Likely[a] (mixed trends for heat waves in several regions)	Very likely[a]	Heat wave	Daily-weekly maximum	Electric reliability, human settlements	Health, life, property, business interruption
			Heat wave, droughts	Monthly-seasonal maximum	Forests (tree health), natural resources, agriculture, water resources, electricity demand and reliability, industry, health, tourism	Health, crop, business interruption
Higher (increasing) minimum temperatures, fewer cold days, frost days, and cold waves[b] over nearly all land areas	Very likely[a] (cold waves not treated by WGI)	Very likely[a]	Frost, frost heave	Daily-monthly minimum	Agriculture, energy demand, health, transport, human settlements	Health, crop, property, business interruption, vehicle
Rainfall/Precipitation Extremes						
More intense precipitation events	Likely[a] over many Northern Hemisphere mid- to high-latitude land areas	Very likely[a] over many areas	Flash flood	Hourly-daily maximum	Human settlements	Property, flood, vehicle, business interruption, life, health
			Flood, inundation, mudslide	Weekly-monthly maximum	Agriculture, forests, transport, water quality, human settlements, tourism	Property, flood, crop, marine, business interruption
Increased summer drying and associated risk of drought	Likely[a] in a few areas	Likely[a] over most mid-latitude continental interiors (lack of consistent projections in other areas)	Summer drought, land subsidence, wildfire	Monthly-seasonal minimum	Forests (tree health), natural resources, agriculture, water resources, (hydro)energy supply, human settlements	Crop, property, health
Increased intensity of mid-latitude storms[c]	Medium likelihood[a] of increase in Northern Hemisphere, decrease in Southern Hemisphere	Little agreement among current models	Snowstorm, ice storm, avalanche	Hourly-weekly	Forests, agriculture, energy distribution and reliability, human settlements, mortality, tourism	Property, crop, vehicle, aviation, life, business interruption
			Hailstorm	Hourly	Agriculture, property	Crop, vehicle, property, aviation
Intensified droughts and floods associated with El Niño events in many different regions (see also droughts and extreme precipitation events)	Inconclusive information	Likely[a]	Drought and floods	Various	Forests (tree health), natural resources, agriculture, water resources, (hydro)energy supply, human settlements	Property, flood, vehicle, crop, marine, business interruption, life, health

Table TS-4 (continued)

Changes in Extreme Climate Phenomena	Observed Changes	Projected Changes	Type of Event Relevant to Insurance Sector	Relevant Time Scale	Sensitive Sectors/ Activities	Sensitive Insurance Branches
	Likelihood					
Wind Extremes						
Increased intensity of mid-latitude storms[b]	No compelling evidence for change	Little agreement among current models	Mid-latitude windstorm	Hourly-daily	Forests, electricity distribution and reliability, human settlements	Property, vehicle, aviation, marine, business interruption, life
			Tornadoes	Hourly	Forests, electricity distribution and reliability, human settlements	Property, vehicle, aviation, marine, business interruption
Increase in tropical cyclone peak wind intensities, mean and peak precipitation intensities[c]	Wind extremes not observed in the few analyses available; insufficient data for precipitation	Likely[a] over some areas	Tropical storms, including cyclones, hurricanes, and typhoons	Hourly-weekly	Forests, electricity distribution and reliability, human settlements, agriculture	Property, vehicle, aviation, marine, business interruption, life
Other Extremes						
Refer to entries above for higher temperatures, increased tropical and mid-latitude storms	Refer to relevant entries above	Refer to relevant entries above	Lightning	Instan-taneous	Electricity distribution and reliability, human settlements, wildfire	Life, property, vehicle, aviation, marine, business interruption
Refer to entries above for increased tropical cyclones, Asian summer monsoon, and intensity of mid-latitude storms	Refer to relevant entries above	Refer to relevant entries above	Tidal surge (associated with onshore gales), coastal inundation	Daily	Coastal zone infrastructure, agriculture and industry, tourism	Life, marine, property, crop
Increased Asian summer monsoon precipitation variability	Not treated by WGI	Likely[a]	Flood and drought	Seasonal	Agriculture, human settlements	Crop, property, health, life

[a] Likelihood refers to judgmental estimates of confidence used by Working Group I: very likely (90-99% chance); likely (66-90% chance). Unless otherwise stated, information on climate phenomena is taken from Working Group I's Summary for Policymakers and Technical Summary. These likelihoods refer to observed and projected changes in extreme climate phenomena and likelihood shown in first three columns of table.
[b] Information from Working Group I, Technical Summary, Section F.5.
[c] Changes in regional distribution of tropical cyclones are possible but have not been established.

insurance and government sectors and complicate adaptation and disaster relief efforts under climate change. [8.3, 15.2.7]

The financial services sector as a whole is expected to be able to cope with the impacts of future climate change, although the historic record shows that low-probability, high-impact events or multiple closely spaced events severely affect parts of the sector, especially if adaptive capacity happens to be simultaneously depleted by nonclimate factors (e.g., adverse market conditions that can deplete insurer loss reserves by eroding the value of securities and other insurer assets). There is high confidence that climate change and anticipated changes in weather-related events that are perceived to be linked to climate change would increase actuarial uncertainty in risk assessment and thus in the functioning of insurance markets. Such developments would place upward pressure on premiums and/or could cause certain risks to be reclassified as uninsurable, with subsequent withdrawal of coverage. This, in turn, would place increased pressure on government-based insurance and relief systems, which already are showing strain in many regions and are attempting to limit their exposures (e.g., by raising deductibles and/or placing caps on maximum claims payable).

Trends toward increasing firm size, diversification, and integration of insurance with other financial services, as well as improved tools to transfer risk, all potentially contribute to robustness. However, the property/casualty insurance and reinsurance segments have greater sensitivity, and individual companies already have experienced catastrophe-related bankruptcies triggered by weather events. Under some conditions and in some regions, the banking industry as a provider of loans also may be vulnerable to climate change. In many cases, however, the banking sector transfers risk back to insurers, who often purchase their debt products. [8.3, 8.4, 15.2.7]

Adaptation[2] to climate change presents complex challenges, as well as opportunities, for the financial services sector. Regulatory involvement in pricing, tax treatment of reserves, and the (in)ability of firms to withdraw from at-risk markets are examples of factors that influence the resilience of the sector. Management of climate-related risk varies by country and region. Usually it is a mixture of commercial and public arrangements and self-insurance. In the face of climate change, the relative role of each can be expected to change. Some potential response options offer co-benefits that support sustainable development and climate change mitigation objectives (e.g., energy-efficiency measures that also make buildings more resilient to natural disasters, in addition to helping the sector adapt to climate changes). [8.3.4, 8.4.2]

The effects of climate change are expected to be greatest in developing countries (especially those that rely on primary production as a major source of income) in terms of loss of life, effects on investment, and effects on the economy. Damages from natural disasters have been as high as half of the gross domestic product (GDP) in one case. Weather disasters set back development, particularly when funds are redirected from development projects to disaster-recovery efforts. [8.5]

Equity issues and development constraints would arise if weather-related risks become uninsurable, insurance prices increase, or the availability of insurance or financing becomes limited. Thus, increased uncertainty could constrain development. Conversely, more extensive penetration of or access to insurance and disaster preparedness/recovery resources would increase the ability of developing countries to adapt to climate change. More widespread introduction of microfinancing schemes and development banking also could be an effective mechanism to help developing countries and communities adapt. [8.3]

This assessment of financial services has identified some areas of improved knowledge and has corroborated and further augmented conclusions reached in the SAR. It also has highlighted many areas where greater understanding is needed— in particular, better analysis of economic losses to determine their causation, assessment of financial resources involved in dealing with climate change damage and adaptation, evaluation of alternative methods to generate such resources, deeper investigation of the sector's vulnerability and resilience to a range of extreme weather event scenarios, and more research into how the sector (private and public elements) could innovate to meet the potential increase in demand for adaptation funding in developed and developing countries, to spread and reduce risks from climate change. [8.7]

4.7. Human Health

Global climate change will have diverse impacts on human health—some positive, most negative. Changes in the frequencies of extreme heat and cold, the frequencies of floods and droughts, and the profile of local air pollution and aeroallergens would affect population health directly. Other health impacts would result from the impacts of climate change on ecological and social systems. These impacts would include changes in infectious disease occurrence, local food production and undernutrition, and various health consequences of population displacement and economic disruption.

There is little published evidence that changes in population health status actually have occurred in response to observed trends in climate over recent decades. A recurring difficulty in identifying such impacts is that the causation of most human health disorders is multifactorial, and the "background" socioeconomic, demographic, and environmental context changes significantly over time.

Studies of the health impacts associated with interannual climate variability (particularly those related to the El Niño cycle) have provided new evidence of human health sensitivity to climate, particularly for mosquito-borne diseases. The combination of existing research-based knowledge, resultant theoretical understandings, and the output of predictive modeling leads to several conclusions about the future impacts of climate change on human population health.

If heat waves increase in frequency and intensity, the risk of death and serious illness would increase, principally in older

---

[2] The term "mitigation" often is used in the insurance and financial services sectors in much the same way as the term "adaptation" is used in the climate research and policy communities.

age groups and the urban poor (high confidence). The effects of an increase in heat waves often would be exacerbated by increased humidity and urban air pollution. The greatest increases in thermal stress are forecast for mid- to high-latitude (temperate) cities, especially in populations with nonadapted architecture and limited air conditioning. Modeling of heat wave impacts in urban populations, allowing for acclimatization, suggests that a number of U.S. cities would experience, on average, several hundred extra deaths each summer. Although the impact of climate change on thermal stress-related mortality in developing country cities may be significant, there has been little research in such populations. Warmer winters and fewer cold spells will decrease cold-related mortality in many temperate countries (high confidence). Limited evidence indicates that in at least some temperate countries, reduced winter deaths would outnumber increased summer deaths (medium confidence). [9.4]

Any increases in the frequency and intensity of extreme events such as storms, floods, droughts, and cyclones would adversely impact human health through a variety of pathways. These natural hazards can cause direct loss of life and injury and can affect health indirectly through loss of shelter, population displacement, contamination of water supplies, loss of food production (leading to hunger and malnutrition), increased risk of infectious disease epidemics (including diarrheal and respiratory disease), and damage to infrastructure for provision of health services (very high confidence). If cyclones were to increase regionally, devastating impacts often would occur, particularly in densely settled populations with inadequate resources. Over recent years, major climate-related disasters have had major adverse effects on human health, including floods in China, Bangladesh, Europe, Venezuela, and Mozambique, as well as Hurricane Mitch, which devastated Central America. [9.5]

Climate change will decrease air quality in urban areas with air pollution problems (medium confidence). An increase in temperature (and, in some models, ultraviolet radiation) increases the formation of ground-level ozone, a pollutant with well-established adverse effects on respiratory health. Effects of climate change on other air pollutants are less well established. [9.6]

Higher temperatures, changes in precipitation, and changes in climate variability would alter the geographic ranges and seasonality of transmission of vector-borne infectious diseases— extending the range and season for some infectious diseases and contracting them for others. Vector-borne infectious diseases are transmitted by blood-feeding organisms such as mosquitoes and ticks. Such organisms depend on the complex interaction of climate and other ecological factors for survival. Currently, 40% of the world population lives in areas with malaria. In areas with limited or deteriorating public health infrastructure, increased temperatures will tend to expand the geographic range of malaria transmission to higher altitudes (high to medium confidence) and higher latitudes (medium to low confidence). Higher temperatures, in combination with conducive patterns of rainfall and surface water, will extend the transmission season in some locations (high confidence). Changes in climate, including changes in climate variability, would affect many other vector-borne infections (such as dengue, leishmaniasis, various types of mosquito-borne encephalitis, Lyme disease, and tick-borne encephalitis) at the margins of their current distributions (medium/high confidence). For some vector-borne diseases in some locations, climate change will decrease transmission via reductions in rainfall or temperatures that are too high for transmission (medium confidence). A range of mathematical models indicate, with high consistency, that climate change scenarios over the coming century would cause a small net increase in the proportion of the world's population living in regions of potential transmission of malaria and dengue (medium to high confidence). A change in climatic conditions will increase the incidence of various types of water- and food-borne infectious diseases. [9.7]

Climate change may cause changes in the marine environment that would alter risks of biotoxin poisoning from human consumption of fish and shellfish. Biotoxins associated with warmer waters, such as ciguatera in tropical waters, could extend their range to higher latitudes (low confidence). Higher SSTs also would increase the occurrence of toxic algal blooms (medium confidence), which have complex relationships with human poisoning and are ecologically and economically damaging. Changes in surface water quantity and quality will affect the incidence of diarrheal diseases (medium confidence). [9.8]

Changes in food supply resulting from climate change could affect the nutrition and health of the poor in some regions of the world. Studies of climate change impacts on food production indicate that, globally, impacts could be positive or negative, but the risk of reduced food yields is greatest in developing countries—where 790 million people are estimated to be undernourished at present. Populations in isolated areas with poor access to markets will be particularly vulnerable to local decreases or disruptions in food supply. Undernourishment is a fundamental cause of stunted physical and intellectual development in children, low productivity in adults, and susceptibility to infectious disease. Climate change would increase the number of undernourished people in the developing world (medium confidence), particularly in the tropics. [9.9, 5.3]

In some settings, the impacts of climate change may cause social disruption, economic decline, and population displacement that would affect human health. Health impacts associated with population displacement resulting from natural disasters or environmental degradation are substantial (high confidence). [9.10]

Table TS-5: Options for adaptation to reduce health impacts of climate change.

Health Outcome	Legislative	Technical	Educational Advisory	Cultural and Behavioral
Thermal stress	- Building guidelines	- Housing, public buildings, urban planning to reduce heat island effects, air conditioning	- Early warning systems	- Clothing, siesta
Extreme weather events	- Planning laws - Building guidelines - Forced migration - Economic incentives for building	- Urban planning - Storm shelters	- Early warning systems	- Use of storm shelters
Air quality	- Emission controls - Traffic restrictions	- Improved public transport, catalytic converters, smoke stacks	- Pollution warnings	- Carpooling
Vector-borne diseases		- Vector control - Vaccination, impregnated bednets - Sustainable surveillance, prevention and control programs	- Health education	- Water storage practices
Water-borne diseases	- Watershed protection laws - Water quality regulation	- Genetic/molecular screening of pathogens - Improved water treatment (e.g., filters) - Improved sanitation (e.g., latrines)	- Boil water alerts	- Washing hands and other hygiene behavior - Use of pit latrines

For each anticipated adverse health impact there is a range of social, institutional, technological, and behavioral adaptation options to lessen that impact (see Table TS-5). Overall, the adverse health impacts of climate change will be greatest in vulnerable lower income populations, predominately within tropical/subtropical countries. There is a basic and general need for public health infrastructure (programs, services, surveillance systems) to be strengthened and maintained. The ability of affected communities to adapt to risks to health also depends on social, environmental, political, and economic circumstances. [9.11]

5. Regional Analysis

The vulnerability of human populations and natural systems to climate change differs substantially across regions and across populations within regions. Regional differences in baseline climate and expected climate change give rise to different exposures to climate stimuli across regions. The natural and social systems of different regions have varied characteristics, resources, and institutions and are subject to varied pressures that give rise to differences in sensitivity and adaptive capacity. From these differences emerge different key concerns for each of the major regions of the world. Even within regions, however, impacts, adaptive capacity, and

vulnerability will vary. Because available studies have not employed a common set of climate scenarios and methods and because of uncertainties regarding the sensitivities and adaptability of natural and social systems, assessment of regional vulnerabilities is necessarily qualitative.

5.1. Africa

Africa is highly vulnerable to climate change. Impacts of particular concern to Africa are related to water resources, food production, human health, desertification, and coastal zones, especially in relation to extreme events. A synergy of land-use and climate change will exacerbate desertification. Selected key impacts in Africa are highlighted in Figure TS-6.

5.1.1. Water Resources

Water resources are a key area of vulnerability in Africa, affecting water supply for household use, agriculture, and industry. In shared river basins, regional cooperation protocols minimize adverse impacts and potential for conflicts. Trends in regional per capita water availability in Africa over the past half century show that water availability has diminished by 75%. Although the past 2 decades have experienced reductions in river flows, especially in sub-Saharan West Africa, the trend mainly reflects the impact of population growth—which, for

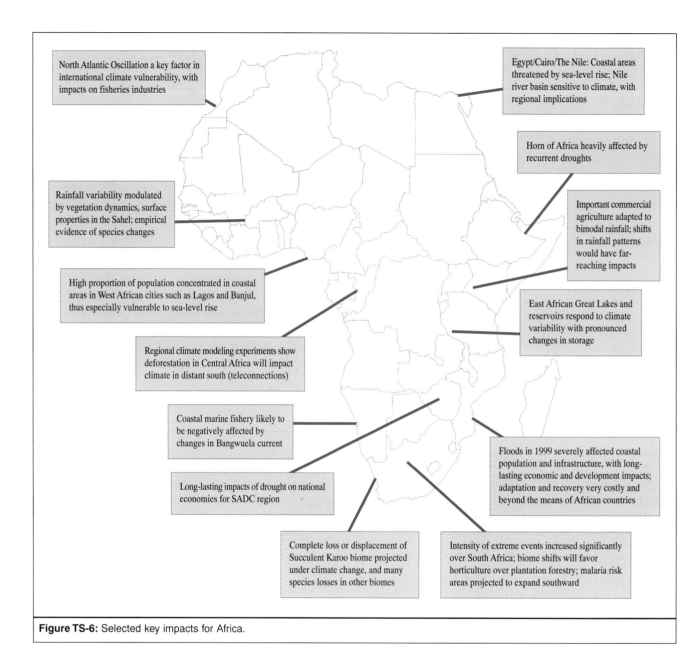

Figure TS-6: Selected key impacts for Africa.

most countries, quadrupled in the same period. Population growth and degradation of water quality are significant threats to water security in many parts of Africa, and the combination of continued population increases and global warming impacts is likely to accentuate water scarcity in subhumid regions of Africa.

Africa is the continent with the lowest conversion factor of precipitation to runoff, averaging 15%. Although the equatorial region and coastal areas of eastern and southern Africa are humid, the rest of the continent is dry subhumid to arid. The dominant impact of global warming will be a reduction in soil moisture in subhumid zones and a reduction in runoff. Current trends in major river basins indicate decreasing runoff of about 17% over the past decade.

Most of Africa has invested significantly in hydroelectric power facilities to underpin economic development. Reservoir storage shows marked sensitivity to variations in runoff and periods of drought. Lake storage and major dams have reached critical levels, threatening industrial activity. Model results and some reservoirs and lakes indicate that global warming will increase the frequency of such low storage as a result of flooding or drought conditions that are related to ENSO. [10.2.1]

5.1.2. Food Security

There is wide consensus that climate change will worsen food security, mainly through increased extremes and temporal/ spatial shifts. The continent already experiences a major deficit in food production in many areas, and potential declines in soil moisture will be an added burden. Food-insecure countries are at a greater risk of adverse impacts of climate change. Inland and marine fisheries provide a significant contribution to protein intake in many African countries. As a result of water stress and land degradation, inland fisheries will be rendered more vulnerable to episodic drought and habitat destruction. Ocean warming is likely to impact coastal marine fisheries. [10.2.2]

5.1.3. Natural Resource Management and Biodiversity

Irreversible losses of biodiversity could be accelerated with climate change. Climate change is expected to lead to drastic shifts of biodiversity-rich biomes such as the Succulent Karoo in South Africa and many losses in species in other biomes. Changes in the frequency, intensity, and extent of vegetation fires and habitat modification from land-use change may negate natural adaptive processes and lead to extinctions. Changes in ecosystems will affect water supply, fuelwood, and other services. [10.2.3.2]

5.1.4. Human Health

Changes in temperature and rainfall will have many negative impacts on human health. Temperature increases will extend disease vector habitats. Where sanitary infrastructure is inadequate, droughts and flooding will result in increased frequency of water-borne diseases. Increased rainfall could lead to more frequent outbreaks of Rift Valley fever. Poor sanitation in urban locations and increased temperatures of coastal waters could aggravate cholera epidemics. [10.2.4.1.1, 10.2.4.4]

5.1.5. Settlements and Infrastructure

Although the basic infrastructure for development—transport, housing, and services—is inadequate in many instances, it nevertheless represents substantial investment by governments. An increase in the frequency of damaging floods, heat waves, dust storms, hurricanes, and other extreme events could degrade the integrity of such critical infrastructures at rates the economies may not be able to tolerate, leading to a serious deterioration of social, health, and economic services delivery systems. This condition will greatly compromise general human welfare. [10.2.5.3]

Sea-level rise, coastal erosion, saltwater intrusion, and flooding will have significant impacts for African communities and economies. Most of Africa's largest cities are along coasts and are highly vulnerable to extreme events, sea-level rise, and coastal erosion because of inadequate physical planning and escalating urban drift. Rapid unplanned expansion is likely to predispose large populations to infectious diseases from climate-related factors such as flooding. [10.2.5.2]

5.1.6. Desertification

Alteration of spatial and temporal patterns in temperature, rainfall, solar radiation, and winds from a changing climate will exacerbate desertification. Desertification is a critical threat to sustainable resource management in arid, semi-arid, and dry subhumid regions of Africa, undermining food and water security. [10.2.6]

5.1.7. Adaptive Capacity

Given the diversity of constraints facing many nations, the overall capacity for Africa to adapt to climate change currently is very low. National action plans that incorporate long-term changes and pursue "no regrets" strategies could increase the adaptive capacity of the region. Seasonal forecasting—for example, linking SSTs to outbreaks of major diseases—is a promising adaptive strategy that will help save lives. Current technologies and approaches, especially in agriculture and water, are unlikely to be adequate to meet projected demands, and increased climate variability will be an additional stress. It is unlikely that African countries on their own will have sufficient resources to respond effectively.

Climate change also offers some opportunities. The processes of adapting to global climate change, including technology transfer and carbon sequestration, offer new development pathways that could take advantage of Africa's resources and human potential. Regional cooperation in science, resource management, and development already are increasing, and access to international markets will diversify economies and increase food security.

This assessment of vulnerability to climate change is marked by uncertainty. The diversity of African climates, high rainfall variability, and a very sparse observational network make predictions of future climate change difficult at the subregional and local level. Underlying exposure and vulnerability to climatic changes are well established. Sensitivity to climatic variations is established but incomplete. However, uncertainty over future conditions means that there is low confidence in projected costs of climate change. This assessment can create the framework for individual states to begin to construct methodologies for estimating such costs, based on their individual circumstances.

5.2. Asia

Climate change will impose significant stress on resources throughout the Asian region. Asia has more than 60% of the world's population; natural resources already are under stress, and the resilience of most sectors in Asia to climate change is poor. Many countries are socioeconomically dependent on natural resources such as water, forests, grassland and rangeland, and fisheries. The magnitude of changes in climate variables

Table TS-6: Sensitivity of selected Asian regions to climate change.

Change in Climatic Elements and Sea-Level Rise	Vulnerable Region	Primary Change	Impacts	
			Primary	*Secondary*
0.5-2°C (10- to 45-cm sea-level rise)	Bangladesh Sundarbans	- Inundation of about 15% (~750 km^2) - Increase in salinity	- Loss of plant species - Loss of wildlife	- Economic loss - Exacerbated insecurity and loss of employment
4°C (+10% rainfall)	Siberian permafrosts	- Reduction in continuous permafrost - Shift in southern limit of Siberian permafrost by ~100-200 km northward	- Change in rock strength - Change in bearing capacity - Change in compressibility of frozen rocks - Thermal erosion	- Effects on construction industries - Effects on mining industry - Effects on agricultural development
>3°C (>+20% rainfall)	Water resources in Kazakhstan	- Change in runoff	- Increase in winter floods - Decrease in summer flows	- Risk to life and property - Summer water stress
~2°C (−5 to 10% rainfall; 45-cm sea-level rise)	Bangladesh lowlands	- About 23-29% increase in extent of inundation	- Change in flood depth category - Change in monsoon rice cropping pattern	- Risk to life and property - Increased health problems - Reduction in rice yield

Table TS-7: Vulnerability of key sectors to impacts of climate change for select subregions in Asia. Key to confidence-level rankings is provided in Section 1.4 of Technical Summary.

Regions	Food and Fiber	Biodiversity	Water Resources	Coastal Ecosystems	Human Health	Settlements
Boreal Asia	Slightly resilient ****	Highly vulnerable ***	Slightly resilient ***	Slightly resilient **	Moderately vulnerable **	Slightly or not vulnerable ***
Arid and Semi-Arid Asia						
- Central Asia	Highly vulnerable ****	Moderately vulnerable **	Highly vulnerable ****	Moderately vulnerable **	Moderately vulnerable ***	Moderately vulnerable ***
- Tibetan Plateau	Slightly or not vulnerable **	Highly vulnerable ***	Moderately vulnerable **	Not applicable	No information	No information
Temperate Asia	Highly vulnerable ****	Moderately vulnerable ***	Highly vulnerable ****	Highly vulnerable ****	Highly vulnerable ***	Highly vulnerable ****
Tropical Asia						
- South Asia	Highly vulnerable ****	Highly vulnerable ***	Highly vulnerable ****	Highly vulnerable ****	Moderately vulnerable ***	Highly vulnerable ***
- Southeast Asia	Highly vulnerable ****	Highly vulnerable ***	Highly vulnerable ****	Highly vulnerable ****	Moderately vulnerable ***	Highly vulnerable ***

would differ significantly across Asian subregions and countries. The climate change sensitivity of a few vulnerable sectors in Asia and the impacts of these limits are presented in Table TS-6.

The region's vulnerability to climate change is captured in Table TS-7 for selected categories of regions/issues.

5.2.1. Agriculture and Food Security

Food insecurity appears to be the primary concern for Asia. Crop production and aquaculture would be threatened by thermal and water stresses, sea-level rise, increased flooding, and strong winds associated with intense tropical cyclones (high confidence). In general, it is expected that areas in mid- and high latitudes will experience increases in crop yield; yields in lower latitudes generally will decrease. A longer duration of the summer season should lead to a northward shift of the agroecosystem boundary in boreal Asia and favor an overall increase in agriculture productivity (medium confidence). Climatic variability and change also will affect scheduling of the cropping season, as well as the duration of the growing period of the crop. In China, yields of several major crops are expected to decline as a result of climate change. Acute water shortages combined with thermal stress should adversely affect wheat and, more severely, rice productivity in India even under the positive effects of elevated CO_2 in the future. Crop diseases such as wheat scab, rice blast, and sheath and culm blight of rice also could become more widespread in temperate and tropical regions of Asia if the climate becomes warmer and wetter. Adaptation measures to reduce the negative effects of climatic variability may include changing the cropping calendar to take advantage of the wet period and to avoid the extreme weather events (e.g., typhoons and strong winds) during the growing season. [11.2.2.1]

Asia dominates world aquaculture, producing 80% of all farmed fish, shrimp, and shellfish. Many wild stocks are under stress as a result of overexploitation, trawling on sea-bottom habitats, coastal development, and pollution from land-based activities. Moreover, marine productivity is greatly affected by plankton shift, such as seasonal shifting of sardine in the Sea of Japan, in response to temperature changes induced during ENSO. Storm surges and cyclonic conditions also routinely lash the coastline, adding sediment loads to coastal waters. Effective conservation and sustainable management of marine and inland fisheries are needed at the regional level so that living aquatic resources can continue to meet regional and national nutritional needs. [11.2.4.4]

5.2.2. Ecosystems and Biodiversity

Climate change would exacerbate current threats to biodiversity resulting from land-use/cover change and population pressure in Asia (medium confidence). Risks to Asia's rich array of living species are climbing. As many as 1,250 of 15,000 higher plant species are threatened in India. Similar trends are evident in China, Malaysia, Myanmar, and Thailand. Many species and a large population of many other species in Asia are likely to be exterminated as a result of the synergistic effects of climate change and habitat fragmentation. In desert ecosystems, increased frequency of droughts may result in a decline in local forage around oases, causing mass mortality among local fauna and threatening their existence. With a 1-m rise in sea level, the Sundarbans (the largest mangrove ecosystems) of Bangladesh will completely disappear. [11.2.1, 11.2.1.6]

Permafrost degradation resulting from global warming would increase the vulnerability of many climate-dependent sectors affecting the economy in boreal Asia (medium confidence). Pronounced warming in high latitudes of the northern hemisphere could lead to thinning or disappearance of permafrost in locations where it now exists. Large-scale shrinkage of the permafrost region in boreal Asia is likely. Poleward movement of the southern boundary of the sporadic zone also is likely in Mongolia and northeast China. The boundary between continuous and discontinuous (intermittent or seasonal) permafrost areas on the Tibetan Plateau is likely to shift toward the center of the plateau along the eastern and western margins. [11.2.1.5]

The frequency of forest fires is expected to increase in boreal Asia (medium confidence). Warmer surface air temperatures, particularly during summer, may create favorable conditions for thunderstorms and associated lightening, which could trigger forest fires in boreal forests more often. Forest fire is expected to occur more frequently in northern parts of boreal Asia as a result of global warming. [11.2.1.3]

5.2.3. Water Resources

Freshwater availability is expected to be highly vulnerable to anticipated climate change (high confidence). Surface runoff increases during winter and summer periods would be pronounced in boreal Asia (medium confidence). Countries in which water use is more than 20% of total potential water resources available are expected to experience severe water stress during drought periods. Surface runoff is expected to decrease drastically in arid and semi-arid Asia under projected climate change scenarios. Climate change is likely to change streamflow volume, as well as the temporal distribution of streamflows throughout the year. With a 2°C increase in air temperature accompanied by a 5-10% decline in precipitation during summer, surface runoff in Kazakhstan would be substantially reduced, causing serious implications for agriculture and livestocks. Water would be a scarce commodity in many south and southeast Asian countries, particularly where reservoir facilities to store water for irrigation are minimal. Growing populations and concentration of populations in urban areas will exert increasing pressures on water availability and water quality. [11.2.3.1]

5.2.4. Extreme Weather Events

Developing countries of temperate and tropical Asia already are quite vulnerable to extreme climate events such as typhoons/ cyclones, droughts, and floods. Climate change and variability would exacerbate these vulnerabilities (high confidence). Extreme weather events are known to cause adverse effects in

Table TS-8: Potential land loss and population exposed in Asian countries for selected magnitudes of sea-level rise, assuming no adaptation.

Country	Sea-Level Rise (cm)	Potential Land Loss (km²)	Potential Land Loss (%)	Population Exposed (million)	Population Exposed (%)
Bangladesh	45	15,668	10.9	5.5	5.0
	100	29,846	20.7	14.8	13.5
India	100	5,763	0.4	7.1	0.8
Indonesia	60	34,000	1.9	2.0	1.1
Japan	50	1,412	0.4	2.9	2.3
Malaysia	100	7,000	2.1	>0.05	>0.3
Pakistan	20	1,700	0.2	n.a.	n.a.
Vietnam	100	40,000	12.1	17.1	23.1

widely separated areas of Asia. There is some evidence of increases in the intensity or frequency of some of these extreme events on regional scales throughout the 20th century. [11.1.2.2, 11.1.2.3, 11.4.1]

Increased precipitation intensity, particularly during the summer monsoon, could increase flood-prone areas in temperate and tropical Asia. There is potential for drier conditions in arid and semi-arid Asia during summer, which could lead to more severe droughts (medium confidence). Many countries in temperate and tropical Asia have experienced severe droughts and floods frequently in the 20th century. Flash floods are likely to become more frequent in many regions of temperate and tropical Asia in the future. A decrease in return period for extreme precipitation events and the possibility of more frequent floods in parts of India, Nepal, and Bangladesh is projected. [11.1.3.3, 11.2.2.2, 11.1.2.3, 11.4.1]

Conversion of forestland to cropland and pasture already is a prime force driving forest loss in tropical and temperate Asian countries. With more frequent floods and droughts, these actions will have far-reaching implications for the environment (e.g., soil erosion, loss of soil fertility, loss of genetic variability in crops, and depletion of water resources). [11.1.4.1]

Tropical cyclones and storm surges continue to take a heavy toll on life and property in India and Bangladesh. An increase in the intensity of cyclones combined with sea-level rise would result in more loss of life and property in low-lying coastal areas in cyclone-prone countries of Asia (medium confidence). The expected increase in the frequency and intensity of climatic extremes will have significant potential effects on crop growth and agricultural production, as well as major economic and environmental implications (e.g., tourism, transportation). [11.2.4.5, 11.2.6.3, 11.3]

A wide range of precautionary measures at regional and national levels, including awareness and acceptance of risk factors among regional communities, is warranted to avert or reduce the impacts of disasters associated with more extreme weather events on economic and social structures of countries in temperate and tropical Asia. [11.3.2]

5.2.5. Deltas and Coastal Zones

The large deltas and low-lying coastal areas of Asia would be inundated by sea-level rise (high confidence). Climate-related stresses in coastal areas include loss and salinization of agricultural land as a result of change in sea level and changing frequency and intensity of tropical cyclones. Estimates of potential land loss resulting from sea-level rise and risk to population displacement provided in Table TS-8 demonstrate the scale of the issue for major low-lying regions of coastal Asia. Currently, coastal erosion of muddy coastlines in Asia is not a result of sea-level rise; it is triggered largely by annual river-borne suspended sediments transported into the ocean by human activities and delta evolution. These actions could exacerbate the impacts of climate change in coastal regions of Asia. [11.2.4.2]

5.2.6. Human Health

Warmer and wetter conditions would increase the potential for higher incidence of heat-related and infectious diseases in tropical and temperate Asia (medium confidence). The rise in surface air temperature and changes in precipitation in Asia will have adverse effects on human health. Although warming would result in a reduction in wintertime deaths in temperate countries, there could be greater frequency and duration of heat stress, especially in megalopolises during summer. Global warming also will increase the incidence of respiratory and cardiovascular diseases in parts of arid and semi-arid Asia and temperate and tropical Asia. Changes in environmental temperature and precipitation could expand vector-borne diseases into temperate and arid Asia. The spread of vector-borne diseases into more northern latitudes may pose a serious threat to human health. Warmer SSTs along Asian coastlines would support higher phytoplankton blooms. These blooms are habitats for infectious bacterial diseases. Water-borne diseases—including cholera and the suite of diarrheal diseases caused by organisms such as giardia, salmonella, and cryptosporidium—could become more common in many countries of south Asia in warmer climate. [11.2.5.1, 11.2.5.2, 11.2.5.4]

5.2.7. Adaptive Capacity

Adaptation to climate change in Asian countries depends on the affordability of adaptive measures, access to technology, and biophysical constraints such as land and water resource availability, soil characteristics, genetic diversity for crop breeding (e.g., crucial development of heat-resistant rice cultivars), and topography. Most developing countries of Asia are faced with increasing population, spread of urbanization, lack of adequate water resources, and environmental pollution, which hinder socioeconomic activities. These countries will have to individually and collectively evaluate the tradeoffs between climate change actions and nearer term needs (such as hunger, air and water pollution, energy demand). Coping strategies would have to be developed for three crucial sectors: land resources, water resources, and food productivity. Adaptation measures that are designed to anticipate the potential effects of climate change can help offset many of the negative effects. [11.3.1]

5.3. Australia and New Zealand

The Australia/New Zealand region spans the tropics to mid-latitudes and has varied climates and ecosystems, including deserts, rainforests, coral reefs, and alpine areas. The climate is strongly influenced by the surrounding oceans. Australia has significant vulnerability to the drying trend projected over much of the country for the next 50-100 years (Figure TS-3) because substantial agricultural areas currently are adversely affected by periodic droughts, and there already are large areas of arid and semi-arid land. New Zealand—a smaller, more mountainous country with a generally more temperate, maritime climate—may be more resilient to climate changes than Australia, although considerable vulnerability remains (medium confidence). Table TS-9 shows key vulnerabilities and adaptability to climate change impacts for Australia and New Zealand. [12.9.5]

Comprehensive cross-sectoral estimates of net climate change impact costs for various GHG emission scenarios and different societal scenarios are not yet available. Confidence remains very low in the IPCC *Special Report on Regional Impacts of Climate Change* estimate for Australia and New Zealand of −1.2 to −3.8% of GDP for an equivalent doubling of CO_2 concentrations. This estimate did not account for many of the effects and adaptations currently identified. [12.9]

Extreme events are a major source of current climate impacts, and changes in extreme events are expected to dominate the impacts of climate change. Return periods for heavy rains, floods,

and sea-level surges of a given magnitude at particular locations would be modified by possible increases in intensity of tropical cyclones and heavy rain events and changes in the location-specific frequency of tropical cyclones. Scenarios of climate change that are based on recent coupled atmosphere-ocean (A-O) models suggest that large areas of mainland Australia will experience significant decreases in rainfall during the 21st century. The ENSO phenomenon leads to floods and prolonged droughts, especially in inland Australia and parts of New Zealand. The region would be sensitive to a change towards a more El Niño-like mean state. [12.1.5]

Before stabilization of GHG concentrations, the north-south temperature gradient in mid-southern latitudes is expected to increase (medium to high confidence), strengthening the westerlies and the associated west-to-east gradient of rainfall across Tasmania and New Zealand. Following stabilization of GHG concentrations, these trends would be reversed (medium confidence). [12.1.5.1]

Climate change will add to existing stresses on achievement of sustainable land use and conservation of terrestrial and aquatic biodiversity. These stresses include invasion by exotic animal and plant species, degradation and fragmentation of natural ecosystems through agricultural and urban development, dryland salinization (Australia), removal of forest cover (Australia and New Zealand), and competition for scarce water resources. Within both countries, economically and socially disadvantaged groups of people, especially indigenous peoples, are particularly vulnerable to stresses on health and living conditions induced by climate change. Major exacerbating problems include rapid population and infrastructure growth in vulnerable coastal areas, inappropriate use of water resources, and complex institutional arrangements. [12.3.2, 12.3.3, 12.4.1, 12.4.2, 12.6.4, 12.8.5]

5.3.1. Water Resources

Water resources already are stressed in some areas and therefore are highly vulnerable, especially with respect to salinization (parts of Australia) and competition for water supply between agriculture, power generation, urban areas, and environmental flows (high confidence). Increased evaporation and possible decreases in rainfall in many areas would adversely affect water supply, agriculture, and the survival and reproduction of key species in parts of Australia and New Zealand (medium confidence). [12.3.1, 12.3.2, 12.4.6, 12.5.2, 12.5.3, 12.5.6]

5.3.2. Ecosystems

A warming of 1°C would threaten the survival of species that currently are growing near the upper limit of their temperature range, notably in marginal alpine regions and in the southwest of Western Australia. Species that are unable to migrate or relocate because of land clearing, soil differences, or topography could

Table TS-9: Main areas of vulnerability and adaptability to climate change impacts in Australia and New Zealand. Degree of confidence that tabulated impacts will occur is indicated by stars in second column (see Section 1.4 of Technical Summary for key to confidence-level rankings). Confidence levels, and assessments of vulnerability and adaptability, are based on information reviewed in Chapter 12, and assume continuation of present population and investment growth patterns.

Sector	Impact	Vulnerability	Adaptation	Adaptability	Section
Hydrology and water supply	- Irrigation and metropolitan supply constraints, and increased salinization—****	High in some areas	- Planning, water allocation, and pricing	Medium	12.3.1, 12.3.2
	- Saltwater intrusion into some island and coastal aquifers—****	High in limited areas	- Alternative water supplies, retreat	Low	12.3.3
Terrestrial ecosystems	- Increased salinization of dryland farms and some streams (Australia)—***	High	- Changes in land-use practices	Low	12.3.3
	- Biodiversity loss notably in fragmented regions, Australian alpine areas, and southwest of WA—****	Medium to high in some areas	- Landscape management; little possible in alpine areas	Medium to low	12.4.2, 12.4.4, 12.4.8
	- Increased risk of fires—***	Medium	- Land management, fire protection	Medium	12.1.5.3, 12.5.4, 12.5.10
	- Weed invasion—***	Medium	- Landscape management	Medium	12.4.3
Aquatic ecosystems	- Salinization of some coastal freshwater wetlands—***	High	- Physical intervention	Low	12.4.7
	- River and inland wetland ecosystem changes—***	Medium	- Change water allocations	Low	12.4.5, 12.4.6
	- Eutrophication—***	Medium in inland Aus. waters	- Change water allocations, reduce nutrient inflows	Medium to low	12.3.4
Coastal ecosystems	- Coral bleaching, especially Great Barrier Reef—****	High	- Seed coral?	Low	12.4.7
	- More toxic algal blooms?—*	Unknown	—	—	12.4.7
Agriculture, grazing, and forestry	- Reduced productivity, increased stress on rural communities if droughts increase, increased forest fire risk—***	Location-dependent, worsens with time	- Management and policy changes, fire prevention, seasonal forecasts	Medium	12.5.2, 12.5.3, 12.5.4
	- Changes in global markets due to climate changes elsewhere—***, but sign uncertain	High, but sign uncertain	- Marketing, planning, niche and fuel crops, carbon trading	Medium	12.5.9
	- Increased spread of pests and diseases—****	Medium	- Exclusion, spraying	Medium	12.5.7
	- Increased CO_2 initially increases productivity but offset by climate changes later—**	Changes with time	- Change farm practices, change industry		12.5.3, 12.5.4
Horticulture	- Mixed impacts (+ and -), depends on species and location—****	Low overall	- Relocate	High	12.5.3
Fish	- Recruitment changes (some species)—**	Unknown net effect	- Monitoring, management	—	12.5.5

Table TS-9 (continued)

Sector	Impact	Vulnerability	Adaptation	Adaptability	Section
Settlements and industry	- Increased impacts of flood, storm, storm surge, sea-level rise—***	High in some places	- Zoning, disaster planning	Moderate	12.6.1, 12.6.4
Human health	- Expansion and spread of vector-borne diseases—****	High	- Quarantine, eradication, or control	Moderate to high	12.7.1, 12.7.4
	- Increased photochemical air pollution—****	Moderate (some cities)	- Emission controls	High	12.7.1

become endangered or extinct. Other Australian ecosystems that are particularly vulnerable include coral reefs and arid and semi-arid habitats. Freshwater wetlands in coastal zones in Australia and New Zealand are vulnerable, and some New Zealand ecosystems are vulnerable to accelerated spread of weeds. [12.4.2, 12.4.3, 12.4.4, 12.4.5, 12.4.7]

5.3.3. Food Production

Agricultural activities are particularly vulnerable to regional reductions in rainfall in southwest and inland Australia (medium confidence). Drought frequency and consequent stresses on agriculture are likely to increase in parts of Australia and New Zealand as a result of higher temperatures and El Niño changes (medium confidence). Enhanced plant growth and water-use efficiency (WUE) resulting from CO_2 increases may provide initial benefits that offset any negative impacts from climate change (medium confidence), although the balance is expected to become negative with warmings in excess of 2-4°C and associated rainfall changes (medium confidence). This is illustrated in Figure TS-7 for wheat production in Australia, for a range of climate change scenarios. Reliance on exports of agricultural and forest products makes the region very sensitive to changes in production and commodity prices that are induced by changes in climate elsewhere. [12.5.2, 12.5.3, 12.5.6, 12.5.9, 12.8.7]

Australian and New Zealand fisheries are influenced by the extent and location of nutrient upwellings governed by prevailing winds and boundary currents. In addition, ENSO influences recruitment of some fish species and the incidence of toxic algal blooms. [12.5.5]

5.3.4. Settlements, Industry, and Human Health

Marked trends toward greater population and investment in exposed regions are increasing vulnerability to tropical cyclones and storm surges. Thus, projected increases in tropical cyclone intensity and possible changes in their location-specific frequency, along with sea-level rise, would have major impacts— notably, increased storm-surge heights for a given return period (medium to high confidence). Increased frequency of high-

intensity rainfall would increase flood damages to settlements and infrastructure (medium confidence). [12.1.5.1, 12.1.5.3, 12.6.1, 12.6.4]

There is high confidence that projected climate changes will enhance the spread of some disease vectors, thereby increasing

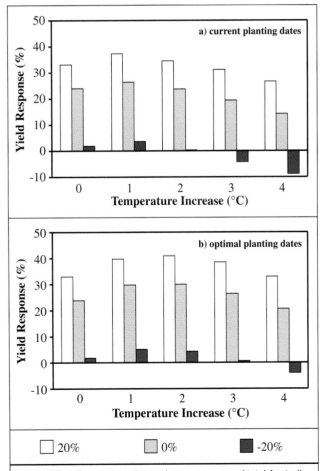

Figure TS-7: Percentage change in average annual total Australian wheat yield for CO_2 (levels of 700 ppm) and a range of changes in temperature and rainfall: a) current planting dates, and b) optimal planting dates. Yield response is shown for rainfall changes of +20% (white), 0 (light blue), and –20% (dark blue), for warmings of 0-4°C.

the potential for disease outbreaks such as mosquito-borne Ross River virus and Murray Valley encephalitis, despite existing biosecurity and health services. [12.7.1]

5.3.5. Key Adaptation Options

Key adaptation options include improved WUE and effective trading mechanisms for water; more appropriate land-use policies; provision of climate information and seasonal forecasts to land users to help them manage for climate variability and change; improved crop cultivars; revised engineering standards and zoning for infrastructure development; and improved biosecurity and health services. However, many natural ecosystems in Australia and New Zealand have only a limited capacity to adapt, and many managed systems will face limits on adaptation imposed by cost, acceptability, and other factors. [12.3.2, 12.3.3, 12.5.6, 12.7.4, 12.8.4, 12.8.5]

5.4. Europe

Present-day weather conditions affect natural, social, and economic systems in Europe in ways that reveal sensitivities and vulnerabilities to climate change in these systems. Climate change may aggravate such effects (very high confidence). Vulnerability to climate change in Europe differs substantially between subregions. Southern Europe and the European Arctic are more vulnerable than other parts of Europe. More-marginal and less-wealthy areas will be less able to adapt, which leads to important equity implications (very high confidence). Findings in the TAR relating to key vulnerabilities in Europe are broadly consistent with those expressed in the IPCC *Special Report on Regional Impacts of Climate Change* and the SAR, but are more specific about subregional effects and include new information concerning adaptive capacity. [13.1.1, 13.1.4, 13.4]

5.4.1. Water Resources

Water resources and their management in Europe are under pressure now, and these pressures are likely to be exacerbated by climate change (high confidence). Flood hazard is likely to increase across much of Europe—except where snowmelt peak has been reduced—and the risk of water shortage is projected to increase, particularly in southern Europe (medium to high confidence). Climate change is likely to widen water resource differences between northern and southern Europe (high confidence). Half of Europe's alpine glaciers could disappear by the end of the 21st century. [13.2.1]

5.4.2. Ecosystems

Natural ecosystems will change as a result of increasing temperature and atmospheric concentration of CO_2. Permafrost will decline; trees and shrubs will encroach into current northern tundra; and broad-leaved trees may encroach into current coniferous forest areas. Net primary productivity in ecosystems is likely to increase (also as a result of nitrogen deposition), but increases in decomposition resulting from increasing temperature may negate any additional carbon storage. Diversity in nature reserves is under threat from rapid change. Loss of important habitats (wetlands, tundra, and isolated habitats) would threaten some species (including rare/endemic species and migratory birds). Faunal shifts as a result of ecosystem changes are expected in marine, aquatic, and terrestrial ecosystems (high confidence; established but incomplete evidence). [13.2.1.4, 13.2.2.1, 13.2.2.3-5]

Soil properties will deteriorate under warmer and drier climate scenarios in southern Europe. The magnitude of this effect will vary markedly between geographic locations and may be modified by changes in precipitation (medium confidence; established but incomplete evidence). [13.2.1.2]

In mountain regions, higher temperatures will lead to an upward shift of biotic zones. There will be a redistribution of species with, in some instances, a threat of extinction (high confidence). [13.2.1.4]

Timber harvest will increase in commercial forests in northern Europe (medium confidence; established but incomplete evidence), although forest pests and disease may increase. Reductions are likely in the Mediterranean, with increased drought and fire risk (high confidence; well-established evidence). [13.2.2.1]

5.4.3. Agriculture and Food Security

Agricultural yields will increase for most crops as a result of increasing atmospheric CO_2 concentration. This increase in yields would be counteracted by the risk of water shortage in southern and eastern Europe and by shortening of the duration of growth in many grain crops because of increasing temperature. Northern Europe is likely to experience overall positive effects, whereas some agricultural production systems in southern Europe may be threatened (medium confidence; established but incomplete evidence).

Changes in fisheries and aquaculture production resulting from climate change embrace faunal shifts that affect freshwater and marine fish and shellfish biodiversity. These changes will be aggravated by unsustainable exploitation levels and environmental change (high confidence).

5.4.4. Human Settlements and Financial Services

The insurance industry faces potentially costly climate change impacts through the medium of property damage, but there is great scope for adaptive measures if initiatives are taken soon (high confidence). Transport, energy, and other industries will face changing demand and market opportunities. The concentration of industry on the coast exposes it to sea-level

Table TS-10: Estimates of flood exposure and incidence for Europe's coasts in 1990 and the 2080s. Estimates of flood incidence are highly sensitive to assumed protection standard and should be interpreted in indicative terms only (former Soviet Union excluded).

Region	1990 Exposed Population (millions)	Flood Incidence	
		1990 Average Number of People Experiencing Flooding (thousands yr^1)	2080s Increase due to Sea-Level Rise, Assuming No Adaptation (%)
Atlantic Coast	19.0	19	50 to 9,000
Baltic Coast	1.4	1	0 to 3,000
Mediterranean Coast	4.1	3	260 to 12,000

rise and extreme events, necessitating protection or removal (high confidence). [13.2.4]

Recreational preferences are likely to change with higher temperatures. Heat waves are likely to reduce the traditional peak summer demand at Mediterranean holiday destinations. Less-reliable snow conditions will impact adversely on winter tourism (medium confidence). [13.2.4.4]

The risk of flooding, erosion, and wetland loss in coastal areas will increase substantially, with implications for human settlement, industry, tourism, agriculture, and coastal natural habitats. Southern Europe appears to be more vulnerable to these changes, although the North Sea coast already has a high exposure to flooding (high confidence). Table TS-10 provides estimates of flood exposure and risk for Europe's coasts. [13.2.1.3]

5.4.5. Human Health

A range of risks is posed for human health through increased exposure to heat episodes (exacerbated by air pollution in urban areas), extension of some vector-borne diseases, and coastal and riverine flooding. Cold-related risks will be reduced (medium confidence; competing explanations). [13.2.5]

5.4.6. Adaptive Capacity

The adaptation potential of socioeconomic systems in Europe is relatively high because of economic conditions [high gross national product (GNP) and stable growth], a stable population (with the capacity to move within the region), and well-developed political, institutional, and technological support systems. However, the adaptation potential for natural systems generally is low (very high confidence). [13.3]

5.5. Latin America

There is ample evidence of climate variability at a wide range of time scales all over Latin America, from intraseasonal to long-term. In many subregions of Latin America, this variability

in climate normally is associated with phenomena that already produce impacts with important socioeconomic and environmental consequences that could be exacerbated by global warming and its associated weather and climate changes.

Variations in precipitation have a strong effect on runoff and streamflow, which are simultaneously affected by melting of glaciers and snow. Precipitation variations and their sign depend on the geographical subregion under consideration. Temperature in Latin America also varies among subregions. Although these variations might depend on the origin and quality of the source data as well as on the record periods used for studies and analyses, some of these variations could be attributed to a climate change condition (low confidence). [14.1.2.1]

ENSO is responsible for a large part of the climate variability at interannual scales in Latin America (high confidence). The region is vulnerable to El Niño, with impacts varying across the continent. For example, El Niño is associated with dry conditions in northeast Brazil, northern Amazonia, the Peruvian-Bolivian Altiplano, and the Pacific coast of Central America. The most severe droughts in Mexico in recent decades have occurred during El Niño years, whereas southern Brazil and northwestern Peru have exhibited anomalously wet conditions. La Niña is associated with heavy precipitation and flooding in Colombia and drought in southern Brazil. If El Niño or La Niña were to increase, Latin America would be exposed to these conditions more often. [14.1.2]

Some subregions of Latin America frequently experience extreme events, and these extraordinary combinations of hydrological and climatic conditions historically have produced disasters in Latin America. Tropical cyclones and associated heavy rain, flooding, and landslides are very common in Central America and southern Mexico. In northwestern South America and northeastern Brazil, many of the extremes that occur are strongly related to El Niño. [14.1.2]

5.5.1. Water Resources

It has been well established that glaciers in Latin America have receded in the past several decades. Warming in high mountain regions could lead to disappearance of significant snow and ice surface (medium confidence), which could affect mountain sport and tourist activities. Because these areas contribute to river streamflow, this trend also would reduce water availability for irrigation, hydropower generation, and navigation. [14.2.4]

5.5.2. Ecosystems

It is well established that Latin America accounts for one of the Earth's largest concentrations of biodiversity, and the impacts of climate change can be expected to increase the risk of biodiversity loss (high confidence). Observed population declines in frogs and small mammals in Central America can be related to regional climate change. The remaining Amazonian forest is threatened by the combination of human disturbance, increases in fire frequency and scale, and decreased precipitation from evapotranspiration loss, global warming, and El Niño. Neotropical seasonally dry forest should be considered severely threatened in Mesoamerica.

Tree mortality increases under dry conditions that prevail near newly formed edges in Amazonian forests. Edges, which affect an increasingly large portion of the forest because of increased deforestation, would be especially susceptible to the effects of reduced rainfall. In Mexico, nearly 50% of the deciduous tropical forest would be affected. Heavy rain during the 1997-1998 ENSO event generated drastic changes in dry ecosystems of northern Peru's coastal zone. Global warming would expand the area suitable for tropical forests as equilibrium vegetation types. However, the forces driving deforestation make it unlikely that tropical forests will be permitted to occupy these increased areas. Land-use change interacts with climate through positive-feedback processes that accelerate loss of humid tropical forests. [14.2.1]

5.5.3. Sea-Level Rise

Sea-level rise will affect mangrove ecosystems by eliminating their present habitats and creating new tidally inundated areas to which some mangrove species may shift. This also would affect the region's fisheries because most commercial shellfish and finfish use mangroves for nurseries and refuge. Coastal inundation that stems from sea-level rise and riverine and flatland flooding would affect water availability and agricultural land, exacerbating socioeconomic and health problems in these areas. [14.2.3]

5.5.4. Agriculture

Studies in Argentina, Brazil, Chile, Mexico, and Uruguay—based on GCMs and crop models—project decreased yields for numerous crops (e.g., maize, wheat, barley, grapes) even when the direct effects of CO_2 fertilization and implementation of moderate adaptation measures at the farm level are considered (high confidence). Predicted increases in temperature will reduce crops yields in the region by shortening the crop cycle. Over the past 40 years, the contribution of agriculture to the GDP of Latin American countries has been on the order of 10%. Agriculture remains a key sector in the regional economy because it employs 30-40% of the economically active population. It also is very important for the food security of the poorest sectors of the population. Subsistence farming could be severely threatened in some parts of Latin America, including northeastern Brazil.

It is established but incomplete that climate change would reduce silvicultural yields because lack of water often limits growth during the dry season, which is expected to become longer and more intense in many parts of Latin America. Table TS-11 summarizes studies undertaken on the region for different crops and management conditions, all under rainfed conditions; most of these results predict negative impacts, particularly for maize. [14.2.2]

5.5.5. Human Health

The scale of health impacts from climate change in Latin America would depend primarily on the size, density, location, and wealth of populations. Exposure to heat or cold waves has impacts on mortality rates in risk groups in the region (medium confidence).

Increases in temperature would affect human health in polluted cities such as Mexico City and Santiago, Chile. It is well established that ENSO causes changes in disease vector populations and in the incidence of water-borne diseases in Brazil, Peru, Bolivia, Argentina, and Venezuela. Studies in Peru and Cuba indicate that increases in temperature and precipitation would change the geographical distribution of infectious diseases such as cholera and meningitis (high confidence), although there is speculation about what the changes in patterns of diseases would be in different places. It is well established that extreme events tend to increase death and morbidity rates (injuries, infectious diseases, social problems, and damage to sanitary infrastructure), as shown in Central America with Hurricane Mitch in 1998, heavy rains in Mexico and Venezuela in 1999, and in Chile and Argentina in 2000. [14.2.5]

5.6. North America

North America will experience both positive and negative climate change impacts (high confidence). Varying impacts on ecosystems and human settlements will exacerbate subregional differences in climate-sensitive resource production and vulnerability to extreme events. Opportunities and challenges

Table TS-11: Assessments of climate change impacts on annual crops in Latin America.

Study[a]	Climate Scenario	Scope	Crop	Yield Impact (%)
Downing, 1992	+3°C −25% precipitation	Norte Chico, Chile	Wheat Maize Potato Grapes	decrease increase increase decrease
Baethgen, 1994	GISS, GFDL, UKMO	Uruguay	Wheat Barley	−30 −40 to −30
de Siqueira et al., 1994	GISS, GFDL, UKMO	Brazil	Wheat Maize Soybeans	−50 to −15 −25 to −2 −10 to +40
Liverman and O'Brien, 1991	GFDL, GISS	Tlaltizapan, Mexico	Maize	−20 −24 −61
Liverman et al., 1994	GISS, GFDL, UKMO	Mexico	Maize	−61 to −6
Sala and Paruelo, 1994	GISS, GFDL, UKMO	Argentina	Maize	−36 to −17
Baethgen and Magrin, 1995	UKMO	Argentina Uruguay (9 sites)	Wheat	−5 to −10
Conde et al., 1997a	CCCM, GFDL	Mexico (7 sites)	Maize	increase-decrease
Magrin et al., 1997a	GISS, UKMO, GFDL, MPI	Argentina (43 sites)	Maize Wheat Sunflower Soybean	−16 to +2 −8 to +7 −8 to +13 −22 to +21
Hofstadter et al., 1997	Incremental	Uruguay	Barley Maize	−10[b] −8 to +5[c] −15[d] −13 to +10[c]

[a] See WGII TAR Chapter 14 reference list for complete source information.
[b] For 1°C increase.
[c] Change of −20 to +20% in precipitation.
[d] For 2°C increase.

to adaptation will arise, frequently involving multiple stresses (Table TS-12). Some innovative adaptation strategies are being tested as a response to current climate-related challenges (e.g., water banks), but few cases have examined how these strategies could be implemented as regional climates continue to change. Shifting patterns in temperature, precipitation, disease vectors, and water availability will require adaptive responses—including, for example, investments in storm protection and water supply infrastructure, as well as community health services. [15.3.2, 15.4]

5.6.1. Communities and Urban Infrastructure

Potential changes in the frequency, severity, and duration of extreme events are among the most important risks associated with climate change in North America. Potential impacts of climate change on cities include fewer periods of extreme winter cold; increased frequency of extreme heat; rising sea levels and risk of storm surge; and changes in timing, frequency, and severity of flooding associated with storms and precipitation extremes. These events—particularly increased heat waves and changes in extreme events—will be accompanied by effects on health.

Communities can reduce their vulnerability to adverse impacts through investments in adaptive infrastructure, which can be expensive. Rural, poor, and indigenous communities may not be able to make such investments. Furthermore, infrastructure investment decisions are based on a variety of needs beyond climate change, including population growth and aging of existing systems. [15.2.5]

Table TS-12: Climate change adaptation issues in North American subregions. Some unique issues for certain locations also are indicated.

North American Subregions	Development Context	Climate Change Adaptation Options and Challenges
Most or all subregions	- Changing commodity markets - Intensive water resources development over large areas—domestic and transboundary - Lengthy entitlement/land claim/treaty agreements—domestic and transboundary - Urban expansion - Transportation expansion	- Role of water/environmental markets - Changing design and operations of water and energy systems - New technology/practices in agriculture and forestry - Protection of threatened ecosystems or adaptation to new landscapes - Increased role for summer (warm weather) tourism - Risks to water quality from extreme events - Managing community health for changing risk factors - Changing roles of public emergency assistance and private insurance
Arctic border	- Winter transport system - Indigenous lifestyles	- Design for changing permafrost and ice conditions - Role of two economies and co-management bodies
Coastal regions	- Declines in some commercial marine resources (cod, salmon) - Intensive coastal zone development	- Aquaculture, habitat protection, fleet reductions - Coastal zone planning in high demand areas
Great Lakes	- Sensitivity to lake level fluctuations	- Managing for reduction in mean levels without increased shoreline encroachment

5.6.2. Water Resources and Aquatic Ecosystems

Uncertain changes in precipitation lead to little agreement regarding changes in total annual runoff across North America. Modeled impacts of increased temperatures on lake evaporation lead to consistent projections of reduced lake levels and outflows for the Great Lakes-St. Lawrence system under most scenarios (medium confidence). Increased incidence of heavy precipitation events will result in greater sediment and non-point-source pollutant loadings to watercourses (medium confidence). In addition, *in regions where seasonal snowmelt is an important aspect of the annual hydrologic regime (e.g., California, Columbia River Basin), warmer temperatures are likely to result in a seasonal shift in runoff, with a larger proportion of total runoff occurring in winter, together with possible reductions in summer flows (high confidence).* This could adversely affect the availability and quality of water for instream and out-of-stream water uses during the summer (medium confidence). Figure TS-8 shows possible impacts. [15.2.1]

Adaptive responses to such seasonal runoff changes include altered management of artificial storage capacity, increased reliance on coordinated management of groundwater and surface water supplies, and voluntary water transfers between various water users. Such actions could reduce the impacts of reduced summer flows on water users, but it may be difficult or impossible to offset adverse impacts on many aquatic ecosystems, and it may not be possible to continue to provide current levels of reliability and quality for all water users. Some regions (e.g., the western United States) are likely to see increased market transfers of available water supplies from irrigated agriculture

to urban and other relatively highly valued uses. Such reallocations raise social priority questions and entail adjustment costs that will depend on the institutions in place.

5.6.3. Marine Fisheries

Climate-related variations in marine/coastal environments are now recognized as playing an important role in determining the productivity of several North American fisheries in the Pacific, North Atlantic, Bering Sea, and Gulf of Mexico regions. There are complex links between climatic variations and changes in processes that influence the productivity and spatial distribution of marine fish populations (high confidence), as well as uncertainties linked to future commercial fishing patterns. Recent experience with Pacific salmon and Atlantic cod suggests that sustainable fisheries management will require timely and accurate scientific information on environmental conditions affecting fish stocks, as well as institutional and operational flexibility to respond quickly to such information. [15.2.3.3]

5.6.4. Agriculture

Small to moderate climate change will not imperil food and fiber production (high confidence). There will be strong regional production effects, with some areas suffering significant loss of comparative advantage to other regions (medium confidence). Overall, this results in a small net effect. The agricultural welfare of consumers and producers would increase with modest warming. However, the benefit would decline at an increasing rate—possibly becoming a net loss—with further warming. There is potential for increased drought in the U.S.

I. Alaska, Yukon, and Coastal British Columbia
Lightly settled/water-abundant region; potential ecological, hydropower, and flood impacts:
　　Increased spring flood risks
　　Glacial retreat/disappearance in south, advance in north; impacts on flows, stream ecology
　　Increased stress on salmon, other fish species
　　Flooding of coastal wetlands
　　Changes in estuary salinity/ecology

V. Sub-Arctic and Arctic
Sparse population (many dependent on natural systems); winter ice cover important feature of hydrologic cycle:
　　Thinner ice cover, 1- to 3-month increase in ice-free season, increased extent of open water
　　Increased lake-level variability, possible complete drying of some delta lakes
　　Changes in aquatic ecology and species distribution as a result of warmer temperatures and longer growing season

VI. Midwest USA and Canadian Prairies
Agricultural heartland mostly rainfed, with some areas relying heavily on irrigation:
　　Annual streamflow decreasing/increasing; possible large declines in summer streamflow
　　Increasing likelihood of severe droughts
　　Possible increasing aridity in semi-arid zones
　　Increases or decreases in irrigation demand and water availability uncertain impacts on farm-sector income, groundwater levels, streamflows, and water quality

II. Pacific Coast States (USA)
Large and rapidly growing population; water abundance decreases north to south; intensive irrigated agriculture; massive water-control infrastructure; heavy reliance on hydropower; endangered species issues; increasing competition for water:
　　More winter rainfall/less snowfall earlier seasonal peak in runoff, increased fall/winter flooding, decreased summer water supply
　　Possible increases in annual runoff in Sierra Nevada and Cascades
　　Possible summer salinity increase in San Francisco Bay and Sacramento/San Joaquin Delta
　　Changes in lake and stream ecology warmwater species benefitting; damage to coldwater species (e.g., trout and salmon)

VII. Great Lakes
Heavily populated and industrialized region; variations in lake levels/flows now affect hydropower, shipping, shoreline structures:
　　Possible precipitation increases coupled with reduced runoff and lake-level declines
　　Reduced hydropower production; reduced channel depths for shipping
　　Decreases in lake ice extent some years w/out ice cover
　　Changes in phytoplankton/zooplankton biomass, northward migration of fish species, possible extirpations of coldwater species

III. Rocky Mountains (USA and Canada)
Lightly populated in north, rapid population growth in south; irrigated agriculture, recreation, urban expansion increasingly competing for water; headwaters area for other regions:
　　Rise in snow line in winter-spring, possible increases in snowfall, earlier snowmelt, more frequent rain on snow changes in seasonal streamflow, possible reductions in summer streamflow, reduced summer soil moisture
　　Stream temperature changes affecting species composition; increased isolation of coldwater stream fish

VIII. Northeast USA and Eastern Canada
Large, mostly urban population generally adequate water supplies, large number of small dams, but limited total reservoir capacity; heavily populated floodplains:
　　Decreased snow cover amount and duration
　　Possible large reductions in streamflow
　　Accelerated coastal erosion, saline intrusion into coastal aquifers
　　Changes in magnitude, timing of ice freeze-up/break-up, with impacts on spring flooding
　　Possible elimination of bog ecosystems
　　Shifts in fish species distributions, migration patterns

IV. Southwest
Rapid population growth, dependence on limited groundwater and surface water supplies, water quality concerns in border region, endangered species concerns, vulnerability to flash flooding:
　　Possible changes in snowpacks and runoff
　　Possible declines in groundwater recharge reduced water supplies
　　Increased water temperatures further stress on aquatic species
　　Increased frequency of intense precipitation events increased risk of flash floods

IX. Southeast, Gulf, and Mid-Atlantic USA
Increasing population especially in coastal areas, water quality/non-point source pollution problems, stress on aquatic ecosystems:
　　Heavily populated coastal floodplains at risk to flooding from extreme precipitation events, hurricanes
　　Possible lower base flows, larger peak flows, longer droughts
　　Possible precipitation increase possible increases or decreases in runoff/river discharge, increased flow variability
　　Major expansion of northern Gulf of Mexico hypoxic zone possible other impacts on coastal systems related to changes in precipitation/non-point source pollutant loading
　　Changes in estuary systems and wetland extent, biotic processes, species distribution

Figure TS-8: Possible water resources impacts in North America.

Great Plains/Canadian Prairies and opportunities for a limited northward shift in production areas in Canada.

Increased production from direct physiological effects of CO_2, and farm- and agricultural market-level adjustments (e.g., behavioral, economic, and institutional) are projected to offset losses. Economic studies that include farm- and agricultural market-level adjustments indicate that the negative effects of climate change on agriculture probably have been overestimated by studies that do not account for these adjustments (medium confidence). However, the ability of farmers to adapt their input and output choices is difficult to forecast and will depend on market and institutional signals. [15.2.3.1]

5.6.5. Forests and Protected Areas

Climate change is expected to increase the areal extent and productivity of forests over the next 50-100 years (medium confidence). However, climate change is likely to cause changes in the nature and extent of several "disturbance factors" (e.g., fire, insect outbreaks) (medium confidence). Extreme or long-term climate change scenarios indicate the possibility of widespread forest decline (low confidence).

There is strong evidence that climate change can lead to the loss of specific ecosystem types—such as high alpine areas and specific coastal (e.g., salt marshes) and inland (e.g., prairie "potholes") wetland types (high confidence). There is moderate potential for adaptation to prevent these losses by planning conservation programs to identify and protect particularly threatened ecosystems. Lands that are managed for timber production are likely to be less susceptible to climate change than unmanaged forests because of the potential for adaptive management. [15.2.2]

5.6.6. Human Health

Vector-borne diseases, including malaria and dengue fever, may expand their ranges in the United States and may develop in Canada. Tick-borne Lyme disease also may see its range expanded in Canada. However, socioeconomic factors such as public health measures will play a large role in determining the existence or extent of such infections. Diseases associated with water may increase with warming of air and water temperatures, combined with heavy runoff events from agricultural and urban surfaces. Increased frequency of convective storms could lead to more cases of thunderstorm-associated asthma. [15.2.4]

5.6.7. Public and Private Insurance Systems

Inflation-corrected catastrophe losses have increased eight-fold in North America over the past 3 decades (high confidence). The exposures and surpluses of private insurers (especially property insurers) and reinsurers have been growing, and weather-related profit losses and insolvencies have been observed.

Insured losses in North America (59% of the global total) are increasing with affluence and as populations continue to move into vulnerable areas. Insurer vulnerability to these changes varies considerably by region.

Recent extreme events have led to several responses by insurers, including increased attention to building codes and disaster preparedness. Insurers' practices traditionally have been based primarily on historic climatic experience; only recently have they begun to use models to predict future climate-related losses, so the potential for surprise is real. Governments play a key role as insurers or providers of disaster relief, especially in cases in which the private sector deems risks to be uninsurable. [15.2.7]

5.7. Polar Regions

Climate change in the polar region is expected to be among the greatest of any region on Earth. Twentieth century data for the Arctic show a warming trend of as much as 5°C over extensive land areas (very high confidence), while precipitation has increased (low confidence). There are some areas of cooling in eastern Canada. The extent of sea ice has decreased by 2.9% per decade, and it has thinned over the 1978-1996 period (high confidence). There has been a statistically significant decrease in spring snow extent over Eurasia since 1915 (high confidence). The area underlain by permafrost has been reduced and has warmed (very high confidence). The layer of seasonally thawed ground above permafrost has thickened in some areas, and new areas of extensive permafrost thawing have developed. *In the Antarctic, a marked warming trend is evident in the Antarctic Peninsula, with spectacular loss of ice shelves (high confidence).* The extent of higher terrestrial vegetation on the Antarctic Peninsula is increasing (very high confidence). Elsewhere, warming is less definitive. There has been no significant change in the Antarctic sea ice since 1973, although it apparently retreated by more than 3° of latitude between the mid-1950s and the early 1970s (medium confidence). [16.1.3.2.]

The Arctic is extremely vulnerable to climate change, and major physical, ecological, and economic impacts are expected to appear rapidly. A variety of feedback mechanisms will cause an amplified response, with consequent impacts on other systems and people. There will be different species compositions on land and sea, poleward shifts in species assemblages, and severe disruptions for communities of people who lead traditional lifestyles. *In developed areas of the Arctic and where the permafrost is ice-rich, special attention will be required to mitigate the detrimental impacts of thawing, such as severe damage to buildings and transport infrastructure (very high confidence).* There also will be beneficial consequences of climatic warming, such as reduced demand for heating energy. Substantial loss of sea ice in the Arctic Ocean will be favorable for opening of Arctic sea routes and ecotourism, which may have large implications for trade and for local communities. [16.2.5.3, 16.2.7.1, 16.2.8.1, 16.2.8.2]

In the Antarctic, projected climate change will generate impacts that will be realized slowly (high confidence). Because the impacts will occur over a long period, however, they will continue long after GHG emissions have stabilized. For example, there will be slow but steady impacts on ice sheets and circulation patterns of the global ocean, which will be irreversible for many centuries into the future and will cause changes elsewhere in the world, including a rise of sea level. Further substantial loss of ice shelves is expected around the Antarctic Peninsula. Warmer temperatures and reduced sea-ice extent are likely to produce long-term changes in the physical oceanography and ecology of the Southern Ocean, with intensified biological activity and increased growth rates of fish. [16.2.3.4, 16.2.4.2]

Polar regions contain important drivers of climate change. The Southern Ocean's uptake of carbon is projected to reduce substantially as a result of complex physical and biological processes. GHG emissions from tundra caused by changes in water content, decomposition of exposed peat, and thawing of permafrost are expected to increase. Reductions in the extent of highly reflective snow and ice will magnify warming (very high confidence). Freshening of waters from increased Arctic runoff and increased rainfall, melt of Antarctic ice shelves, and reduced sea-ice formation will slow the thermohaline circulations of the North Atlantic and Southern Oceans and reduce the ventilation of deep ocean waters. [16.3.1]

Adaptation to climate change will occur in natural polar ecosystems, mainly through migration and changing mixes of species. Some species may become threatened (e.g., walrus, seals, and polar bears), whereas others may flourish (e.g., caribou and fish). Although such changes may be disruptive to many local ecological systems and particular species, the possibility remains that predicted climate change eventually may increase the overall productivity of natural systems in polar regions. [16.3.2]

For indigenous communities who follow traditional lifestyles, opportunities for adaptation to climate change are limited (very high confidence). Changes in sea ice, seasonality of snow, habitat, and diversity of food species will affect hunting and gathering practices and could threaten longstanding traditions and ways of life. Technologically developed communities are likely to adapt quite readily to climate change by adopting altered modes of transport and by increased investment to take advantage of new commercial and trade opportunities. [16.3.2]

5.8. Small Island States

Climate change and sea-level rise pose a serious threat to the small island states, which span the ocean regions of the Pacific, Indian, and Atlantic Oceans as well as the Caribbean and Mediterranean Seas. Characteristics of small island states that increase their vulnerability include their small physical size relative to large expanses of ocean; limited natural resources;

relative isolation; extreme openness of small economies that are highly sensitive to external shocks and highly prone to natural disasters and other extreme events; rapidly growing populations with high densities; poorly developed infrastructure; and limited funds, human resources, and skills. These characteristics limit the capacity of small island states to mitigate and adapt to future climate change and sea-level rise. [17.1.2]

Many small island states already are experiencing the effects of current large interannual variations in oceanic and atmospheric conditions. As a result, the most significant and more immediate consequences for small island states are likely to be related to changes in rainfall regimes, soil moisture budgets, prevailing winds (speed and direction), short-term variations in regional and local sea levels, and patterns of wave action. These changes are manifest in past and present trends of climate and climate variability, with an upward trend in average temperature by as much as $0.1°C$ per decade and sea-level rise of 2 mm yr^{-1} in the tropical ocean regions in which most of the small island states are located. Analysis of observational data from various regions indicates an increase in surface air temperature that has been greater than global rates of warming, particularly in the Pacific Ocean and Caribbean Sea. Much of the variability in the rainfall record of the Pacific and Caribbean islands appears to be closely related to the onset of ENSO. However, part of the variability also may be attributable to shifts in the Intertropical and South Pacific Convergence Zone, whose influence on rainfall variability patterns must be better understood. The interpretation of current sea-level trends also is constrained by limitations of observational records, particularly from geodetic-controlled tide gauges. [17.1.3]

5.8.1. Equity and Sustainable Development

Although the contribution of small island states to global emissions of GHG is insignificant, projected impacts of climate change and sea-level rise on these states are likely to be serious. The impacts will be felt for many generations because of small island states' low adaptive capacity, high sensitivity to external shocks, and high vulnerability to natural disasters. Adaptation to these changing conditions will be extremely difficult for most small island states to accomplish in a sustainable manner. [17.2.1]

5.8.2. Coastal Zone

Much of the coastal change currently experienced in small island states is attributed to human activities on the coast. Projected sea-level rise of 5 mm yr^{-1} over the next 100 years, superimposed on further coastal development, will have negative impacts on the coasts (high confidence). This in turn will increase the vulnerability of coastal environments by reducing natural resilience and increasing the cost of adaptation. Given that severity will vary regionally, the most serious considerations for some small island states will be whether they will have adequate potential to adapt to sea-level rise within their own national boundaries. [17.2.2.1, 17.2.3]

5.8.3. Ecosystems and Biodiversity

Projected future climate change and sea-level rise will affect shifts in species composition and competition. It is estimated that one of every three known threatened plants are island endemics while 23% of bird species found on islands are threatened. [17.2.5]

Coral reefs, mangroves, and seagrass beds that often rely on stable environmental conditions will be adversely affected by rising air and sea temperature and sea levels (medium confidence). Episodic warming of the sea surface has resulted in greatly stressed coral populations that are subject to widespread coral bleaching. Mangroves, which are common on low-energy, nutrient/sediment-rich coasts and embayments in the tropics, have been altered by human activities. Changes in sea levels are likely to affect landward and alongshore migration of remnants of mangrove forests that provide protection for coasts and other resources. An increase in SST would adversely affect seagrass communities, which already are under stress from land-based pollution and runoff. Changes in these systems are likely to negatively affect fishery populations that depend on them for habitat and breeding grounds. [17.2.4]

5.8.4. Water Resources, Agriculture, and Fisheries

Water resources and agriculture are of critical concern because most small island states possess limited arable land and water resources. Communities rely on rainwater from catchments and a limited freshwater lens. In addition, arable farming, especially on low islands and atolls, is concentrated at or near the coast. Changes in the height of the water table and soil salinization as a consequence of sea-level rise would be stressful for many staple crops, such as taro.

Although fishing is largely artisinal or small-scale commercial, it is an important activity on most small islands and makes a significant contribution to the protein intake of island inhabitants. Many breeding grounds and habitats for fish and shellfish—such as mangroves, coral reefs, seagrass beds, and salt ponds—will face increasing threats from likely impacts of projected climate change. Water resources, agriculture, and fisheries already are sensitive to currently observed variability in oceanic and atmospheric conditions in many small island states, and the impacts are likely to be exacerbated by future climate and sea-level change (high confidence). [17.2.6, 17.2.8.1]

5.8.5. Human Health, Settlement, Infrastructure and Tourism

Several human systems are likely to be affected by projected changes in climate and sea levels in many small island states.

Human health is a major concern given that many tropical islands are experiencing high incidences of vector- and water-borne diseases that are attributable to changes in temperature and rainfall, which may be linked to the ENSO phenomenon, droughts, and floods. Climate extremes also create a huge burden on some areas of human welfare, and these burdens are likely to increase in the future. Almost all settlements, socioeconomic infrastructure, and activities such as tourism are located at or near coastal areas in small island states. Tourism provides a major source of revenue and employment for many small island states (Table TS-13). Changes in temperature and rainfall regimes, as well as loss of beaches, could devastate the economies of many small island states (high confidence). Because these areas are very vulnerable to future climate change and sea-level rise, it is important to protect and nourish beaches and sites by implementing programs that constitute wise use resources. Integrated coastal management has been identified as one approach that would be useful for many small island states for a sustainable tourism industry. [17.2.7, 17.2.9]

5.8.6. Sociocultural and Traditional Assets

Certain traditional island assets (good and services) also will be at risk from climate change and sea-level rise. These assets include subsistence and traditional technologies (skills and knowledge) and cohesive community structures that, in the past, have helped to buttress the resilience of these islands to various forms of shock. Sea-level rise and climate changes, combined with other environmental stresses, already have destroyed unique cultural and spiritual sites, traditional heritage assets, and important coastal protected areas in many Pacific island states. [17.2.10]

6. Adaptation, Sustainable Development, and Equity

Adaptation to climate change has the potential to substantially reduce many of the adverse impacts of climate change and enhance beneficial impacts, though neither without cost nor without leaving residual damage. In natural systems, adaptation is reactive, whereas in human systems it also can be anticipatory. Figure TS-9 presents types and examples of adaptation to climate change. Experience with adaptation to climate variability and extremes shows that in the private and public sectors there are constraints to achieving the potential of adaptation. The adoption and effectiveness of private, or market-driven, adaptations in sectors and regions are limited by other forces, institutional conditions, and various sources of market failure. There is little evidence to suggest that private adaptations will be employed to offset climate change damages in natural environments. In some instances, adaptation measures may have inadvertent consequences, including environmental damage. The ecological, social, and economic costs of relying on reactive, autonomous adaptation to the

Table TS-13: Estimates of flood exposure and incidence for Europe's coasts in 1990 and the 2080s. Estimates of flood incidence are highly sensitive to assumed protection standard and should be interpreted in indicative terms only (former Soviet Union excluded).

Country	Number of Tourists (000s)[a]	Tourists as % of Population[a]	Tourist Receipts[b] as % of GNP	Tourist Receipts[b] as % of Exports
Antigua and Barbuda	232	364	63	74
Bahamas	1618	586	42	76
Barbados	472	182	39	56
Cape Verde	45	11	12	37
Comoros	26	5	11	48
Cuba	1153	11	9	n/a
Cyprus	2088	281	24	49
Dominica	65	98	16	33
Dominican Republic	2211	28	14	30
Fiji	359	45	19	29
Grenada	111	116	27	61
Haiti	149	2	4	51
Jamaica	1192	46	32	40
Maldives	366	131	95	68
Malta	1111	295	23	29
Mauritius	536	46	16	27
Papua New Guinea	66	2	2	3
St. Kitts and Nevis	88	211	31	64
St. Lucia	248	165	41	67
St. Vincent	65	55	24	46
Samoa	68	31	20	49
Seychelles	130	167	35	52
Singapore	7198	209	6	4
Solomon Islands	16	4	3	4
Trinidad and Tobago	324	29	4	8
Vanuatu	49	27	19	41

[a] Data on tourist inflows and ratio to population pertain to 1997.

[b] Data for tourist receipts pertain to 1997 for the Bahamas, Cape Verde, Jamaica, the Maldives, Malta, Mauritius, Samoa, Seychelles, Singapore, and Solomon Islands; 1996 for Antigua and Barbuda, Cuba, Dominica, Dominican Republic, Fiji, Grenada, Haiti, Papua New Guinea, St. Lucia, and St. Vincent; 1995 for Barbados, Comoros, Cyprus, Trinidad and Tobago, and Vanuatu; and 1994 for St. Kitts and Nevis.

cumulative effects of climate change are substantial. Many of these costs can be avoided through planned, anticipatory adaptation. Designed appropriately, many adaptation strategies could provide multiple benefits in the near and longer terms. However, there are limits on their implementation and effectiveness. Enhancement of adaptive capacity reduces the vulnerability of sectors and regions to climate change, including variability and extremes, and thereby promotes sustainable development and equity. [18.2.4, 18.3.4]

Planned anticipatory adaptation has the potential to reduce vulnerability and realize opportunities associated with climate change, regardless of autonomous adaptation. Adaptation facilitated by public agencies is an important part of societal response to climate change. Implementation of adaptation policies, programs, and measures usually will have immediate and future benefits. Adaptations to current climate and climate-related risks (e.g., recurring droughts, storms, floods, and other extremes) generally are consistent with adaptation to changing and changed climatic conditions. Adaptation measures are likely to be implemented only if they are consistent with or integrated with decisions or programs that address nonclimatic stresses. Vulnerabilities associated with climate change are rarely experienced independently of nonclimatic conditions. Impacts of climatic stimuli are felt via economic or social stresses,

	Anticipatory	Reactive
Natural Systems		• Changes in length of growing season • Changes in ecosystem composition • Wetland migration
Human Systems *Private*	• Purchase of insurance • Construction of houses on stilts • Redesign of oil rigs	• Changes in farm practices • Changes in insurance premiums • Purchase of air-conditioning
Human Systems *Public*	• Early-warning systems • New building codes, design standards • Incentives for relocation	• Compensatory payments, subsidies • Enforcement of building codes • Beach nourishment

Figure TS-9: Types of adaptation to climate change, including examples.

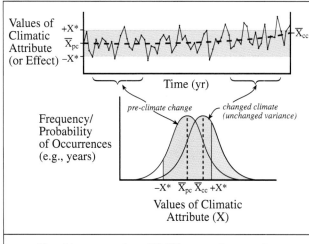

- - - Trend in mean value of X (20-yr running mean)

$\overline{X}_{pc}$ = Mean value of climatic attribute (X) at start of time series (pre-climate change)

$\overline{X}_{cc}$ = Mean value of climatic attribute (X) at end of time series (climate change)

+X* = Upper critical value of X for system of interest: values < –X* are problematic and considered "extreme" or beyond "damage threshold"

–X* = Lower critical value of X for system of interest: values < –X* are problematic and considered "extreme" or beyond "damage threshold"

☐ Coping range or zone of minimal hazard potential for system of interest

☐ Probability of "extreme" events (i.e., climatic attribute values > +X*)

Figure TS-10: Climate change, variability, extremes, and coping range.

and adaptations to climate (by individuals, communities, and governments) are evaluated and undertaken in light of these conditions. The costs of adaptation often are marginal to other management or development costs. To be effective, climate change adaptation must consider nonclimatic stresses and be consistent with existing policy criteria, development objectives, and management structures. [18.3.5, 18.4]

The key features of climate change for vulnerability and adaptation are related to variability and extremes, not simply changed average conditions (Figure TS-10). Societies and economies have been making adaptations to climate for centuries. Most sectors, regions, and communities are reasonably adaptable to changes in average conditions, particularly if the changes are gradual. However, losses from climatic variations and extremes are substantial and, in some sectors, increasing. These losses indicate that autonomous adaptation has not been sufficient to offset damages associated with temporal variations in climatic conditions. Communities therefore are more

vulnerable and less adaptable to changes in the frequency and/ or magnitude of conditions other than average, especially extremes, which are inherent in climate change. The degree to which future adaptations are successful in offsetting adverse impacts of climate change will be determined by success in adapting to climate change, variability, and extremes. [18.2.2]

6.1. Adaptive Capacity

The capacity to adapt varies considerably among regions, countries, and socioeconomic groups and will vary over time. Table TS-14 summarizes adaptation measures and capacities by sector, and Table TS-15 provides this information for each region covered by the TAR. The most vulnerable regions and communities are highly exposed to hazardous climate change effects and have limited adaptive capacity. The ability to adapt and cope with climate change impacts is a function of wealth, scientific and technical knowledge, information, skills, infrastructure, institutions, and equity. Countries with limited economic resources, low levels of technology, poor information and skills, poor infrastructure, unstable or weak institutions, and inequitable empowerment and access to resources have little capacity to adapt and are highly vulnerable. Groups and regions with adaptive capacity that is limited along any of these dimensions are more vulnerable to climate change damages, just as they are more vulnerable to other stresses. [18.5, 18.7]

6.2. Development, Sustainability, and Equity

Activities required for enhancement of adaptive capacity are essentially equivalent to those promoting sustainable development. Enhancement of adaptive capacity is a necessary condition for reducing vulnerability, particularly for the most vulnerable regions, nations, and socioeconomic groups. Many sectors and regions that are vulnerable to climate change also are under pressure from forces such as population growth and resource depletion. Climate adaptation and sustainability goals can be jointly advanced by changes in policies that lessen pressure on resources, improve management of environmental risks, and enhance adaptive capacity. Climate adaptation and equity goals can be jointly pursued through initiatives that promote the welfare of the poorest members of society—for example, by improving food security, facilitating access to safe water and health care, and providing shelter and access to other resources. Development decisions, activities, and programs play important roles in modifying the adaptive capacity of communities and regions, yet they tend not to take into account risks associated with climate variability and change. Inclusion of climatic risks in the design and implementation of development initiatives is necessary to reduce vulnerability and enhance sustainability. [18.6.1]

Table TS-14: Adaptation and adaptive capacity in sectors (key findings from Chapters 4 through 9).

Sector	Key Findings
Water Resources	- Water managers have experience with adapting to change. Many techniques exist to assess and implement adaptive options. However, the pervasiveness of climate change may preclude some traditional adaptive strategies, and available adaptations often are not used. - Adaptation can involve management on the supply side (e.g., altering infrastructure or institutional arrangements) and on the demand side (changing demand or risk reduction). Numerous no-regret policies exist, which will generate net social benefits regardless of climate change. - Climate change is just one of numerous pressures facing water managers. Nowhere are water management decisions taken solely to cope with climate change, although it is increasingly considered for future resource management. Some vulnerabilities are outside the conventional responsibility of water managers. - Estimates of the economic costs of climate change impacts on water resources depend strongly on assumptions made about adaptation. Economically optimum adaptation may be prevented by constraints associated with uncertainty, institutions, and equity. - Extreme events often are catalysts for change in water management, by exposing vulnerabilities and raising awareness of climate risks. Climate change modifies indicators of extremes and variability, complicating adaptation decisions. - Ability to adapt is affected by institutional capacity, wealth, management philosophy, planning time scale, organizational and legal framework, technology, and population mobility. - Water managers need research and management tools aimed at adapting to uncertainty and change, rather than improving climate scenarios.
Ecosystems and Their Services	- Adaptation to loss of some ecosystem services may be possible, especially in managed ecosystems. However, adaptation to losses in wild ecosystems and biodiversity may be difficult or impossible. - There is considerable capacity for adaptation in agriculture, including crop changes and resource substitutions, but adaptation to evolving climate change and interannual variability is uncertain. - Adaptations in agriculture are possible, but they will not happen without considerable transition costs and equilibrium (or residual) costs. - Greater adverse impacts are expected in areas where resource endowments are poorest and the ability of farmers to adapt is most limited. - In many countries where rangelands are important, lack of infrastructure and investment in resource management limit options for adaptation. - Commercial forestry is adaptable, reflecting a history of long-term management decisions under uncertainty. Adaptations are expected in land-use management (species-selection silviculture) and product management (processing-marketing). - Adaptation in developed countries will fare better, while developing countries and countries in transition, especially in the tropics and subtropics, will fare worse.
Coastal Zones	- Without adaptations, the consequences of global warming and sea-level rise would be disastrous. - Coastal adaptation entails more than just selecting one of the technical options to respond to sea-level rise (strategies can aim to protect, accommodate, or retreat). It is a complex and iterative process rather than a simple choice. - Adaptation options are more acceptable and effective when they are incorporated into coastal zone management, disaster mitigation programs, land-use planning, and sustainable development strategies. - Adaptation choices will be conditioned by existing policies and development objectives, requiring researchers and policymakers to work toward a commonly acceptable framework for adaptation. - The adaptive capacity of coastal systems to perturbations is related to coastal resilience, which has morphological, ecological, and socioeconomic components. Enhancing resilience—including the technical, institutional, economic, and cultural capability to cope with impacts—is a particularly appropriate adaptive strategy given future uncertainties and the desire to maintain development opportunities. - Coastal communities and marine-based economic sectors with low exposure or high adaptive capacity will be least affected. Communities with lower economic resources, poorer infrastructure, less-developed communications and transportation systems, and weak social support systems have less access to adaptation options and are more vulnerable.
Human Settlements, Energy, and Industry	- Larger and more costly impacts of climate change occur through changed probabilities of extreme weather events that overwhelm the design resiliency of human systems. - Many adaptation options are available to reduce the vulnerability of settlements. However, urban managers, especially in developing countries, have so little capacity to deal with current problems (housing, sanitation, water, and power) that dealing with climate change risks is beyond their means. - Lack of financial resources, weak institutions, and inadequate or inappropriate planning are major barriers to adaptation in human settlements. - Successful environmental adaptation cannot occur without locally based, technically competent, and politically supported leadership. - Uncertainty with respect to capacity and the will to respond hinder assessment of adaptation and vulnerability.

Table TS-14 (continued)

Sector	Key Findings
Insurance and Other Financial Services	- Adaptation in financial and insurance services in the short term is likely to be to changing frequencies and intensities of extreme weather events. - Increasing risk could lead to a greater volume of traditional business and development of new financial risk management products, but increased variability of loss events would heighten actuarial uncertainty. - Financial services firms have adaptability to external shocks, but there is little evidence that climate change is being incorporated into investment decisions. - The adaptive capacity of the financial sector is influenced by regulatory involvement, the ability of firms to withdraw from at-risk markets, and fiscal policy regarding catastrophe reserves. - Adaptation will involve changes in the roles of private and public insurance. Changes in the timing, intensity, frequency, and/or spatial distribution of climate-related losses will generate increased demand on already overburdened government insurance and disaster assistance programs. - Developing countries seeking to adapt in a timely manner face particular difficulties, including limited availability of capital, poor access to technology, and absence of government programs. - Insurers' adaptations include raising prices, non-renewal of policies, cessation of new policies, limiting maximum claims, and raising deductibles—actions that can seriously affect investment in developing countries. - Developed countries generally have greater adaptive capacity, including technology and economic means to bear costs.
Human Health	- Adaptation involves changes in society, institutions, technology, or behavior to reduce potential negative impacts or increase positive ones. There are numerous adaptation options, which may occur at the population, community, or personal levels. - The most important and cost-effective adaptation measure is to rebuild public health infrastructure—which, in much of the world, has declined in recent years. Many diseases and health problems that may be exacerbated by climate change can be effectively prevented with adequate financial and human public health resources, including training, surveillance and emergency response, and prevention and control programs. - Adaptation effectiveness will depend on timing. "Primary" prevention aims to reduce risks before cases occur, whereas "secondary" interventions are designed to prevent further cases. - Determinants of adaptive capacity to climate-related threats include level of material resources, effectiveness of governance and civil institutions, quality of public health infrastructure, and preexisting burden of disease. - Capacity to adapt also will depend on research to understand associations between climate, weather, extreme events, and vector-borne diseases.

7. Global Issues and Synthesis

7.1. Detection of Climate Change Impacts

Observational evidence indicates that climate changes in the 20th century already have affected a diverse set of physical and biological systems. Examples of observed changes with linkages to climate include shrinkage of glaciers; thawing of permafrost; shifts in ice freeze and break-up dates on rivers and lakes; increases in rainfall and rainfall intensity in most mid- and high latitudes of the Northern Hemisphere; lengthening of growing seasons; and earlier flowering dates of trees, emergence of insects, and egg-laying in birds. Statistically significant associations between changes in regional climate and observed changes in physical and biological systems have been documented in freshwater, terrestrial, and marine environments on all continents. [19.2]

The presence of multiple causal factors (e.g., land-use change, pollution) makes attribution of many observed impacts to regional climate change a complex challenge. Nevertheless, studies of systems subjected to significant regional climate change—and with known sensitivities to that change—find changes that are consistent with well-established relationships between climate and physical or biological processes (e.g., shifts in the energy balance of glaciers, shifts in the ranges of animals and plants when temperatures exceed physiological thresholds) in about 80% of biological cases and about 99% of physical cases. Table TS-16 shows ~450 changes in processes or species that have been associated with regional temperature changes. Figure TS-11 illustrates locations at which studies have documented regional temperature change impacts. These consistencies enhance our confidence in the associations between changes in regional climate and observed changes in physical and biological systems. Based on observed changes, there is high confidence that 20th century climate changes have had a discernible impact on many physical and biological systems. Changes in biota and physical systems observed in the 20th century indicate that these systems are sensitive to climatic changes that are small relative to changes that have been projected for the 21st century. High sensitivity of biological systems to long-term climatic change also is demonstrated by paleorecords. [19.2.2.]

Table TS-15: Adaptation and capacity in regions (key findings from Chapters 10 through 17).

Sector	Key Findings
Africa	- Adaptive measures would enhance flexibility and have net benefits in water resources (irrigation and water reuse, aquifer and groundwater management, desalinization), agriculture (crop changes, technology, irrigation, husbandry), and forestry (regeneration of local species, energy-efficient cook stoves, sustainable community management). - Without adaptation, climate change will reduce the wildlife reserve network significantly by altering ecosystems and causing species' emigrations and extinctions. This represents an important ecological and economic vulnerability in Africa. - A risk-sharing approach between countries will strengthen adaptation strategies, including disaster management, risk communication, emergency evacuation, and cooperative water resource management. - Most countries in Africa are particularly vulnerable to climate change because of limited adaptive capacity as a result of widespread poverty, recurrent droughts, inequitable land distribution, and dependence on rainfed agriculture. - Enhancement of adaptive capacity requires local empowerment in decisionmaking and incorporation of climate adaptation within broader sustainable development strategies.
Asia	- Priority areas for adaptation are land and water resources, food productivity, and disaster preparedness and planning, particularly for poorer, resource-dependent countries. - Adaptations already are required to deal with vulnerabilities associated with climate variability, in human health, coastal settlements, infrastructure, and food security. Resilience of most sectors in Asia to climate change is very poor. Expansion of irrigation will be difficult and costly in many countries. - For many developing countries in Asia, climate change is only one of a host of problems to deal with, including nearer term needs such as hunger, water supply and pollution, and energy. Resources available for adaptation to climate are limited. Adaptation responses are closely linked to development activities, which should be considered in evaluating adaptation options. - Early signs of climate change already have been observed and may become more prominent over 1 or 2 decades. If this time is not used to design and implement adaptations, it may be too late to avoid upheavals. Long-term adaptation requires anticipatory actions. - A wide range of precautionary measures are available at the regional and national level to reduce economic and social impacts of disasters. These strategies include awareness-building and expansion of the insurance industry. - Development of effective adaptation strategies requires local involvement, inclusion of community perceptions, and recognition of multiple stresses on sustainable management of resources. - Adaptive capacities vary between countries, depending on social structure, culture, economic capacity, and level of environmental disruptions. Limiting factors include poor resource and infrastructure bases, poverty and disparities in income, weak institutions, and limited technology. - The challenge in Asia lies in identifying opportunities to facilitate sustainable development with strategies that make climate-sensitive sectors resilient to climate variability. - Adaptation strategies would benefit from taking a more systems-oriented approach, emphasizing multiple interactive stresses, with less dependence on climate scenarios.
Australia and New Zealand	- Adaptations are needed to manage risks from climatic variability and extremes. Pastoral economies and communities have considerable adaptability but are vulnerable to any increase in the frequency or duration of droughts. - Adaptation options include water management, land-use practices and policies, engineering standards for infrastructure, and health services. - Adaptations will be viable only if they are compatible with the broader ecological and socioeconomic environment, have net social and economic benefits, and are taken up by stakeholders. - Adaptation responses may be constrained by conflicting short- and long-term planning horizons. - Poorer communities, including many indigenous settlements, are particularly vulnerable to climate-related hazards and stresses on health because they often are in exposed areas and have less adequate housing, health care, and other resources for adaptation.
Europe	- Adaptation potential in socioeconomic systems is relatively high because strong economic conditions, stable population (with capacity to migrate), and well-developed political, institutional, and technological support systems. - The response of human activities and the natural environment to current weather perturbations provides a guide to critical sensitivities under future climate change. - Adaptation in forests requires long-term planning; it is unlikely that adaptation measures will be put in place in a timely manner. - Farm-level analyses show that if adaptation is fully implemented large reductions in adverse impacts are possible. - Adaptation for natural systems generally is low. - More marginal and less wealthy areas will be less able to adapt; thus, without appropriate policies of response, climate change may lead to greater inequities.

Table TS-15 (continued)

Sector	Key Findings
Latin America	- Adaptation measures have potential to reduce climate-related losses in agriculture and forestry. - There are opportunities for adapting to water shortages and flooding through water resource management. - Adaptation measures in the fishery sector include changing species captured and increasing prices to reduce losses.
North America	- Strain on social and economic systems from rapid climate and sea-level changes will increase the need for explicit adaptation strategies. In some cases, adaptation may yield net benefits, especially if climate change is slow. - Stakeholders in most sectors believe that technology is available to adapt, although at some social and economic cost. - Adaptation is expected to be more successful in agriculture and forestry. However, adaptations for water, health, food, energy, and cities are likely to require substantial institutional and infrastructure changes. - In the water sector, adaptations to seasonal runoff changes include storage, conjunctive supply management, and transfer. It may not be possible to continue current high levels of reliability of water supply, especially with transfers to high-valued uses. Adaptive measures such as "water markets" may lead to concerns about accessibility and conflicts over allocation priorities. - Adaptations such as levees and dams often are successful in managing most variations in weather but can increase vulnerability to the most extreme events. - There is moderate potential for adaptation through conservation programs that protect particularly threatened ecosystems, such as high alpines and wetlands. It may be difficult or impossible to offset adverse impacts on aquatic systems.
Polar Regions	- Adaptation will occur in natural polar ecosystems through migration and changing mixes of species. Species such as walrus, seals, and polar bears will be threatened; while others, such as fish, may flourish. - Potential for adaptation is limited in indigenous communities that follow traditional lifestyles. - Technologically developed communities are likely to adapt quite readily, although the high capital investment required may result in costs in maintaining lifestyles. - Adaptation depends on technological advances, institutional arrangements, availability of financing, and information exchange.
Small Island States	- The need for adaptation has become increasingly urgent, even if swift implementation of global agreements to reduce future emissions occurs. - Most adaptation will be carried out by people and communities that inhabit island countries; support from governments is essential for implementing adaptive measures. - Progress will require integration of appropriate risk-reduction strategies with other sectoral policy initiatives in areas such as sustainable development planning, disaster prevention and management, integrated coastal zone management, and health care planning. - Strategies for adaptation to sea-level rise are retreat, accommodate, and protect. Measures such as retreat to higher ground, raising of the land, and use of building set-backs appear to have little practical utility, especially when hindered by limited physical size. - Measures for reducing the severity of health threats include health education programs, health care facilities, sewerage and solid waste management, and disaster preparedness plans. - Islanders have developed some capacity to adapt by application of traditional knowledge, locally appropriate technology, and customary practice. Overall adaptive capacity is low, however, because of the physical size of nations, limited access to capital and technology, shortage of human resource skills, lack of tenure security, overcrowding, and limited access to resources for construction. - Many small islands require external financial, technical, and other assistance to adapt. Adaptive capacity may be enhanced by regional cooperation and pooling of limited resources.

Signals of regional climate change impacts are expected to be clearer in physical and biotic systems than in social and economic systems, which are simultaneously undergoing many complex non-climate-related stresses, such as population growth and urbanization. Preliminary indications suggest that some social and economic systems have been affected in part by 20th century regional climate changes (e.g., increased damages by floods and droughts in some locations, with apparent increases in insurance impacts). Coincident or alternative explanations for such observed regional impacts result in only low to medium confidence about determining whether climate change is affecting these systems. [19.2.2.4]

7.2. Five Reasons for Concern

Some of the current knowledge about climate change impacts, vulnerability, and adaptation is synthesized here along five reasons for concern: unique and threatened systems, global aggregate impacts, distribution of impacts, extreme weather events, and large-scale singular events. Consideration of these reasons

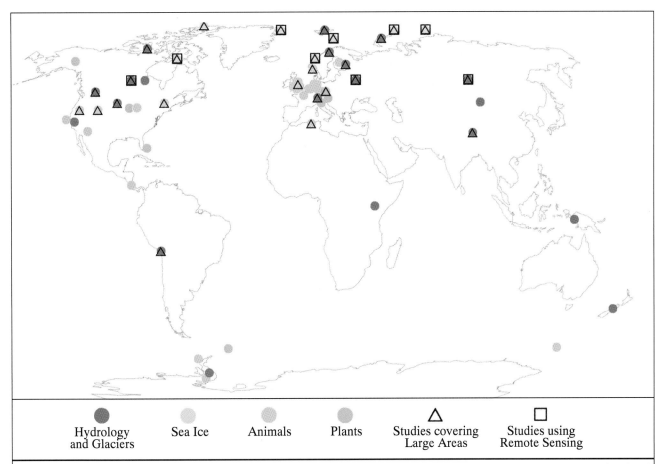

Figure TS-11: Locations at which systematic long-term studies meet stringent criteria documenting recent temperature-related regional climate change impacts on physical and biological systems. Hydrology, glacial retreat, and sea-ice data represent decadal to century trends. Terrestrial and marine ecosystem data represent trends of at least 2 decades. Remote-sensing studies cover large areas. Data are for single or multiple impacts that are consistent with known mechanisms of physical/biological system responses to observed regional temperature-related changes. For reported impacts spanning large areas, a representative location on the map was selected.

for concern contribute to understanding of vulnerabilities and potential benefits associated with human-induced climate change that can aid deliberations by policymakers of what could constitute dangerous interference with the climate system in the context of Article 2 of the UNFCCC. No single dimension is paramount.

Figure TS-12 presents qualitative findings about climate change impacts related to the reasons for concern. At a small increase in global mean temperature,[3] some of the reasons for concern show the potential for negative impacts, whereas others show little adverse impact or risk. At higher temperature increases, all lines of evidence show a potential for adverse impacts, with impacts in each reason for concern becoming more negative at increasing temperatures. There is high confidence in this general relationship between impacts and temperature change, but confidence generally is low in estimates of temperature change thresholds at which different categories of impacts would happen. [19.8]

7.2.1. Unique and Threatened Systems

Small increases in global average temperature may cause significant and irreversible damage to some systems and species, including possible local, regional, or global loss. Some plant and animal species, natural systems, and human settlements are highly sensitive to climate and are likely to be adversely affected by climate changes associated with scenarios of <1°C mean global warming. Adverse impacts to species and systems would become more numerous and more serious for climatic changes that would accompany a global mean warming of 1-2°C and are highly likely to become even more numerous and serious at higher temperatures. The greater the rate and magnitude of temperature and other climatic changes, the greater the likelihood that critical

---

[3] Intervals of global mean temperature increase of 0-2, 2-3, and >3°C relative to 1990 are labeled small, moderate, and large, respectively. The relatively large range for the "small" designation results because the literature does adequately address a warming of 1-2°C. These magnitudes of change in global mean temperature should be taken as an approximate indicator of when impacts might occur; they are not intended to define absolute thresholds or to describe all relevant aspects of climate change impacts, such as rate of change in climate and changes in precipitation, extreme climate events, or lagged (latent) effects such as rising sea levels.

Table TS-16: Processes and species found in studies to be associated with regional temperature change.[a]

Region	Glaciers, Snow Cover/Melt, Lake/Stream Ice[b]		Vegetation		Invertebrates		Amphibians and Reptiles		Birds		Mammals	
Africa	1	0	—	—	—	—	—	—	—	—	—	—
Antarctica	3	2	2	0	—	—	—	—	2	0	—	—
Asia	14	0	—	—	—	—	—	—	—	—	—	—
Australia	1	0	—	—	—	—	—	—	—	—	—	—
Europe	29	4	13	1	46	1	7	0	258	92	7	0
North America	36	4	32	11	—	—	—	—	17	4	3	0
Latin America	3	0	—	—	—	—	22	0	15	0	—	—
Total	87	10	47	12	46	1	29	0	292	96	10	0

[a] The columns represent the number of species and processes in each region that were found in each particular study to be associated with regional temperature change. For inclusion in the table, each study needed to show that the species or process was changing over time and that the regional temperature was changing over time; most studies also found a significant association between how the temperature and species or processes were changing. The first number indicates the number of species or processes changing in the manner predicted with global warming. The second number is the number of species or processes changing in a manner opposite to that predicted with a warming planet. Empty cells indicate that no studies were found for this region and category.

[b] Sea ice not included.

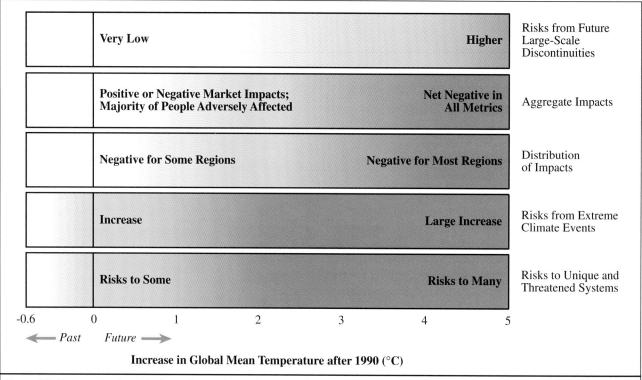

Figure TS-12: Impacts of or risks from climate change, by reason for concern. Each row corresponds to a reason for concern, and shades correspond to severity of impact or risk. White means no or virtually neutral impact or risk, yellow means somewhat negative impacts or low risks, and red means more negative impacts or higher risks. Global-averaged temperatures in the 20th century increased by 0.6°C and led to some impacts. Impacts are plotted against increases in global mean temperature after 1990. This figure addresses only how impacts or risks change as thresholds of increase in global mean temperature are crossed, not how impacts or risks change at different rates of change in climate. These temperatures should be taken as approximate indications of impacts, not as absolute thresholds.

thresholds of systems would be surpassed. Many of these threatened systems are at risk from climate change because they face nonclimate pressures such as those related to human land use, land-use change, and pollution. [19.3]

Species that may be threatened with local or global extinction by changes in climate that may accompany a small mean global temperature increase include critically endangered species generally, species with small ranges and low population densities, species with restricted habitat requirements, and species for which suitable habitat is patchy in distribution, particularly if under pressure from human land-use and land-cover change. Examples of species that may be threatened by small changes include forest birds in Tanzania, the Resplendent Quetzal in Central America, the mountain gorilla in Africa, amphibians that are endemic to cloud forests of the neotropics, the spectacled bear of the Andes, the Bengal tiger and other species that are endemic to the Sundarban wetlands, and rainfall-sensitive plant species that are endemic to the Cape Floral Kingdom of South Africa. Natural systems that may be threatened include coral reefs, mangroves, and other coastal wetlands; montane ecosystems that are restricted to the upper 200-300 m of mountainous areas; prairie wetlands; remnant native grasslands; coldwater and some coolwater fish habitat; ecosystems overlying permafrost; and ice edge ecosystems that provide habitat for polar bears and penguins. Human settlements that may be placed at serious risk by changes in climate and sea level that may be associated with medium to large mean warming include some settlements of low-lying coastal areas and islands, floodplains, and hillsides—particularly those of low socioeconomic status such as squatter and other informal settlements. Other potentially threatened settlements include traditional peoples that are highly dependent on natural resources that are sensitive to climate change. [19.3]

7.2.2. Aggregate Impacts

With a small temperature increase, aggregate market-sector impacts could amount to plus or minus a few percent of world GDP (medium confidence); aggregate nonmarket impacts could be negative (low confidence). The small net impacts are mainly the result of the fact that developed economies, many of which could have positive impacts, contribute the majority of global production. Applying more weight to impacts in poorer countries to reflect equity concerns, however, can result in net aggregate impacts that are negative even at medium warming. It also is possible that a majority of people will be negatively affected by climate change scenarios in this range, even if the net aggregate monetary impact is positive. With medium to higher temperature increases, benefits tend to decrease and damages increase, so the net change in global economic welfare becomes negative—and increasingly negative with greater warming (medium confidence). Some sectors, such as coastal and water resources, could have negative impacts in developed and developing countries. Other sectors, such as agriculture and human health, could have net positive impacts in some countries and net negative impacts in other countries. [19.5]

Results are sensitive to assumptions about changes in regional climate, levels of development, adaptive capacity, rates of change, valuation of impacts, and methods used for aggregating losses and gains, including the choice of discount rate. In addition, these studies do not consider potentially important factors such as changes in extreme events, advantageous and complementary responses to the threat of non-climate-driven

extreme events, rapid change in regional climate (e.g., resulting from changes in ocean circulation), compounding effects of multiple stresses, or conflicting or complementary reaction to those stresses. Because these factors have yet to be accounted for in estimates of aggregate impacts and estimates do not include all possible categories of impacts, particularly nonmarket impacts, estimates of aggregate economic welfare impacts of climate change are considered to be incomplete. Given the uncertainties about aggregate estimates, the possibility of negative effects at a small increase in temperature cannot be excluded. [19.5]

7.2.3. Distribution of Impacts

Developing countries tend to be more vulnerable to climate change than developed countries (high confidence). Developing countries are expected to suffer more adverse impacts than developed countries (medium confidence). A small temperature increase would have net negative impacts on market sectors in many developing countries (medium confidence) and net positive impacts on market sectors in many developed countries (medium confidence). The different results are attributable partly to differences in exposures and sensitivities (e.g., present temperatures are below optimal in mid- and high latitudes for many crops but at or above optimal in low latitudes) and partly to lesser adaptive capacity in developing countries relative to developed countries. At a medium temperature increase, net positive impacts would start to turn negative and negative impacts would be exacerbated (high confidence). The results of these studies do not fully take into account nonmarket impacts of climate change such as impacts on natural systems, which may be sensitive to small amounts of warming. Particularly vulnerable regions include deltaic regions, low-lying small island states, and many arid regions where droughts and water availability are problematic even without climate change. Within regions or countries, impacts are expected to fall most heavily, in relative terms, on impoverished persons. The poorest members of society can be inferred to be most vulnerable to climate change because of their lack of resources with which to cope and adapt to impacts, but few studies have explicitly examined the distribution of impacts on the poor relative to other segments of society. [19.4]

Impacts on unmanaged systems are likely to increase in severity with time, but impacts on managed systems could increase or decrease through the 21st century. The distribution of impacts over the 21st century is influenced by several factors. As GHG concentrations increase, the magnitude of exposure to change in climate stimuli also would increase. Nonclimate pressures on natural and social systems, which increase the vulnerability of systems, also may grow through time as a result of population growth and increased demands for land, water, public infrastructure, and other resources. Increased population, incomes, and wealth also mean that more people and human-made resources potentially would be exposed to climate change, which would tend to increase market-sector damages in absolute dollar terms; this has been the case historically. Counteracting these tendencies are factors such as increased wealth and technology and improved institutions, which can raise adaptive capacity and reduce vulnerability to climate change. [8, 19.4]

Whether impacts and vulnerability increase or decrease with time is likely to depend in part on the rates of climate change and development and may differ for managed and unmanaged systems. The more rapid the rate of climate change, the greater would be future exposure to potentially adverse changes and the greater the potential for exceeding system thresholds. The more rapid the rate of development, the more resources would be exposed to climate change in the future—but so too would the adaptive capacity of future societies. The benefits of increased adaptive capacity are likely to be greater for intensively managed systems than for systems that presently are unmanaged or lightly managed. For this reason, and because of the possibility that nonclimate pressures on natural systems may increase in the future, the vulnerability of natural systems is expected to increase with time (medium confidence). [19.4.2, 19.4.3]

Future development paths, sustainable or otherwise, will shape future vulnerability to climate change, and climate change impacts may affect prospects for sustainable development in different parts of the world. Climate change is one of many stresses that confront human and natural systems. The severity of many of these stresses will be determined in part by the development paths followed by human societies; paths that generate lesser stresses are expected to lessen the vulnerability of human and natural systems to climate change. Development also can influence future vulnerability by enhancing adaptive capacity through accumulation of wealth, technology, information, skills, and appropriate infrastructure; development of effective institutions; and advancement of equity. Climate change impacts could affect prospects for sustainable development by changing the capacity to produce food and fiber, the supply and quality of water, and human health and by diverting financial and human resources to adaptation. [18]

7.2.4. Extreme Weather Events

Many climatic impacts are related to extreme weather events, and the same will hold for the impacts of climate change. The large damage potential of extreme events arises from their severity, suddenness, and unpredictability, which makes them difficult to adapt to. Development patterns can increase vulnerability to extreme events. For example, large development along coastal regions increases exposure to storm surges and tropical cyclones, increasing vulnerability.

The frequency and magnitude of many extreme climate events increase even with a small temperature increase and will become greater at higher temperatures (high confidence). Extreme events include, for example, floods, soil moisture deficits, tropical cyclones, storms, high temperatures, and fires.

The impacts of extreme events often are large locally and could strongly affect specific sectors and regions. Increases in extreme events can cause critical design or natural thresholds to be exceeded, beyond which the magnitude of impacts increases rapidly (high confidence). Multiple nonextreme consecutive events also can be problematic because they can lessen adaptive capacity by depleting reserves of insurance and reinsurance companies. [8, 19.6.3.1]

An increase in the frequency and magnitude of extreme events would have adverse effects throughout sectors and regions. Agriculture and water resources may be particularly vulnerable to changes in hydrological and temperature extremes. Coastal infrastructure and ecosystems may be adversely affected by changes in the occurrence of tropical cyclones and storm surges. Heat-related mortality is likely to increase with higher temperatures; cold-related mortality is likely to decrease. Floods may lead to the spread of water-related and vector-borne diseases, particularly in developing countries. Many of the monetary damages from extreme events will have repercussions on a broad scale of financial institutions, from insurers and reinsurers to investors, banks, and disaster relief funds. Changes in the statistics of extreme events have implications for the design criteria of engineering applications (e.g., levee banks, bridges, building design, and zoning), which are based on estimates of return periods, and for assessment of the economic performance and viability of particular enterprises that are affected by weather. [19.6.3.1]

7.2.5. Large-Scale Singular Events

Human-induced climate change has the potential to trigger large-scale changes in Earth systems that could have severe consequences at regional or global scales. The probabilities of triggering such events are poorly understood but should not be ignored, given the severity of their consequences. Events of this type that might be triggered include complete or partial shutdown of the North Atlantic and Antarctic Deep Water formation, disintegration of the West Antarctic and Greenland Ice Sheets, and major perturbations of biosphere-regulated carbon dynamics. Determining the timing and probability of occurrence of large-scale discontinuities is difficult because these events are triggered by complex interactions between components of the climate system. The actual discontinuous impact could lag the trigger by decades to centuries. These triggers are sensitive to the magnitude and rate of climate change. Large temperature increases have the potential to lead to large-scale discontinuities in the climate system (medium confidence).

These discontinuities could cause severe impacts on the regional and even global scale, but indepth impact analyses are still lacking. Several climate model simulations show complete shutdown of the North Atlantic thermohaline circulation with high warming. Although complete shutdown may take several centuries to occur, regional shutdown of convection and significant weakening of the thermohaline circulation may take place within the next century. If this were to occur, it could lead to a rapid regional climate change in the North Atlantic region, with major societal and ecosystem impacts. Collapse of the West Antarctic Ice Sheet would lead to a global sea-level rise of several meters, which may be very difficult to adapt to. Although the disintegration might take many hundreds of years, this process could be triggered irreversibly in the next century. The relative magnitude of feedback processes involved in cycling of carbon through the oceans and the terrestrial biosphere is shown to be distorted by increasing temperatures. Saturation and decline of the net sink effect of the terrestrial biosphere—which is projected to occur over the next century—in step with similar processes, could lead to dominance of positive feedbacks over negative ones and strong amplification of the warming trend. [19.6.3.2]

8. Information Needs

Although progress has been made, considerable gaps in knowledge remain regarding exposure, sensitivity, adaptability, and vulnerability of physical, ecological, and social systems to climate change. Advances in these areas are priorities for advancing understanding of potential consequences of climate change for human society and the natural world, as well as to support analyses of possible responses.

Exposure. Advances in methods for projecting exposures to climate stimuli and other nonclimate stresses at finer spatial scales are needed to improve understanding of potential consequences of climate change, including regional differences, and stimuli to which systems may need to adapt. Work in this area should draw on results from research on system sensitivity, adaptability, and vulnerability to identify the types of climate stimuli and nonclimate stresses that affect systems most. This research is particularly needed in developing countries, many of which lack historical data, adequate monitoring systems, and research and development capabilities. Developing local capacity in environmental assessment and management will increase investment effectiveness. Methods of investigating possible changes in the frequency and intensity of extreme climate events, climate variability, and large-scale, abrupt changes in the Earth system such as slowing or shutdown of thermohaline circulation of oceans are priorities. Work also is needed to advance understanding of how social and economic factors influence the exposures of different populations.

Sensitivity. Sensitivity to climate stimuli is still poorly quantified for many natural and human systems. Responses of systems to climate change are expected to include strong nonlinearities, discontinuous or abrupt responses, time-varying responses, and complex interactions with other systems. However, quantification of the curvature, thresholds, and

interactions of system responses is poorly developed for many systems. Work is needed to develop and improve process-based, dynamic models of natural, social, and economic systems; to estimate model parameters of system responses to climate variables; and to validate model simulation results. This work should include use of observational evidence, paleo-observations where applicable, and long-term monitoring of systems and forces acting on them. Continued efforts to detect impacts of observed climate change is a priority for further investigation that can provide empirical information for understanding of system sensitivity to climate change

Adaptability. Progress has been made in the investigation of adaptive measures and adaptive capacity. However, work is needed to better understand the applicability of adaptation experiences with climate variability to climate change, to use this information to develop empirically based estimates of the effectiveness and costs of adaptation, and to develop predictive models of adaptive behavior that take into account decision making under uncertainty. Work also is needed to better understand the determinants of adaptive capacity and to use this information to advance understanding of differences in adaptive capacity across regions, nations, and socioeconomic groups, as well as how capacity may change through time. Advances in these areas are expected to be useful for identifying successful strategies for enhancing adaptation capacity in ways that can be complementary to climate change mitigation, sustainable development, and equity goals.

Vulnerability. Assessments of vulnerability to climate change are largely qualitative and address the sources and character of vulnerability. Further work is needed to integrate information about exposures, sensitivity, and adaptability to provide more detailed and quantitative information about the potential impacts of climate change and the relative degree of vulnerability of different regions, nations, and socioeconomic groups. Advances will require development and refinement of multiple measures or indices of vulnerability such as the number or percentage of persons, species, systems, or land area negatively or positively affected; changes in productivity of systems; the monetary value of economic welfare change in absolute and relative terms; and measures of distributional inequities.

Uncertainty. Large gaps remain in refining and applying methods for treating uncertainties, particularly with respect to providing scientific information for decisionmaking. Improvements are required in ways of expressing the likelihood, confidence, and range of uncertainty for estimates of outcomes, as well as how such estimates fit into broader ranges of uncertainty. Methods for providing "traceable accounts" of how any aggregated estimate is made from disaggregated information must be refined. More effort is needed to translate judgments into probability distributions in integrated assessment models.

Climate Change 2001:
Mitigation

Working Group III Summaries

Summary for Policymakers
A Report of Working Group III of the Intergovernmental Panel on Climate Change

Technical Summary of the Working Group III Report
A Report accepted by Working Group III of the Intergovernmental Panel on Climate Change but not approved in detail

Part of the Working Group III contribution to the Third Assessment Report
of the Intergovernmental Panel on Climate Change

Contents

Climate Change 2001:
Mitigation

Summary for Policymakers

A Report of Working Group III of the Intergovernmental Panel on Climate Change

This summary, approved in detail at the Sixth Session of IPCC Working Group III (Accra, Ghana, 28 February-3 March 2001), represents the formally agreed statement of the IPCC concerning climate change mitigation.

Based on a draft prepared by:

Tariq Banuri, Terry Barker, Igor Bashmakov, Kornelis Blok, Daniel Bouille, Renate Christ, Ogunlade Davidson, Jae Edmonds, Ken Gregory, Michael Grubb, Kirsten Halsnaes, Tom Heller, Jean-Charles Hourcade, Catrinus Jepma, Pekka Kauppi, Anil Markandya, Bert Metz, William Moomaw, Jose Roberto Moreira, Tsuneyuki Morita, Nebojsa Nakicenovic, Lynn Price, Richard Richels, John Robinson, Hans Holger Rogner, Jayant Sathaye, Roger Sedjo, Priyaradshi Shukla, Leena Srivastava, Rob Swart, Ferenc Toth, John Weyant

Introduction

1. *This report assesses the scientific, technological, environmental, economic and social aspects of the mitigation of climate change.* Research in climate change mitigation[1] has continued since the publication of the IPCC Second Assessment Report (SAR), taking into account political changes such as the agreement on the Kyoto Protocol to the United Nations Framework Convention on Climate Change (UNFCCC) in 1997, and is reported on here. The Report also draws on a number of IPCC Special Reports, notably the Special Report on Aviation and the Global Atmosphere, the Special Report on Methodological and Technological Issues in Technology Transfer (SRTT), the Special Report on Emissions Scenarios (SRES), and the Special Report on Land Use, Land Use Change and Forestry (SRLULUCF).

The Nature of the Mitigation Challenge

2. *Climate change[2] is a problem with unique characteristics.* It is global, long-term (up to several centuries), and involves complex interactions between climatic, environmental, economic, political, institutional, social and technological processes. This may have significant international and intergenerational implications in the context of broader societal goals such as equity and sustainable development. Developing a response to climate change is characterized by decision-making under uncertainty and risk, including the possibility of non-linear and/or irreversible changes (Sections 1.2.5, 1.3, 10.1.2, 10.1.4, 10.4.5).[3]

3. *Alternative development paths[4] can result in very different greenhouse gas emissions.* The SRES and the mitigation scenarios assessed in this report suggest that the type, magnitude, timing and costs of mitigation depend on different national circumstances and socio-economic, and technological development paths and the desired level of greenhouse gas concentration stabilization in the atmosphere (see Figure SPM-1 for an example for total CO_2 emissions). Development paths leading to low emissions depend on a wide range of policy choices and require major policy changes in areas other than climate change (Sections 2.2.2, 2.3.2, 2.4.4, 2.5).

4. *Climate change mitigation will both be affected by, and have impacts on, broader socio-economic policies and trends, such as those relating to development, sustainability and equity.* Climate mitigation policies may promote sustainable development when they are consistent with such broader societal objectives. Some mitigation actions may yield extensive benefits in areas outside of climate change: for example, they may reduce health problems; increase employment; reduce negative environmental impacts (like air pollution); protect and enhance forests, soils and watersheds; reduce those subsidies and taxes which enhance greenhouse gas emissions; and induce technological change and diffusion, contributing to wider goals of sustainable development. Similarly, development paths that meet sustainable development objectives may result in lower levels of greenhouse gas emissions (Sections 1.3, 1.4, 2.2.3, 2.4.4, 2.5, 7.2.2, 8.2.4).

5. *Differences in the distribution of technological, natural and financial resources among and within nations and regions, and between generations, as well as differences in mitigation costs, are often key considerations in the analysis of climate change mitigation options.* Much of the debate about the future differentiation of contributions of countries to mitigation and related equity issues also considers these circumstances[5]. The challenge of addressing climate change raises an important issue of equity, namely the extent to which the impacts of climate change or mitigation policies create or exacerbate inequities both within and across nations and regions. Greenhouse gas stabilization scenarios assessed in this report (except those where stabilization occurs without new climate policies, e.g. B1) assume that developed countries and countries with economies in transition limit and reduce their greenhouse gas emissions first.[6]

6. *Lower emissions scenarios require different patterns of energy resource development.* Figure SPM-2 compares the cumulative carbon emissions between 1990 and 2100 for various SRES scenarios to carbon contained in global fossil fuel reserves and resources[7]. This figure shows that there are abundant fossil fuel resources that will not limit carbon emissions during the 21st century. However, different from the relatively large coal and unconventional oil and gas deposits, the carbon in proven

---

[1] Mitigation is defined here as an anthropogenic intervention to reduce the sources of greenhouse gases or enhance their sinks.

[2] *Climate change* in IPCC usage refers to any change in climate over time, whether due to natural variability or as a result of human activity. This usage differs from that in the UNFCCC, where *climate change* refers to a change of climate that is attributed directly or indirectly to human activity that alters the composition of the global atmosphere and that is in addition to natural climate variability observed over comparable time periods.

[3] Section numbers refer to WGIII TAR.

[4] In this report "alternative development paths" refer to a variety of possible scenarios for societal values and consumption and production patterns in all countries, including but not limited to a continuation of today's trends. These paths do not include additional climate initiatives which means that no scenarios are included that explicitly assume implementation of the UNFCCC or the emission targets of the Kyoto Protocol, but do include assumptions about other policies that influence greenhouse gas emissions indirectly.

[5] Approaches to equity have been classified into a variety of categories, including those based on allocation, outcome, process, rights, liability, poverty, and opportunity, reflecting the diverse expectations of fairness used to judge policy processes and the corresponding outcomes (Sections 1.3, 10.2).

[6] Emissions from all regions diverge from baselines at some point. Global emissions diverge earlier and to a greater extent as stabilization levels are lower or underlying scenarios are higher. Such scenarios are uncertain, do not provide information on equity implications and how such changes may be achieved or who may bear any costs incurred.

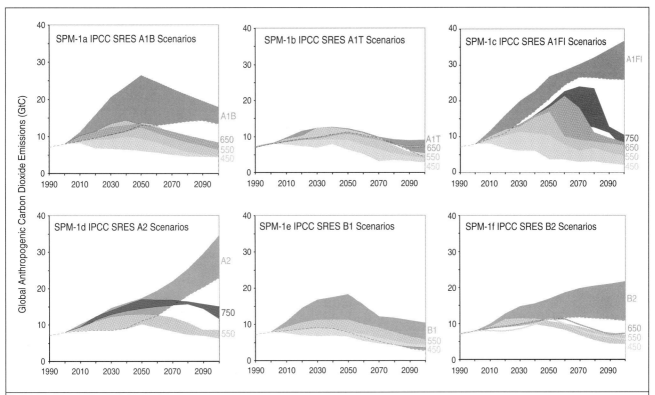

Figure SPM-1: Comparison of reference and stabilization scenarios. The figure is divided into six parts, one for each of the reference scenario groups from the Special Report on Emissions Scenarios (SRES, see Box SPM-1). Each part of the figure shows the range of total global CO$_2$ emissions (gigatonnes of carbon (GtC)) from all anthropogenic sources for the SRES reference scenario group (shaded in grey) and the ranges for the various mitigation scenarios assessed in the TAR leading to stabilization of CO$_2$ concentrations at various levels (shaded in colour). Scenarios are presented for the A1 family subdivided into three groups (the balanced A1B group (Figure SPM-1a), non-fossil fuel A1T (Figure SPM-1b) and the fossil intensive A1FI (Figure SPM-1c)) with stabilization of CO$_2$ concentrations at 450, 550, 650 and 750 ppmv; for the A2 group with stabilization at 550 and 750 ppmv in Figure SPM-1d, the B1 group with stabilization at 450 and 550 ppmv in Figure SPM-1e, and the B2 group with stabilization at 450, 550 and 650 ppmv in Figure SPM-1f. The literature is not available to assess 1000 ppmv stabilization scenarios. The figure illustrates that the lower the stabilization level and the higher the baseline emissions, the wider the gap. The difference between emissions in different scenario groups can be as large as the gap between reference and stabilization scenarios within one scenario group. The dotted lines depict the boundaries of the ranges where they overlap.

conventional oil and gas reserves, or in conventional oil resources, is much less than the cumulative carbon emissions associated with stabilization of carbon dioxide at levels of 450 ppmv or higher (the reference to a particular concentration level does not imply an agreed-upon desirability of stabilization at this level). These resource data may imply a change in the energy mix and the introduction of new sources of energy during the 21st century. The choice of energy mix and associated investment will determine whether,

and if so, at what level and cost, greenhouse concentrations can be stabilized. Currently most such investment is directed towards discovering and developing more conventional and unconventional fossil resources (Sections 2.5.1, 2.5.2, 3.8.3, 8.4).

Options to Limit or Reduce Greenhouse Gas Emissions and Enhance Sinks

7. *Significant technical progress relevant to greenhouse gas emissions reduction has been made since the SAR in 1995 and has been faster than anticipated.* Advances are taking place in a wide range of technologies at different stages of development, e.g., the market introduction of wind turbines, the rapid elimination of industrial by-product gases such as N$_2$O from adipic acid production and perfluorocarbons from aluminium production, efficient hybrid engine cars, the advancement of fuel cell technology, and the demonstration of underground carbon dioxide storage. Technological options for emissions reduction include

---

[7] Reserves are those occurrences that are identified and measured as economically and technically recoverable with current technologies and prices. Resources are those occurrences with less certain geological and/or economic characteristics, but which are considered potentially recoverable with foreseeable technological and economic developments. The resource base includes both categories. On top of that, there are additional quantities with unknown certainty of occurrence and/or with unknown or no economic significance in the foreseeable future, referred to as "additional occurrences" (SAR, Working Group II). Examples of unconventional fossil fuel resources include tar sands, shale oil, other heavy oil, coal bed methane, deep geopressured gas, gas in acquifers, *etc.*

Box SPM-1. The Emissions Scenarios of the IPCC Special Report on Emissions Scenarios (SRES)

A1. The A1 storyline and scenario family describes a future world of very rapid economic growth, global population that peaks in mid-century and declines thereafter, and the rapid introduction of new and more efficient technologies. Major underlying themes are convergence among regions, capacity building and increased cultural and social interactions, with a substantial reduction in regional differences in per capita income. The A1 scenario family develops into three groups that describe alternative directions of technological change in the energy system. The three A1 groups are distinguished by their technological emphasis: fossil intensive (A1FI), non-fossil energy sources (A1T), or a balance across all sources (A1B) (where balanced is defined as not relying too heavily on one particular energy source, on the assumption that similar improvement rates apply to all energy supply and end use technologies).

A2. The A2 storyline and scenario family describes a very heterogeneous world. The underlying theme is self-reliance and preservation of local identities. Fertility patterns across regions converge very slowly, which results in continuously increasing population. Economic development is primarily regionally oriented and per capita economic growth and technological change more fragmented and slower than other storylines.

B1. The B1 storyline and scenario family describes a convergent world with the same global population, that peaks in mid-

century and declines thereafter, as in the A1 storyline, but with rapid change in economic structures toward a service and information economy, with reductions in material intensity and the introduction of clean and resource-efficient technologies. The emphasis is on global solutions to economic, social and environmental sustainability, including improved equity, but without additional climate initiatives.

B2. The B2 storyline and scenario family describes a world in which the emphasis is on local solutions to economic, social and environmental sustainability. It is a world with continuously increasing global population, at a rate lower than A2, intermediate levels of economic development, and less rapid and more diverse technological change than in the B1 and A1 storylines. While the scenario is also oriented towards environmental protection and social equity, it focuses on local and regional levels.

An illustrative scenario was chosen for each of the six scenario groups A1B, A1FI, A1T, A2, B1 and B2. All should be considered equally sound.

The SRES scenarios do not include additional climate initiatives, which means that no scenarios are included that explicitly assume implementation of the United Nations Framework Convention on Climate Change or the emissions targets of the Kyoto Protocol.

---

improved efficiency of end use devices and energy conversion technologies, shift to low-carbon and renewable biomass fuels, zero-emissions technologies, improved energy management, reduction of industrial by-product and process gas emissions, and carbon removal and storage (Section 3.1, 4.7).

Table SPM-1 summarizes the results from many sectoral studies, largely at the project, national and regional level with some at the global levels, providing estimates of potential greenhouse gas emission reductions in the 2010 to 2020 timeframe. Some key findings are:

- Hundreds of technologies and practices for end-use energy efficiency in buildings, transport and manufacturing industries account for more than half of this potential (Sections 3.3, 3.4, 3.5).
- At least up to 2020, energy supply and conversion will remain dominated by relatively cheap and abundant fossil fuels. Natural gas, where transmission is economically feasible, will play an important role in emission reduction together with conversion efficiency improvement, and greater use of combined cycle and/or co-generation plants (Section 3.8.4).
- Low-carbon energy supply systems can make an important contribution through biomass from forestry and agricultural

by-products, municipal and industrial waste to energy, dedicated biomass plantations, where suitable land and water are available, landfill methane, wind energy and hydropower, and through the use and lifetime extension of nuclear power plants. After 2010, emissions from fossil and/or biomass-fueled power plants could be reduced substantially through pre- or post-combustion carbon removal and storage. Environmental, safety, reliability and proliferation concerns may constrain the use of some of these technologies (Section 3.8.4).
- In agriculture, methane and nitrous oxide emissions can be reduced, such as those from livestock enteric fermentation, rice paddies, nitrogen fertilizer use and animal wastes (Section 3.6).
- Depending on application, emissions of fluorinated gases can be minimized through process changes, improved recovery, recycling and containment, or avoided through the use of alternative compounds and technologies (Section 3.5 and Chapter 3 Appendix).

The potential emissions reductions found in Table SPM-1 for sectors were aggregated to provide estimates of global potential emissions reductions taking account of potential overlaps between and within sectors and technologies to the extent possible given

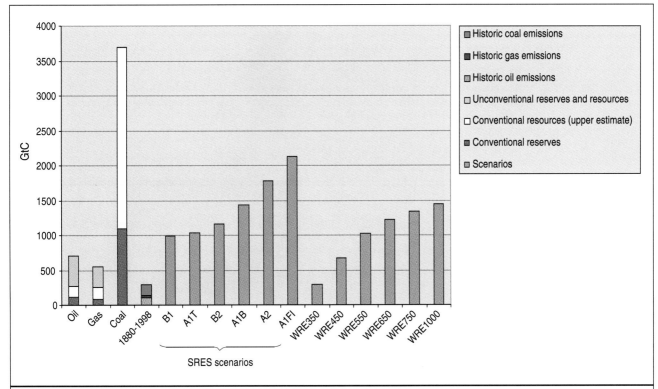

Figure SPM-2: Carbon in oil, gas and coal reserves and resources compared with historic fossil fuel carbon emissions 1860–1998, and with cumulative carbon emissions from a range of SRES scenarios and TAR stabilization scenarios up until 2100. Data for reserves and resources are shown in the left hand columns (Section 3.8.2). Unconventional oil and gas includes tar sands, shale oil, other heavy oil, coal bed methane, deep geopressured gas, gas in acquifers, *etc*. Gas hydrates (clathrates) that amount to an estimated 12,000GtC are not shown. The scenario columns show both SRES reference scenarios as well as scenarios which lead to stabilization of CO_2 concentrations at a range of levels. Note that if by 2100 cumulative emissions associated with SRES scenarios are equal to or smaller than those for stabilization scenarios, this does not imply that these scenarios equally lead to stabilization.

the information available in the underlying studies. Half of these potential emissions reductions may be achieved by 2020 with direct benefits (energy saved) exceeding direct costs (net capital, operating, and maintenance costs), and the other half at a net direct cost of up to US$100/t$C_{eq}$ (at 1998 prices). These cost estimates are derived using discount rates in the range of 5% to 12%, consistent with public sector discount rates. Private internal rates of return vary greatly, and are often significantly higher, affecting the rate of adoption of these technologies by private entities.

Depending on the emissions scenario this could allow global emissions to be reduced below 2000 levels in 2010–2020 at these net direct costs. Realizing these reductions involve additional implementation costs, which in some cases may be substantial, the possible need for supporting policies (such as those described in Paragraph 18), increased research and development, effective technology transfer and overcoming other barriers (Paragraph 17). These issues, together with costs and benefits not included in this evaluation are discussed in Paragraphs 11, 12 and 13.

The various global, regional, national, sector and project studies assessed in this report have different scopes and assumptions.

Studies do not exist for every sector and region. The range of emissions reductions reported in Table SPM-1 reflects the uncertainties (see Box SPM-2) of the underlying studies on which they are based (Sections 3.3-3.8)

8. *Forests, agricultural lands, and other terrestrial ecosystems offer significant carbon mitigation potential. Although not necessarily permanent, conservation and sequestration of carbon may allow time for other options to be further developed and implemented.* Biological mitigation can occur by three strategies: (a) conservation of existing carbon pools, (b) sequestration by increasing the size of carbon pools, and (c) substitution of sustainably produced biological products, e.g. wood for energy intensive construction products and biomass for fossil fuels (Sections 3.6, 4.3). Conservation of threatened carbon pools may help to avoid emissions, if leakage can be prevented, and can only become sustainable if the socio-economic drivers for deforestation and other losses of carbon pools can be addressed. Sequestration reflects the biological dynamics of growth, often starting slowly, passing through a maximum, and then declining over decades to centuries.

Table SPM-1: Estimates of potential global greenhouse gas emission reductions in 2010 and in 2020 (Sections 3.3-3.8 and Chapter 3 Appendix).

Sector	Historic emissions in 1990 (MtC_{eq}/yr)	Historic C_{eq} annual growth rate over 1990-1995 (%)	Potential emission reductions in 2010 (MtC_{eq}/yr)	Potential emission reductions in 2020 (MtC_{eq}/yr)	Net direct costs per tonne of carbon avoided
Buildings[a] CO_2 only	1,650	1.0	700–750	1,000–1,100	Most reductions are available at negative net direct costs.
Transport CO_2 only	1,080	2.4	100–300	300–700	Most studies indicate net direct costs less than US$25/tC but two suggest net direct costs will exceed US$50/tC.
Industry CO_2 only – Energy efficiency – Material efficiency	2,300	0.4	300–500 ~200	700–900 ~600	More than half available at net negative direct costs. Costs are uncertain.
Industry Non-CO_2 gases	170		~100	~100	N_2O emissions reduction costs are US$0–10/tC$_{eq}$.
Agriculture[b] CO_2 only Non-CO_2 gases	210 1,250–2,800	n.a.	150–300	350–750	Most reductions will cost between US$0–100/tC$_{eq}$ with limited opportunities for negative net direct cost options.
Waste[b] CH_4 only	240	1.0	~200	~200	About 75% of the savings as methane recovery from landfills at net negative direct cost; 25% at a cost of US$20/tC$_{eq}$.
Montreal Protocol replacement applications Non-CO_2 gases	0	n.a.	~100	n.a.	About half of reductions due to difference in study baseline and SRES baseline values. Remaining half of the reductions available at net direct costs below $200/tC$_{eq}$.
Energy supply and conversion[c] CO_2 only	(1,620)	1.5	50–150	350–700	Limited net negative direct cost options exist; many options are available for less than US$100/tC$_{eq}$.
Total	**6,900–8,400[d]**		**1,900–2,600[e]**	**3,600–5,050[e]**	

[a] Buildings include appliances, buildings, and the building shell.

[b] The range for agriculture is mainly caused by large uncertainties about CH_4, N_2O, and soil-related emissions of CO_2. Waste is dominated by methane landfill and the other sectors could be estimated with more precision as they are dominated by fossil CO_2.

[c] Included in sector values above. Reductions include electricity generation options only (fuel switching to gas/nuclear, CO_2 capture and storage, improved power station efficiencies, and renewables).

[d] Total includes all sectors reviewed in WGIII TAR Chapter 3 for all six gases. It excludes non-energy related sources of CO_2 (cement production, 160 Mt C; gas flaring, 60 Mt C; and land-use change, 600–1,400 Mt C) and energy used for conversion of fuels in the end-use sector totals (630 Mt C). If petroleum refining and coke oven gas were added, global year 1990 CO_2 emissions of 7,100 Mt C would increase by 12%. Note that forestry emissions and their carbon sink mitigation options are not included.

[e] The baseline SRES scenarios (for six gases included in the Kyoto Protocol) project a range of emissions of 11,500–14,000 Mt C$_{eq}$ for the year 2010 and of 12,000–16,000 Mt C$_{eq}$ for the year 2020. The emissions reduction estimates are most compatible with baseline emissions trends in the SRES B2 scenario. The potential reductions take into account regular turnover of capital stock. They are not limited to cost-effective options, but exclude options with costs above US$100 t C$_{eq}$ (except for Montreal Protocol gases) or options that will not be adopted through the use of generally accepted policies

Conservation and sequestration result in higher carbon stocks, but can lead to higher future carbon emissions if these ecosystems are severely disturbed by either natural or direct/indirect human-induced disturbances. Even though natural disturbances are normally followed by re-sequestration, activities to manage such disturbances can play an important role in limiting carbon emissions. Substitution benefits can, in principle, continue indefinitely. Appropriate management of land for crop, timber and sustainable bio-energy production, may increase benefits for climate change mitigation. Taking into account competition for land use and the SAR and SRLULUCF assessments, the estimated global potential of biological mitigation options is in the order of 100GtC (cumulative), although there are substantial uncertainties associated with this estimate, by 2050, equivalent to about 10% to 20% of potential fossil fuel emissions during that period. Realization of this potential depends upon land and water availability as well as the rates of adoption of different land management practices. The largest biological potential for atmospheric carbon mitigation is in subtropical and tropical regions. Cost estimates reported to date of biological mitigation vary significantly from US$0.1/tC to about US$20/tC in several tropical countries and from US$20/tC to US$100/tC in non-tropical countries. Methods of financial analysis and carbon accounting have not been comparable. Moreover, the cost calculations do not cover, in many instances, *inter alia*, costs for infrastructure, appropriate discounting, monitoring, data collection and implementation costs, opportunity costs of land and maintenance, or other recurring costs, which are often excluded or overlooked. The lower end of the ranges are biased downwards, but understanding and treatment of costs is improving over time. These biological mitigation options may have social, economic and environmental benefits beyond reductions in atmospheric CO_2, if implemented appropriately. (e.g., biodiversity, watershed protection, enhancement of sustainable land management and rural employment). However, if implemented inappropriately, they may pose risks of negative impacts (e.g., loss of biodiversity, community disruption and ground-water pollution). Biological mitigation options may reduce or increase non-CO_2 greenhouse gas emissions (Sections 4.3, 4.4).

Box SPM-2. Approaches to Estimating Costs and Benefits, and their Uncertainties

For a variety of factors, significant differences and uncertainties surround specific quantitative estimates of the costs and benefits of mitigation options. The SAR described two categories of approaches to estimating costs and benefits: bottom-up approaches, which build up from assessments of specific technologies and sectors, such as those described in Paragraph 7, and top-down modelling studies, which proceed from macroeconomic relationships, such as those discussed in Paragraph 13. These two approaches lead to differences in the estimates of costs and benefits, which have been narrowed since the SAR. Even if these differences were resolved, other uncertainties would remain. The potential impact of these uncertainties can be usefully assessed by examining the effect of a change in any given assumption on the aggregate cost results, provided any correlation between variables is adequately dealt with.

9. *There is no single path to a low emission future and countries and regions will have to choose their own path. Most model results indicate that known technological options[8] could achieve a broad range of atmospheric CO_2 stabilization levels, such as 550ppmv, 450ppmv or below over the next 100 years or more, but implementation would require associated socio-economic and institutional changes.* To achieve stabilization at these levels, the scenarios suggest that a very significant reduction in world carbon emissions per unit of GDP from 1990 levels will be necessary. Technological improvement and technology transfer play a critical role in the stabilization scenarios assessed in this report. For the crucial energy sector, almost all greenhouse gas mitigation and concentration stabilization scenarios are characterized by the introduction of efficient technologies for both energy use and supply, and of low- or no-carbon energy. However, no single technology option will provide all of the emissions reductions needed. Reduction options in non-energy sources and non-CO_2 greenhouse gases will also provide significant potential for reducing emissions. Transfer of technologies between countries and regions will widen the choice of options at the regional level and economies of scale and learning will lower the costs of their adoption (Sections 2.3.2, 2.4, 2.5).

10. *Social learning and innovation, and changes in institutional structure could contribute to climate change mitigation.* Changes in collective rules and individual behaviours may have significant effects on greenhouse gas emissions, but take place within a complex institutional, regulatory and legal setting. Several studies suggest that current incentive systems can encourage resource intensive production and consumption patterns that increase greenhouse gas emissions in all sectors, e.g. transport and housing. In the shorter term, there are opportunities to influence through social innovations individual and organizational behaviours. In the longer term such innovations, in combination with technological change, may further enhance socio-economic potential, particularly if preferences and cultural norms shift towards lower emitting and sustainable behaviours. These innovations frequently meet with resistance, which may be addressed by encouraging greater public participation in the decision-making processes. This can help contribute to new approaches to sustainability and equity (Sections 1.4.3, 5.3.8, 10.3.2, 10.3.4).

---

[8] "Known technological options" refer to technologies that exist in operation or pilot plant stage today, as referenced in the mitigation scenarios discussed in this report. It does not include any new technologies that will require drastic technological breakthroughs. In this way it can be considered to be a conservative estimate, considering the length of the scenario period.

The Costs and Ancillary[9] Benefits of Mitigation Actions

11. *Estimates of cost and benefits of mitigation actions differ because of (i) how welfare is measured, (ii) the scope and methodology of the analysis, and (iii) the underlying assumptions built into the analysis. As a result, estimated costs and benefits may not reflect the actual costs and benefits of implementing mitigation actions.* With respect to (i) and (ii), costs and benefits estimates, *inter alia*, depend on revenue recycling, and whether and how the following are considered: implementation and transaction cost, distributional impacts, multiple gases, land-use change options, benefits of avoided climate change, ancillary benefits, no regrets opportunities[10] and valuation of externalities and non-market impacts. Assumptions include, *inter alia*:

- Demographic change, the rate and structure of economic growth; increases in personal mobility, technological innovation such as improvements in energy efficiency and the availability of low-cost energy sources, flexibility of capital investments and labour markets, prices, fiscal distortions in the no-policy (baseline) scenario.
- The level and timing of the mitigation target.
- Assumptions regarding implementation measures, e.g. the extent of emissions trading, the Clean Development Mechanism (CDM) and Joint Implementation (JI), regulation, and voluntary agreements[11] and the associated transaction costs.
- Discount rates: the long time scales make discounting assumptions critical and there is still no consensus on appropriate long-term rates, though the literature shows increasing attention to rates that decline over time and hence give more weight to benefits that occur in the long term. These discount rates should be distinguished from the higher rates that private agents generally use in market transactions.

(Sections 7.2, 7.3, 8.2.1, 8.2.2, 9.4)

12. *Some sources of greenhouse gas emissions can be limited at no or negative net social cost to the extent that policies can exploit no regrets opportunities* (Sections 7.3.4, 9.2.1):

- *Market imperfections.* Reduction of existing market or institutional failures and other barriers that impede adoption of cost-effective emission reduction measures, can lower private costs compared to current practice. This can also reduce private costs overall.
- *Ancillary benefits.* Climate change mitigation measures will have effects on other societal issues. For example, reducing carbon emissions in many cases will result in the simultaneous reduction in local and regional air pollution. It is likely that mitigation strategies will also affect transportation, agriculture, land-use practices and waste management and will have an impact on other issues of social concern, such as employment, and energy security. However, not all of the effects will be positive; careful policy selection and design can better ensure positive effects and minimize negative impacts. In some cases, the magnitude of ancillary benefits of mitigation may be comparable to the costs of the mitigating measures, adding to the no regrets potential, although estimates are difficult to make and vary widely (Sections 7.3.3, 8.2.4, 9.2.2-9.2.8, 9.2.10).
- *Double dividend.* Instruments (such as taxes or auctioned permits) provide revenues to the government. If used to finance reductions in existing distortionary taxes ("revenue recycling"), these revenues reduce the economic cost of achieving greenhouse gas reductions. The magnitude of this offset depends on the existing tax structure, type of tax cuts, labour market conditions, and method of recycling. Under some circumstances, it is possible that the economic benefits may exceed the costs of mitigation (Sections 7.3.3, 8.2.2, 9.2.1)

13. *The cost estimates for Annex B countries to implement the Kyoto Protocol vary between studies and regions as indicated in Paragraph 11, and depend strongly upon the assumptions regarding the use of the Kyoto mechanisms, and their interactions with domestic measures.* The great majority of global studies reporting and comparing these costs use international energy-economic models. Nine of these studies suggest the following GDP impacts[12] (Sections 7.3.5, 8.3.1, 9.2.3, 10.4.4):

Annex II countries[13]: *In the absence of emissions trading between Annex B countries*[14], *the majority of global studies show reductions in projected GDP of about 0.2% to 2% in 2010 for different Annex II regions. With full emissions trading between Annex B countries, the estimated reductions in 2010*

---

[9] Ancillary benefits are the ancillary, or side effects, of policies aimed exclusively at climate change mitigation. Such policies have an impact not only on greenhouse gas emissions, but also on resource use efficiency, like reduction in emissions of local and regional air pollutants associated with fossil fuel use, and on issues such as transportation, agriculture, land-use practices, employment, and fuel security. Sometimes these benefits are referred to as "ancillary impacts" to reflect that in some cases the benefits may be negative.

[10] In this report, as in the SAR, no regrets opportunities are defined as those options whose benefits such as reduced energy costs and reduced emissions of local/regional pollutants equal or exceed their costs to society, excluding the benefits of avoided climate change.

[11] A voluntary agreement is an agreement between a government authority and one or more private parties, as well as a unilateral commitment that is recognized by the public authority, to achieve environmental objectives or to improve environmental performance beyond compliance.

[12] Many other studies incorporating more precisely the country specifics and diversity of targeted policies provide a wider range of net cost estimates (Section 8.2.2).

[13] Annex II countries: Group of countries included in Annex II to the UNFCCC, including all developed countries in the Organisation of Economic Co-operation and Development.

[14] Annex B countries: Group of countries included in Annex B in the Kyoto Protocol that have agreed to a target for their greenhouse gas emissions, including all the Annex I countries (as amended in 1998) but Turkey and Belarus.

are between 0.1% and 1.1% of projected GDP[15]. These studies encompass a wide range of assumptions as listed in Paragraph 11. Models whose results are reported in this paragraph assume full use of emissions trading without transaction cost. Results for cases that do not allow Annex B trading assume full domestic trading within each region. Models do not include sinks or non-CO_2 greenhouse gases. They do not include the CDM, negative cost options, ancillary benefits, or targeted revenue recycling.

For all regions costs are also influenced by the following factors:

- Constraints on the use of Annex B trading, high transaction costs in implementing the mechanisms, and inefficient domestic implementation could raise costs.
- Inclusion in domestic policy and measures of the no regrets possibilities[10] identified in Paragraph 12, use of the CDM, sinks, and inclusion of non-CO_2 greenhouse gases, could lower costs. Costs for individual countries can vary more widely.

The models show that the Kyoto mechanisms are important in controlling risks of high costs in given countries, and thus can complement domestic policy mechanisms. Similarly, they can minimize risks of inequitable international impacts and help to level marginal costs. The global modelling studies reported above show national marginal costs to meet the Kyoto targets from about US$20/tC up to US$600/tC without trading, and a range from about US$15/tC up to US$150/tC with Annex B trading. The cost reductions from these mechanisms may depend on the details of implementation, including the compatibility of domestic and international mechanisms, constraints, and transaction costs.

Economies in transition: For most of these countries, GDP effects range from negligible to a several per cent increase. This reflects opportunities for energy efficiency improvements not available to Annex II countries. Under assumptions of drastic energy efficiency improvement and/or continuing economic recessions in some countries, the assigned amounts may exceed projected emissions in the first commitment period. In this case, models show increased GDP due to revenues from trading assigned amounts. However, for some economies in transition, implementing the Kyoto Protocol will have similar impact on GDP as for Annex II countries.

14. *Cost-effectiveness studies with a century timescale estimate that the costs of stabilizing CO_2 concentrations in the*

atmosphere increase as the concentration stabilization level declines. Different baselines can have a strong influence on absolute costs. While there is a moderate increase in the costs when passing from a 750ppmv to a 550ppmv concentration stabilization level, there is a larger increase in costs passing from 550ppmv to 450ppmv unless the emissions in the baseline scenario are very low. These results, however, do not incorporate carbon sequestration, gases other than CO_2 and did not examine the possible effect of more ambitious targets on induced technological change[16]. Costs associated with each concentration level depend on numerous factors including the rate of discount, distribution of emission reductions over time, policies and measures employed, and particularly the choice of the baseline scenario: for scenarios characterized by a focus on local and regional sustainable development for example, total costs of stabilizing at a particular level are significantly lower than for other scenarios[17] (Sections 2.5.2, 8.4.1, 10.4.6).

15. *Under any greenhouse gas mitigation effort, the economic costs and benefits are distributed unevenly between sectors; to a varying degree, the costs of mitigation actions could be reduced by appropriate policies.* In general, it is easier to identify activities, which stand to suffer economic costs compared to those which may benefit, and the economic costs are more immediate, more concentrated and more certain. Under mitigation policies, coal, possibly oil and gas, and certain energy-intensive sectors, such as steel production, are most likely to suffer an economic disadvantage. Other industries including renewable energy industries and services can be expected to benefit in the long term from price changes and the availability of financial and other resources that would otherwise have been devoted to carbon-intensive sectors. Policies such as the removal of subsidies from fossil fuels may increase total societal benefits through gains in economic efficiency, while use of the Kyoto mechanisms could be expected to reduce the net economic cost of meeting Annex B targets. Other types of policies, for example exempting carbon-intensive industries, redistribute the costs but increase total societal costs at the same time. Most studies show that the distributional effects of a carbon tax can have negative income effects on low-income groups unless the tax revenues are used directly or indirectly to compensate such effects (Section 9.2.1).

---

[15] Many metrics can be used to present costs. For example, if the annual costs to developed countries associated with meeting Kyoto targets with full Annex B trading are in the order of 0.5% of GDP, this represents US$125 billion (1000 million) per year, or US$125 per person per year by 2010 in Annex II (SRES assumptions). This corresponds to an impact on economic growth *rates* over ten years of less than 0.1 percentage point.

[16] Induced technological change is an emerging field of inquiry. None of the literature reviewed in TAR on the relationship between the century-scale CO_2 concentrations and costs, reported results for models employing induced technological change. Models with induced technological change under some circumstances show that century-scale concentrations can differ, with similar GDP growth but under different policy regimes (Section 8.4.1.4).

[17] See Figure SPM-1 for the influence of reference scenarios on the magnitude of the required mitigation effort to reach a given stabilization level.

16. *Emission constraints in Annex I countries have well established, albeit varied "spillover" effects[18] on non-Annex I countries* (Sections 8.3.2, 9.3).

- *Oil-exporting, non-Annex I countries: Analyses report costs differently, including, inter alia, reductions in projected GDP and reductions in projected oil revenues[19].* The study reporting the lowest costs shows reductions of 0.2% of projected GDP with no emissions trading, and less than 0.05% of projected GDP with Annex B emissions trading in 2010[20]. The study reporting the highest costs shows reductions of 25% of projected oil revenues with no emissions trading, and 13% of projected oil revenues with Annex B emissions trading in 2010. These studies do not consider policies and measures[21] other than Annex B emissions trading, that could lessen the impact on non-Annex I, oil-exporting countries, and therefore tend to overstate both the costs to these countries and overall costs. The effects on these countries can be further reduced by removal of subsidies for fossil fuels, energy tax restructuring according to carbon content, increased use of natural gas, and diversification of the economies of non-Annex I, oil-exporting countries.

- *Other non-Annex I countries: They may be adversely affected by reductions in demand for their exports to OECD nations and by the price increase of those carbon-intensive and other products they continue to import. These countries may benefit from the reduction in fuel prices, increased exports of carbon-intensive products and the transfer of environmentally sound technologies and know-how.* The net balance for a given country depends on which of these factors dominates. Because of these complexities, the breakdown of winners and losers remains uncertain.

- *Carbon leakage[22]. The possible relocation of some carbon-intensive industries to non-Annex I countries and wider impacts on trade flows in response to changing prices may lead to leakage in the order of 5%-20%* (Section 8.3.2.2). Exemptions, for example for energy-intensive industries,

make the higher model estimates for carbon leakage unlikely, but would raise aggregate costs. The transfer of environmentally sound technologies and know-how, not included in models, may lead to lower leakage and especially on the longer term may more than offset the leakage.

Ways and Means for Mitigation

17. *The successful implementation of greenhouse gas mitigation options needs to overcome many technical, economic, political, cultural, social, behavioural and/or institutional barriers which prevent the full exploitation of the technological, economic and social opportunities of these mitigation options.* The potential mitigation opportunities and types of barriers vary by region and sector, and over time. This is caused by the wide variation in mitigation capacity. The poor in any country are faced with limited opportunities to adopt technologies or change their social behaviour, particularly if they are not part of a cash economy, and most countries could benefit from innovative financing and institutional reform and removing barriers to trade. In the industrialized countries, future opportunities lie primarily in removing social and behavioural barriers; in countries with economies in transition, in price rationalization; and in developing countries, in price rationalization, increased access to data and information, availability of advanced technologies, financial resources, and training and capacity building. Opportunities for any given country, however, might be found in the removal of any combination of barriers (Sections 1.5, 5.3, 5.4).

18. *National responses to climate change can be more effective if deployed as a portfolio of policy instruments to limit or reduce greenhouse gas emissions.* The portfolio of national climate policy instruments may include - according to national circumstances - emissions/carbon/energy taxes, tradable or non-tradable permits, provision and/or removal of subsidies, deposit/refund systems, technology or performance standards, energy mix requirements, product bans, voluntary agreements, government spending and investment, and support for research and development. Each government may apply different evaluation criteria, which may lead to different portfolios of instruments. The literature in general gives no preference for any particular policy instrument. Market based instruments may be cost-effective in many cases, especially where capacity to administer them is developed. Energy efficiency standards and performance regulations are widely used, and may be effective in many countries, and sometimes precede market based instruments. Voluntary agreements have recently been used more frequently, sometimes preceding the introduction of more stringent measures. Information campaigns, environmental labelling, and green marketing, alone or in combination with incentive subsidies, are increasingly emphasized to inform and shape consumer or producer behaviour. Government and/or privately supported research and development is important in advancing the long-term application and transfer of mitigation

---

[18] Spillover effects incorporate only economic effects, not environmental effects.

[19] Details of the six studies reviewed are found in Table 9-4 of the underlying report.

[20] These estimated costs can be expressed as differences in GDP growth rates over the period 2000–2010. With no emissions trading, GDP growth rate is reduced by 0.02 percentage points/year; with Annex B emissions trading, growth rate is reduced by less than 0.005 percentage points/year.

[21] These policies and measures include: those for non-CO_2 gases and non-energy sources of all gases; offsets from sinks; industry restructuring (e.g., from energy producer to supplier of energy services); use of OPEC's market power; and actions (e.g. of Annex B Parties) related to funding, insurance, and the transfer of technology. In addition, the studies typically do not include the following policies and effects that can reduce the total cost of mitigation: the use of tax revenues to reduce tax burdens or finance other mitigation measures; environmental ancillary benefits of reductions in fossil fuel use; and induced technological change from mitigation policies.

[22] Carbon leakage is defined here as the increase in emissions in non-Annex B countries due to implementation of reductions in Annex B, expressed as a percentage of Annex B reductions.

technologies beyond the current market or economic potential (Section 6.2).

19. *The effectiveness of climate change mitigation can be enhanced when climate policies are integrated with the non-climate objectives of national and sectorial policy development and be turned into broad transition strategies to achieve the long-term social and technological changes required by both sustainable development and climate change mitigation.* Just as climate policies can yield ancillary benefits that improve wellbeing, non-climate policies may produce climate benefits. It may be possible to significantly reduce greenhouse gas emissions by pursuing climate objectives through general socio-economic policies. In many countries, the carbon intensity of energy systems may vary depending on broader programmmes for energy infrastructure development, pricing, and tax policies. Adopting state-of-the-art environmentally sound technologies may offer particular opportunity for environmentally sound development while avoiding greenhouse gas intensive activities. Specific attention can foster the transfer of those technologies to small and medium size enterprises. Moreover, taking ancillary benefits into account in comprehensive national development strategies can lower political and institutional barriers for climate-specific actions (Sections 2.2.3, 2.4.4, 2.4.5, 2.5.1, 2.5.2, 10.3.2, 10.3.4).

20. *Co-ordinated actions among countries and sectors may help to reduce mitigation cost, address competitiveness concerns, potential conflicts with international trade rules, and carbon leakage. A group of countries that wants to limit its collective greenhouse gas emissions could agree to implement well-designed international instruments.* Instruments assessed in this report and being developed in the Kyoto Protocol are emissions trading; Joint Implementation (JI); the Clean Development Mechanism (CDM); other international instruments also assessed in this report include co-ordinated or harmonized emission/carbon/energy taxes; an emission/carbon/ energy tax; technology and product standards; voluntary agreements with industries; direct transfers of financial resources and technology; and co-ordinated creation of enabling environments such as reduction of fossil fuel subsidies. Some of these have been considered only in some regions to date (Sections 6.3, 6.4.2, 10.2.7, 10.2.8).

21. *Climate change decision-making is essentially a sequential process under general uncertainty.* The literature suggests that a prudent risk management strategy requires a careful consideration of the consequences (both environmental and economic), their likelihood and society's attitude toward risk. The latter is likely to vary from country to country and perhaps even from generation to generation. This report therefore confirms the SAR finding that the value of better information about climate change processes and impacts and society's responses to them is likely to be great. Decisions about near-term climate policies are in the process of being made while the concentration stabilization target is

still being debated. The literature suggests a step-by-step resolution aimed at stabilizing greenhouse gas concentrations. This will also involve balancing the risks of either insufficient or excessive action. The relevant question is not "what is the best course for the next 100 years", but rather "what is the best course for the near term given the expected long-term climate change and accompanying uncertainties" (Section 10.4.3).

22. *This report confirms the finding in the SAR that earlier actions, including a portfolio of emissions mitigation, technology development and reduction of scientific uncertainty, increase flexibility in moving towards stabilization of atmospheric concentrations of greenhouse gases. The desired mix of options varies with time and place.* Economic modelling studies completed since the SAR indicate that a gradual near-term transition from the world's present energy system towards a less carbon-emitting economy minimizes costs associated with premature retirement of existing capital stock. It also provides time for technology development, and avoids premature lock-in to early versions of rapidly developing low-emission technology. On the other hand, more rapid near-term action would decrease environmental and human risks associated with rapid climatic changes.

It would also stimulate more rapid deployment of existing low-emission technologies, provide strong near-term incentives to future technological changes that may help to avoid lock-in to carbon-intensive technologies, and allow for later tightening of targets should that be deemed desirable in light of evolving scientific understanding (Sections 2.3.2, 2.5.2, 8.4.1, 10.4.2, 10.4.3).

23. *There is an inter-relationship between the environmental effectiveness of an international regime, the cost-effectiveness of climate policies and the equity of the agreement.* Any international regime can be designed in a way that enhances both its efficiency and its equity. The literature assessed in this report on coalition formation in international regimes presents different strategies that support these objectives, including how to make it more attractive to join a regime through appropriate distribution of efforts and provision of incentives. While analysis and negotiation often focus on reducing system costs, the literature also recognizes that the development of an effective regime on climate change must give attention to sustainable development and non-economic issues (Sections 1.3, 10.2).

Gaps in Knowledge

24. *Advances have been made since previous IPCC assessments in the understanding of the scientific, technical, environmental, and economic and social aspects of mitigation of climate change. Further research is required, however, to strengthen future assessments and to reduce uncertainties as far as possible in order that sufficient information is available for policy*

making about responses to climate change, including research in developing countries.

The following are high priorities for further narrowing gaps between current knowledge and policy making needs:

· *Further exploration of the regional, country and sector specific potentials of technological and social innovation options.* This includes research on the short, medium and long-term potential and costs of both CO_2 and non-CO_2, non-energy mitigation options; understanding of technology diffusion across different regions; identifying opportunities in the area of social innovation leading to decreased greenhouse gas emissions; comprehensive analysis of the impact of mitigation measures on carbon flows in and out of the terrestrial system; and some basic inquiry in the area of geo-engineering.

· *Economic, social and institutional issues related to climate change mitigation in all countries.* Priority areas include: analysis of regionally specific mitigation options and barriers; the implications of equity assessments; appropriate methodologies and improved data sources for climate change mitigation and capacity building in the area of integrated assessment; strengthening future research and assessments, especially in the developing countries.

· *Methodologies for analysis of the potential of mitigation options and their cost, with special attention to comparability of results.* Examples include: characterizing and measuring barriers that inhibit greenhouse gas-reducing action; making mitigation modelling techniques more consistent, reproducible, and accessible; modelling technology learning; improving analytical tools for evaluating ancillary benefits, e.g. assigning the costs of abatement to greenhouse gases and to other pollutants; systematically analyzing the dependency of costs on baseline assumptions for various greenhouse gas stabilization scenarios; developing decision analytical frameworks for dealing with uncertainty as well as socio-economic and ecological risk in climate policy making; improving global models and studies, their assumptions and their consistency in the treatment and reporting of non-Annex I countries and regions.

· *Evaluating climate mitigation options in the context of development, sustainability and equity.* Examples include: exploration of alternative development paths, including sustainable consumption patterns in all sectors, including the transportation sector; integrated analysis of mitigation and adaptation; identifying opportunities for synergy between explicit climate policies and general policies promoting sustainable development; integration of intra- and inter-generational equity in climate change mitigation analysis; implications of equity assessments; analysis of scientific, technical and economic implications of options under a wide variety of stabilization regimes.

Climate Change 2001: Mitigation

Technical Summary

A Report Accepted by Working Group III of the IPCC but not Approved in Detail

"Acceptance" of IPCC Reports at a Session of the Working Group or Panel signifies that the material has not been subject to line by line discussion and agreement, but nevertheless presents a comprehensive, objective, and balanced view of the subject matter.

Lead Authors
Tariq Banuri (Pakistan), Terry Barker (UK), Igor Bashmakov (Russian Federation), Kornelis Blok (Netherlands), John Christensen (Denmark), Ogunlade Davidson (Sierra Leone), Michael Grubb (UK), Kirsten Halsnæs (Denmark), Catrinus Jepma (Netherlands), Eberhard Jochem (Germany), Pekka Kauppi (Finland), Olga Krankina (Russian Federation), Alan Krupnick (USA), Lambert Kuijpers (Netherlands), Snorre Kverndokk (Norway), Anil Markandya (UK), Bert Metz (Netherlands), William R. Moomaw (USA), Jose Roberto Moreira (Brazil), Tsuneyuki Morita (Japan), Jiahua Pan (China), Lynn Price (USA), Richard Richels (USA), John Robinson (Canada), Jayant Sathaye (USA), Rob Swart (Netherlands), Kanako Tanaka (Japan), Tomihiro Taniguchi (Japan), Ferenc Toth (Germany), Tim Taylor (UK), John Weyant (USA)

Review Editors
Rajendra Pachauri (India)

1. Scope of the Report

1.1. Background

In 1998, Working Group (WG) III of the Intergovernmental Panel on Climate Change (IPCC) was charged by the IPCC Plenary for the Panel's Third Assessment Report (TAR) to assess the scientific, technical, environmental, economic, and social aspects of the mitigation of climate change. Thus, the mandate of the Working Group was changed from a predominantly disciplinary assessment of the economic and social dimensions on climate change (including adaptation) in the Second Assessment Report (SAR), to an interdisciplinary assessment of the options to control the emissions of greenhouse gases (GHGs) and/or enhance their sinks.

After the publication of the SAR, continued research in the area of mitigation of climate change, which was partly influenced by political changes such as the adoption of the Kyoto Protocol to the United Nations Framework Convention on Climate Change (UNFCCC) in 1997, has been undertaken and is reported on here. The report also draws on a number of IPCC Special Reports[1] and IPCC co-sponsored meetings and Expert Meetings that were held in 1999 and 2000, particularly to support the development of the IPCC TAR. This summary follows the 10 chapters of the report.

1.2. Broadening the Context of Climate Change Mitigation

This chapter places climate change mitigation, mitigation policy, and the contents of the rest of the report in the broader context of development, equity, and sustainability. This context reflects the explicit conditions and principles laid down by the UNFCCC on the pursuit of the ultimate objective of stabilizing greenhouse gas concentrations. The UNFCCC imposes three conditions on the goal of stabilization: namely that it should take place within a time-frame sufficient to "allow ecosystems to adapt naturally to climate change, to ensure that food production is not threatened and to enable economic development to proceed in a sustainable manner" (Art. 2). It also specifies several principles to guide this process: equity, common but differentiated responsibilities, precaution, cost-effective measures, right to sustainable development, and support for an open international economic system (Art. 3).

Previous IPCC assessment reports sought to facilitate this pursuit by comprehensively describing, cataloguing, and comparing technologies and policy instruments that could be used to achieve mitigation of greenhouse gas emissions in a cost-

effective and efficient manner. The present assessment advances this process by including recent analyses of climate change that place policy evaluations in the context of sustainable development. This expansion of scope is consistent both with the evolution of the literature on climate change and the importance accorded by the UNFCCC to sustainable development - including the recognition that "Parties have a right to, and should promote sustainable development" (Art. 3.4). It therefore goes some way towards filling the gaps in earlier assessments.

Climate change involves complex interactions between climatic, environmental, economic, political, institutional, social, and technological processes. It cannot be addressed or comprehended in isolation of broader societal goals (such as equity or sustainable development), or other existing or probable future sources of stress. In keeping with this complexity, a multiplicity of approaches have emerged to analyze climate change and related challenges. Many of these incorporate concerns about development, equity, and sustainability (DES) (albeit partially and gradually) into their framework and recommendations. Each approach emphasizes certain elements of the problem, and focuses on certain classes of responses, including for example, optimal policy design, building capacity for designing and implementing policies, strengthening synergies between climate change mitigation and/or adaptation and other societal goals, and policies to enhance societal learning. These approaches are therefore complementary rather than mutually exclusive.

This chapter brings together three broad classes of analysis, which differ not so much in terms of their ultimate goals as of their points of departure and preferred analytical tools. The three approaches start with concerns, respectively, about efficiency and cost-effectiveness, equity and sustainable development, and global sustainability and societal learning. The difference between the three approaches selected lies in their starting point not in their ultimate goals. Regardless of the starting point of the analysis, many studies try in their own way to incorporate other concerns. For example, many analyses that approach climate change mitigation from a cost-effectiveness perspective try to bring in considerations of equity and sustainability through their treatment of costs, benefits, and welfare. Similarly, the class of studies that are motivated strongly by considerations of inter-country equity tend to argue that equity is needed to ensure that developing countries can pursue their internal goals of sustainable development—a concept that includes the implicit components of sustainability and efficiency. Likewise, analysts focused on concerns of global sustainability have been compelled by their own logic to make a case for global efficiency—often modelled as the decoupling of production from material flows—and social equity. In other words, each of the three perspectives has led writers to search for ways to incorporate concerns that lie beyond their initial starting point. All three classes of analyses look at the relationship of climate change mitigation with all three goals—development, equity, and sustainability—albeit in different and often highly complementary

---

[1] Notably the Special Report on Aviation and the Global Atmosphere, the Special Report on Methodological and Technological Issues in Technology Transfer, the Special Report on Emissions Scenarios, and the Special Report on Land Use, Land-Use Change and Forestry.

ways. Nevertheless, they frame the issues differently, focus on different sets of causal relationships, use different tools of analysis, and often come to somewhat different conclusions.

There is no presumption that any particular perspective for analysis is most appropriate at any level. Moreover, the three perspectives are viewed here as being highly synergistic. The important changes have been primarily in the types of questions being asked and the kinds of information being sought. In practice, the literature has expanded to add new issues and new tools, subsuming rather than discarding the analyses included in the other perspectives. The range and scope of climate policy analyses can be understood as a gradual broadening of the types and extent of uncertainties that analysts have been willing and able to address.

The first perspective on climate policy analysis is cost effectiveness. It represents the field of conventional climate policy analysis that is well represented in the First through Third Assessments. These analyses have generally been driven directly or indirectly by the question of what is the most cost-effective amount of mitigation for the global economy starting from a particular baseline GHG emissions projection, reflecting a specific set of socio-economic projections. Within this framework, important issues include measuring the performance of various technologies and the removal of barriers (such as existing subsidies) to the implementation of those candidate policies most likely to contribute to emissions reductions. In a sense, the focus of analysis here has been on identifying an efficient pathway through the interactions of mitigation policies and economic development, conditioned by considerations of equity and sustainability, but not primarily guided by them. At this level, policy analysis has almost always taken the existing institutions and tastes of individuals as given; assumptions that might be valid for a decade or two, but may become more questionable over many decades.

The impetus for the expansion in the scope of the climate policy analysis and discourse to include equity considerations was to address not simply the impacts of climate change and mitigation policies on global welfare as a whole, but also of the effects of climate change and mitigation policies on existing inequalities among and within nations. The literature on equity and climate change has advanced considerably over the last two decades, but there is no consensus on what constitutes fairness. Once equity issues were introduced into the assessment agenda, though, they became important components in defining the search for efficient emissions mitigation pathways. The considerable literature that indicated how environmental policies could be hampered or even blocked by those who considered them unfair became relevant. In light of these results, it became clear how and why any widespread perception that a mitigation strategy is unfair would likely engender opposition to that strategy, perhaps to the extent of rendering it non-optimal (or even infeasible, as could be the case if non-Annex I countries never participate). Some cost-effectiveness analyses had, in fact, laid the groundwork

for applying this literature by demonstrating the sensitivity of some equity measures to policy design, national perspective, and regional context. Indeed, cost-effectiveness analyses had even highlighted similar sensitivities for other measures of development and sustainability. As mentioned, the analyses that start from equity concerns have by and large focused on the needs of developing countries, and in particular on the commitment expressed in Article 3.4 of the UNFCCC to the pursuit of sustainable development. Countries differ in ways that have dramatic implications for scenario baselines and the range of mitigation options that can be considered. The climate policies that are feasible, and/or desirable, in a particular country depend significantly on its available resources and institutions, and on its overall objectives including climate change as but one component. Recognizing this heterogeneity may, thus, lead to a different range of policy options than has been considered likely thus far and may reveal differences in the capacities of different sectors that may also enhance appreciation of what can be done by non-state actors to improve their ability to mitigate.

The third perspective is global sustainability and societal learning. While sustainability has been incorporated in the analyses in a number of ways, a class of studies takes the issue of global sustainability as their point of departure. These studies focus on alternative pathways to pursue global sustainability and address issues like decoupling growth from resource flows, for example through eco-intelligent production systems, resource light infrastructure and appropriate technologies, and decoupling wellbeing from production, for example through intermediate performance levels, regionalization of production systems, and changing lifestyles. One popular method for identifying constraints and opportunities within this perspective is to identify future sustainable states and then examine possible transition paths to those states for feasibility and desirability. In the case of developing countries this leads to a number of possible strategies that can depart significantly from those which the developed countries pursued in the past.

1.3. Integrating the Various Perspectives

Extending discussions of how nations might respond to the mitigation challenge so that they include issues of cost-effectiveness and efficiency, distribution narrowly defined, equity more broadly defined, and sustainability, adds enormous complexity to the problem of uncovering how best to respond to the threat of climate change. Indeed, recognizing that these multiple domains are relevant complicates the task assigned to policymakers and international negotiators by opening their deliberations to issues that lie beyond the boundaries of the climate change problem, *per se*. Their recognition thereby underlines the importance of integrating scientific thought across a wide range of new policy-relevant contexts, but not simply because of some abstract academic or narrow parochial interest advanced by a small set of researchers or nations. Cost-effectiveness, equity, and sustainability have all been identified as critical issues by the drafters of the UNFCCC, and they are an integral part of the charge given to the drafters of the TAR. Integration across the domains of cost-effectiveness, equity, and sustainability is therefore profoundly relevant to policy deliberations according to the letter as well as the spirit of the UNFCCC itself.

The literature being brought to bear on climate change mitigation increasingly shows that policies lying beyond simply reducing GHG emissions from a specified baseline to minimize costs can be extremely effective in abating the emission of GHGs. Therefore, a portfolio approach to policy and analysis would be more effective than exclusive reliance on a narrow set of policy instruments or analytical tools. Besides the flexibility that an expanded range of policy instruments and analytical tools can provide to policymakers for achieving climate objectives, the explicit inclusion of additional policy objectives also increases the likelihood of "buy-in" to climate policies by more participants. In particular, it will expand the range of no regrets[2] options. Finally, it could assist in tailoring policies to short-, medium-, and long-term goals.

In order to be effective, however, a portfolio approach requires weighing the costs and impacts of the broader set of policies according to a longer list of objectives. Climate deliberations need to consider the climate ramifications of policies designed primarily to address a wide range of issues including DES, as well as the likely impacts of climate policies on the achievement of these objectives. As part of this process the opportunity costs and impacts of each instrument are measured against the multiple criteria defined by these multiple objectives. Furthermore, the number of decision makers or stakeholders to be considered is increased beyond national policymakers and international negotiators to include state, local, community, and household agents, as well as non-government organizations (NGOs).

The term "ancillary benefits" is often used in the literature for the ancillary, or secondary, effects of climate change mitigation policies on problems other than GHG emissions, such as reductions in local and regional air pollution, associated with the reduction of fossil fuels, and indirect effects on issues such as transportation, agriculture, land use practices, biodiversity preservation, employment, and fuel security. Sometimes these are referred to as "ancillary impacts", to reflect the fact that in some cases the benefits may be negative[3]. The concept of

"mitigative capacity" is also introduced as a possible way to integrate results derived from the application of the three perspectives in the future. The determinants of the capacity to mitigate climate change include the availability of technological and policy options, and access to resources to underwrite undertaking those options. These determinants are the focus of much of the TAR. The list of determinants is, however, longer than this. Mitigative capacity also depends upon nation-specific characteristics that facilitate the pursuit of sustainable development – e.g., the distribution of resources, the relative empowerment of various segments of the population, the credibility of empowered decision makers, the degree to which climate objectives complement other objectives, access to credible information and analyses, the will to act on that information, the ability to spread risk intra- and inter-generationally, and so on. Given that the determinants of mitigative capacity are essentially the same as those of the analogous concept of adaptive capacity introduced in the WGII Report, this approach may provide an integrated framework for assessing both sets of options.

2. Greenhouse Gas Emissions Scenarios

2.1. Scenarios

A long-term view of a multiplicity of future possibilities is required to consider the ultimate risks of climate change, assess critical interactions with other aspects of human and environmental systems, and guide policy responses. Scenarios offer a structured means of organizing information and gleaning insight on the possibilities.

Each mitigation scenario describes a particular future world, with particular economic, social, and environmental characteristics, and they therefore implicitly or explicitly contain information about DES. Since the difference between reference case scenarios and stabilization and mitigation scenarios is simply the addition of deliberate climate policy, it can be the case that the differences in emissions among different reference case scenarios are greater than those between any one such scenario and its stabilization or mitigation version.

This section presents an overview of three scenario literatures: general mitigation scenarios produced since the SAR, narrative-based scenarios found in the general futures literature, and mitigation scenarios based on the new reference scenarios developed in the IPCC SRES.

2.2. Greenhouse Gas Emissions Mitigation Scenarios

This report considers the results of 519 quantitative emissions scenarios from 188 sources, mainly produced after 1990. The review focuses on 126 mitigation scenarios that cover global emissions and have a time horizon encompassing the coming

---

[2] In this report, as in the SAR, no regrets options are defined as those options whose benefits such as reduced energy costs and reduced emissions of local/regional pollutants equal or exceed their costs to society, excluding the benefits of avoided climate change. They are also known as negative cost options.

[3] In this report sometimes the term "co-benefits" is also used to indicate the additional benefits of policy options that are implemented for various reasons at the same time, acknowledging that most policies designed to address GHG mitigation also have other, often at least equally important, rationales, e.g., related to objectives of development, sustainability and equity. The benefits of avoided climate change are not covered in ancillary or co-benefits. See also Section 7.2.

century. Technological improvement is a critical element in all the general mitigation scenarios.

Based on the type of mitigation, the scenarios fall into four categories: concentration stabilization scenarios, emission stabilization scenarios, safe emission corridor scenarios, and other mitigation scenarios. All the reviewed scenarios include energy-related carbon dioxide (CO_2) emissions; several also include CO_2 emissions from land-use changes and industrial processes, and other important GHGs.

Policy options used in the reviewed mitigation scenarios take into account energy systems, industrial processes, and land use, and depend on the underlying model structure. Most of the scenarios introduce simple carbon taxes or constraints on emissions or concentration levels. Regional targets are introduced in the models with regional disaggregation. Emission permit trading is introduced in more recent work. Some models employ policies of supply-side technology introduction, while others emphasize efficient demand-side technology.

Allocation of emission reduction among regions is a contentious issue. Only some studies, particularly recent ones, make explicit assumptions about such allocations in their scenarios. Some studies offer global emission trading as a mechanism to reduce mitigation costs.

Technological improvement is a critical element in all the general mitigation scenarios.

Detailed analysis of the characteristics of 31 scenarios for stabilization of CO_2 concentrations at 550ppmv[4] (and their baseline scenarios) yielded several insights:

- There is a wide range in baselines, reflecting a diversity of assumptions, mainly with respect to economic growth and low-carbon energy supply. High economic growth scenarios tend to assume high levels of progress in the efficiency of end-use technologies; however, carbon intensity reductions were found to be largely independent of economic growth assumptions. The range of future trends shows greater divergence in scenarios that focus on developing countries than in scenarios that look at developed nations. There is little consensus with respect to future directions in developing regions.
- The reviewed 550ppmv stabilization scenarios vary with respect to reduction time paths and the distribution of emission reductions among regions. Some scenarios suggested that emission trading may lower the overall

mitigation cost, and could lead to more mitigation in the non-OECD countries. The range of assumed mitigation policies is very wide. In general, scenarios in which there is an assumed adoption of high-efficiency measures in the baseline show less scope for further introduction of efficiency measures in the mitigation scenarios. In part this results from model input assumptions, which do not assume major technological breakthroughs. Conversely, baseline scenarios with high carbon intensity reductions show larger carbon intensity reductions in their mitigation scenarios.

Only a small set of studies has reported on scenarios for mitigating non-CO_2 gases. This literature suggests that small reductions of GHG emissions can be accomplished at lower cost by including non-CO_2 gases; that both CO_2 and non-CO_2 emissions would have to be controlled in order to slow the increase of atmospheric temperature sufficiently to achieve climate targets assumed in the studies; and that methane (CH_4) mitigation can be carried out more rapidly, with a more immediate impact on the atmosphere, than CO_2 mitigation.

Generally, it is clear that mitigation scenarios and mitigation policies are strongly related to their baseline scenarios, but no systematic analysis has been published on the relationship between mitigation and baseline scenarios.

2.3. Global Futures Scenarios

Global futures scenarios do not specifically or uniquely consider GHG emissions. Instead, they are more general "stories" of possible future worlds. They can complement the more quantitative emissions scenario assessments, because they consider dimensions that elude quantification, such as governance and social structures and institutions, but which are nonetheless important to the success of mitigation policies. Addressing these issues reflects the different perspectives presented in Section 1: cost-effectiveness and/or efficiency, equity, and sustainability.

A survey of this literature has yielded a number of insights that are relevant to GHG emissions scenarios and sustainable development. First, a wide range of future conditions has been identified by futurists, ranging from variants of sustainable development to collapse of social, economic, and environmental systems. Since future values of the underlying socio-economic drivers of emissions may vary widely, it is important that climate policies should be designed so that they are resilient against widely different future conditions.

Second, the global futures scenarios that show falling GHG emissions tend to show improved governance, increased equity and political participation, reduced conflict, and improved environmental quality. They also tend to show increased energy efficiency, shifts to non-fossil energy sources, and/or shifts to a post-industrial (service-based) economy; population tends

---

[4] The reference to a particular concentration level does not imply an agreed-upon desirability of stabilization at this level. The selection of 550ppmv is based on the fact that the majority of studies in the literature analyze this level, and does not imply any endorsement of this level as a target for climate change mitigation policies.

to stabilize at relatively low levels, in many cases thanks to increased prosperity, expanded provision of family planning, and improved rights and opportunities for women. A key implication is that sustainable development policies can make a significant contribution to emission reduction.

Third, different combinations of driving forces are consistent with low emissions scenarios, which agrees with the SRES findings. The implication of this seems to be that it is important to consider the linkage between climate policy and other policies and conditions associated with the choice of future paths in a general sense.

2.4. Special Report on Emissions Scenarios

Six new GHG emission reference scenario groups (not including specific climate policy initiatives), organized into 4 scenario "families", were developed by the IPCC and published as the Special Report on Emissions Scenarios (SRES). Scenario families A1 and A2 emphasize economic development but differ with respect to the degree of economic and social convergence; B1 and B2 emphasize sustainable development but also differ in terms of degree of convergence (see Box TS-1). In all, six models were used to generate the 40 scenarios that comprise

the six scenario groups. Six of these scenarios, which should be considered equally sound, were chosen to illustrate the whole set of scenarios. These six scenarios include marker scenarios for each of the worlds as well as two scenarios, A1FI and A1T, which illustrate alternative energy technology developments in the A1 world (see Figure TS.1).

The SRES scenarios lead to the following findings:
- Alternative combinations of driving-force variables can lead to similar levels and structure of energy use, land-use patterns, and emissions.
- Important possibilities for further bifurcations in future development trends exist within each scenario family.
- Emissions profiles are dynamic across the range of SRES scenarios. They portray trend reversals and indicate possible emissions cross-over among different scenarios.
- Describing potential future developments involves inherent ambiguities and uncertainties. One and only one possible development path (as alluded to, for instance, in concepts such as "business-as-usual scenario") simply does not exist. The multi-model approach increases the value of the SRES scenario set, since uncertainties in the choice of model input assumptions can be more explicitly separated from the specific model behaviour and related modelling uncertainties.

Box TS-1. The Emissions Scenarios of the IPCC Special Report on Emissions Scenarios (SRES)

A1. The A1 storyline and scenario family describes a future world of very rapid economic growth, global population that peaks in mid-century and declines thereafter, and the rapid introduction of new and more efficient technologies. Major underlying themes are convergence among regions, capacity building and increased cultural and social interactions, with a substantial reduction in regional differences in per capita income. The A1 scenario family develops into three groups that describe alternative directions of technological change in the energy system. The three A1 groups are distinguished by their technological emphasis: fossil intensive (A1FI), non-fossil energy sources (A1T), or a balance across all sources (A1B) (where balanced is defined as not relying too heavily on one particular energy source, on the assumption that similar improvement rates apply to all energy supply and end use technologies).

A2. The A2 storyline and scenario family describes a very heterogeneous world. The underlying theme is self-reliance and preservation of local identities. Fertility patterns across regions converge very slowly, which results in continuously increasing population. Economic development is primarily regionally oriented and per capita economic growth and technological change more fragmented and slower than other storylines.

B1. The B1 storyline and scenario family describes a convergent world with the same global population, that peaks in mid-

century and declines thereafter, as in the A1 storyline, but with rapid change in economic structures toward a service and information economy, with reductions in material intensity and the introduction of clean and resource-efficient technologies. The emphasis is on global solutions to economic, social and environmental sustainability, including improved equity, but without additional climate initiatives.

B2. The B2 storyline and scenario family describes a world in which the emphasis is on local solutions to economic, social and environmental sustainability. It is a world with continuously increasing global population, at a rate lower than A2, intermediate levels of economic development, and less rapid and more diverse technological change than in the B1 and A1 storylines. While the scenario is also oriented towards environmental protection and social equity, it focuses on local and regional levels.

An illustrative scenario was chosen for each of the six scenario groups A1B, A1FI, A1T, A2, B1 and B2. All should be considered equally sound.

The SRES scenarios do not include additional climate initiatives, which means that no scenarios are included that explicitly assume implementation of the United Nations Framework Convention on Climate Change or the emissions targets of the Kyoto Protocol.

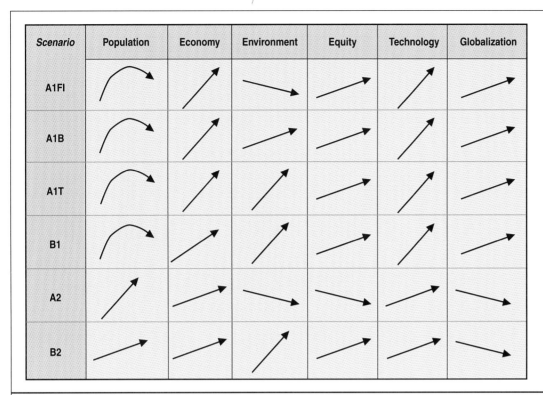

Figure TS-1: Qualitative directions of SRES scenarios for different indicators.

2.5. Review of Post-SRES Mitigation Scenarios

Recognizing the importance of multiple baselines in evaluating mitigation strategies, recent studies analyze and compare mitigation scenarios using as their baselines the new SRES scenarios. This allows for the assessment in this report of 76 "post-SRES mitigation scenarios" produced by nine modelling teams. These mitigation scenarios were quantified on the basis of storylines for each of the six SRES scenarios that describe the relationship between the kind of future world and the capacity for mitigation.

Quantifications differ with respect to the baseline scenario, including assumed storyline, the stabilization target, and the model that was used. The post-SRES scenarios cover a very wide range of emission trajectories, but the range is clearly below the SRES range. All scenarios show an increase in CO_2 reduction over time. Energy reduction shows a much wider range than CO_2 reduction, because in many scenarios a decoupling between energy use and carbon emissions takes place as a result of a shift in primary energy sources.

In general, the lower the stabilization target and the higher the level of baseline emissions, the larger the CO_2 divergence from the baseline that is needed, and the earlier that it must occur. The A1FI, A1B, and A2 worlds require a wider range of and more strongly implemented technology and/or policy measures than A1T, B1, and B2. The 450ppmv stabilization case requires more drastic emission reduction to occur earlier than under

the 650ppmv case, with very rapid emission reduction over the next 20 to 30 years (see Figure TS-2).

A key policy question is what kind of emission reductions in the medium term (after the Kyoto Protocol commitment period) would be needed. Analysis of the post-SRES scenarios (most of which assume developing country emissions to be below baselines by 2020) suggests that stabilization at 450ppmv will require emissions reductions in Annex I countries after 2012 that go significantly beyond their Kyoto Protocol commitments. It also suggests that it would not be necessary to go much beyond the Kyoto commitments for Annex I by 2020 to achieve stabilization at 550ppmv or higher. However, it should be recognized that several scenarios indicate the need for significant Annex I emission reductions by 2020 and that none of the scenarios introduces other constraints such as a limit to the rate of temperature change.

An important policy question already mentioned concerns the participation of developing countries in emission mitigation. A preliminary finding of the post-SRES scenario analysis is that, if it is assumed that the CO_2 emission reduction needed for stabilization occurs in Annex I countries only, Annex I per capita CO_2 emissions would fall below non-Annex I per capita emissions during the 21st century in nearly all of the stabilization scenarios, and before 2050 in two-thirds of the scenarios, if developing countries emissions follow the baseline scenarios. This suggests that the stabilization target and the baseline emission level are both important determinants of

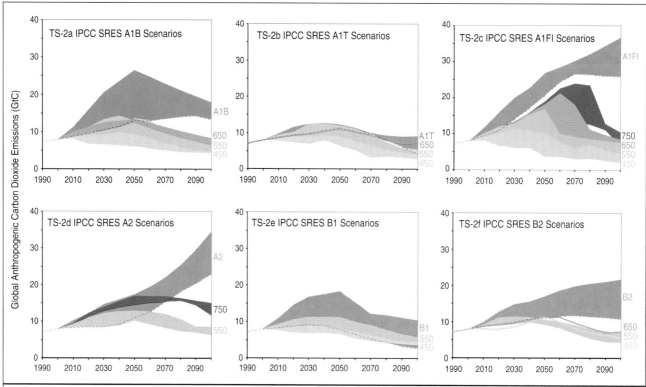

Figure TS-2: Comparison of reference and stabilization scenarios. The figure is divided into six parts, one for each of the reference scenario groups from the Special Report on Emissions Scenarios (SRES). Each part of the figure shows the range of total global CO₂ emissions (gigatonnes of carbon (GtC)) from all anthropogenic sources for the SRES reference scenario group (shaded in grey) and the ranges for the various mitigation scenarios assessed in the TAR leading to stabilization of CO₂ concentrations at various levels (shaded in colour). Scenarios are presented for the A1 family subdivided into three groups (the balanced A1B group (Figure TS-2a), non-fossil fuel A1T (Figure TS-2b), and the fossil intensive A1FI (Figure TS-2c)) and stabilization of CO₂ concentrations at 450, 550, 650 and 750ppmv; for the A2 group with stabilization at 550 and 750ppmv in Figure TS-2d, the B1 group and stabilization at 450 and 550ppmv in Figure TS-2e, and the B2 group including stabilization at 450, 550, and 650ppmv in Figure TS-2f. The literature is not available to assess 1000ppmv stabilization scenarios. The figure illustrates that the lower the stabilization level and the higher the baseline emissions, the wider the gap. The difference between emissions in different scenario groups can be as large as the gap between reference and stabilization scenarios within one scenario group. The dotted lines depict the boundaries of the ranges where they overlap (see Box TS-1).

the timing when developing countries emissions might need to diverge from their baseline.

Climate policy would reduce per capita final energy use in the economy-emphasized worlds (A1FI, A1B, and A2), but not in the environment-emphasized worlds (B1 and B2). The reduction in energy use caused by climate policies would be larger in Annex I than in non-Annex I countries. However, the impact of climate policies on equity in per capita final energy use would be much smaller than that of the future development path.

There is no single path to a low emission future and countries and regions will have to choose their own path. Most model results indicate that known technological options[5] could

achieve a broad range of atmospheric CO₂ stabilization levels, such as 550ppmv, 450ppmv or, below over the next 100 years or more, but implementation would require associated socio-economic and institutional changes.

Assumed mitigation options differ among scenarios and are strongly dependent on the model structure. However, common features of mitigation scenarios include large and continuous energy efficiency improvements and afforestation as well as low-carbon energy, especially biomass over the next 100 years and natural gas in the first half of the 21st century. Energy conservation and reforestation are reasonable first steps, but innovative supply-side technologies will eventually be required. Possible robust options include using natural gas and combined-cycle technology to bridge the transition to more advanced fossil fuel and zero-carbon technologies, such as hydrogen fuel cells. Solar energy as well as either nuclear energy or carbon removal and storage would become increasingly important for a higher emission world or lower stabilization target.

[5] "Known technological options" refer to technologies that exist in operation or pilot plant stage today, as referenced in the mitigation scenarios discussed in this report. It does not include any new technologies that will require drastic technological breakthroughs. In this way it can be considered to be a conservative estimate, considering the length of the scenario period.

Integration between global climate policies and domestic air pollution abatement policies could effectively reduce GHG emissions in developing regions for the next two or three decades. However, control of sulphur emissions could amplify possible climate change, and partial trade-offs are likely to persist for environmental policies in the medium term.

Policies governing agriculture, land use and energy systems could be linked for climate change mitigation. Supply of biomass energy as well as biological CO_2 sequestration would broaden the available options for carbon emission reductions, although the post-SRES scenarios show that they cannot provide the bulk of the emission reductions required. That has to come from other options.

3. Technological and Economic Potential of Mitigation Options

3.1. Key Developments in Knowledge about Technological Options to Mitigate GHG Emissions in the Period up to 2010-2020 since the Second Assessment Report

Technologies and practices to reduce GHG emissions are continuously being developed. Many of these technologies focus on improving the efficiency of fossil fuel energy or electricity use and the development of low carbon energy sources, since the majority of GHG emissions (in terms of CO_2 equivalents) are related to the use of energy. Energy intensity (energy consumed divided by gross domestic product (GDP)) and carbon intensity (CO_2 emitted from burning fossil fuels divided by the amount of energy produced) have been declining for more than 100 years in developed countries without explicit government policies for decarbonization, and have the potential to decline further. Much of this change is the result of a shift away from high carbon fuels such as coal towards oil and natural gas, through energy conversion efficiency improvements and the introduction of hydro and nuclear power. Other non-fossil fuel energy sources are also being developed and rapidly implemented and have a significant potential for reducing GHG emissions. Biological sequestration of CO_2 and CO_2 removal and storage can also play a role in reducing GHG emissions in the future (see also Section 4 below). Other technologies and measures focus on the non-energy sectors for reducing emissions of the remaining major GHGs: CH4, nitrous oxide (N^2O), hydrofluorocarbons (HFCs), perfluorocarbons (PFCs), and sulphur hexafluoride (SF_6).

Since the SAR several technologies have advanced more rapidly than was foreseen in the earlier analysis. Examples include the market introduction of efficient hybrid engine cars, rapid advancement of wind turbine design, demonstration of underground carbon dioxide storage, and the near elimination of N_2O emissions from adipic acid production. Greater energy efficiency opportunities for buildings, industry, transportation, and energy supply are available, often at a lower cost than was expected. By the year 2010 most of the opportunities to reduce emissions will still come from energy efficiency gains in the end-use sectors, by switching to natural gas in the electric power sector, and by reducing the release of process GHGs from industry, e.g., N_2O, perfluoromethane (CF_4), and HFCs. By the year 2020, when a proportion of the existing power plants will have been replaced in developed countries and countries with economies in transition (EITs), and when many new plants will become operational in developing countries, the use of renewable sources of energy can begin contributing to the reduction of CO_2 emissions. In the longer term, nuclear energy technologies − with inherent passive characteristics meeting stringent safety, proliferation, and waste storage goals − along with physical carbon removal and storage from fossil fuels and biomass, followed by sequestration, could potentially become available options.

Running counter to the technological and economic potential for GHG emissions reduction are rapid economic development and accelerating change in some socio-economic and behavioural trends that are increasing total energy use, especially in developed countries and high-income groups in developing countries. Dwelling units and vehicles in many countries are growing in size, and the intensity of electrical appliance use is increasing. Use of electrical office equipment in commercial buildings is increasing. In developed countries, and especially the USA, sales of larger, heavier, and less efficient vehicles are also increasing. Continued reduction or stabilization in retail energy prices throughout large portions of the world reduces incentives for the efficient use of energy or the purchase of energy efficient technologies in all sectors. With a few important exceptions, countries have made little effort to revitalize policies or programmes to increase energy efficiency or promote renewable energy technologies. Also since the early 1990s, there has been a reduction in both public and private resources devoted to R&D (research and development) to develop and implement new technologies that will reduce GHG emissions.

In addition, and usually related to technological innovation options, there are important possibilities in the area of social innovation. In all regions, many options are available for lifestyle choices that may improve quality of life, while at the same time decreasing resource consumption and associated GHG emissions. Such choices are very much dependent on local and regional cultures and priorities. They are very closely related to technological changes, some of which can be associated with profound lifestyle changes, while others do not require such changes. While these options were hardly noted in the SAR, this report begins to address them.

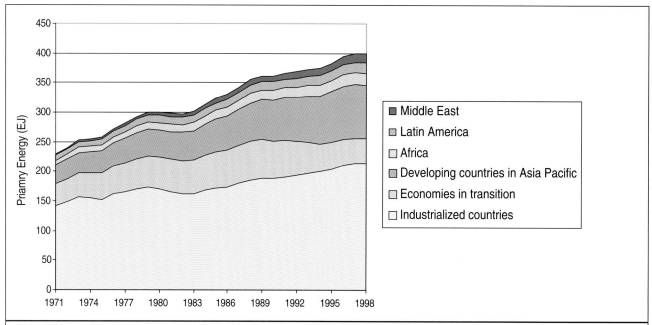

Figure TS-3: World primary energy use by region from 1971 to 1998.
Note: Primary energy calculated using the IEA's physical energy content method based on the primary energy sources used to produce heat and electricity.

3.2. Trends in Energy Use and Associated Greenhouse Gas Emissions

Global consumption of energy and associated emission of CO_2 continue an upward trend in the 1990s (Figures TS-3 and TS-4). Fossil fuels remain the dominant form of energy utilized in the world, and energy use accounts for more than two thirds of the GHG emissions addressed by the Kyoto Protocol. In 1998, 143 exajoules (EJ) of oil, 82EJ of natural gas, and 100EJ of coal were consumed by the world's economies. Global primary energy consumption grew an average of 1.3% annually between 1990 and 1998. Average annual growth rates were 1.6% for developed countries and 2.3% to 5.5% for developing countries between 1990 and 1998. Primary energy use for the EITs declined at an annual rate of 4.7% between 1990 and 1998 owing to the loss of heavy industry, the decline in overall economic activity, and restructuring of the manufacturing sector.

Average global carbon dioxide emissions grew – approximately at the same rate as primary energy – at a rate of 1.4% per year between 1990 and 1998, which is much slower than the 2.1% per year growth seen in the 1970s and 1980s. This was in large measure because of the reductions from the EITs and structural changes in the industrial sector of the developed countries. Over the longer term, global growth in CO_2 emissions from energy use was 1.9% per year between 1971 and 1998. In 1998, developed countries were responsible for over 50% of energy-related CO_2 emissions, which grew at a rate of 1.6% annually from 1990. The EITs accounted for 13% of 1998 emissions, and their emissions have been declining at an annual

rate of 4.6% per year since 1990. Developing countries in the Asia-Pacific region emitted 22% of the global total carbon dioxide, and have been the fastest growing with increases of 4.9% per year since 1990. The rest of the developing countries accounted for slightly more than 10% of total emissions, growing at an annual rate of 4.3% since 1990.

During the period of intense industrialization from 1860 to 1997, an estimated 13,000EJ of fossil fuel were burned, releasing 290GtC into the atmosphere, which along with land-use change has raised atmospheric concentrations of CO_2 by 30%. By comparison, estimated natural gas resources[6] are comparable to those for oil, being approximately 35,000EJ. The coal resource base is approximately four times as large. Methane clathrates (not counted in the resource base) are estimated to be approximately 780,000EJ. Estimated fossil fuel reserves contain 1,500GtC, being more than 5 times the carbon already released, and if estimated resources are added, there is a total of 5,000GtC remaining in the ground. The scenarios modelled by the SRES without any specific GHG

---

[6] Reserves are those occurrences that are identified and measured as economically and technically recoverable with current technologies and prices. Resources are those occurrences with less certain geological and/or economic characteristics, but which are considered poten-tially recoverable with foreseeable technological and economic developments. The resource base includes both categories. On top of that there are additional quantities with unknown certainty of occurrence and/or with unknown or no economic significance in the foreseeable future, referred to as "additional occurrences" (SAR). Examples of unconventional fossil fuel resources are tar sands and shale oils, geo-pressured gas, and gas in aquifers.

emission policies foresee cumulative release ranging from approximately 1,000 GtC to 2,100 GtC from fossil fuel consumption between 2,000 and 2,100. Cumulative carbon emissions for stabilization profiles of 450 to 750ppmv over that same period are between 630 and 1,300GtC (see Figure TS-5).

Fossil-fuel scarcity, at least at the global level, is therefore not a significant factor in considering climate change mitigation. On the contrary, different from the relatively large coal and unconventional oil and gas deposits, the carbon in conventional oil and gas reserves or in conventional oil resources is much

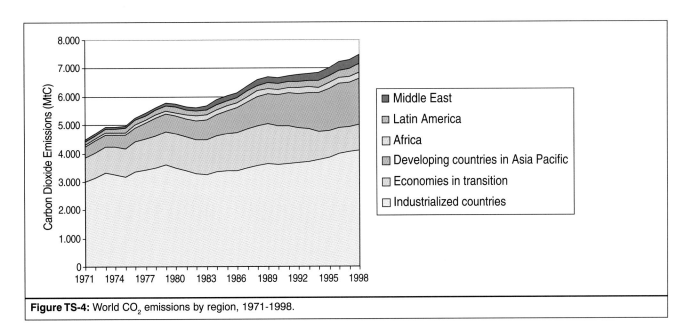

Figure TS-4: World CO_2 emissions by region, 1971-1998.

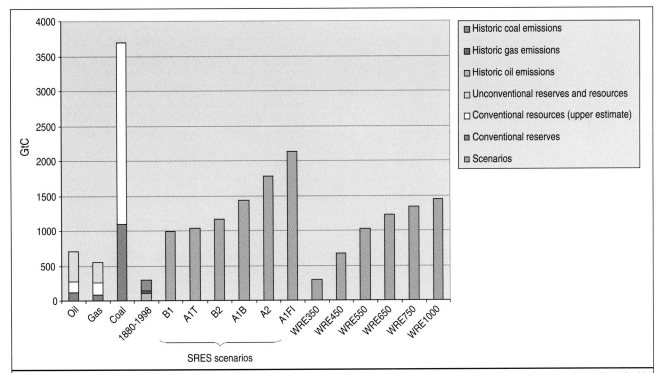

Figure TS-5: Carbon in oil, gas and coal reserves and resources compared with historic fossil fuel carbon emissions 1860-1998, and with cumulative carbon emissions from a range of SRES scenarios and TAR stabilization scenarios up until 2100. Data for reserves and resources are shown in the left hand columns. Unconventional oil and gas includes tar sands, shale oil, other heavy oil, coal bed methane, deep geopressured gas, gas in acquifers, etc. Gas hydrates (clathrates) that amount to an estimated 12,000 GtC are not shown. The scenario columns show both SRES reference scenarios as well as scenarios which lead to stabilization of CO_2 concentrations at a range of levels. Note that if by 2100 cumulative emissions associated with SRES scenarios are equal to or smaller than those for stabilization scenarios, this does not imply that these scenarios equally lead to stabilization.

less than the cumulative carbon emissions associated with stabilisation at 450ppmv or higher (Figure TS-5). In addition, there is the potential to contribute large quantities of other GHGs as well. At the same time it is clear from Figure TS-5 that the conventional oil and gas reserves are only a small fraction of the total fossil fuel resource base. These resource data may imply a change in the energy mix and the introduction of new sources of energy during the 21st century. The choice of energy mix and associated investment will determine whether, and if so at what level and cost, greenhouse concentrations can be stabilized. Currently most such investment is directed towards discovering and developing more conventional and unconventional fossil resources.

3.3. Sectoral Mitigation Technological Options[7]

The potential[8] for major GHG emission reductions is estimated for each sector for a range of costs (Table TS-1). In the industrial sector, costs for carbon emission abatement are estimated to range from negative (i.e., no regrets, where reductions can be made at a profit), to around US$300/tC[9]. In the buildings sector, aggressive implementation of energy-efficient technologies and measures can lead to a reduction in CO_2 emissions from residential buildings in 2010 by 325MtC/yr in developed and EIT countries at costs ranging from –US$250 to –US$150/tC and by 125MtC in developing countries at costs of –US$250 to US$50/tC. Similarly, CO_2 emissions from commercial buildings in 2010 can be reduced by 185MtC in developed and EIT countries at costs ranging from –US$400 to –US$250/tC avoided and by 80MtC in developing countries at costs ranging from –US$400 to US$0/tC. In the transport sector costs range from –US$200/tC to US$300/tC, and in the agricultural sector from –US$100/tC to US$300/tC. Materials management, including recycling and landfill gas recovery, can also produce savings at negative to modest costs under US$100/tC. In the energy supply sector a number of fuel switching and technological substitutions are possible at costs from –US$100 to more than US$200/tC. The realization of this potential will be determined by the market conditions as influenced by human and societal preferences and government interventions.

Table TS-2 provides an overview and links with barriers and mitigation impacts. Sectoral mitigation options are discussed in more detail below.

3.3.1. The Main Mitigation Options in the Buildings Sector

The buildings sector contributed 31% of global energy-related CO_2 emissions in 1995, and these emissions have grown at an annual rate of 1.8% since 1971. Building technology has continued on an evolutionary trajectory with incremental gains during the past five years in the energy efficiency of windows, lighting, appliances, insulation, space heating, refrigeration, and air conditioning. There has also been continued development of building controls, passive solar design, integrated building design, and the application of photovoltaic systems in buildings. Fluorocarbon emissions from refrigeration and air conditioning applications have declined as chlorofluorocarbons (CFCs) have been phased out, primarily thanks to improved containment and recovery of the fluorocarbon refrigerant and, to a lesser extent, owing to the use of hydrocarbons and other non-fluorocarbon refrigerants. Fluorocarbon use and emission from insulating foams have declined as CFCs have been phased out, and are projected to decline further as HCFCs are phased out. R&D effort has led to increased efficiency of refrigerators and cooling and heating systems. In spite of the continued improvement in technology and the adoption of improved technology in many countries, energy use in buildings has grown more rapidly than total energy demand from 1971 through 1995, with commercial building energy registering the greatest annual percentage growth (3.0% compared to 2.2% in residential buildings). This is largely a result of the increased amenity that consumers demand – in terms of increased use of appliances, larger dwellings, and the modernization and expansion of the commercial sector – as economies grow. There presently exist significant cost-effective technological opportunities to slow this trend. The overall technical potential for reducing energy-related CO_2 emissions in the buildings sector using existing technologies combined with future technical advances is 715MtC/yr in 2010 for a base case with carbon emissions of 2,600MtC/yr (27%), 950MtC/yr in 2020 for a base case with carbon emissions of 3,000MtC/yr (31%), and 2,025MtC/yr in 2050 for a base case with carbon emissions of 3,900MtC/yr (52%). Expanded R&D can assure continued technology improvement in this sector.

3.3.2. The Main Mitigation Options in the Transport Sector

In 1995, the transport sector contributed 22% of global energy-related carbon dioxide emissions; globally, emissions from this sector are growing at a rapid rate of approximately 2.5% annually. Since 1990, principal growth has been in the developing countries (7.3% per year in the Asia–Pacific region) and is actually declining at a rate of 5.0% per year for the EITs. Hybrid gasoline-electric vehicles have been introduced on a commercial basis with fuel economies 50%-100% better than those of comparably sized four-passenger vehicles. Biofuels produced from wood, energy

---

[7] International Energy Statistics (IEA) report sectoral data for the industrial and transport sectors, but not for buildings and agriculture, which are reported as "other". In this section, information on energy use and CO_2 emissions for these sectors has been estimated using an allocation scheme and based on a standard electricity conversion factor of 33%. In addition, values for the EIT countries are from a different source (British Petroleum statistics). Thus, the sectoral values can differ from the aggregate values presented in section 3.2, although general trends are the same. In general, there is uncertainty in the data for the EITs and for the commercial and residential sub-categories of the buildings sector in all regions.

[8] The potential differs in different studies assessed but the aggregate potential reported in Sections 3 and 4 refers to the socio-economic potential as indicated in Figure TS-7.

[9] All costs in US$.

Table TS-1: Estimations of greenhouse gas emission reductions and cost per tonne of carbon equivalent avoided following the anticipated socio-economic potential uptake by 2010 and 2020 of selected energy efficiency and supply technologies, either globally or by region and with varying degrees of uncertainty.

	Region	US$/tC avoided (−400 −200 0 +200)	2010 Potential[a]	2010 Probability[b]	2020 Potential[a]	2020 Probability[b]	References, comments, and relevant section in Chapter 3 of the main report
Buildings/appliances Residential sector	OECD/EIT		◆◆◆◆◆	◇◇◇◇◇	◆◆◆◆◆	◇◇◇◇◇	Acosta Moreno *et al.*, 1996; Brown *et al.*, 1998 Wang and Smith, 1999
	Dev. cos.		◆◆◆◆	◇◇◇	◆◆◆◆	◇◇◇◇◇	
Commercial sector	OECD/EIT		◆◆◆◆	◇◇◇◇	◆◆◆◆◆	◇◇◇◇◇	
	Dev. cos.		◆◆◆	◇◇◇	◆◆◆◆	◇◇◇◇◇	
Transport Automobile efficiency improvements	USA		◆◆◆◆◆	◇◇◇◇	◆◆◆◆◆	◇◇◇	Interlab. Working Group, 1997 Brown *et al.*, 1998 US DOE/ EIA, 1998 ECMT, 1997 (8 countries only) Kashiwagi *et al.*, 1999 Denis and Koopman, 1998 Worrell *et al.*, 1997b
	Europe		◆◆◆◆◆	◇◇	◆◆◆◆◆	◇◇	
	Japan		◆◆◆◆	◇◇	◆◆◆◆◆	◇◇	
	Dev. cos.		◆◆◆◆	◇◇	◆◆◆◆◆	◇◇	
Manufacturing CO_2 removal − fertilizer; refineries	Global		◆	◇◇◇◇	◆	◇◇◇◇	Table 3.21
Material efficiency improvement	Global		◆◆◆◆	◇◇◇	◆◆◆◆	◇◇◇	Table 3.21
Blended cements	Global		◆	◇◇◇	◆	◇◇◇	Table 3.21
N_2O reduction by chem. indus.	Global		◆	◇◇◇◇	◆	◇◇◇	Table 3.21
PFC reduction by Al industry	Global		◆	◇◇◇	◆	◇◇◇	Table 3.21
HFC-23 reduction by chem. industry	Global		◆◆	◇◇◇	◆◆	◇◇◇	Table 3.21
Energy efficient improvements	Global		◆◆◆◆◆	◇◇◇◇◇	◆◆◆◆◆	◇◇◇◇	Table 3.19
Agriculture Increased uptake of conservation tillage and cropland management	Dev. cos.		◆	◇◇	◆	◇◇	Zhou, 1998; *Table 3.27* Dick *et al.*, 1998 IPCC, 2000
	Global		◆◆◆	◇◇	◆◆◆◆	◇◇◇	
Soil carbon sequestration	Global		◆◆◆	◇◇	◆◆◆◆	◇◇◇	Lal and Bruce, 1999 *Table 3.27*
Nitrogenous fertilizer management	OECD		◆	◇◇◇	◆	◇◇◇	Kroeze & Mosier, 1999 *Table 3.27*
	Global		◆	◇◇◇	◆◆◆	◇◇◇◇◇	OECD, 1999; IPCC, 2000
Enteric methane reduction	OECD		◆◆	◇◇	◆◆	◇◇◇	Kroeze & Mosier, 1999 *Table 3.27* OECD, 1998 Reimer & Freund, 1999 Chipato, 1999
	USA		◆	◇◇	◆	◇◇◇	
	Dev. cos.		◆	◇	◆	◇◇	
Rice paddy irrigation and fertilizers	Global		◆◆◆	◇◇	◆◆◆◆	◇◇◇	Riemer & Freund, 1999 IPCC, 2000
Wastes Landfill methane capture	OECD		◆◆◆◆	◇◇◇	◆◆◆◆	◇◇◇◇	Landfill methane USEPA, 1999

Table TS-1 (continued)

	Region	US$/tC avoided (−400 −200 0 +200)	2010 Potential[a]	2010 Probability[b]	2020 Potential[a]	2020 Probability[b]	References, comments, and relevant section in Chapter 3 of the main report
Energy supply							
Nuclear for coal	Global		♦♦♦♦	◇◇	♦♦♦♦♦	◇◇◇◇	Totals[c] – See Section 3.8.6
	Annex I		♦♦	◇◇	♦♦♦♦	◇◇	Table 3.35a
	Non-Annex I		♦♦	◇◇◇	♦♦♦♦♦	◇◇◇	Table 3.35b
Nuclear for gas	Annex I		♦♦♦	◇	♦♦♦♦	◇	Table 3.35c
	Non-Annex I		♦	◇	♦♦♦	◇	Table 3.35d
Gas for coal	Annex I		♦	◇◇◇	♦♦♦♦	◇◇◇◇	Table 3.35a
	Non-Annex I		♦	◇◇◇◇	♦♦♦♦	◇◇◇◇	Tables 3.35b
CO_2 capture from coal	Global		♦	◇◇	♦♦	◇◇	Tables 3.35a + b
CO_2 capture from gas	Global		♦	◇◇	♦♦	◇◇	Tables 3.35c + d
Biomass for coal	Global		♦	◇◇◇◇	♦♦♦	◇◇◇◇	Tables 3.35a + b Moore, 1998; Interlab w. gp. 1997
Biomass for gas	Global		♦	◇	♦	◇◇◇	Tables 3.35c + d
Wind for coal or gas	Global		♦♦♦	◇◇◇		◇◇◇◇	Tables 3.35a - d BTM Cons 1999; Greenpeace, 1999
Co-fire coal with 10% biomass	USA		♦	◇◇◇	♦♦	◇◇◇	Sulilatu, 1998
Solar for coal	Annex I		♦	◇	♦	◇	Table 3.35a
	Non-Annex I		♦	◇	♦	◇	Table 3.35b
Hydro for coal	Global		♦♦	◇	♦♦♦	◇◇	Tables 3.35a + b
Hydro for gas	Global		♦	◇	♦♦	◇◇	Tables 3.35c + d

Notes:

[a] Potential in terms of tonnes of carbon equivalent avoided for the cost range of US$/tC given.

♦ = <20 MtC/yr   ♦♦ = 20-50 MtC/yr   ♦♦♦ = 50-100MtC/yr   ♦♦♦♦ = 100-200MtC/yr   ♦♦♦♦♦ = >200 MtC/yr

[b] Probability of realizing this level of potential based on the costs as indicated from the literature.

◇ = Very unlikely   ◇◇ = Unlikely   ◇◇◇ = Possible   ◇◇◇◇ = Probable   ◇◇◇◇◇ = Highly probable

[c] Energy supply total mitigation options assumes that not all the potential will be realized for various reasons including competition between the individual technologies as listed below the totals.

crops, and waste may also play an increasingly important role in the transportation sector as enzymatic hydrolysis of cellulosic material to ethanol becomes more cost effective. Meanwhile, biodiesel, supported by tax exemptions, is gaining market share in Europe. Incremental improvements in engine design have, however, largely been used to enhance performance rather than to improve fuel economy, which has not increased since the SAR. Fuel cell powered vehicles are developing rapidly, and are scheduled to be introduced to the market in 2003. Significant improvements in the fuel economy of aircraft appear to be both technically and economically possible for the next generation fleet. Nevertheless, most evaluations of the technological efficiency improvements (Table TS-3) show that because of growth in demand for transportation, efficiency improvement alone is not enough to avoid GHG emission growth. Also, there is evidence that, other things being equal, efforts to improve fuel efficiency have only partial effects in emission reduction because of resulting increases in driving distances caused by lower specific operational costs.

Table TS-2: Technological options, abrriers, opportunities, and impacts on production in various sectors.

Technological options	Barriers and opportunities	Implications of mitigation policies on sectors
Buildings, households and services: Hundreds of technologies and measures exist that can improve the energy efficiency of appliances and equipment as well as building structures in all regions of the world. It is estimated that CO_2 emissions from residential buildings in 2010 can be reduced by 325MtC in developed countries and the EIT region at costs ranging from −US$ 250 to −US$ 150/tC and by 125MtC in developing countries at costs of −US$ 250 to US$ 50/tC. Similarly, CO_2 emissions from commercial buildings in 2010 can be reduced by 185MtC in industrialized countries and the EIT region at costs ranging from −US$ 400 to −US$ 250/tC and by 80MtC in developing countries at costs ranging from −US$ 400 to US$ 0/tC. These savings represent almost 30% of buildings, CO_2 emissions in 2010 and 2020 compared to a central scenario such as the SRES B2 Marker scenario.	**Barriers:** In developed countries a market structure not conducive to efficiency improvements, misplaced incentives, and lack of information; and in developing countries lack of financing and skills, lack of information, traditional customs, and administered pricing. **Opportunities:** Developing better marketing approaches and skills, information-based marketing, voluntary programmes and standards have been shown to overcome barriers in developed countries. Affordable credit skills, capacity building, information base and consumer awareness, standards, incentives for capacity building, and deregulation of the energy industry are ways to address the aforementioned barriers in the developing world.	**Service industries:** Many will gain output and employment depending on how mitigation policies are implemented, however in general the increases are expected to be small and diffused. **Households and the informal sector:** The impact of mitigation on households comes directly through changes in the technology and price of household's use of energy and indirectly through macroeconomic effects on income and employment. An important ancillary benefit is the improvement in indoor and outdoor air quality, particularly in developing countries and cities all over the world.
Transportation: Transportation technology for light-duty vehicles has advanced more rapidly than anticipated in the SAR, as a consequence of international R&D efforts. Hybrid-electric vehicles have already appeared in the market and introduction of fuel cell vehicles by 2003 has been announced by most major manufacturers. The GHG mitigation impacts of technological efficiency improvements will be diminished to some extent by the rebound effect, unless counteracted by policies that effectively increase the price of fuel or travel. In countries with high fuel prices, such as Europe, the rebound effect may be as large as 40%; in countries with low fuel prices, such as the USA, the rebound appears to be no larger than 20%. Taking into account rebound effects, technological measures can reduce GHG emissions by 5%-15% by 2010 and 15%-35% by 2020, in comparison to a baseline of continued growth.	**Barriers:** Risk to manufacturers of transportation equipment is an important barrier to more rapid adoption of energy efficient technologies in transport. Achieving significant energy efficiency improvements generally requires a "clean sheet" redesign of vehicles, along with multibillion dollar investments in new production facilities. On the other hand, the value of greater efficiency to customers is the difference between the present value of fuel savings and increased purchase price, which net can often be a small quantity. Although markets for transport vehicles are dominated by a very small number of companies in the technical sense, they are nonetheless highly competitive in the sense that strategic errors can be very costly. Finally, many of the benefits of increased energy efficiency accrue in the form of social rather than private benefits. For all these reasons, the risk to manufacturers of sweeping technological change to improve energy efficiency is generally perceived to outweigh the direct market benefits. Enormous public and private investments in transportation infrastructure and a built environment adapted to motor vehicle travel pose significant barriers to changing the modal structure of transportation in many countries. **Opportunities:** Information technologies are creating new opportunities for pricing some of the external costs of transportation, from congestion to environmental pollution. Implementation of more efficient pricing can provide greater incentives for energy efficiency in both equipment and modal structure. The factors that hinder the adoption of fuel-efficient technologies in transport vehicle markets create conditions under which energy efficiency regulations, voluntary or mandatory, can be effective. Well-formulated regulations eliminate much of the risk of making sweeping technological changes, because all competitors face the same regulations. Study after study has demonstrated the existence of technologies capable of reducing vehicle carbon intensities by up to 50% or in the longer run 100%, approximately cost-effectively. Finally, intensive R&D efforts for light-duty road vehicles have achieved dramatic improvements in hybrid power-train and fuel cell technologies. Similar efforts could be directed at road freight, air, rail, and marine transport technologies, with potentially dramatic pay-offs.	**Transportation:** Growth in transportation demand is projected to remain, influenced by GHG mitigation policies only in a limited way. Only limited opportunities for replacing fossil carbon based fuels exist in the short to medium term. The main effect of mitigation policies will be to improve energy efficiency in all modes of transportation.

Table TS-2 (continued).

Technological options	*Barriers and opportunities*	*Implications of mitigation policies on sectors*
Industry: Energy efficiency improvement is the main emission reduction option in industry. Especially in industrialized countries much has been done already to improve energy efficiency, but options for further reductions remain. 300–500MtC/yr and 700–1,100MtC/yr can be reduced by 2010 and 2020, respectively, as compared to a scenario like SRES B2. The larger part of these options has net negative costs. Non-CO_2 emissions in industry are generally relatively small and can be reduced by over 85%, most at moderate or sometimes even negative costs.	**Barriers:** lack of full-cost pricing, relatively low contribution of energy to production costs, lack of information on part of the consumer and producer, limited availability of capital and skilled personnel are the key barriers to the penetration of mitigation technology in the industrial sector in all, but most importantly in developing countries. **Opportunities:** legislation to address local environmental concerns; voluntary agreements, especially if complemented by government efforts; and direct subsidies and tax credits are approaches that have been successful in overcoming the above barriers. Legislation, including standards, and better marketing are particularly suitable approaches for light industries.	**Industry:** Mitigation is expected to lead to structural change in manufacturing in Annex I countries (partly caused by changing demands in private consumption), with those sectors supplying energy-saving equipment and low-carbon technologies benefitting and energy-intensive sectors having to switch fuels, adopt new technologies, or increase prices. However, rebound effects may lead to unexpected negative results.
Land-use change and forestry: There are three fundamental ways in which land use or management can mitigate atmospheric CO_2 increases: protection, sequestration, and substitutiona. These options show different temporal patterns; consequently, the choice of options and their potential effectiveness depend on the target time frame as well as on site productivity and disturbance history. The SAR estimated that globally these measures could reduce atmospheric C by about 83 to 131GtC by 2050 (60 to 87GtC in forests and 23 to 44GtC in agricultural soils). Studies published since then have not substantially revised these estimates. The costs of terrestrial management practices are quite low compared to alternatives, and range from 0 ('win-win' opportunities) to US$12/tC.	**Barriers:** to mitigation in land-use change and forestry include lack of funding and of human and institutional capacity to monitor and verify, social constraints such as food supply, people living off the natural forest, incentives for land clearing, population pressure, and switch to pastures because of demand for meat. In tropical countries, forestry activities are often dominated by the state forest departments with a minimal role for local communities and the private sector. In some parts of the tropical world, particularly Africa, low crop productivity and competing demands on forests for crop production and fuelwood are likely to reduce mitigation opportunities. **Opportunities:** in land use and forestry, incentives and policies are required to realize the technical potential. There may be in the form of government regulations, taxes, and subsidies, or through economic incentives in the form of market payments for capturing and holding carbon as suggested in the Kyoto Protocol, depending on its implementation following decisions by CoP.	GHG mitigation policies can have a large effect on land use, especially through carbon sequestration and biofuel production. In tropical countries, large-scale adoption of mitigation activities could lead to biodiversity conservation, rural employment generation and watershed protection contributing to sustainable development. To achieve this, institutional changes to involve local communities and industry and necessary thereby leading to a reduced role for governments in managing forests.
Agriculture and waste management: Energy inputs are growing by <1% per year globally with the highest increases in non-OECD countries but they have reduced in the EITs. Several options already exist to decrease GHG emissions for investments of –US$ 50 to 150/tC. These include increasing carbon stock by cropland management (125MtC/yr by 2010); reducing CH_4 emissions from better livestock management (>30MtC/yr) and rice production (7MtC/yr); soil carbon sequestration (50-100MtC/yr) and reducing N_2O emissions from animal wastes and application of N measures are feasible in most regions given appropriate technology transfer and incentives for farmers to change their traditional methods. Energy cropping to displace fossil fuels has good prospects if the costs can be made more competitive and the crops are produced sustainably. Improved waste management can decrease GHG emissions by 200MtC$_{eq}$ in 2010 and 320MtC$_{eq}$ in 2020 as compared to 240MtC$_{eq}$ emissions in 1990.	**Barriers:** In agriculture and waste management, these include inadequate R& D funding, lack of intellectual property rights, lack of national human and institutional capacity and information in the developing countries, farm-level adoption constraints, lack of incentives and information for growers in developed countries to adopt new husbandry techniques, (need other benefits, not just greenhouse gas reduction). **Opportunities:** Expansion of credit schemes, shifts in research priorities, development of institutional linkages across countries, trading in soil carbon, and integration of food, fibre, and energy products are ways by which the barriers may be overcome. Measures should be linked with moves towards sustainable production methods. Energy cropping provides benefits of land use diversification where suitable land is currently under utilized for food and fibre production and water is readily available.	**Energy:** forest and land management can provide a variety of solid, liquid, or gaseous fuels that are renewable and that can substitute for fossil fuels. **Materials:** products from forest and other biological materials are used for construction, packaging, papers, and many other uses and are often less energy-intensive than are alternative materials that provide the same service. **Agriculture/land use:** commitment of large areas to carbon sequestration or carbon management may compliment or conflict with other demands for land, such as agriculture. GHG mitigation will have an impact on agriculture through increased demand for biofuel production in many regions. Increasing competition for arable land may increase prices of food and other agricultural products.

Table TS-2 (continued).

Technological options	Barriers and opportunities	Implications of mitigation policies on sectors
Waste management: Utilization of methane from landfills and from coal beds. The use of landfill gas for heat and electric power is also growing. In several industrial countries and especially in Europe and Japan, waste-to-energy facilities have become more efficient with lower air pollution emissions, paper and fibre recycling, or by utilizing waste paper as a biofuel in waste to energy facilities.	**Barriers:** Little is being done to manage landfill gas or to reduce waste in rapidly growing markets in much of the developing world. **Opportunities:** countries like the US and Germany have specific policies to either reduce methane producing waste, and/or requirements to utilize methane from landfills as an energy source. Costs of recovery are negative for half of landfill methane.	
Energy sector: In the energy sector, options are available both to increase conversion efficiency and to increase the use of primary energy with less GHGs per unit of energy produced, by sequestering carbon, and reducing GHG leakages. Win-win options such as coal bed methane recovery and improved energy efficiency in coal and gas fired power generation as well as co-production of heat and electricity can help to reduce emissions. With economic development continuing, efficiency increases alone will be insufficient to control GHG emissions from the energy sector. Options to decrease emissions per unit energy produced include new renewable forms of energy, which are showing strong growth but still account for less than 1% of energy produced worldwide. Technologies for CO_2 capture and disposal to achieve "clean fossil" energy have been proposed and could contribute significantly at costs competitive with renewable energy although considerable research is still needed on the feasibility and possible environmental impacts of such methods to determine their application and usage. Nuclear power and, in some areas, larger scale hydropower could make a substantially increased contribution but face problems of costs and acceptability. Emerging fuel cells are expected to open opportunities for increasing the average energy conversion efficiency in the decades to come.	**Barriers:** key barriers are human and institutional capacity, imperfect capital markets that discourage investment in small decentralized systems, more uncertain rates of return on investment, high trade tariffs, lack of information, and lack of intellectual property rights for mitigation technologies. For renewable energy, high first costs, lack of access to capital, and subsidies for fossil fuels and key barriers. **Opportunities** for developing countries include promotion of leapfrogs in energy supply and demand technology, facilitating technology transfer through creating an enabling environment, capacity building, and appropriate mechanisms for transfer of clean and efficient energy technologies. Full cost pricing and information systems provide opportunities in developed countries. Ancillary benefits associated with improved technology, and with reduced production and use of fossil fuels, can be substantial.	**Coal:** Coal production, use, and employment are likely to fall as a result of greenhouse gas mitigation policies, compared with projections of energy supply without additional climate policies. However, the costs of adjustment will be much lower if policies for new coal production also encourage clean coal technology. **Oil:** Global mitigation policies are likely to lead to reductions in oil production and trade, with energy exporters likely to face reductions in real incomes as compared to a situation without such policies. The effect on the global oil price of achieving the Kyoto targets, however, may be less severe than many of the models predict, because of the options to include non-CO_2 gases and the flexible mechanisms in achieving the target, which are often not included in the models. **Gas:** Over the next 20 years mitigation may influence the use of natural gas may positively or negatively, depending on regional and local conditions. In the Annex I countries any switch that takes place from coal or oil would be towards natural gas and renewable sources for power generation. In the case of the non-Annex 1 countries, the potential for switching to natural gas is much higher, however energy security and the availability of domestic resources are considerations, particularly for countries such as China and India with large coal reserves. **Renewables:** Renewable sources are very diverse and the mitigation impact would depend on technological development. It would vary from region to region depending on resource endowment. However, mitigation is very likely to lead to larger markets for the renewables industry. In that situation, R&D for cost reduction and enhanced performance and increased flow of funds to renewables could increase their application leading to cost reduction. **Nuclear:** There is substantial technical potential for nuclear power development to reduce greenhouse gas emissions; whether this is realized will depend on relative costs, political factors, and public acceptance.
Halocarbons: Emissions of HFCs are growing as HFCs are being used to replace some of the ozone-depleting substances being phased out. Compared to SRES projections for HFCs in 2010, it is estimated that emissions could be lower by as much as 100MtC$_{eq}$ at costs below US\$ 200/tC$_{eq}$. About half of the estimated reduction is an artifact caused by the SRES baseline values being higher than the study baseline for this report. The remainder could be accomplished by reducing emissions through containment, recovering and recycling refrigerants, and through use of alternative fluids and technologies.	**Barriers:** uncertainty with respect to the future of HFC policy in relation to global warming and ozone depletion. **Opportunities:** capturing new technological developments.	

Table TS-2 (continued).		
Technological options	***Barriers and opportunities***	***Implications of mitigation policies on sectors***
Geo-engineering: Regarding mitigation opportunities in marine ecosystems and geo-engineering[b], human understanding of biophysical systems, as well as many ethical, legal, and equity assessments are still rudimentary.	**Barriers:** In geo-engineering, the risks for unanticipated consequences are large and it may not even be possible to engineer the regional distribution of temperature and precipitation. **Opportunities:** Some basic inquiry appears appropriate.	**Sector not yet in existence:** not applicable.

[a] 'Protection' refers to active measures that maintain and preserve existing C reserves, including those in vegetation, soil organic matter, and products exported from the ecosystem (e. g., preventing the conversion of tropical forests for agricultural purposes and avoiding drainage of wetlands). 'Sequestration' refers to measures, deliberately undertaken, that increase C stocks above those already present (e. g., afforestation, revised forest management, enhanced C storage in wood products, and altered cropping systems, including more forage crops, reduced tillage). "Substitution" refers to practices that substitute renewable biological products for fossil fuels or energy-intensive products, thereby avoiding the emission of CO_2 from combustion of fossil fuels.
[b] Geo-engineering involves efforts to stabilize the climate system by directly managing the energy balance of the earth, thereby overcoming the enhanced greenhouse effect.

Table TS-3: Projected energy intensities for transportation from 5-Laboratory Study in the USA.[a]

			2010	
Determinants	***1997***	***BAU***	***Energy efficiency***	***HE/LC***
New passenger car l/100km	8.6	8.5	6.3	5.5
New light truck l/100km	11.5	11.4	8.7	7.6
Light-duty fleet l/100km[b]	12.0	12.1	10.9	10.1
Aircraft efficiency (seat-l/100km)	4.5	4.0	3.8	3.6
Freight truck fleet l/100km	42.0	39.2	34.6	33.6
Rail efficiency (tonne-km/MJ)	4.2	4.6	5.5	6.2

[a] BAU, Business as usual; HE/LC, high-energy/low-carbon.
[b] Includes existing passenger cars and light trucks.

3.3.3. The Main Mitigation Options in the Industry Sector

Industrial emissions account for 43% of carbon released in 1995. Industrial sector carbon emissions grew at a rate of 1.5% per year between 1971 and 1995, slowing to 0.4% per year since 1990. Industries continue to find more energy efficient processes and reductions of process-related GHGs. This is the only sector that has shown an annual decrease in carbon emissions in OECD economies (−0.8%/yr between 1990 and 1995). The CO_2 from EITs declined most strongly (−6.4% per year between 1990 and 1995 when total industrial production dropped).

Differences in the energy efficiency of industrial processes between different developed countries, and between developed and developing countries remain large, which means that there are substantial differences in relative emission reduction potentials between countries.

Improvement of the energy efficiency of industrial processes is the most significant option for lowering GHG emissions. This potential is made up of hundreds of sector-specific technologies. The worldwide potential for energy efficiency improvement − compared to a baseline development − for the year 2010 is estimated to be 300-500MtC and for the year 2020 700-900MtC. In the latter case continued technological development is necessary to realize the potential. The majority of energy efficiency improvement options can be realized at net negative costs.

Another important option is material efficiency improvement (including recycling, more efficient product design, and material substitution); this may represent a potential of 600MtC in the year 2020. Additional opportunities for CO_2 emissions reduction exist through fuel switching, CO_2 removal and storage, and the application of blended cements.

A number of specific processes not only emit CO_2, but also non-CO_2 GHGs. The adipic acid manufacturers have strongly reduced their N_2O emissions, and the aluminium industry has made major gains in reducing the release of PFCs (CF_4, C_2F_6). Further reduction of non-CO_2 GHGs from manufacturing industry to low levels is often possible at relatively low costs per tonne of C-equivalent (tC_{eq}) mitigated.

Sufficient technological options are known today to reduce GHG emissions from industry in absolute terms in most developed countries by 2010, and to limit growth of emissions in this sector in developing countries significantly.

3.3.4. The Main Mitigation Options in the Agricultural Sector

Agriculture contributes only about 4% of global carbon emissions from energy use, but over 20% of anthropogenic GHG emissions (in terms of MtC_{eq}/yr) mainly from CH_4 and N_2O as well as carbon from land clearing. There have been modest gains in energy efficiency for the agricultural sector since the SAR, and biotechnology developments related to plant and animal production could result in additional gains, provided concerns about adverse environmental effects can be adequately addressed. A shift from meat towards plant production for human food purposes, where feasible, could increase energy efficiency and decrease GHG emissions (especially N_2O and CH_4 from the agricultural sector). Significant abatement of GHG emissions can be achieved by 2010 through changes in agricultural practices, such as:

- soil carbon uptake enhanced by conservation tillage and reduction of land use intensity;
- CH_4 reduction by rice paddy irrigation management, improved fertilizer use, and lower enteric CH_4 emissions from ruminant animals;
- avoiding anthropogenic agricultural N_2O emissions (which for agriculture exceeds carbon emission from fossil fuel use) through the use of slow release fertilizers, organic manure, nitrification inhibitors, and potentially genetically-engineered leguminous plants. N_2O emissions are greatest in China and the USA, mainly from fertilizer use on rice paddy soils and other agricultural soils. More significant contributions can be made by 2020 when more options to control N_2O emissions from fertilized soils are expected to become available.

Uncertainties on the intensity of use of these technologies by farmers are high, since they may have additional costs involved in their uptake. Economic and other barriers may have to be removed through targetted policies.

3.3.5. The Main Mitigation Options in the Waste Management Sector

There has been increased utilization of CH_4 from landfills and from coal beds. The use of landfill gas for heat and electric power is also growing because of policy mandates in countries like Germany, Switzerland, the EU, and USA. Recovery costs are negative for half of landfill CH_4. Requiring product life management in Germany has been extended from packaging to vehicles and electronics goods. If everyone in the USA increased per capita recycling rates from the national average to the per capita recycling rate achieved in Seattle, Washington, the result would be a reduction of 4% of total US GHG emissions. Debate is taking place over whether the greater reduction in lifecycle

GHG emissions occurs through paper and fibre recycling or by utilizing waste paper as a biofuel in waste-to-energy facilities. Both options are better than landfilling in terms of GHG emissions. In several developed countries, and especially in Europe and Japan, waste-to-energy facilities have become more efficient with lower air pollution emissions.

3.3.6. The Main Mitigation Options in the Energy Supply Sector

Fossil fuels continue to dominate heat and electric power production. Electricity generation accounts for 2,100MtC/yr or 37.5% of global carbon emissions[10]. Baseline scenarios without carbon emission policies anticipate emissions of 3,500 and 4,000MtC_{eq} for 2010 and 2020, respectively. In the power sector, low-cost combined cycle gas turbines (CCGTs) with conversion efficiencies approaching 60% for the latest model have become the dominant option for new electric power plants wherever adequate natural gas supply and infrastructure are available. Advanced coal technologies based on integrated gasification combined cycle or supercritical (IGCCS) designs potentially have the capability of reducing emissions at modest cost through higher efficiencies. Deregulation of the electric power sector is currently a major driver of technological choice. Utilization of distributed industrial and commercial combined heat and power (CHP) systems to meet space heating and manufacturing needs could achieve substantial emission reductions. The further implications of the restructuring of the electric utility industry in many developed and developing countries for CO_2 emissions are uncertain at this time, although there is a growing interest in distributed power supply systems based on renewable energy sources and also using fuel cells, micro-turbines and Stirling engines.

The nuclear power industry has managed to increase significantly the capacity factor at existing facilities, which improved their economics sufficiently that extension of facility life has become cost effective. But other than in Asia, relatively few new plants are being proposed or built. Efforts to develop intrinsically safe and less expensive nuclear reactors are proceeding with the goal of lowering socio-economic barriers and reducing public concern about safety, nuclear waste storage, and proliferation. Except for a few large projects in India and China, construction of new hydropower projects has also slowed because of few available major sites, sometimes-high costs, and local environmental and social concerns. Another development is the rapid growth of wind turbines, whose annual growth rate has exceeded 25% per year, and by 2000 exceeded 13GW of installed capacity. Other renewables, including solar and biomass, continue to grow as costs decline, but total contributions from non-hydro renewable

---

[10] Note that the section percentages do not add up to 100% as these emissions have been allocated to the four sectors in the paragraphs above.

sources remain below 2% globally. Fuel cells have the potential to provide highly efficient combined sources of electricity and heat as power densities increase and costs continue to drop. By 2010, co-firing of coal with biomass, gasification of fuel wood, more efficient photovoltaics, off-shore wind farms, and ethanol-based biofuels are some of the technologies that are capable of penetrating the market. Their market share is expected to increase by 2020 as the learning curve reduces costs and capital stock of existing generation plants is replaced.

Physical removal and storage of CO_2 is potentially a more viable option than at the time of the SAR. The use of coal or biomass as a source of hydrogen with storage of the waste CO_2 represents a possible step to the hydrogen economy. CO_2 has been stored in an aquifer, and the integrity of storage is being monitored. However, long-term storage is still in the process of being demonstrated for that particular reservoir. Research is also needed to determine any adverse and/or beneficial environmental impacts and public health risks of uncontrolled release of the various storage options. Pilot CO_2 capture and storage facilities are expected to be operational by 2010, and may be capable of making major contributions to mitigation by 2020. Along with biological sequestration, physical removal and storage might complement current efforts at improving efficiency, fuel switching, and the development of renewables, but must be able to compete economically with them.

The report considers the potential for mitigation technologies in this sector to reduce CO_2 emissions to 2020 from new power plants. CCGTs are expected to be the largest provider of new capacity between now and 2020 worldwide, and will be a strong competitor to displace new coal-fired power stations where additional gas supplies can be made available. Nuclear power has the potential to reduce emissions if it becomes politically acceptable, as it can replace both coal and gas for electricity production. Biomass, based mainly on wastes and agricultural and forestry by-products, and wind power are also potentially capable of making major contributions by 2020. Hydropower is an established technology and further opportunities exist beyond those anticipated to contribute to reducing CO_2 equivalent emissions. Finally, while costs of solar power are expected to decline substantially, it is likely to remain an expensive option by 2020 for central power generation, but it is likely to make increased contributions in niche markets and off-grid generation. The best mitigation option is likely to be dependent on local circumstances, and a combination of these technologies has the potential to reduce CO_2 emissions by 350-700MtC by 2020 compared to projected emissions of around 4,00MtC from this sector.

3.3.7. The Main Mitigation Options for Hydrofluorocarbons and Perfluorocarbons

HFC and, to a lesser extent, PFC use has grown as these chemicals replaced about 8% of the projected use of CFCs by weight in 1997; in the developed countries the production of CFCs and other ozone depleting substances (ODSs) was halted in 1996 to comply with the Montreal Protocol to protect the stratospheric ozone layer. HCFCs have replaced an additional 12% of CFCs. The remaining 80% have been eliminated through controlling emissions, specific use reductions, or alternative technologies and fluids including ammonia, hydrocarbons, carbon dioxide and water, and not-in-kind technologies. The alternative chosen to replace CFCs and other ODSs varies widely among the applications, which include refrigeration, mobile and stationary air-conditioning, heat pumps, medical and other aerosol delivery systems, fire suppression, and solvents. Simultaneously considering energy efficiency with ozone layer protection is important, especially in the context of developing countries, where markets have just begun to develop and are expected to grow at a fast rate.

Based on current trends and assuming no new uses outside the ODS substitution area, HFC production is projected to be 370kt or 170MtC$_{eq}$/yr by 2010, while PFC production is expected to be less than 12MtC$_{eq}$/yr. For the year 2010, annual emissions are more difficult to estimate. The largest emissions are likely to be associated with mobile air conditioning followed by commercial refrigeration and stationary air conditioning. HFC use in foam blowing is currently low, but if HFCs replaces a substantial part of the HCFCs used here, their use is projected to reach 30MtC$_{eq}$/yr by 2010, with emissions in the order of 5-10MtC$_{eq}$/yr.

3.4. The Technological and Economic Potential of Greenhouse Gas Mitigation: Synthesis

Global emissions of GHGs grew on average by 1.4% per year during the period 1990 to 1998. In many areas, technical progress relevant to GHG emission reduction since the SAR has been significant and faster than anticipated. The total potential for worldwide GHG emissions reductions resulting from technological developments and their adoption amount to 1,900 to 2,600MtC/yr by 2010, and 3,600 to 5,050MtC/yr by 2020. The evidence on which this conclusion is based is extensive, but has several limitations. No comprehensive worldwide study of technological potential has yet been done, and the existing regional and national studies generally have varying scopes and make different assumptions about key parameters. Therefore, the estimates as presented in Table TS-1 should be considered to be indicative only. Nevertheless, the main conclusion in the paragraph above can be drawn with high confidence.

Costs of options vary by technology and show regional differences. Half of the potential emissions reductions may be achieved by 2020 with direct benefits (energy saved) exceeding direct costs (net capital, operating, and maintenance costs), and the other half at a net direct cost of up to US$100/tC$_{eq}$ (at 1998 prices). These cost estimates are derived using discount rates in the range of 5% to 12%, consistent with public sector discount rates. Private internal rates of return vary greatly, and are often significantly higher, which affects the rate of adoption of these technologies by private entities. Depending on the emissions scenario this could allow global emissions to be reduced below 2000 levels in 2010-2020 at these net direct costs. Realizing these reductions will involve additional implementation costs, which in some cases may be substantial, and will possibly need supporting policies (such as those described in Section 6), increased research and development, effective technology transfer, and other barriers to be overcome (Section 5 for details).

Hundreds of technologies and practices exist to reduce GHG emissions from the buildings, transport, and industry sectors. These energy efficiency options are responsible for more than half of the total emission reduction potential of these sectors. Efficiency improvements in material use (including recycling) will also become more important in the longer term. The energy supply and conversion sector will remain dominated by cheap and abundant fossil fuels. However, there is significant emission reduction potential thanks to a shift from coal to natural gas, conversion efficiency improvement of power plants, the expansion of distributed co-generation plants in industry, commercial buildings and institutions, and CO_2 recovery and sequestration. The continued use of nuclear power plants (including their lifetime extension), and the application of renewable energy sources could avoid some additional emissions from fossil fuel use. Biomass from by-products and wastes such as landfill gas are potentially important energy sources that can be supplemented by energy crop production where suitable land and water are available. Wind energy and hydropower will also contribute, more so than solar energy because of its relatively high costs. N_2O and fluorinated GHG reductions have already been achieved through major technological advances. Process changes, improved containment and recovery, and the use of alternative compounds and technologies have been implemented. Potential for future reductions exists, including process-related emissions from insulated foam and semiconductor production and by-product emissions from aluminium and HCFC-22. The potential for energy efficiency improvements connected to the use of fluorinated gases is of a similar magnitude to reductions of direct emissions. Soil carbon sequestration, enteric CH_4 control, and conservation tillage can all contribute to mitigating GHG emissions from agriculture.

Appropriate policies are required to realize these potentials. Furthermore, on-going research and development is expected to significantly widen the portfolio of technologies that provide emission reduction options. Maintaining these R&D activities together with technology transfer actions will be necessary if the longer term potential as outlined in Table TS-1 is to be realized. Balancing mitigation activities in the various sectors with other goals, such as those related to DES, is key to ensuring they are effective.

4. Technological and Economic Potential of Options to Enhance, Maintain and Manage Biological Carbon Reservoirs and Geo-engineering

4.1. Mitigation through Terrestrial Ecosystem and Land Management

Forests, agricultural lands, and other terrestrial ecosystems offer significant, if often temporary, mitigation potential. Conservation and sequestration allow time for other options to be further developed and implemented. The IPCC SAR estimated that about 60 to 87GtC could be conserved or sequestered in forests by the year 2050 and another 23 to 44GtC could be sequestered in agricultural soils. The current assessment of the potential of biological mitigation options is in the order of 100GtC (cumulative) by 2050, equivalent to about 10% to 20% of projected fossil fuel emissions during that period. In this section, biological mitigation measures in terrestrial ecosystems are assessed, focusing on the mitigation potential, ecological and environmental constraints, economics, and social considerations. Also, briefly, the so-called geo-engineering options are discussed.

Increased carbon pools through the management of terrestrial ecosystems can only partially offset fossil fuel emissions. Moreover, larger C stocks may pose a risk for higher CO_2 emissions in the future, if the C-conserving practices are discontinued. For example, abandoning fire control in forests, or reverting to intensive tillage in agriculture may result in a rapid loss of at least part of the C accumulated during previous years. However, using biomass as a fuel or wood to displace more energy-intensive materials can provide permanent carbon mitigation benefits. It is useful to evaluate terrestrial sequestration opportunities alongside emission reduction strategies, as both approaches will likely be required to control atmospheric CO_2 levels.

Carbon reservoirs in most ecosystems eventually approach some maximum level. The total amount of carbon stored and/or carbon emission avoided by a forest management project at any given time is dependent on the specific management practices (see Figure TS-6). Thus, an ecosystem depleted of carbon by past events may have a high potential rate of carbon accumulation, while one with a large carbon pool tends to have a low rate of carbon sequestration. As ecosystems eventually approach their maximum carbon pool, the sink (i.e., the rate of change of the pool) will diminish. Although both the sequestration rate and pool of carbon may be relatively high at some stages, they cannot be maximized simultaneously. Thus, management strategies for an ecosystem may depend on whether the goal is to enhance short-term accumulation or to maintain the carbon reservoirs through time. The ecologically achievable balance between the two goals is constrained by disturbance history, site productivity, and target time frame. For example, options to maximize sequestration by 2010 may not maximize sequestration by 2020 or 2050; in some cases,

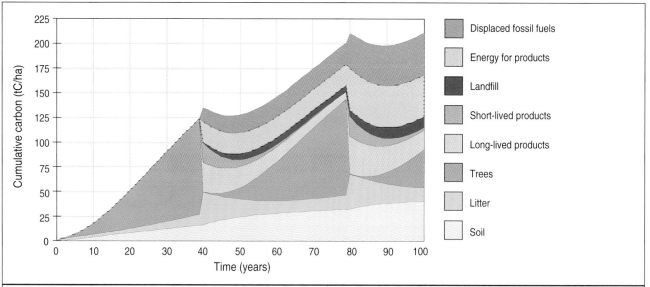

Figure TS-6: Carbon balance from a hypothetical forest management project.

Note: The figure shows cumulative carbon-stock changes for a scenario involving afforestation and harvest for a mix of traditional forest products with some of the harvest being used as a fuel. Values are illustrative of what might be observed in the southeastern USA or Central Europe. Regrowth restores carbon to the forest and the (hypothetical) forest stand is harvested every 40 years, with some litter left on the ground to decay, and products accumulate or are disposed of in landfills. These are net changes in that, for example, the diagram shows savings in fossil fuel emissions with respect to an alternative scenario that uses fossil fuels and alternative, more energy-intensive products to provide the same services.

maximizing sequestration by 2010 may lead to lower carbon storage over time.

The effectiveness of C mitigation strategies, and the security of expanded C pools, will be affected by future global changes, but the impacts of these changes will vary by geographical region, ecosystem type, and local abilities to adapt. For example, increases in atmospheric CO_2, changes in climate, modified nutrient cycles, and altered (either natural or human induced disturbance) regimes can each have negative or positive effects on C pools in terrestrial ecosystems.

In the past, land management has often resulted in reduced C pools, but in many regions like Western Europe, C pools have now stabilized and are recovering. In most countries in temperate and boreal regions forests are expanding, although current C pools are still smaller than those in pre-industrial or pre-historic times. While complete recovery of pre-historic C pools is unlikely, there is potential for substantial increases in carbon stocks. The Food and Agriculture Organization (FAO) and the UN Economic Commission for Europe (ECE)'s statistics suggest that the average net annual increment exceeded timber fellings in managed boreal and temperate forests in the early 1990s. For example, C stocks in live tree biomass have increased by 0.17GtC/yr in the USA and 0.11GtC/yr in Western Europe, absorbing about 10% of global fossil CO_2 emissions for that time period. Though these estimates do not include changes in litter and soils, they illustrate that land surfaces play a significant and changing role in the atmospheric carbon budget. Enhancing these carbon pools provides potentially powerful opportunities for climate mitigation.

In some tropical countries, however, the average net loss of forest carbon stocks continues, though rates of deforestation may have declined slightly in the past decade. In agricultural lands, options are now available to recover partially the C lost during the conversion from forest or grasslands.

4.2. Social and Economic Considerations

Land is a precious and limited resource used for many purposes in every country. The relationship of climate mitigation strategies with other land uses may be competitive, neutral, or symbiotic. An analysis of the literature suggests that C mitigation strategies can be pursued as one element of more comprehensive strategies aimed at sustainable development, where increasing C stocks is but one of many objectives. Often, measures can be adopted within forestry, agriculture, and other land uses to provide C mitigation and, at the same time, also advance other social, economic, and environmental goals. Carbon mitigation can provide additional value and income to land management and rural development. Local solutions and targets can be adapted to priorities of sustainable development at national, regional, and global levels.

A key to making C mitigation activities effective and sustainable is to balance it with other ecological and/or environmental, economic, and social goals of land use. Many biological mitigation strategies may be neutral or favourable for all three goals and become accepted as "no regrets" or "win-win" solutions. In other cases, compromises may be needed. Important potential environmental impacts include effects on biodiversity, effects on amount and quality of water resources (particularly where they are already scarce), and long-term impacts on ecosystem productivity. Cumulative environmental, economic, and social impacts could be assessed in individual projects and also from broader, national and international perspectives. An important issue is "leakage" – an expanded or conserved C pool in one area leading to increased emissions elsewhere. Social acceptance at the local, national, and global levels may also influence how effectively mitigation policies are implemented.

4.3. Mitigation Options

In tropical regions there are large opportunities for C mitigation, though they cannot be considered in isolation of broader policies in forestry, agriculture, and other sectors. Additionally, options vary by social and economic conditions: in some regions slowing or halting deforestation is the major mitigation opportunity; in other regions, where deforestation rates have declined to marginal levels, improved natural forest management practices, afforestation, and reforestation of degraded forests and wastelands are the most attractive opportunities. However, the current mitigative capacity[11] is often weak and sufficient land and water is not always available.

Non-tropical countries also have opportunities to preserve existing C pools, enhance C pools, or use biomass to offset fossil fuel use. Examples of strategies include fire or insect control, forest conservation, establishing fast-growing stands, changing silvicultural practices, planting trees in urban areas, ameliorating waste management practices, managing agricultural lands to store more C in soils, improving management of grazing lands, and re-planting grasses or trees on cultivated lands.

Wood and other biological products play several important roles in carbon mitigation: they act as a carbon reservoir; they can replace construction materials that require more fossil fuel input; and they can be burned in place of fossil fuels for renewable energy. Wood products already contribute somewhat to climate mitigation, but if infrastructures and incentives can be developed, wood and agricultural products may become a vital element of a sustainable economy: they are among the few renewable resources available on a large scale.

4.4. Criteria for Biological Carbon Mitigation Options

To develop strategies that mitigate atmospheric CO_2 and advance other, equally important objectives, the following criteria merit consideration:
- potential contributions to C pools over time;
- sustainability, security, resilience, permanence, and robustness of the C pool maintained or created;

- compatibility with other land-use objectives;
- leakage and additionality issues;
- economic costs;
- environmental impacts other than climate mitigation;
- social, cultural, and cross-cutting issues, as well as issues of equity; and
- the system-wide effects on C flows in the energy and materials sector.

Activities undertaken for other reasons may enhance mitigation. An obvious example is reduced rates of tropical deforestation. Furthermore, because wealthy countries generally have a stable forest estate, it could be argued that economic development is associated with activities that build up forest carbon reservoirs.

4.5. Economic Costs

Most studies suggest that the economic costs of some biological carbon mitigation options, particularly forestry options, are quite modest through a range. Cost estimates of biological mitigation reported to date vary significantly from US\$0.1/tC to about US\$20/tC in several tropical countries and from US\$20 to US\$100/tC in non-tropical countries. Moreover the cost calculations do not cover, in many instances, *inter alia*, costs for infrastructure, appropriate discounting, monitoring, data collection and interpretation, and opportunity costs of land and maintenance, or other recurring costs, which are often excluded or overlooked. The lower end of the ranges are biased downwards, but understanding and treatment of costs is improving over time. Furthermore, in many cases biological mitigation activities may have other positive impacts, such as protecting tropical forests or creating new forests with positive external environmental effects. However, costs rise as more biological mitigation options are exercised and as the opportunity costs of the land increases. Biological mitigation costs appear to be lowest in developing countries and higher in developed countries. If biological mitigation activities are modest, leakage is likely to be small. However, the amount of leakage could rise if biological mitigation activities became large and widespread.

4.6. Marine Ecosystem and Geo-engineering

Marine ecosystems may also offer possibilities for removing CO_2 from the atmosphere. The standing stock of C in the marine biosphere is very small, however, and efforts could focus, not on increasing biological C stocks, but on using biospheric processes to remove C from the atmosphere and transport it to the deep ocean. Some initial experiments have been performed, but fundamental questions remain about the permanence and stability of C removals, and about unintended consequences of the large-scale manipulations required to have a significant impact on the atmosphere. In addition, the economics of such approaches have not yet been determined.

[11] Mitigative capacity: the social, political, and economic structures and conditions that are required for effective mitigation.

Geo-engineering involves efforts to stabilize the climate system by directly managing the energy balance of the earth, thereby overcoming the enhanced greenhouse effect. Although there appear to be possibilities for engineering the terrestrial energy balance, human understanding of the system is still rudimentary. The prospects of unanticipated consequences are large, and it may not even be possible to engineer the regional distribution of temperature, precipitation, etc. Geo-engineering raises scientific and technical questions as well as many ethical, legal, and equity issues. And yet, some basic inquiry does seem appropriate.

In practice, by the year 2010 mitigation in land use, land-use change, and forestry activities can lead to significant mitigation of CO_2 emissions. Many of these activities are compatible with, or complement, other objectives in managing land. The overall effects of altering marine ecosystems to act as carbon sinks or of applying geo-engineering technology in climate change mitigation remain unresolved and are not, therefore, ready for near-term application.

5. Barriers, Opportunities, and Market Potential of Technologies and Practices

5.1. Introduction

The transfer of technologies and practices that have the potential to reduce GHG emissions is often hampered by barriers[12] that slow their penetration. The opportunity[13] to mitigate GHG concentrations by removing or modifying barriers to or otherwise accelerating the spread of technology may be viewed within a framework of different potentials for GHG mitigation (Figure TS-7). Starting at the bottom, one can imagine addressing barriers (often referred to as market failures) that relate to markets, public policies, and other institutions that inhibit the diffusion of technologies that are (or are projected to be) cost-effective for users without reference to any GHG benefits they may generate. Amelioration of this class of "market and institutional imperfections" would increase GHG mitigation towards the level that is labelled as the "economic potential". The economic potential represents the level of GHG mitigation that could be achieved if all technologies that are cost-effective from the consumers' point of view were implemented. Because economic potential is evaluated from the consumer's point of view, we would evaluate cost-effectiveness using market prices and the private rate of time discounting, and also take into account consumers' preferences regarding the acceptability of the technologies' performance characteristics.

---

[12] A barrier is any obstacle to reaching a potential that can be overcome by a policy, programme, or measure.

[13] An opportunity is a situation or circumstance to decrease the gap between the market potential of a technology or practice and the economic, socio-economic, or technological potential.

Of course, elimination of all these market and institutional barriers would not produce technology diffusion at the level of the "technical potential". The remaining barriers, which define the gap between economic potential and technical potential, are usefully placed in two groups separated by a socio-economic potential. The first group consists of barriers derived from people's preferences and other social and cultural barriers to the diffusion of new technology. That is, even if market and institutional barriers are removed, some GHG-mitigating technologies may not be widely used simply because people do not like them, are too poor to afford them, or because existing social and cultural forces operate against their acceptance. If, in addition to overcoming market and institutional barriers, this second group of barriers could be overcome, what is labelled as the "socio-economic potential" would be achieved. Thus, the socio-economic potential represents the level of GHG mitigation that would be approached by overcoming social and cultural obstacles to the use of technologies that are cost-effective.

Finally, even if all market, institutional, social, and cultural barriers were removed, some technologies might not be widely used simply because they are too expensive. Elimination of this requirement would therefore take us up to the level of "technological potential", the maximum technologically feasible extent of GHG mitigation through technology diffusion.

An issue arises as to how to treat the relative environmental costs of different technologies within this framework. Because the purpose of the exercise is ultimately to identify opportunities for global climate change policies, the technology potentials are defined without regard to GHG impacts. Costs and benefits associated with other environmental impacts would be part of the cost-effectiveness calculation underlying economic potential only insofar as existing environmental regulations or policies internalize these effects and thereby impose them on consumers. Broader impacts might be ignored by consumers, and hence not enter into the determination of economic potential, but they would be incorporated into a social cost-effectiveness calculation. Thus, to the extent that other environmental benefits make certain technologies socially cost-effective, even if they are not cost-effective from a consumer's point of view, the GHG benefits of diffusion of such technologies would be incorporated in the socio-economic potential.

5.2. Sources of Barriers and Opportunities

Technological and social innovation is a complex process of research, experimentation, learning, and development that can contribute to GHG mitigation. Several theories and models have been developed to understand its features, drivers, and implications. New knowledge and human capital may result from R&D spending, through learning by doing, and/or in an evolutionary process. Most innovations require some social or behavioural change on the part of users. Rapidly changing economies, as well as social and institutional structures offer opportunities for locking

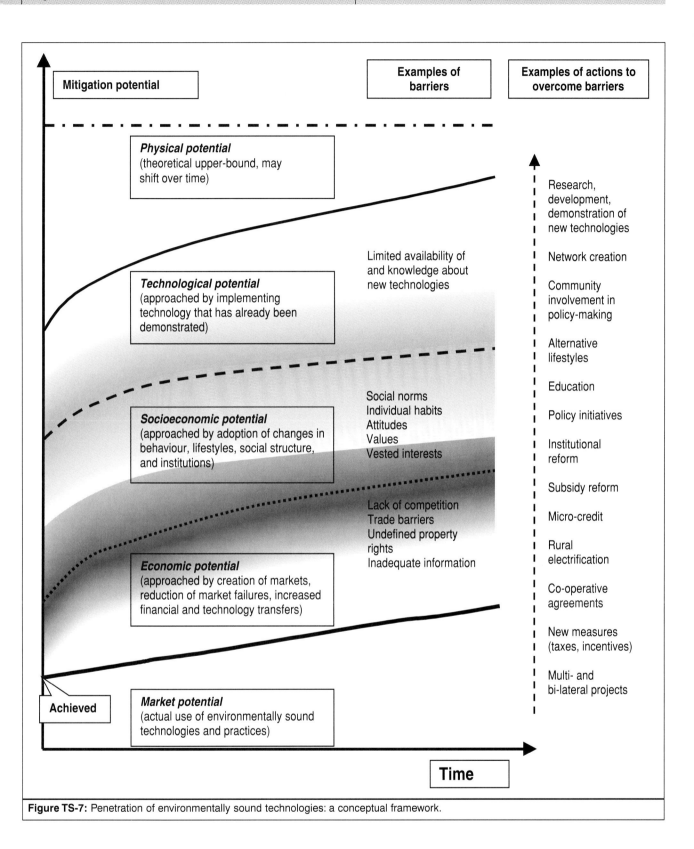

Figure TS-7: Penetration of environmentally sound technologies: a conceptual framework.

in to GHG-mitigative technologies that may lead countries on to sustainable development pathways. The pathways will be influenced by the particular socio-economic context that reflects prices, financing, international trade, market structure, institutions, the provision of information, and social, cultural, and behavioural factors; key elements of these are described below.

Unstable macroeconomic conditions increase risk to private investment and finance. Unsound government borrowing and fiscal policy lead to chronic public deficits and low liquidity in the private sector. Governments may also create perverse micro-economic incentives that the encourage rent-seeking and corruption, rather than the efficient use of resources. Trade barriers that favour inefficient technologies, or prevent access to foreign technology, slow technology diffusion. Tied aid still dominates in official development assistance. It distorts the efficiency of technology choice, and may crowd-out viable business models.

Commercial financing institutions face high risks with developing "green" financial products. Environmentally sound technologies with relatively small project sizes and long repayment periods deter banks with their high transaction costs. Small collateral value makes it difficult to use financing instruments, such as project finance. Innovative approaches in the private sector to address these issues include leasing, environmental and ethical banks, micro-credits or small grants facilities targetted at low income households, environmental funds, energy service companies (ESCOs), and green venture capital. The insurance industry has already begun to react to risks of climate change. New green financial institutions, such as forestry investment funds, have tapped market opportunities by working towards capturing values of standing forests.

Distorted or incomplete prices are also important barriers. The absence of a market price for certain impacts (externalities), such as environmental harm, constitutes a barrier to the diffusion of environmentally beneficial technologies. Distortion of prices because of taxes, subsidies, or other policy interventions that make resource consumption more or less expensive to consumers also impedes the diffusion of resource-conserving technologies.

Network externalities can generate barriers. Some technologies operate in such a way that a given user's equipment interacts with the equipment of other users so as to create "network externalities". For example, the attractiveness of vehicles using alternative fuels depends on the availability of convenient refuelling sites. On the other hand, the development of a fuel distribution infrastructure depends on there being a demand for alternative fuel vehicles.

Misplaced incentives result between landlords and tenants when the tenant is responsible for the monthly cost of fuel and/or electricity, and the landlord is prone to provide the cheapest-first-cost equipment without regard to its monthly energy use. Similar problems are encountered when vehicles are purchased by companies for the use of their employees.

Vested interests: A major barrier to the diffusion of technical progress lies in the vested interests who specialize in conventional technologies and who may, therefore, be tempted to collude and exert political pressure on governments to impose administrative procedures, taxes, trade barriers, and regulations in order to delay or even prevent the arrival of new innovations that might destroy their rents.

Lack of effective regulatory agencies impedes the introduction of environmentally sound technologies. Many countries have excellent constitutional and legal provisions for environmental protection but the latter are not enforced. However, "informal regulation" under community pressure from, for example, non-governmental organizations (NGOs), trade unions, neighbourhood organizations, *etc.* may substitute for formal regulatory pressure.

Information is often considered *as a public good*. Generic information regarding the availability of different kinds of technologies and their performance characteristics may have the attributes of a "public good" and hence may be underprovided by the private market. This problem is exacerbated by the fact that even after a technology is in place and being used, it is often difficult to quantify the energy savings that resulted from its installation owing to measurement errors and the difficulty with baseline problems. Knowing that this uncertainty will prevail can itself inhibit technology diffusion.

Current lifestyles, behaviours, and consumption patterns have developed within current and historical socio-cultural contexts. Changes in behaviour and lifestyles may result from a number of intertwined processes, such as:

- scientific, technological, and economic developments;
- developments in dominant world views and public discourse;
- changes in the relationships among institutions, political alliances, or actor networks;
- changes in social structures or relationships within firms and households; and
- changes in psychological motivation (e.g., convenience, social prestige, career, *etc.*).

Barriers take various forms in association with each of the above processes.

In some situations policy development is based on a model of human psychology that has been widely criticized. People are assumed to be rational welfare-maximizers and to have a fixed set of values. Such a model does not explain processes, such as learning, habituation, value formation, or the bounded rationality, observed in human choice. Social structures can affect consumption, for example, through the association of objects with status and class. Individuals' adoption of more sustainable consumption patterns depends not only on the match between those patterns and their perceived needs, but also on the extent to which they understand their consumption options, and are able to make choices.

Uncertainty: Another important barrier is uncertainty. A consumer may be uncertain about future energy prices and, therefore, future energy savings. Also, there may be uncertainty about the next generation of equipment – will next year bring a cheaper or better model? In practical decision making, a barrier is often associated with the issue of sunk cost and long lifetimes of infrastructure, and the associated irreversibilities of investments of the non-fungible infrastructure capita.

5.3. Sector- and Technology-specific Barriers and Opportunities

The following sections describe barriers and opportunities particular to each mitigation sector (see also Table TS-2).

Buildings: The poor in every country are affected far more by barriers in this sector than the rich, because of inadequate access to financing, low literacy rates, adherence to traditional customs, and the need to devote a higher fraction of their income to satisfy basic needs, including fuel purchases. Other barriers in this sector are lack of skills and social barriers, misplaced incentives, market structure, slow stock turnover, administratively set prices, and imperfect information. Integrated building design for residential construction could lead to energy saving by 40%-60%, which in turn could reduce the cost of living (Section 3.3.4).

Policies, programmes, and measures to remove barriers and reduce energy costs, energy use, and carbon emissions in residential and commercial buildings fall into ten general categories: voluntary programmes, building efficiency standards, equipment efficiency standards, state market transformation programmes, financing, government procurement, tax credits, energy planning (production, distribution, and end-use), and accelerated R&D. Affordable credit financing is widely recognized in Africa as one of the critical measures to remove the high first-cost barrier. Poor macroeconomic management captured by unstable economic conditions often leads to financial repression and higher barriers. As many of several obstacles can be observed simultaneously in the innovation chain of an energy-efficient investment or organizational measure, policy measures usually have to be applied as a bundle to realize the economic potential of a particular technology.

Transport: The car has come to be widely perceived in modern societies as a means of freedom, mobility and safety, a symbol of personal status and identity, and as one of the most important products in the industrial economy. Several studies have found that people living in denser and more compact cities rely less on cars, but it is not easy, even taking congestion problems into account, to motivate the shift away from suburban sprawl to compact cities as advocated in some literature. An integrated approach to town and transport planning and the use of incentives are key to energy efficiency and saving in the transport sector. This is an area, where lock-in effects are very important: when land-use patterns have been chosen there is hardly a way back. This represents an opportunity in particular for the developing world.

Transport fuel taxes are commonly used, but have proved very unpopular in some countries, especially where they are seen as revenue-raising measures. Charges on road users have been accepted where they are earmarked to cover the costs of transport provision. Although trucks and cars may be subject to different barriers and opportunities because of differences in their purpose of use and travel distance, a tax policy that assesses the full cost of GHG emissions would result in a similar impact on CO_2 reductions in road transport. Several studies have explored the potential for adjusting the way existing road taxes, licence fees, and insurance premiums are levied and have found potential emissions reductions of around 10% in OECD countries. Inadequate development and provision of convenient and efficient mass transport systems encourage the use of more energy consuming private vehicles. It is the combination of policies protecting road transport interest, however, that poses the greatest barrier to change, rather than any single type of instrument.

New and used vehicles and/or their technologies mostly flow from the developed to developing countries. Hence, a global approach to reducing emissions that targets technology in developed countries would have a significant impact on future emissions from developing countries.

Industry: In industry, barriers may take many forms, and are determined by the characteristics of the firm (size and structure) and the business environment. Cost-effective energy efficiency measures are often not undertaken as a result of lack of information and high transaction costs for obtaining reliable information. Capital is used for competing investment priorities, and is subject to high hurdle rates for energy efficiency investments. Lack of skilled personnel, especially for small and medium-sized enterprises (SMEs), leads to difficulties installing new energy-efficient equipment compared to the simplicity of buying energy. Other barriers are the difficulty of quantifying energy savings and slow diffusion of innovative technology into markets, while at the same time firms typically underinvest in R&D, despite the high rates of return on investment.

A wide array of policies to reduce barriers, or the perception of barriers, has been used and tested in the industrial sector in developed countries, with varying success rates. Information programmes are designed to assist energy consumers in understanding and employing technologies and practices to use energy more efficiently. Forms of environmental legislation have been a driving force in the adoption of new technologies. New approaches to industrial energy efficiency improvement in developed countries include voluntary agreements (VAs).

In the energy supply sector virtually all the generic barriers cited in Section 5.2 restrict the introduction of environmentally sound technologies and practices. The increasing deregulation of energy supply, while making it more efficient, has raised particular concerns. Volatile spot and contract prices, short-term outlook of private investors, and the perceived risks of nuclear and hydropower plants have shifted fuel and technology choice towards natural gas and oil plants, and away from renewable energy, including − to a lesser extent − hydropower, in many countries.

Co-generation or combined production of power and heat (CHP) is much more efficient than the production of energy for each of these uses alone. The implementation of CHP is closely linked to the availability and density of industrial heat loads, district heating, and cooling networks. Yet, its implementation is hampered by lack of information, the decentralized character of the technology, the attitude of grid operators, the terms of grid connection, and a lack of policies that foster long-term planning. Firm public policy and regulatory authority is necessary to install and safeguard harmonized conditions, transparency, and unbundling of the main power supply functions.

Agriculture and Forestry: Lack of adequate capacity for research and provision of extension services will hamper the spread of technologies that suit local conditions, and the declining Consultative Group on International Agricultural Research (CGIAR) system has exacerbated this problem in the developing world. Adoption of new technology is also limited by small farm size, credit constraints, risk aversion, lack of access to information and human capital, inadequate rural infrastructure and tenurial arrangements, and unreliable supply of complementary inputs. Subsidies for critical inputs to agriculture, such as fertilizers, water supply, and electricity and fuels, and to outputs in order to maintain stable agricultural systems and an equitable distribution of wealth distort markets for these products.

Measures to address the above barriers include:
- The expansion of credit and savings schemes;
- Shifts in international research funding towards water-use efficiency, irrigation design, irrigation management, adaptation to salinity, and the effect of increased CO_2 levels on tropical crops;
- The improvement of food security and disaster early warning systems;
- The development of institutional linkages between countries; and
- The rationalization of input and output prices of agricultural commodities, taking DES issues into consideration.

The forestry sector faces land-use regulation and other macroeconomic policies that usually favour conversion to other land uses such as agriculture, cattle ranching, and urban industry. Insecure land tenure regimes and tenure rights and subsidies favouring agriculture or livestock are among the most important barriers for ensuring sustainable management of forests as well as sustainability of carbon abatement. In relation to climate change mitigation, other issues, such as lack of technical capability, lack of credibility about the setting of project baselines, and monitoring of carbon stocks, poses difficult challenges.

Waste Management: Solid waste and wastewater disposal and treatment represent about 20% of human-induced methane emissions. The principal barriers to technology transfer in this sector include limited financing and institutional capability, jurisdictional complexity, and the need for community involvement. Climate change mitigation projects face further barriers resulting from unfamiliarity with CH_4 capture and potential electricity generation,

unwillingness to commit additional human capacity for climate mitigation, and the additional institutional complexity required not only by waste treatment but also by energy generation and supply. The lack of clear regulatory and investment frameworks can pose significant challenges for project development.

To overcome the barriers and to avail the opportunities in waste management, it is necessary to have a multi-project approach, the components of which include the following:
- Building databases on availability of wastes, their characteristics, distribution, accessibility, current practices of utilization and/ or disposal technologies, and economic viability;
- Institutional mechanism for technology transfer though a co-ordinated programme involving the R&D institutions, financing agencies, and industry; and
- Defining the role of stakeholders including local authorities, individual householders, industries, R&D institutions, and the government.

Regional Considerations: Changing global patterns provide an opportunity for introducing GHG mitigation technologies and practices that are consistent with DES goals. A culture of energy subsidies, institutional inertia, fragmented capital markets, vested interests, etc., however, presents major barriers to their implementation, and may be particular issues in developing and EIT countries. Situations in these two groups of countries call for a more careful analysis of trade, institutional, financial, and income barriers and opportunities, distorted prices, and information gaps. In the developed countries, other barriers such as the current carbon-intensive lifestyle and consumption patterns, social structures, network externalities, and misplaced incentives offer opportunities for intervention to control the growth of GHG emissions. Lastly, new and used technologies mostly flow from the developed to developing and transitioning countries. A global approach to reducing emissions that targets technology that is transferred from developed to developing countries could have a significant impact on future emissions.

6. Policies, Measures, and Instruments

6.1. Policy Instruments and Possible Criteria for their Assessment

The purpose of this section is to examine the major types of policies and measures that can be used to implement options to mitigate net concentrations of GHGs in the atmosphere. In keeping within the defined scope of this Report, policies and measures that can be used to implement or reduce the costs of adaptation to climate change are not examined. Alternative policy instruments are discussed and assessed in terms of specific criteria, all on the basis of the most recent literature. There is naturally some emphasis on the instruments mentioned in the Kyoto Protocol (the Kyoto mechanisms), because they are new and focus on achieving GHG emissions limits, and the extent of their envisaged international application is unprecedented. In addition to economic dimensions, political economy, legal, and institutional elements are discussed insofar as they are relevant to these policies and measures.

Any individual country can choose from a large set of possible policies, measures, and instruments, including (in arbitrary order): emissions, carbon, or energy taxes, tradable permits, subsidies, deposit-refund systems, voluntary agreements, non-tradable permits, technology and performance standards, product bans, and direct government spending, including R&D investment. Likewise, a group of countries that wants to limit its collective GHG emissions could agree to implement one, or a mix, of the following instruments (in arbitrary order): tradable quotas, joint implementation, clean development mechanism, harmonized emissions or carbon or energy taxes, an international emissions, carbon, or energy tax, non-tradable quotas, international technology and product standards, voluntary agreements, and direct international transfers of financial resources and technology.

Possible criteria for the assessment of policy instruments include: environmental effectiveness; cost effectiveness; distributional considerations including competitiveness concerns; administrative and political feasibility; government revenues; wider economic effects including implications for international trade rules; wider environmental effects including carbon leakage; and effects on changes in attitudes, awareness, learning, innovation, technical progress, and dissemination of technology. Each government may apply different weights to various criteria when evaluating GHG mitigation policy options depending on national and sector level circumstances. Moreover, a government may apply different sets of weights to the criteria when evaluating national (domestic) versus international policy instruments. Co-ordinated actions could help address competitiveness concerns, potential conflicts with international trade rules, and carbon leakage.

The economics literature on the choice of policies adopted has emphasized the importance of interest group pressures, focusing on the demand for regulation. But it has tended to neglect the "supply side" of the political equation, emphasized in the political science literature: the legislators and government and party officials who design and implement regulatory policy, and who ultimately decide which instruments or mix of instruments will be used. However, the point of compliance of alternative policy instruments, whether they are applied to fossil fuel users or manufacturers, for example, is likely to be politically crucial to the choice of policy instrument. And a key insight is that some forms of regulation actually can benefit the regulated industry, for example, by limiting entry into the industry or imposing higher costs on new entrants. A policy that imposes costs on industry as a whole might still be supported by firms who would fare better than their competitors. Regulated firms, of course, are not the only group with a stake in regulation: opposing interest groups will fight for their own interests.

6.2. National Policies, Measures, and Instruments

In the case of countries in the process of structural reform, it is important to understand the new policy context to develop reasonable assessments of the feasibility of implementing GHG mitigation policies. Recent measures taken to liberalize energy markets have been inspired for the most part by desires to increase competition in energy and power markets, but they also can have significant emission implications, through their impact on the production and technology pattern of energy or power supply. In the long run, the consumption pattern change might be more important than the sole implementation of climate change mitigation measures.

Market-based instruments – principally domestic taxes and domestic tradable permit systems – will be attractive to governments in many cases because they are efficient. They will frequently be introduced in concert with conventional regulatory measures. When implementing a domestic emissions tax, policymakers must consider the collection point, the tax base, the variation among sectors, the association with trade, employment, revenue, and the exact form of the mechanism. Each of these can influence the appropriate design of a domestic emissions tax, and political or other concerns are likely to play a role as well. For example, a tax levied on the energy content of fuels could be much more costly than a carbon tax for equivalent emissions reduction, because an energy tax raises the price of all forms of energy, regardless of their contribution to CO_2 emissions. Yet, many nations may choose to use energy taxes for reasons other than cost effectiveness, and much of the analysis in this section applies to energy taxes, as well as carbon taxes.

A country committed to a limit on its GHG emissions also can meet this limit by implementing a tradable permit system that directly or indirectly limits emissions of domestic sources. Like a tax, a tradable permit system poses a number of design issues, including type of permit, ways to allocate permits, sources included, point of compliance, and use of banking. To be able to cover all sources with a single domestic permit regime is unlikely. The certainty provided by a tradable permit system of achieving a given emissions level for participating sources comes at the cost of the uncertainty of permit prices (and hence compliance costs). To address this concern, a hybrid policy that caps compliance costs could be adopted, but the level of emissions would no longer be guaranteed.

For a variety of reasons, in most countries the management of GHG emissions will not be addressed with a single policy instrument, but with a portfolio of instruments. In addition to one or more market-based policies, a portfolio might include standards and other regulations, voluntary agreements, and information programmes:

- Energy efficiency standards have been effective in reducing energy use in a growing number of countries. They may be especially effective in many countries where the capacity to administer market instruments is relatively limited, thereby helping to develop this administrative infrastructure. They need updating to remain effective. The main disadvantage of standards is that they can be inefficient, but efficiency can be improved if the standard focuses on the desired results and leaves as much flexibility as possible in the choice of how to achieve the results.

- Voluntary agreements (VAs) may take a variety of forms. Proponents of VAs point to low transaction costs and consensus elements, while sceptics emphasize the risk of "free riding", and the risk that the private sector will not pursue real emissions reduction in the absence of monitoring and enforcement. Voluntary agreements sometimes precede the introduction of more stringent measures.
- Imperfect information is widely recognized as a key market failure that can have significant effects on improved energy efficiency, and hence emissions. Information instruments include environmental labelling, energy audits, and industrial reporting requirements, and information campaigns are marketing elements in many energy-efficiency programmes.

A growing literature has demonstrated theoretically, and with numerical simulation models, that the economics of addressing GHG reduction targets with domestic policy instruments depend strongly on the choice of those instruments. Price-based policies tend to lead to positive marginal and positive total mitigation costs. In each case, the interaction of these abatement costs with the existing tax structure and, more generally, with existing factor prices is important. Price-based policies that generate revenues can be coupled with measures to improve market efficiency. However, the role of non-price policies, which affect the sign of the change in the unit price of energy services, often remains decisive.

6.3. International Policies and Measures

Turning to international policies and measures, the Kyoto Protocol defines three international policy instruments, the so-called Kyoto mechanisms: international emissions trading (IET), joint implementation (JI), and the Clean Development Mechanism (CDM). Each of these international policy instruments provides opportunities for Annex I Parties to fulfil their commitments cost-effectively. IET essentially would allow Annex I Parties to exchange part of their assigned national emission allowances (targets). IET implies that countries with high marginal abatement costs (MACs) may acquire emission reductions from countries with low MACs. Similarly, JI would allow Annex I Parties to exchange emission reduction units among themselves on a project-by-project basis. Under the CDM, Annex I Parties would receive credit – on a project-by-project basis – for reductions accomplished in non-Annex I countries.

Economic analyses indicate that the Kyoto mechanisms could reduce significantly the overall cost of meeting the Kyoto emissions limitation commitments. However, achievement of the potential cost savings requires the adoption of domestic policies that allow individual entities to use the mechanisms to meet their national emissions limitation obligations. If domestic policies limit the use of the Kyoto mechanisms, or international rules governing the mechanisms limit their use, the cost savings may be reduced.

In the case of JI, host governments have incentives to ensure that emission reduction units (ERUs) are issued only for real emission reductions, assuming that they face strong penalties for non-compliance with national emissions limitation commitments. In the case of CDM, a process for independent certification of emission reductions is crucial, because host governments do not have emissions limitation commitments and hence may have less incentive to ensure that certified emission reductions (CERs) are issued only for real emission reductions. The main difficulty in implementing project-based mechanisms, both JI and CDM, is determining the net additional emission reduction (or sink enhancement) achieved; baseline definition may be extremely complex. Various other aspects of these Kyoto mechanisms are awaiting further decision making, including: monitoring and verification procedures, financial additionality (assurance that CDM projects will not displace traditional development assistance flows), and possible means of standardizing methodologies for project baselines.

The extent to which developing country (non-Annex I) Parties will effectively implement their commitments under the UNFCCC may depend, among other factors, on the transfer of environmentally sound technologies (ESTs).

6.4. Implementation of National and International Policy Instruments

Any international or domestic policy instrument can be effective only if accompanied by adequate systems of monitoring and enforcement. There is a linkage between compliance enforcement and the amount of international co-operation that will actually be sustained. Many multilateral environmental agreements address the need to co-ordinate restrictions on conduct taken in compliance with obligations they impose and the expanding legal regime under the WTO and/or GATT umbrella. Neither the UNFCCC nor the Kyoto Protocol now provides for specific trade measures in response to non-compliance. But several domestic policies and measures that might be developed and implemented in conjunction with the Kyoto Protocol could conflict with WTO provisions. International differences in environmental regulation may have trade implications.

One of the main concerns in environmental agreements (including the UNFCCC and the Kyoto Protocol) has been with reaching wider participation. The literature on international environmental agreements predicts that participation will be incomplete, and incentives may be needed to increase participation (see also Section 10).

7. Costing Methodologies

7.1. Conceptual Basis

Using resources to mitigate greenhouse gases (GHGs) generates opportunity costs that should be considered to help guide reasonable policy decisions. Actions taken to abate GHG emissions or to increase carbon sinks divert resources from other alternative uses. Assessing the costs of these actions should ideally consider the total value that society attaches to the goods and services forgone because of the diversion of resources to climate protection. In some cases, the sum of benefits and costs will be negative, meaning that society gains from undertaking the mitigation action.

This section addresses the methodological issues that arise in the estimation of the monetary costs of climate change. The focus is on the correct assessment of the costs of mitigation measures to reduce the emissions of GHGs. The assessment of costs and benefits should be based on a systematic analytical framework to ensure comparability and transparency of estimates. One well-developed framework assesses costs as changes in social welfare based on individual values. These individual values are reflected by the willingness to pay (WTP) for environmental improvements or the willingness to accept (WTA) compensation. From these value measures can be derived measures such as the social surpluses gained or lost from a policy, the total resource costs, and opportunity costs.

While the underlying measures of welfare have limits and using monetary values remains controversial, the view is taken that the methods to "convert" non-market inputs into monetary terms provide useful information for policymakers. These methods should be pursued when and where appropriate. It is also considered useful to supplement this welfare-based cost methodology with a broader assessment that includes equity and sustainability dimensions of climate change mitigation policies. In practice, the challenge is to develop a consistent and comprehensive definition of the key impacts to be measured.

A frequent criticism of this costing method is that it is inequitable, as it gives greater weight to the "well off". This is because, typically, a well-off person has a greater WTP or WTA than a less well-off person and hence the choices made reflect more the preferences of the better off. This criticism is valid, but there is no coherent and consistent method of valuation that can replace the existing one in its entirety. Concerns about, for example, equity can be addressed along with the basic cost estimation. The estimated costs are one piece of information in the decision-making process for climate change that can be supplemented with other information on other social objectives, for example impacts on key stakeholders and the meeting of poverty objectives.

In this section the costing methodology is overviewed, and issues involved in using these methods addressed.

7.2. Analytical Approaches

Cost assessment is an input into one or more rules for decision-making, including cost-benefit analysis (CBA), cost-effectiveness analysis (CEA), and multi-attribute analysis. The analytical approaches differ primarily by how the objectives of the decision-making framework are selected, specified, and valued. Some objectives in mitigation policies can be specified in economic units (e.g., costs and benefits measured in monetary units), and some in physical units (e.g., the amount of pollutants dispersed in tonnes of CO_2). In practice, however, the challenge is in developing a consistent and comprehensive definition of every important impact to be measured.

7.2.1. Co-Benefits and Costs and Ancillary Benefits and Costs

The literature uses a number of terms to depict the associated benefits and costs that arise in conjunction with GHG mitigation policies. These include co-benefits, ancillary benefits, side benefits, secondary benefits, collateral benefits, and associated benefits. In the current discussion, the term "co-benefits" refers to the non-climate benefits of GHG mitigation policies that are explicitly incorporated into the initial creation of mitigation policies. Thus, the term co-benefits reflects that most policies designed to address GHG mitigation also have other, often at least equally important, rationales involved at the inception of these policies (e.g., related to objectives of development, sustainability, and equity). In contrast, the term ancillary benefits connotes those secondary or side effects of climate change mitigation policies on problems that arise subsequent to any proposed GHG mitigation policies.

Policies aimed at mitigating GHGs, as stated earlier, can yield other social benefits and costs (here called ancillary or co-benefits and costs), and a number of empirical studies have made a preliminary attempt to assess these impacts. It is apparent that the actual magnitude of the ancillary benefits or co-benefits assessed critically depends on the scenario structure of the analysis, in particular on the assumptions about policy management in the baseline case. This implies that whether a particular impact is included or not depends on the primary objective of the programme. Moreover, something that is seen as a GHG reduction programme from an international perspective may be seen, from a national perspective, as one in which local pollutants and GHGs are equally important.

7.2.2. Implementation Costs

All climate change policies necessitate some costs of implementation, that is costs of changes to existing rules and regulations, making sure that the necessary infrastructure is available, training and educating those who are to implement the policy as well those affected by the measures, etc. Unfortunately, such costs are not fully covered in conventional cost analyses.

Implementation costs in this context are meant to reflect the more permanent institutional aspects of putting a programme into place and are different to those costs conventionally considered as transaction costs. The latter, by definition, are temporary costs. Considerable work needs to be done to quantify the institutional and other costs of programmes, so that the reported figures are a better representation of the true costs that will be incurred if programmes are actually implemented.

7.2.3. Discounting

There are broadly two approaches to discounting–an ethical or prescriptive approach based on what rates of discount should be applied, and a descriptive approach based on what rates of discount people (savers as well as investors) actually apply in their day-to-day decisions. For mitigation analysis, the country must base its decisions at least partly on discount rates that reflect the opportunity cost of capital. Rates that range from 4% to 6% would probably be justified in developed countries. The rate could be 10–12% or even higher in developing countries. It is more of a challenge to argue that climate change mitigation projects should face different rates, unless the mitigation project is of very long duration. The literature shows increasing attention to rates that decline over time and hence give more weight to benefits that occur in the long term. Note that these rates do not reflect private rates of return, which typically must be greater to justify a project, at around 10–25%.

7.2.4. Adaptation and Mitigation Costs and the Link Between Them

While most people appreciate that adaptation choices affect the costs of mitigation, this obvious point is often not addressed in climate policymaking. Policy is fragmented - with mitigation being seen as addressing climate change and adaptation seen as a means of reacting to natural hazards. Usually mitigation and adaptation are modelled separately as a necessary simplification to gain traction on an immense and complex issue. As a consequence, the costs of risk reduction action are frequently estimated separately, and therefore each measure is potentially biased. This realization suggests that more attention to the interaction of mitigation and adaptation, and its empirical ramification, is worthwhile, though uncertainty about the nature and timing of impacts, including surprises, will constrain the extent to which the associated costs can be fully internalized.

7.3. System Boundaries: Project, Sector, and Macro

Researchers make a distinction between project, sector, and economywide analyses. Project level analysis considers a "stand-alone" investment assumed to have insignificant secondary impacts on markets. Methods used for this level include CBA, CEA, and life-cycle analysis. Sector level analysis examines sectoral policies in a "partial-equilibrium" context in which all other variables are assumed to be exogenous. Economy-wide analysis explores how policies affect all sectors and markets, using various macroeconomic and general equilibrium models. A trade-off exists between the level of detail in the assessment and complexity of the system considered. This section presents some of the key assumptions made in cost analysis.

A combination of different modelling approaches is required for an effective assessment of climate change mitigation options. For example, detailed project assessment has been combined with a more general analysis of sectoral impacts, and macroeconomic carbon tax studies have been combined with the sectoral modelling of larger technology investment programmes.

7.3.1. Baselines

The baseline case, which by definition gives the emissions of GHGs in the absence of the climate change interventions being considered, is critical to the assessment of the costs of climate change mitigation. This is because the definition of the baseline scenario determines the potential for future GHG emissions reduction, as well as the costs of implementing these reduction policies. The baseline scenario also has a number of important implicit assumptions about future economic policies at the macroeconomic and sectoral levels, including sectoral structure, resource intensity, prices, and thereby technology choice.

7.3.2. Consideration of No Regrets Options

No regrets options are by definition actions to reduce GHG emissions that have negative net costs. Net costs are negative because these options generate direct or indirect benefits, such as those resulting from reductions in market failures, double dividends through revenue recycling and ancillary benefits, large enough to offset the costs of implementing the options. The no regrets issue reflects specific assumptions about the working and the efficiency of the economy, especially the existence and stability of a social welfare function, based on a social cost concept:

- Reduction of existing market or institutional failures and other barriers that impede adoption of cost-effective emission reduction measures can lower private costs compared to current practice. This can also reduce private costs overall.
- A double dividend related to recycling of the revenue of carbon taxes in such a way that it offsets distortionary taxes.
- Ancillary benefits and costs (or ancillary impacts), which can be synergies or trade-offs in cases in which the reduction of GHG emissions has joint impacts on other environmental policies (i.e., relating to local air pollution, urban congestion, or land and natural resource degradation).

Market Imperfections

The existence of a no regrets potential implies that market and institutions do not behave perfectly, because of market imperfections such as lack of information, distorted price signals, lack of competition, and/or institutional failures related to inadequate regulation, inadequate delineation of property rights, distortion-inducing fiscal systems, and limited financial markets. Reduction of market imperfections suggests it is possible to identify and implement policies that can correct these market and institutional failures without incurring costs larger than the benefits gained.

Double Dividend

The potential for a double dividend arising from climate mitigation policies was extensively studied during the 1990s. In addition to the primary aim of improving the environment (the first dividend), such policies, if conducted through revenue-raising instruments such as carbon taxes or auctioned emission permits, yield a second dividend, which can be set against the gross costs of these policies. All domestic GHG policies have an indirect economic cost from the interactions of the policy instruments with the fiscal system, but in the case of revenue-raising policies this cost is partly offset (or more than offset) if, for example, the revenue is used to reduce existing distortionary taxes. Whether these revenue-raising policies can reduce distortions in practice depends on whether revenues can be "recycled" to tax reduction.

Ancillary Benefits and Costs (Ancillary Impacts)

The definition of ancillary impacts is given above. As noted there, these can be positive as well as negative. It is important to recognize that gross and net mitigation costs cannot be established as a simple summation of positive and negative impacts, because the latter are interlinked in a very complex way. Climate change mitigation costs (gross and well as net costs) are only valid in relation to a comprehensive specific scenario and policy assumption structure.

The existence of no regrets potentials is a necessary, but not a sufficient, condition for the potential implementation of these options. The actual implementation also requires the development of a policy strategy that is complex as comprehensive enough to address these market and institutional failures and barriers.

7.3.3. Flexibility

For a wide variety of options, the costs of mitigation depend on what regulatory framework is adopted by national governments to reduce GHGs. In general, the more flexibility the framework allows, the lower the costs of achieving a given reduction. More flexibility and more trading partners can reduce costs. The opposite is expected with inflexible rules and few trading partners. Flexibility can be measured as the ability to reduce carbon emissions at the lowest cost, either domestically or internationally.

7.3.4. Development, Equity, and Sustainability Issues

Climate change mitigation policies implemented at a national level will, in most cases, have implications for short-term economic and social development, local environmental quality,

and intra-generational equity. Mitigation cost assessments that follow this line can address these impacts on the basis of a decision-making framework that includes a number of side-impacts to the GHG emissions reduction policy objective. The goal of such an assessment is to inform decision makers about how different policy objectives can be met efficiently, given priorities of equity and other policy constraints (natural resources, environmental objectives). A number of international studies have applied such a broad decision-making framework to the assessment of development implications of CDM projects.

There are a number of key linkages between mitigation costing issues and broader development impacts of the policies, including macroeconomic impacts, employment creation, inflation, the marginal costs of public funds, capital availability, spillovers, and trade.

7.4. Special Issues Relating to Developing Countries and EITs

A number of special issues related to technology use should be considered as the critical determinants of climate change mitigation potential and related costs for developing countries. These include current technological development levels, technology transfer issues, capacity for innovation and diffusion, barriers to efficient technology use, institutional structure, human capacity aspects, and foreign exchange earnings.

Climate change studies in developing countries and EITs need to be strengthened in terms of methodology, data, and policy frameworks. Although a complete standardization of the methods is not possible, to achieve a meaningful comparison of results it is essential to use consistent methodologies, perspectives, and policy scenarios in different nations.

The following modifications to conventional approaches are suggested:

- Alternative development pathways should be analyzed with different patterns of investment in infrastructure, irrigation, fuel mix, and land-use policies.
- Macroeconomic studies should consider market transformation processes in the capital, labour, and power markets.
- Informal and traditional sector transactions should be included in national macroeconomic statistics. The value of non-commercial energy consumption and the unpaid work of household labour for non-commercial energy collection is quite significant and needs to be considered explicitly in economic analysis.
- The costs of removing market barriers should be considered explicitly.

7.5. Modelling Approaches to Cost Assessment

The modelling of climate mitigation strategies is complex and a number of modelling techniques have been applied including input-output models, macroeconomic models, computable general equilibrium (CGE) models, and energy sector based models. Hybrid models have also been developed to provide more detail on the structure of the economy and the energy sector. The appropriate use of these models depends on the subject of the evaluation and the availability of data.

As discussed in Section 6, the main categories of climate change mitigation policies include: market-oriented policies, technology-oriented policies, voluntary policies, and research and development policies. Climate change mitigation policies can include all four of the above policy elements. Most analytical approaches, however, only consider some of the four elements. Economic models, for example, mainly assess market-oriented policies and in some cases technology policies primarily those related to energy supply options, while engineering approaches mainly focus on supply and demand side technology policies. Both of these approaches are relatively weak in the representation of research and development and voluntary agreement policies.

8. Global, Regional, and National Costs and Ancillary Benefits

8.1. Introduction

The UNFCCC (Article 2) has as its ultimate goal the "stabilisation of greenhouse gas concentrations in the atmosphere at a level that will prevent dangerous anthropogenic interference with the climate system"[14]. In addition, the Convention (Article 3.3) states that "policies and measures to deal with climate change should be cost-effective so as to ensure global benefits at the lowest possible costs"[15]. This section reports on literature on the costs of greenhouse gas mitigation policies at the national, regional, and global levels. Net welfare gains or losses are reported, including (when available) the ancillary benefits of mitigation policies. These studies employ the full range of analytical tools described in the previous chapter. These range from technologically detailed bottom-up models to more aggregate top-down models, which link the energy sector to the rest of the economy.

8.2. Gross Costs of GHG Abatement in Technology-Detailed Models

In technology-detailed "bottom-up" models and approaches, the cost of mitigation is derived from the aggregation of technological

and fuel costs such as: investments, operation and maintenance costs, and fuel procurement, but also (and this is a recent trend) revenues and costs from import and exports.

Models can be ranked along two classification axes. First, they range from simple engineering-economics calculations effected technology-by-technology, to integrated partial equilibrium models of whole energy systems. Second, they range from the strict calculation of direct technical costs of reduction to the consideration of observed technology-adoption behaviour of markets, and of the welfare losses due to demand reductions and revenue gains and losses due to changes in trade.

This leads to contrasting two generic approaches, namely the engineering-economics approach and least-cost equilibrium modelling. In the first approach, each technology is assessed independently via an accounting of its costs and savings. Once these elements have been estimated, a unit cost can be calculated for each action, and each action can be ranked according to its costs. This approach is very useful to point out the potentials for negative cost abatements due to the 'efficiency gap' between the best available technologies and technologies currently in use. However, its most important limitation is that studies neglect or do not treat in a systematic way the interdependence of the various actions under examination.

Partial equilibrium least-costs models have been constructed to remedy this defect, by considering all actions simultaneously and selecting the optimal bundle of actions in all sectors and at all time periods. These more integrated studies conclude higher total costs of GHG mitigation than the strict technology by technology studies. Based on an optimization framework they give very easily interpretable results that compare an optimal response to an optimal baseline; however, their limitation is that they rarely calibrate the base year of the model to the existing non optimal situation and implicitly assume an optimal baseline. They consequently provide no information about the negative cost potentials.

Since the publication of the SAR, the bottom-up approaches have produced a wealth of new results for both Annex I and non-Annex I countries, as well as for groups of countries. Furthermore, they have extended their scope much beyond the

---

[14] "The ultimate objective of this Convention and any related legal instruments that the Conference of Parties may adopt is to achieve, in accordance with the relevant provisions of the Convention, stabilization of greenhouse gas concentrations in the atmosphere at such a level that would prevent dangerous interference with the climate system. Such a level should be achieved within a timeframe sufficient to allow ecosystems to adapt naturally to climate change, to ensure that food production is not threatened, and to enable economic development to proceed in a sustainable manner."

[15] "The Parties should take precautionary measures to anticipate, prevent, or minimise the causes of climate change and mitigate its adverse effects. Where there are threats of serious irreversible damage, lack of full scientific certainty should not be used as a reason for postponing such measures, taking into account that polices and measures to deal with climate change should be cost-effective so as to ensure global benefits at the lowest possible costs. To achieve this, such policies and measures should take into account different socio-economic contexts, be comprehensive, cover all relevant sources, sinks and reservoirs of greenhouse gases and adaptation, and comprise all economic sectors. Efforts to address climate change may be carried out co-operatively by interested Parties."

classical computations of direct abatement costs by inclusion of demand effects and some trade effects.

However, the modelling results show considerable variations from study to study, which are explained by a number of factors, some of which reflect the widely differing conditions that prevail in the countries studied (e.g., energy endowment, economic growth, energy intensity, industrial and trade structure), and others reflect modelling assumptions and assumptions about negative cost potentials.

However, as in the SAR, there is agreement on a no regrets potential resulting from the reduction of existing market imperfections, consideration of ancillary benefits, and inclusion of double dividends. This means that some mitigation actions can be realized at negative costs. The no regrets potential results from existing market or institutional imperfections that prevent cost-effective emission reduction measures from being taken. The key question is whether such imperfections can be removed cost-effectively by policy measures.

The second important policy message is that short and medium term marginal abatement costs, which govern most of the macroeconomic impacts of climate policies, are very sensitive to uncertainty regarding baseline scenarios (rate of growth and energy intensity) and technical costs. Even with significant negative cost options, marginal costs may rise quickly beyond a certain anticipated mitigation level. This risk is far lower in models allowing for carbon trading. Over the long term this risk is reduced as technical change curbs down the slope of marginal cost curves.

8.3. Costs of Domestic Policy to Mitigate Carbon Emissions

Particularly important for determining the gross mitigation costs is the magnitude of emissions reductions required in order to meet a given target, thus the emissions baseline is a critical factor. The growth rate of CO_2 depends on the growth rate in GDP, the rate of decline of energy use per unit of output, and the rate of decline of CO_2 emissions per unit of energy use.

In a multi-model comparison project that engaged more than a dozen modelling teams internationally, the gross costs of complying with the Kyoto Protocol were examined, using energy sector models. Carbon taxes are implemented to lower emissions and the tax revenue is recycled lump sum. The magnitude of the carbon tax provides a rough indication of the amount of market intervention that would be needed and equates the marginal abatement cost to meet a prescribed emissions target. The size of the tax required to meet a specific target will be determined by the marginal source of supply (including conservation) with and without the target. This in turn will depend on such factors as the size of the necessary emissions reductions, assumptions about the cost and availability of carbon-based and carbon-free technologies, the fossil fuel resource base, and short- and long-term price elasticities.

With no international emission trading, the carbon taxes necessary to meet the Kyoto restrictions in 2010 vary a lot among the models. Note from Table TS-4[16] that for the USA they are calculated to be in the range US$76 to US$322, for OECD Europe between US$20 and US$665, for Japan between US$97 and US$645, and finally for the rest of OECD (CANZ) between US$46 and US$425. All numbers are reported in 1990 dollars. Marginal abatement costs are in the range of US$20–US$135/tC if international trading is allowed. These models do not generally include no regrets measures or take account of the mitigation potential of CO_2 sinks and of greenhouse gases other than CO_2.

However, there is no strict correlation between the level of the carbon tax and GDP variation and welfare because of the influence of the country specifics (countries with a low share of fossil energy in their final consumption suffer less than others for the same level of carbon tax) and because of the content of the policies.

The above studies assume, to allow an easy comparison across countries, that the revenues from carbon taxes (or auctioned emissions permits) are recycled in a lump-sum fashion to the economy. The net social cost resulting from a given marginal cost of emissions constraint can be reduced if the revenues are targetted to finance cuts in the marginal rates of pre-existing distortionary taxes, such as income, payroll, and sales taxes. While recycling revenues in a lump-sum fashion confers no efficiency benefit, recycling through marginal rate cuts helps avoid some of the efficiency costs or dead-weight loss of existing taxes. This raises the possibility that revenue-neutral carbon taxes might offer a double dividend by (1) improving the environment and (2) reducing the costs of the tax system.

One can distinguish a weak and a strong form of the double dividend. The weak form asserts that the costs of a given revenue-neutral environmental reform, when revenues are devoted to cuts in marginal rates of prior distortionary taxes, are reduced relative to the costs when revenues are returned in lump-sum fashion to households or firms. The strong form of the double-dividend assertion is that the costs of the revenue-neutral environmental tax reform are zero or negative. While the weak form of the double-dividend claim receives virtually universal support, the strong form of the double dividend assertion is controversial.

Where to recycle revenues from carbon taxes or auctioned permits depends upon the country specifics. Simulation results show that in economies that are especially inefficient or distorted along non-environmental lines, the revenue-recycling effect can indeed be strong enough to outweigh the primary

---

[16] The highest figures cited in this sentence are all results from one model: the ABARE-GTEM model.

Table TS-4: Energy Modelling Forum main results. Marginal abatement costs (in 1990 US$/tC; 2010 Kyoto target).

Model	No trading				Annex I trading	Global trading
	US	OECD-E	Japan	CANZ		
ABARE-GTEM	322	665	645	425	106	23
AIM	153	198	234	147	65	38
CETA	168				46	26
Fund					14	10
G-Cubed	76	227	97	157	53	20
GRAPE		204	304		70	44
MERGE3	264	218	500	250	135	86
MIT-EPPA	193	276	501	247	76	
MS-MRT	236	179	402	213	77	27
Oxford	410	966	1074		224	123
RICE	132	159	251	145	62	18
SGM	188	407	357	201	84	22
WorldScan	85	20	122	46	20	5
Administration	154				43	18
EIA	251				110	57
POLES	135.8	135.3	194.6	131.4	52.9	18.4

Note: The results of the Oxford model are not included in the ranges cited in the TS and SPM because this model has not been subject to substantive academic review (and hence is inappropriate for IPCC assessment), and relies on data from the early 1980s for a key parametization that determines the model results. This model is entirely unrelated to the CLIMOX model, from the Oxford Institutes of Energy Studies, referred to in Table TS-6.
EMF-16. GDP losses (as a percentage of total GDP) associated with complying with the prescribed targets under the Kyoto Protocol. Four regions include USA, OECD Europe (OECD-E), Japan, and Canada, Australia and New Zealand (CANZ). Scenarios include no trading, Annex B trading only, and full global trading.

cost and tax-interaction effect so that the strong double dividend may materialize. Thus, in several studies involving European economies, where tax systems may be highly distorted in terms of the relative taxation of labour, the strong double dividend can be obtained, in any case more frequently than in other recycling options. In contrast, most studies of carbon taxes or permits policies in the USA demonstrate that recycling through lower labour taxation is less efficient than through capital taxation; but they generally do not find a strong double dividend. Another conclusion is that even in cases of no strong double-dividend effect, one fares considerably better with a revenue-recycling policy in which revenues are used to cut marginal rates of prior taxes, than with a non-revenue recycling policy, like for example grandfathered quotas.

In all countries where CO_2 taxes have been introduced, some sectors have been exempted by the tax, or the tax is differentiated across sectors. Most studies conclude that tax exemptions raise economic costs relative to a policy involving uniform taxes. However, results differ in the magnitude of the costs of exemptions.

8.4. Distributional Effects of Carbon Taxes

As well as the total costs, the distribution of the costs is important for the overall evaluation of climate policies. A policy that leads to an efficiency gain may not be welfare improving overall if some people are in a worse position than before, and vice versa. Notably, if there is a wish to reduce the income

differences in the society, the effect on the income distribution should be taken into account in the assessment.

The distributional effects of a carbon tax appear to be regressive unless the tax revenues are used either directly or indirectly in favour of the low-income groups. Recycling the tax revenue by reducing the labour tax may have more attractive distributional consequences than a lump-sum recycling, in which the recycled revenue is directed to both wage earners and capital owners. Reduced taxation of labour results in increased wages and favours those who earn their income mainly from labour. However, the poorest groups in the society may not even earn any income from labour. In this regard, reducing labour taxes may not always be superior to recycling schemes that distribute to all groups of a society and might reduce the regressive character of carbon taxes.

8.5. Aspects of International Emission Trading

It has long been recognized that international trade in emission quota can reduce mitigation costs. This will occur when countries with high domestic marginal abatement costs purchase emission quota from countries with low marginal abatement costs. This is often referred to as "where flexibility". That is, allowing reductions to take place where it is cheapest to do so regardless of geographical location. It is important to note that where the reductions take place is independent of who pays for the reductions.

Table TS-5: Energy Modeling Forum main results. GDP loss in 2010 (in % of GDP; 2010 Kyoto target).

Model	No trading				Annex I trading				Global trading			
	US	OECD-E	Japan	CANZ	US	OECD-E	Japan	CANZ	US	OECD-E	Japan	CANZ
ABARE-GTEM	1.96	0.94	0.72	1.96	0.47	0.13	0.05	0.23	0.09	0.03	0.01	0.04
AIM	0.45	0.31	0.25	0.59	0.31	0.17	0.13	0.36	0.20	0.08	0.01	0.35
CETA	1.93				0.67				0.43			
G-Cubed	0.42	1.50	0.57	1.83	0.24	0.61	0.45	0.72	0.06	0.26	0.14	0.32
GRAPE		0.81	0.19			0.81	0.10			0.54	0.05	
MERGE3	1.06	0.99	0.80	2.02	0.51	0.47	0.19	1.14	0.20	0.20	0.01	0.67
MS-MRT	1.88	0.63	1.20	1.83	0.91	0.13	0.22	0.88	0.29	0.03	0.02	0.32
Oxford	1.78	2.08	1.88		1.03	0.73	0.52		0.66	0.47	0.33	
RICE	0.94	0.55	0.78	0.96	0.56	0.28	0.30	0.54	0.19	0.09	0.09	0.19

Note: The results of the Oxford model are not included in the ranges cited in the TS and SPM because this model has not been subject to substantive academic review (and hence is inappropriate for IPCC assessment), and relies on data from the early 1980s for a key parametization that determines the model results. This model is entirely unrelated to the CLIMOX model, from the Oxford Institutes of Energy Studies, referred to in Table TS-6.

"Where flexibility" can occur on a number of scales. It can be global, regional or at the country level. In the theoretical case of full global trading, all countries agree to emission caps and participate in the international market as buyers or sellers of emission allowances. The CDM may allow some of these cost reductions to be captured. When the market is defined at the regional level (e.g., Annex B countries), the trading market is more limited. Finally, trade may take place domestically with all emission reductions occurring in the country of origin.

Table TS-5 shows the cost reductions from emission trading for Annex B and full global trading compared to a no-trading case. The calculation is made by various models with both global and regional detail. In each instance, the goal is to meet the emission reduction targets contained in the Kyoto Protocol. All of the models show significant gains as the size of the trading market is expanded. The difference among models is due in part to differences in their baseline, the assumptions about the cost and availability of low-cost substitutes on both the supply and demand sides of the energy sector, and the treatment of short-term macro shocks. In general, all calculated gross costs for the non-trading case are below 2% of GDP (which is assumed to have increased significantly in the period considered) and in most cases below 1%. Annex B trading lowers the costs for the OECD region as a whole to less than 0.5% and regional impacts within this vary between 0.1% to 1.1%. Global trading in general would decrease these costs to well below 0.5% of GDP with OECD average below 0.2%.

The issue of the so-called "hot air"[17] also influences the cost of implementing the Kyoto Protocol. The recent decline in economic activity in Eastern Europe and the former Soviet Union has led to a decrease in their GHG emissions. Although this trend is eventually expected to reverse, for some countries emissions are still projected to lie below the constraint imposed by the Kyoto Protocol. If this does occur, these countries will have excess emission quota that may be sold to countries in search of low-cost options for meeting their own targets. The cost savings from trading are sensitive to the magnitude of "hot air".

Numerous assessments of reduction in projected GDP have been associated with complying with Kyoto-type limits. Most economic analyses have focused on gross costs of carbon emitting activities[18], ignoring the cost-saving potential of mitigating non-CO_2 gases and using carbon sequestration and neither taking into account environmental benefits (ancillary benefits and avoided climate change), nor using revenues to remove distortions. Including such possibilities could lower costs.

A constraint would lead to a reallocation of resources away from the pattern that is preferred in the absence of a limit and into potentially costly conservation and fuel substitution. Relative prices will also change. These forced adjustments lead to reductions in economic performance, which impact GDP. Clearly, the broader the permit trading market, the greater the opportunity for reducing overall mitigation costs. Conversely, limits on the extent to which a country can satisfy its obligations through the purchase of emissions quota can increase mitigation costs. Several studies have calculated the magnitude of the increase to be substantial falling in particular on countries with the highest marginal abatement costs. But another parameter likely to limit the savings from carbon trading is the very functioning of trading systems (transaction costs, management costs, insurance against uncertainty, and strategic behaviour in the use of permits).

[17] Hot air: a few countries, notably those with economies in transition, have assigned amount units that appear to be well in excess of their anticipated emissions (as a result of economic downturn). This excess is referred to as hot air.

[18] Although some studies include multi-gas analysis, much research is needed on this potential both intertemporally and regionally.

8.6. Ancillary Benefits of Greenhouse Gas Mitigation

Policies aimed at mitigating greenhouse gases can have positive and negative side effects on society, not taking into account benefits of avoided climate change. This section assesses in particular those studies that evaluate the side effects of climate change mitigation. Therefore the term "ancillary benefits or costs" is used. There is little agreement on the definition, reach, and size of these ancillary benefits, and on methodologies for integrating them into climate policy. Criteria are established for reviewing the growing literature linking specific carbon mitigation policies to monetized ancillary benefits. Recent studies that take an economy-wide, rather than a sectoral, approach to ancillary benefits are described in the report and their credibility is examined (Section 9 presents sectoral analyses). In spite of recent progress in methods development, it remains very challenging to develop quantitative estimates of the ancillary effects, benefits and costs of GHG mitigation policies. Despite these difficulties, in the short term, ancillary benefits of GHG policies under some circumstances can be a significant fraction of private (direct) mitigation costs and in some cases they can be comparable to the mitigation costs. According to the literature, ancillary benefits may be of particular importance in developing countries, but this literature is as yet limited.

The exact magnitude, scale, and scope of these ancillary benefits and costs will vary with local geographical and baseline conditions. In some circumstances, where baseline conditions involve relatively low carbon emissions and population density, benefits may be low. The models most in use for ancillary benefit estimation – the computable general equilibrium (CGE) models – have difficulty in estimating ancillary benefits because they rarely have, and may not be able to have, the necessary spatial detail.

With respect to baseline considerations most of the literature on ancillary benefits systematically treats only government policies and regulations with respect to the environment. In contrast, other regulatory policy baseline issues, such as those relating to energy, transportation, and health, have been generally ignored, as have baseline issues that are not regulatory, such as those tied with technology, demography, and the natural resource base. For the studies reviewed here, the biggest share of the ancillary benefits is related to public health. A major component of uncertainty for modelling ancillary benefits for public health is the link between emissions and atmospheric concentrations, particularly in light of the importance of secondary pollutants. However, it is recognized that there are significant ancillary benefits in addition to those for public health that have not been quantified or monetized. At the same time, it appears that there are major gaps in the methods and models for estimating ancillary costs.

8.7. "Spillover" Effects[19] from Actions Taken in Annex B on Non-Annex B Countries

In a world where economies are linked by international trade and capital flows, abatement of one economy will have welfare impacts on other abating or non-abating economies. These impacts are called spillover effects, and include effects on trade, carbon leakage, transfer and diffusion of environmentally sound technology, and other issues (Figure TS-8).

As to the trade effects, the dominant finding of the effects of emission constraints in Annex B countries on non-Annex B countries in simulation studies prior to the Kyoto Protocol was that Annex B abatement would have a predominantly adverse impact on non-Annex B regions. In simulations of the Kyoto Protocol, the results are more mixed with some non-Annex B regions experiencing welfare gains and other losses. This is mainly due to a milder target in the Kyoto simulations than in pre-Kyoto simulations. It was also universally found that most non-Annex B economies that suffered welfare losses under uniform independent abatement would suffer smaller welfare losses under emissions trading.

A reduction in Annex B emissions will tend to result in an increase in non-Annex B emissions reducing the environmental effectiveness of Annex B abatement. This is called "carbon leakage", and can occur in the order of 5%-20% through a possible relocation of carbon-intensive industries because of reduced Annex B competitiveness in the international marketplace, lower producer prices of fossil fuels in the international market, and changes in income due to better terms of trade.

While the SAR reported that there was a high variance in estimates of carbon leakage from the available models, there has been some reduction in the variance of estimates obtained in the subsequent years. However, this may largely result from the development of new models based on reasonably similar assumptions and data sources. Such developments do not necessarily reflect more widespread agreement about appropriate behavioural assumptions. One robust result seems to be that carbon leakage is an increasing function of the stringency of the abatement strategy. This means that leakage may be a less serious problem under the Kyoto target than under the more stringent targets considered previously. Also emission leakage is lower under emissions trading than under independent abatement. Exemptions for energy-intensive industries found in practice, and other factors, make the higher model estimates for carbon leakage unlikely, but would raise aggregate costs.

Carbon leakage may also be influenced by the assumed degree of competitiveness in the world oil market. While most studies

---

[19] "Spillovers" from domestic mitigation strategies are the effects that these strategies have on other countries. Spillover effects can be positive or negative and include effects on trade, carbon leakage, transfer and diffusion of environmentally sound technology, and other issues.

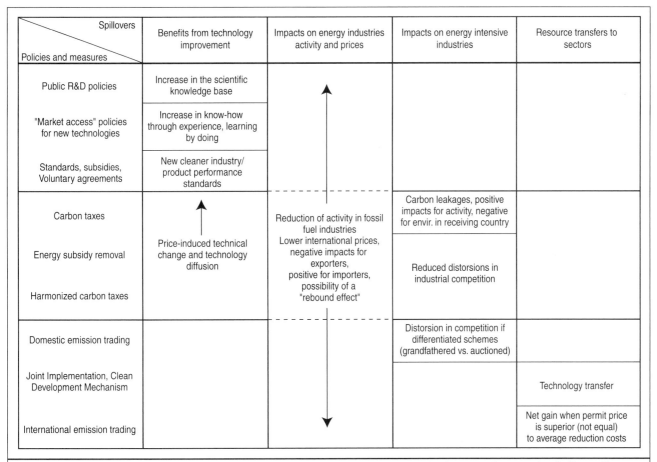

Spillovers / Policies and measures	Benefits from technology improvement	Impacts on energy industries activity and prices	Impacts on energy intensive industries	Resource transfers to sectors
Public R&D policies	Increase in the scientific knowledge base			
"Market access" policies for new technologies	Increase in know-how through experience, learning by doing			
Standards, subsidies, Voluntary agreements	New cleaner industry/ product performance standards			
Carbon taxes	Price-induced technical change and technology diffusion	Reduction of activity in fossil fuel industries Lower international prices, negative impacts for exporters, positive for importers, possibility of a "rebound effect"	Carbon leakages, positive impacts for activity, negative for envir. in receiving country	
Energy subsidy removal			Reduced distorsions in industrial competition	
Harmonized carbon taxes				
Domestic emission trading			Distorsion in competition if differentiated schemes (grandfathered vs. auctioned)	
Joint Implementation, Clean Development Mechanism				Technology transfer
International emission trading				Net gain when permit price is superior (not equal) to average reduction costs

Figure TS-8: "Spillovers" from domestic mitigation strategies are the effects that these strategies have on other countries. Spillover effects can be positive or negative and include effects on trade, carbon leakage, transfer and diffusion of environmentally sound technology, and other issues.

assume a competitive oil market, studies considering imperfect competition find lower leakage if OPEC is able to exercise a degree of market power over the supply of oil and therefore reduce the fall in the international oil price. Whether or not OPEC acts as a cartel can have a reasonably significant effect on the loss of wealth to OPEC and other oil producers and on the level of permit prices in Annex B regions (see also Section 9.2).

The third spillover effect mentioned above, the transfer and diffusion of environmentally sound technology, is related to induced technical change (see Section 8.10). The transfer of environmentally sound technologies and know-how, not included in models, may lead to lower leakage and especially on the longer term may more than offset the leakage.

8.8. Summary of the Main Results for Kyoto Targets

The cost estimates for Annex B countries to implement the Kyoto Protocol vary between studies and regions, and depend strongly upon the assumptions regarding the use of the Kyoto mechanisms, and their interactions with domestic measures. The great majority of global studies reporting and comparing these costs use international energy-economic models. Nine of these studies suggest the following GDP impacts[20]:

Annex II countries[21]: In the absence of emissions trading between Annex B countries[22], the majority of global studies show reductions in projected GDP of about 0.2% to 2% in 2010 for different Annex II regions. With full emissions trading between Annex B countries, the estimated reductions in 2010 are between 0.1% and 1.1% of projected GDP[23]. These studies encompass a wide range of assumptions. Models whose results are reported here assume full use of emissions trading without transaction cost. Results for cases that do not allow Annex B trading assume full domestic trading within each region.

---

[20] Many other studies incorporating more precisely the country specifics and diversity of targetted policies provide a wider range of net cost estimates.

[21] Annex II countries: Group of countries included in Annex II to the UNFCCC, including all developed countries in the Organisation of Economic Co-operation and Development.

[22] Annex B countries: Group of countries included in Annex B in the Kyoto Protocol that have agreed to a target for their greenhouse gas emissions, including all the Annex I countries (as amended in 1998) but Turkey and Belarus.

[23] Many metrics can be used to present costs. For example, if the annual costs to developed countries associated with meeting Kyoto targets with full Annex B trading are in the order of 0.5% of GDP, this represents US$125 billion (1000 million) per year, or US$125 per person per year by 2010 in Annex II (SRES assumptions). This corresponds to an impact on economic growth *rates* over ten years of less than 0.1 percentage point.

Models do not include sinks or non-CO_2 greenhouse gases. They do not include the CDM, negative cost options, ancillary benefits, or targeted revenue recycling.

For all regions costs are also influenced by the following factors:

- Constraints on the use of Annex B trading, high transaction costs in implementing the mechanisms and inefficient domestic implementation could raise costs.
- Inclusion in domestic policy and measures of the no regrets possibilities[2], use of the CDM, sinks, and inclusion of non-CO_2 greenhouse gases, could lower costs. Costs for individual countries can vary more widely.

The models show that the Kyoto mechanisms, are important in controlling risks of high costs in given countries, and thus can complement domestic policy mechanisms. Similarly, they can minimize risks of inequitable international impacts and help to level marginal costs. The global modelling studies reported above show national marginal costs to meet the Kyoto targets from about US\$20/tC up to US\$600/tC without trading, and a range from about US\$15/tC up to US\$150/tC with Annex B trading. The cost reductions from these mechanisms may depend on the details of implementation, including the compatibility of domestic and international mechanisms, constraints, and transaction costs.

Economies in transition: For most of these countries, GDP effects range from negligible to a several percent increase. This reflects opportunities for energy efficiency improvements not available to Annex II countries. Under assumptions of drastic energy efficiency improvement and/or continuing economic recessions in some countries, the assigned amounts may exceed projected emissions in the first commitment period. In this case, models show increased GDP through revenues from trading assigned amounts. However, for some economies in transition, implementing the Kyoto Protocol will have similar impacts on GDP as for Annex II countries.

Non-Annex I countries: Emission constraints in Annex I countries have well established, albeit varied "spillover" effects[24] on non-Annex I countries.

- Oil-exporting, non-Annex I countries: Analyses report costs differently, including, *inter alia*, reductions in projected GDP and reductions in projected oil revenues[25]. The study reporting the lowest costs shows reductions of 0.2% of projected GDP with no emissions trading, and less than 0.05% of projected GDP with Annex B emissions trading in 2010[26]. The study reporting the highest costs

shows reductions of 25% of projected oil revenues with no emissions trading, and 13% of projected oil revenues with Annex B emissions trading in 2010. These studies do not consider policies and measures[27] other than Annex B emissions trading, that could lessen the impact on non-Annex I, oil-exporting countries, and therefore tend to overstate both the costs to these countries and overall costs. The effects on these countries can be further reduced by removal of subsidies for fossil fuels, energy tax restructuring according to carbon content, increased use of natural gas, and diversification of the economies of non-Annex I, oil-exporting countries.

- Other non-Annex I countries: They may be adversely affected by reductions in demand for their exports to OECD nations and by the price increase of those carbon-intensive and other products they continue to import. These countries may benefit from the reduction in fuel prices, increased exports of carbon-intensive products and the transfer of environmentally sound technologies and know-how. The net balance for a given country depends on which of these factors dominates. Because of these complexities, the breakdown of winners and losers remains uncertain.
- Carbon leakage[28]: The possible relocation of some carbon-intensive industries to non-Annex I countries and wider impacts on trade flows in response to changing prices may lead to leakage in the order of 5-20%. Exemptions, for example for energy-intensive industries, make the higher model estimates for carbon leakage unlikely, but would raise aggregate costs. The transfer of environmentally sound technologies and know-how, not included in models, may lead to lower leakage and especially on the longer term may more than offset the leakage.

8.9. The Costs of Meeting a Range of Stabilization Targets

Cost-effectiveness studies with a century timescale estimate that the costs of stabilizing CO_2 concentrations in the atmosphere increase as the concentration stabilization level declines. Different baselines can have a strong influence on absolute costs. While there is a moderate increase in the costs when passing from a 750ppmv to a 550ppmv concentration stabilization

---

[24] Spillover effects here incorporate only economic effects, not environmental effects.

[25] Details of the six studies reviewed are found in Table 9-4 of the underlying report.

[26] These estimated costs can be expressed as differences in GDP growth rates over the period 2000-2010. With no emissions trading, GDP growth rate is reduced by 0.02 percentage points/year; with Annex B emissions trading, growth rate is reduced by less than 0.005 percentage points/year.

[27] These policies and measures include: those for non-CO_2 gases and non-energy sources of all gases; offsets from sinks; industry restructuring (e.g., from energy producer to supplier of energy services); use of OPEC's market power; and actions (e.g. of Annex B Parties) related to funding, insurance, and the transfer of technology. In addition, the studies typically do not include the following policies and effects that can reduce the total cost of mitigation: the use of tax revenues to reduce tax burdens or finance other mitigation measures; environmental ancillary benefits of reductions in fossil fuel use; and induced technological change from mitigation policies.

[28] Carbon leakage is defined here as the increase in emissions in non-Annex B countries resulting from implementation of reductions in Annex B, expressed as a percentage of Annex B reductions.

level, there is a larger increase in costs passing from 550ppmv to 450ppmv unless the emissions in the baseline scenario are very low. These results, however, do not incorporate carbon sequestration and gases other than CO_2, and did not examine the possible effect of more ambitious targets on induced technological change[29]. In particular, the choice of the reference scenario has a strong influence. Recent studies using the IPCC SRES reference scenarios as baselines against which to analyze stabilization clearly show that the average reduction in projected GDP in most of the stabilization scenarios reviewed here is under 3% of the baseline value (the maximum reduction across all the stabilization scenarios reached 6.1% in a given year). At the same time, some scenarios (especially in the A1T group) showed an increase in GDP compared to the baseline because of apparent positive economic feedbacks of technology development and transfer. The GDP reduction (averaged across storylines and stabilization levels) is lowest in 2020 (1%), reaches a maximum in 2050 (1.5%), and declines by 2100 (1.3%). However, in the scenario groups with the highest baseline emissions (A2 and A1FI), the size of the GDP reduction increases throughout the modelling period. Due to their relatively small scale when compared to absolute GDP levels, GDP reductions in the post-SRES stabilization scenarios do not lead to significant declines in GDP growth rates over this century. For example, the annual 1990-2100 GDP growth rate across all the stabilization scenarios was reduced on average by only 0.003% per year, with a maximum reduction reaching 0.06% per year.

The concentration of CO_2 in the atmosphere is determined more by cumulative rather than by year-by-year emissions. That is, a particular concentration target can be reached through a variety of emissions pathways. A number of studies suggest that the choice of emissions pathway can be as important as the target itself in determining overall mitigation costs. The studies fall into two categories: those that assume that the target is known and those that characterize the issue as one of decision making under uncertainty.

For studies that assume that the target is known, the issue is one of identifying the least-cost mitigation pathway for achieving the prescribed target. Here the choice of pathway can be seen as a carbon budget problem. This problem has been so far addressed in terms of CO_2 only and very limited treatment has been given to non-CO_2 GHGs. A concentration target defines an allowable amount of carbon to be emitted into the atmosphere between now and the date at which the target is to be achieved. The issue is how best to allocate the carbon budget over time.

Most studies that have attempted to identify the least-cost pathway for meeting a particular target conclude that such as pathway tends to depart gradually from the model's baseline in the early years with more rapid reductions later on. There are several reasons why this is so. A gradual near-term transition from the world's present energy system minimizes premature retirement of existing capital stock, provides time for technology development, and avoids premature lock-in to early versions of rapidly developing low-emission technology. On the other hand, more aggressive near-term action would decrease environmental risks associated with rapid climatic changes, stimulate more rapid deployment of existing low-emission technologies (see also Section 8.10), provide strong near-term incentives to future technological changes that may help to avoid lock-in to carbon intensive technologies, and allow for later tightening of targets should that be deemed desirable in light of evolving scientific understanding.

It should also be noted that the lower the concentration target, the smaller the carbon budget, and hence the earlier the departure from the baseline. However, even with higher concentration targets, the more gradual transition from the baseline does not negate the need for early action. All stabilization targets require future capital stock to be less carbon-intensive. This has immediate implications for near-term investment decisions. New supply options typically take many years to enter into the marketplace. An immediate and sustained commitment to R&D is required if low-carbon low-cost substitutes are to be available when needed.

The above addresses the issue of mitigation costs. It is also important to examine the environmental impacts of choosing one emission pathway over another. This is because different emission pathways imply not only different emission reduction costs, but also different benefits in terms of avoided environmental impacts (see Section 10).

The assumption that the target is known with certainty is, of course, an oversimplification. Fortunately, the UNFCCC recognizes the dynamic nature of the decision problem. It calls for periodic reviews "in light of the best scientific information on climate change and its impacts." Such a sequential decision making process aims to identify short-term hedging strategies in the face of long-term uncertainties. The relevant question is not "what is the best course of action for the next hundred years" but rather "what is the best course for the near-term given the long-term uncertainties."

Several studies have attempted to identify the optimal near-term hedging strategy based on the uncertainty regarding the long-term objective. These studies find that the desirable amount of hedging depends upon one's assessment of the stakes, the odds, and the cost of mitigation. The risk premium − the amount that society is willing to pay to avoid risk − ultimately is a political decision that differs among countries.

---

[29] Induced technological change is an emerging field of inquiry. None of the literature reviewed in TAR on the relationship between the century-scale CO_2 concentrations and costs reported results for models employing induced technological change. Models with induced technological change under some circumstances show that century-scale concentrations can differ, with similar GDP growth but under different policy regimes.

8.10. The Issue of Induced Technological Change

Most models used to assess the costs of meeting a particular mitigation objective tend to oversimplify the process of technical change. Typically, the rate of technical change is assumed to be independent of the level of emissions control. Such change is referred to as autonomous. In recent years, the issue of induced technical change has received increased attention. Some argue that such change might substantially lower and perhaps even eliminate the costs of CO_2 abatement policies. Others are much less sanguine about the impact of induced technical change.

Recent research suggests that the effect on timing depends on the source of technological change. When the channel for technological change is R&D, the induced technological change makes it preferable to concentrate more abatement efforts in the future. The reason is that technological change lowers the costs of future abatement relative to current abatement, making it more cost-effective to place more emphasis on future abatement. But, when the channel for technological change is learning-by-doing, the presence of induced technological change has an ambiguous impact on the optimal timing of abatement. On the one hand, induced technical change makes future abatement less costly, which suggests emphasizing future abatement efforts. On the other hand, there is an added value to current abatement because such abatement contributes to experience or learning and helps reduce the costs of future abatement. Which of these two effects dominates depends on the particular nature of the technologies and cost functions.

Certain social practices may resist or enhance technological change. Therefore, public awareness-raising and education may help encourage social change to an environment favourable for technological innovation and diffusion. This represents an area for further research.

9. Sectoral Costs and Ancillary Benefits of Mitigation

9.1. Differences between Costs of Climate Change Mitigation Evaluated Nationally and by Sector

Policies adopted to mitigate global warming will have implications for specific sectors, such as the coal industry, the oil and gas industry, electricity, manufacturing, transportation, and households. A sectoral assessment helps to put the costs in perspective, to identify the potential losers and the extent and location of the losses, and to identify the sectors that may benefit. However, it is worth noting that the available literature to make this assessment is limited: there are few comprehensive studies of the sectoral effects of mitigation, compared with those on the macro GDP effects, and they tend to be for Annex I countries and regions.

There is a fundamental problem for mitigation policies. It is well established that, compared to the situation for potential gainers, the potential sectoral losers are easier to identify, and their losses are likely to be more immediate, more concentrated, and more certain. The potential sectoral gainers (apart from the renewables sector and perhaps the natural gas sector) can only expect a small, diffused, and rather uncertain gain, spread over a long period. Indeed many of those who may gain do not exist, being future generations and industries yet to develop.

It is also well established that the overall effects on GDP of mitigation policies and measures, whether positive or negative, conceal large differences between sectors. In general, the energy intensity and the carbon intensity of the economies will decline. The coal and perhaps the oil industries are expected to lose substantial proportions of their traditional output relative to those in the reference scenarios, though the impact of this on the industries will depend on diversification, and other sectors may increase their outputs but by much smaller proportions. Reductions in fossil fuel output below the baseline will not impact all fossil fuels equally. Fuels have different costs and price sensitivities; they respond differently to mitigation policies. Energy-efficiency technology is fuel and combustion device-specific, and reductions in demand can affect imports differently from output. Energy-intensive sectors, such as heavy chemicals, iron and steel, and mineral products, will face higher costs, accelerated technical or organizational change, or loss of output (again relative to the reference scenario) depending on their energy use and the policies adopted for mitigation.

Industries concerned directly with mitigation are likely to benefit from action. These industries include renewable and nuclear electricity, producers of mitigation equipment (incorporating energy- and carbon-saving technologies), agriculture and forestry producing energy crops, and research services producing energy and carbon-saving R&D. They may benefit in the long term from the availability of financial and other resources that would otherwise have been taken up in fossil fuel production. They may also benefit from reductions in tax burdens if taxes are used for mitigation and the revenues recycled as reductions in employer, corporate, or other taxes. Those studies that report reductions in GDP do not always provide a range of recycling options, suggesting that policy packages increasing GDP have not been explored. The extent and nature of the benefits will vary with the policies followed. Some mitigation policies can lead to net overall economic benefits, implying that the gains from many sectors will outweigh the losses for coal and other fossil fuels, and energy-intensive industries. In contrast, other less-well-designed policies can lead to overall losses.

It is worth placing the task faced by mitigation policy in an historical perspective. CO_2 emissions have tended to grow more slowly than GDP in a number of countries over the past

40 years. The reasons for such trends vary but include:
- a shift away from coal and oil and towards nuclear and gas as the source of energy;
- improvements in energy efficiency by industry and households; and
- a shift from heavy manufacturing towards more service and information-based economic activity.

These trends will be encouraged and strengthened by mitigation policies.

9.2. Selected Specific Sectoral Findings on Costs of Climate Change Mitigation

9.2.1. Coal

Within this broad picture, certain sectors will be substantially affected by mitigation. Relative to the reference case, the coal industry, producing the most carbon-intensive of products, faces almost inevitable decline in the long term, relative to the baseline projection. Technologies still under development, such as CO_2 removal and storage from coal-burning plants and in-situ gasification, could play a future role in maintaining the output of coal whilst avoiding CO_2 and other emissions. Particularly large effects on the coal sector are expected from policies such as the removal of fossil fuel subsidies or the restructuring of energy taxes so as to tax the carbon content rather than the energy content of fuels. It is a well-established finding that removal of the subsidies would result in substantial reductions in GHG emissions, as well as stimulating economic growth. However, the effects in specific countries depend heavily on the type of subsidy removed and the commercial viability of alternative energy sources, including imported coal.

9.2.2. Oil

The oil industry also faces a potential relative decline, although this may be moderated by lack of substitutes for oil in transportation, substitution away from solid fuels towards liquid fuels in electricity generation, and the diversification of the industry into energy supply in general.

Table TS-6 shows a number of model results for the impacts of implementation of the Kyoto Protocol on oil exporting countries. Each model uses a different measure of impact, and many use different groups of countries in their definition of oil exporters. However, the studies all show that the use of the flexibility mechanisms will reduce the economic cost to oil producers.

Thus, studies show a wide range of estimates for the impact of GHG mitigation policies on oil production and revenue. Much of these differences are attributable to the assumptions made about: the availability of conventional oil reserves, the degree of mitigation required, the use of emission trading, control of GHGs other than CO_2, and the use of carbon sinks. However, all studies show a net growth in both oil production and revenue to at least 2020, and significantly less impact on the real price of oil than has resulted from market fluctuations over the past 30 years. Figure TS-9 shows the projection of real oil prices to 2010 from the IEA's 1998 *World Energy Outlook*, and the effect of Kyoto implementation from the G-cubed model, the study which shows the largest fall in Organization of Oil Exporting Countries (OPEC) revenues in Table TS-6. The 25% loss in OPEC revenues in the non-trading scenario implies a 17% fall in oil prices shown for 2010 in the figure; this is reduced to a fall of just over 7% with Annex I trading.

These studies typically do not consider some or all of the following policies and measures that could lessen the impact on oil exporters:
- policies and measures for non-CO_2 GHGs or non-energy sources of all GHGs;
- offsets from sinks;
- industry restructuring (e.g., from energy producer to supplier of energy services);
- the use of OPEC's market power; and

Table TS-6: Costs of Kyoto Protocol implementation for oil exporting region/countries.[a]

Model[b]	Without trading[c]	With Annex I trading	With "global trading"
G-Cubed	-25% oil revenue	-13% oil revenue	-7% oil revenue
GREEN	-3% real income	"Substantially reduced loss"	n.a.
GTEM	0.2% GDP loss	<0.05% GDP loss	n.a.
MS-MRT	1.39% welfare loss	1.15% welfare loss	0.36% welfare loss
OPEC Model	-17% OPEC revenue	-10% OPEC revenue	-8% OPEC revenue
CLIMOX	n.a.	-10% some oil exporters' revenues	n.a.

[a] The definition of oil exporting country varies: for G-Cubed and the OPEC model it is the OPEC countries, for GREEN it is a group of oil exporting countries, for GTEM it is Mexico and Indonesia, for MS-MRT it is OPEC + Mexico, and for CLIMOX it is West Asian and North African oil exporters.
[b] The models all considere the global economy to 2010 with mitigation according to the Kyoto Protocol targets (usually in the models, applied to CO_2 mitigation by 2010 rather than GHG emissions for 2008 to 2012) achieved by imposing a carbon tax or auctioned emission permits with revenues recycled through lump-sum payments to consumers; no co-benefits, such as reductions in local air pollution damages, are taken into account in the results.
[c] "Trading" denotes trading in emission permits between countries.

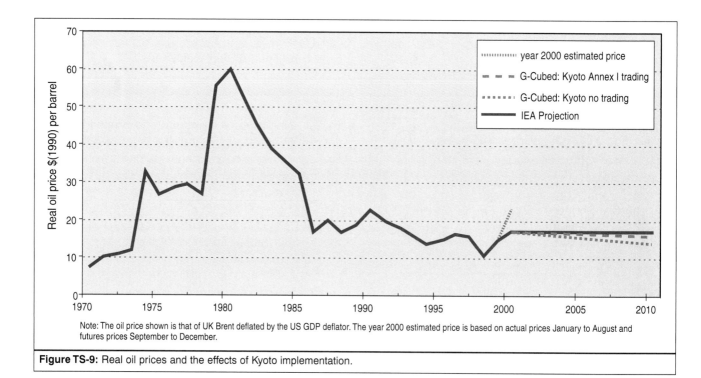

Note: The oil price shown is that of UK Brent deflated by the US GDP deflator. The year 2000 estimated price is based on actual prices January to August and futures prices September to December.

Figure TS-9: Real oil prices and the effects of Kyoto implementation.

actions (e.g., of Annex B Parties) related to funding, insurance, and the transfer of technology.

In addition, the studies typically do not include the following policies and effects that can reduce the total cost of mitigation:
- the use of tax revenues to reduce tax burdens or finance other mitigation measures;
- environmental co- or ancillary benefits of reductions in fossil fuel use; and
- induced technical change from mitigation policies.

As a result, the studies may tend to overstate both the costs to oil exporting countries and overall costs.

9.2.3. Gas

Modelling studies suggest that mitigation policies may have the least impact on oil, the most impact on coal, with the impact on gas somewhere between; these findings are established but incomplete. The high variation across studies for the effects of mitigation on gas demand is associated with the importance of its availability in different locations, its specific demand patterns, and the potential for gas to replace coal in power generation.

These results are different from recent trends, which show natural gas usage growing faster than the use of either coal or oil. They can be explained as follows. In the transport sector, the largest user of oil, current technology and infrastructure will not allow much switching from oil to non-fossil fuel alternatives in Annex I countries before about 2020. Annex B countries can only meet their Kyoto Protocol commitments by reducing overall energy use and this will result in a reduction in natural

gas demand, unless this is offset by a switch towards natural gas for power generation. The modelling of such a switch remains limited in these models.

9.2.4. Electricity

In general as regards the effects on the electricity sector, mitigation policies either mandate or directly provide incentives for increased use of zero-emitting technologies (such as nuclear, hydro, and other renewables) and lower-GHG-emitting generation technologies (such as combined cycle natural gas). Or, second, they drive their increased use indirectly by more flexible approaches that place a tax on or require a permit for emission of GHGs. Either way, the result will be a shift in the mix of fuels used to generate electricity towards increased use of the zero- and lower-emitting generation technologies, and away from the higher-emitting fossil fuels.

Nuclear power would have substantial advantages as a result of GHG mitigation policies, because power from nuclear fuel produces negligible GHGs. In spite of this advantage, nuclear power is not seen as the solution to the global warming problem in many countries. The main issues are (1) the high costs compared to alternative CCGTs, (2) public acceptance involving operating safety and waste, (3) safety of radioactive waste management and recycling of nuclear fuel, (4) the risks of nuclear fuel transportation, and (5) nuclear weapons proliferation.

9.2.5. Transport

Unless highly efficient vehicles (such as fuel cell vehicles) become rapidly available, there are few options available to reduce transport energy use in the short term, which do not involve significant economic, social, or political costs. No government has yet demonstrated policies that can reduce the overall demand for mobility, and all governments find it politically difficult to contemplate such measures. Substantial additional improvements in aircraft energy efficiency are most likely to be accomplished by policies that increase the price of, and therefore reduce the amount of, air travel. Estimated price elasticities of demand are in the range of −0.8 to −2.7. Raising the price of air travel by taxes faces a number of political hurdles. Many of the bilateral treaties that currently govern the operation of the air transport system contain provisions for exemptions of taxes and charges, other than for the cost of operating and improving the system.

9.3. Sectoral Ancillary Benefits of Greenhouse Gas Mitigation

The direct costs for fossil fuel consumption are accompanied by environmental and public health benefits associated with a reduction in the extraction and burning of the fuels. These benefits come from a reduction in the damages caused by these activities, especially a reduction in the emissions of pollutants that are associated with combustion, such as SO_2, NO_x, CO and other chemicals, and particulate matter. This will improve local and regional air and water quality, and thereby lessen damage to human, animal, and plant health, and to ecosystems. If all the pollutants associated with GHG emissions are removed by new technologies or end-of-pipe abatement (for example, flue gas desulphurization on a power station combined with removal of all other non-GHG pollutants), then this ancillary benefit will no longer exist. But such abatement is limited at present and it is expensive, especially for small-scale emissions from dwellings and cars (See also Section 8.6).

9.4. The Effects of Mitigation on Sectoral Competitiveness

Mitigation policies are less effective if they lead to loss of international competitiveness or the migration of GHG-emitting industries from the region implementing the policy (so-called carbon leakage). The estimated effects, reported in the literature, on international price competitiveness are small while those on carbon leakage appear to beat the stage of competing explanations, with large differences depending on the models and the assumptions used. There are several reasons for expecting that such effects will not be substantial. First, mitigation policies actually adopted use a range of instruments and usually include special treatment to minimize adverse industrial effects, such as exemptions for energy-intensive industries. Second, the models assume that any migrating industries will use the average technology of the area to which they will move; however, instead they may adopt newer, lower CO_2-emitting technologies. Third, the mitigation policies also encourage low-emission technologies and these also may migrate, reducing emissions in industries in other countries (see also Section 8.7).

9.5. Why the Results of Studies Differ

The results in the studies assessed come from different approaches and models. A proper interpretation of the results requires an understanding of the methods adopted and the underlying assumptions of the models and studies. Large differences in results can arise from the use of different reference scenarios or baselines. And the characteristics of the baseline can markedly affect the quantitative results of modelling mitigation policy. For example, if air quality is assumed to be satisfactory in the baseline, then the potential for air-quality ancillary benefits in any GHG mitigation scenario is ruled out by assumption. Even with similar or the same baseline assumptions, the studies yield different results.

As regards the costs of mitigation, these differences appear to be largely caused by different approaches and assumptions, with the most important being the type of model adopted. Bottom-up engineering models assuming new technological opportunities tend to show benefits from mitigation. Top-down general equilibrium models appear to show lower costs than top-down time-series econometric models. The main assumptions leading to lower costs in the models are that:

- new flexible instruments, such as emission trading and joint implementation, are adopted;
- revenues from taxes or permit sales are returned to the economy by reducing burdensome taxes; and
- ancillary benefits, especially from reduced air pollution, are included in the results.

Finally, long-term technological progress and diffusion are largely given in the top-down models; different assumptions or a more integrated, dynamic treatment could have major effects on the results.

10. Decision Analytical Frameworks

10.1. Scope for and New Developments in Analyses for Climate Change Decisions

Decision making frameworks (DMFs) related to climate change involve multiple levels ranging from global negotiations to individual choices and a diversity of actors with different resource endowments, and diverging values and aspirations. This explains why it is difficult to arrive at a management strategy that is acceptable for all. The dynamic interplay among economic sectors and related social interest groups makes it difficult to arrive at a national position to be represented at international fora in the first place. The intricacies of international climate negotiations result from the manifold often-ambiguous national positions as well as from the linkages of climate change policy with other socio-economic objectives.

No DMF can reproduce the above diversity in its full richness. Yet analysts have made significant progress in several directions since SAR. First, they integrate an increasing number of issues into a single analytical framework in order to provide an internally consistent assessment of closely related components, processes, and subsystems. The resulting integrated assessment models (IAMs) cited in Chapter 9, and indeed throughout the whole report, provide useful insights into a number of climate policy issues for policymakers. Second, scientists pay increasing attention to the broader context of climate related issues that have been ignored or paid marginal attention previously. Among other factors, this has fostered the integration of development, sustainability and equity issues into the present report.

Climate change is profoundly different from most other environmental problems with which humanity has grappled. A combination of several features lends the climate problem its uniqueness. They include public good issues raising from the concentration of GHGs in the atmosphere that requires collective global action, the multiplicity of decision makers ranging from global down to the micro level of firms and individuals, and the heterogeneity of emissions and their consequences around the world. Moreover, the long-term nature of climate change originates from the fact that it is the concentration of GHGs that matters rather than their annual emissions and this feature raises the thorny issues of intergenerational transfers of wealth and environmental goods and bads. Next, human activities associated with climate change are widespread, which makes narrowly defined technological solutions impossible, and the interactions of climate policy with other broad socio-economic policies are strong. Finally, large uncertainties or in some areas even ignorance characterize many aspects of the problem and require a risk management approach to be adopted in all DMFs that deal with climate change.

Policymakers therefore have to grapple with great uncertainties in choosing the appropriate responses. A wide variety of tools have been applied to help them make fundamental choices. Each of those decision analysis frameworks (DAFs) has its own merits and shortcoming through its ability to address some of the above features well, but other facets less adequately. Recent analyses with well-established tools such as cost–benefit analysis as well as newly developed frameworks like the tolerable windows or safe landing approach provide fresh insights into the problem.

Figure TS-10a shows the results of a cost-effectiveness analysis exploring the optimal hedging strategy when uncertainty with respect to the long-term stabilization target is not resolved until 2020, suggesting that abatement over the next few years would be economically valuable if there is a significant probability of having to stay below ceilings that would be otherwise reached within the characteristic time scales of the systems producing greenhouse gases. The degree of near-term hedging in the above analysis is sensitive to the date of resolution of uncertainty, the inertia in the energy system, and the fact that the ultimate concentration target (once it has been

revealed) must be met at all costs. Other experiments, such as those with cost-benefit models framed as a Bayesian decision analysis problem show that optimal near-term (next two decades) emission paths diverge only modestly under perfect foresight, and hedging even for low-probability, high-consequence scenarios (see Figure TS-10b). However, decisions about near-term climate policies may have to be made while the stabilization target is still being debated. Decision-making therefore should consider appropriate hedging against future resolution of that target and possible revision of the scientific insights in the risks of climate change. There are significant differences in the two approaches. With a cost-effectiveness analysis, the target must be made regardless of costs. With a cost-benefit analysis, costs and benefits are balanced at the margin. Nevertheless, the basic message is quite similar and involves the explicit incorporation of uncertainty and its sequential resolution over time. The desirable amount of hedging depends upon one's assessment of the stakes, the odds, and the costs of policy measures. The risk premium − the amount that society is willing to pay to reduce risk − ultimately is a political decision that differs among countries.

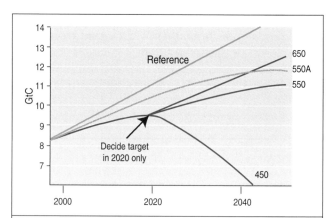

Figure TS-10a: Optimal carbon dioxide emissions strategy, using a cost-effectiveness approach.

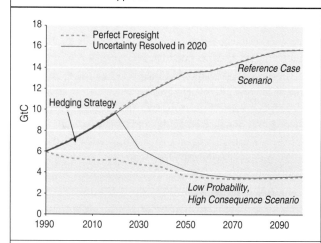

Figure TS-10b: Optimal hedging strategy for low probability, high consequence scenario using a cost-benefits optimization approach.

Cost-effectiveness analyses seek the lowest cost of achieving an environmental target by equalizing the marginal costs of mitigation across space and time. Long-term cost-effectiveness studies estimate the costs of stabilizing atmospheric CO_2 concentrations at different levels and find that the costs of the 450ppmv ceiling are substantially greater than those of the 750ppmv limit. Rather than seeking a single optimal path, the tolerable windows/safe landing approach seeks to delineate the complete array of possible emission paths that satisfy externally defined climate impact and emission cost constraints. Results indicate that delaying near-term effective emission reductions can drastically reduce the future range of options for relatively tight climate change targets, while less tight targets offer more near-term flexibility.

10.2. International Regimes and Policy Options

The structure and characteristics of international agreements on climate change will have a significant influence on the effectiveness and costs and benefits of mitigation. The effectiveness and the costs and benefits of an international climate change regime (such as the Kyoto Protocol or other possible future agreements) depend on the number of signatories to the agreement and their abatement targets and/or policy commitment. At the same time, the number of signatories depends on the question of how equitably the commitments of participants are shared. Economic efficiency (minimizing costs by maximizing participation) and equity (the allocation of emissions limitation commitments) are therefore strongly linked.

There is a three-way relationship between the design of the international regime, the cost-effectiveness/efficiency of climate policies, and the equity of the consequent economic outcomes. As a consequence, it is crucial to design the international regime in a way that is considered both efficient and equitable. The literature presents different theoretical strategies to optimize an international regime. For example, it can be made attractive for countries to join the group that commits to specific targets for limitation and reduction of emissions by increasing the equity of a larger agreement – and therefore its efficiency – through measures like an appropriate distribution of targets over time, the linkage of the climate debate with other issues ("issue linkage"), the use of financial transfers to affected countries ("side payments"), or technology transfer agreements.

Two other important concerns shape the design of an international regime: "implementation" and "compliance". The effectiveness of the regime, which is a function of both implementation and compliance, is related to actual changes of behaviour that promote the goals of the accord. *Implementation* refers to the translation of international accords into domestic law, policy, and regulations by national governments. *Compliance* is related to whether and to what extent countries do in fact adhere to provisions of an accord. *Monitoring*, *reporting*, and *verification* are essential for the effectiveness of international

environmental regimes, as the systematic monitoring, assessment, and handling of implementation failures have been so far relatively rare. Nonetheless, efforts to provide "systems of implementation review" are growing, and are already incorporated into the UNFCCC structure. The challenge for the future is to make them more effective, especially by improving data on national emissions, policies, and measures.

10.3. Linkages to National and Local Sustainable Development Choices

Much of the ambiguity related to sustainable development and climate change arises from the lack of measurements that could provide policymakers with essential information on the alternative choices at stake, how those choices affect clear and recognizable social, economic, and environmental critical issues, and also provide a basis for evaluating their performance in achieving goals and targets. Therefore, indicators are indispensable to make the concept of sustainable development operational. At the national level important steps in the direction of defining and designing different sets of indicators have been undertaken; however, much work remains to be done to translate sustainability objectives into practical terms.

It is difficult to generalize about sustainable development policies and choices. Sustainability implies and requires diversity, flexibility, and innovation. Policy choices are meant to introduce changes in technological patterns of natural resource use, production and consumption, structural changes in the production systems, spatial distribution of population and economic activities, and behavioural patterns. Climate change literature has by and large addressed the first three topics, while the relevance of choices and decisions related to behavioural patterns and lifestyles has been paid scant attention. Consumption patterns in the industrialized countries are an important reason for climate change. If people changed their preferences this could alleviate climate change considerably. To change consumption patterns, however, people must not only change their behaviour but also change themselves because these patterns are an essential element of lifestyles and, therefore, of self-esteem. Yet, apart from climate change there are other reasons to do so as well as indications that this change can be fostered politically.

A critical requirement of sustainable development is a capacity to design policy measures that, without hindering development and consistent with national strategies, could exploit potential synergies between national economic growth objectives and environmentally focused policies. Climate change mitigation strategies offer a clear example of how co-ordinated and harmonized policies can take advantage of the synergies between the implementation of mitigation options and broader objectives. Energy efficiency improvements, including energy conservation, switch to low carbon content fuels, use of renewable energy sources and the introduction of more advanced non conventional energy technologies, are expected to have significant impacts on curbing actual GHG emission tendencies. Similarly, the adoption of new technologies and practices in agriculture and forestry activities as well as the adoption of clean production processes could make substantial contributions to the GHG mitigation effort. Depending on the specific context in which they are applied, these options may entail positive side effects or double dividends, which in some cases are worth undertaking whether or not there are climate-related reasons for doing so.

Sustainable development requires radical technological and related changes in both developed and developing countries. Technological innovation and the rapid and widespread transfer and implementation of individual technological options and choices, as well as overall technological systems, constitute major elements of global strategies to achieve both climate stabilization and sustainable development. However, technology transfer requires more than technology itself. An enabling environment for the successful transfer and implementation of technology plays a crucial role, particularly in developing countries. If technology transfer is to bring about economic and social benefits it must take into account the local cultural traditions and capacities as well as the institutional and organizational circumstances required to handle, operate, replicate, and improve the technology on a continuous basis.

The process of integrating and internalizing climate change and sustainable development policies into national development agendas requires new problem solving strategies and decision-making approaches. This task implies a twofold effort. On one hand, sustainable development discourse needs greater analytical and intellectual rigor (methods, indicators, etc.) to make this concept advance from theory to practice. On the other hand, climate change discourse needs to be aware of both the restrictive set of assumptions underlying the tools and methods applied in the analysis, and the social and political implications of scientific constructions of climate change. Over recent years a good deal of analytical work has addressed the problem in both directions. Various approaches have been explored to transcend the limits of the standard views and decision frameworks in dealing with issues of uncertainty, complexity, and the contextual influences of human valuation and decision making. A common theme emerges: the emphasis on participatory decision making frameworks for articulating new institutional arrangements.

10.4. Key Policy-relevant Scientific Questions

Different levels of globally agreed limits for climate change (or for corresponding atmospheric GHG concentrations), entail different balances of mitigation costs and net damages for individual nations. Considering the uncertainties involved and future learning, climate stabilization will inevitably be an iterative process: nation states determine their own national targets based on their own exposure and their sensitivity to other countries' exposure to climate change. The global target emerges from consolidating national targets, possibly involving side payments, in global negotiations. Simultaneously, agreement on burden sharing and the agreed global target determines national costs. Compared to the expected net damages associated with the global target, nation states might reconsider their own national targets, especially as new information becomes available on global and regional patterns and impacts of climate change. This is then the starting point for the next round of negotiations. It follows from the above that establishing the "magic number" (i.e., the upper limit for global climate change or GHG concentration in the atmosphere) will be a long process and its source will primarily be the policy process, hopefully helped by improving science.

Looking at the key dilemmas in climate change decision making, the following conclusions emerge (see also Table TS-7):

- a carefully crafted portfolio of mitigation, adaptation, and learning activities appears to be appropriate over the next few decades to hedge against the risk of intolerable magnitudes and/or rates of climate change (impact side) and against the need to undertake painfully drastic emission reductions if the resolution of uncertainties reveals that climate change and its impacts might imply high risks;
- emission reduction is an important form of mitigation, but the mitigation portfolio includes a broad range of other activities, including investments to develop low-cost non-carbon, energy efficient and carbon management technologies that will make future CO_2 mitigation less expensive;
- timing and composition of mitigation measures (investment into technological development or immediate emission reductions) is highly controversial because of the technological features of energy systems, and the range of uncertainties involved in the impacts of different emission paths;
- international flexibility instruments help reduce the costs of emission reductions, but they raise a series of implementation and verification issues that need to be balanced against the cost savings;
- while there is a broad consensus to use the Pareto optimality[30] as the efficiency principle, there is no agreement on the best equity principle on wich to build an equitable international regime. Efficiency and equity are important concerns in negotiating emission limitation schemes, and they are not mutually exclusive. Therefore, equity will play an important role in determining the distribution of emissions allowances and/or within compensation schemes following emission trading that could lead to a disproportionately high level of burden on certain countries. Finally, it could be more important to build a regime on the combined implications of the various equity principles rather than to select any one particular equity principle. Diffusing non-carbon, energy-efficient, as

---

[30] Pareto optimum is a requirement or status that an individual's welfare could not be further improved without making others in the society worse off.

Table TS-7: Balancing the near-term mitigation portfolio.

Issue	Favouring modest early abatement	Favouring stringent early abatement
Technology development	· Energy technologies are changing and improved versions of existing technologies are becoming available, even without policy intervention. · Modest early deployment of rapidly improving technologies allows learning-curve cost reductions, without premature lock-in to existing, low-productivity technology. · The development of radically advanced technologies will require investment in basic research.	· Availability of low-cost measures may have substantial impact on emissions trajectories. · Endogenous (market-induced) change could accelerate development of low-cost solutions (learning-by-doing). · Clustering effects highlight the importance of moving to lower emission trajectories. · Induces early switch of corporate energy R&D from fossil frontier developments to low carbon technologies.
Capital stock and inertia	· Beginning with initially modest emissions limits avoids premature retirement of existing capital stocks and takes advantage of the natural rate of capital stock turnover. · It also reduces the switching cost of existing capital and prevents rising prices of investments caused by crowding out effects.	· Exploit more fully natural stock turnover by influencing new investments from the present onwards. · By limiting emissions to levels consistent with low CO_2 concentrations, preserves an option to limit CO_2 concentrations to low levels using current technology. · Reduces the risks from uncertainties in stabilization constraints and hence the risk of being forced into very rapid reductions that would require premature capital retirement later.
Social effects and inertia	· Gradual emission reduction reduces the extent of induced sectoral unemployment by giving more time to retrain the workforce and for structural shifts in the labour market and education. · Reduces welfare losses associated with the need for fast changes in people's lifestyles and living arrangements.	· Especially if lower stabilization targets would be required ultimately, stronger early action reduces the maximum rate of emissions abatement required subsequently and reduces associated transitional problems, disruption, and the welfare losses associated with the need for faster later changes in people's lifestyles and living arrangements.
Discounting and intergenerational equity	· Reduces the present value of future abatement costs (*ceteris paribus*), but possibly reduces future relative costs by furnishing cheap technologies and increasing future income levels.	· Reduces impacts and (*ceteris paribus*) reduces their present value.
Carbon cycle and radiative change	· Small increase in near-term, transient CO_2 concentration. · More early emissions absorbed, thus enabling higher total carbon emissions this century under a given stabilization constraint (to be compensated by lower emissions thereafter).	· Small decrease in near-term, transient CO_2 concentration. · Reduces peak rates in temperature change.
Climate change impacts	· Little evidence on damages from multi-decade episodes of relatively rapid change in the past.	· Avoids possibly higher damages caused by faster rates of climate change.

well as other GHG reducing technologies worldwide could make a significant contribution to reducing emissions over the short term, but many barriers hamper technology transfer, including market imperfections, political problems, and the often-neglected transaction costs;

· some obvious linkages exist between current global and continental environmental problems and attempts of the international community to resolve them, but the potential synergies of jointly tackling several of them have not yet been thoroughly explored, let alone exploited.

Mitigation and adaptation decisions related to anthropogenically induced climate change differ. Mitigation decisions involve many countries, disperse benefits globally over decades to centuries (with some near-term ancillary benefits), are driven by public policy action, based on information available today, and the relevant regulation will require rigorous enforcement. In contrast, adaptation decisions involve a shorter time span between outlays and returns, related costs and benefits accrue locally, and their implementation involves local public policies and private adaptation of the affected social agents, both based on improving information. Local mitigation and adaptive capacities vary significantly across regions and over time. A portfolio of mitigation and adaptation policies will depend on local or national priorities and preferred approaches in combination with international responsibilities.

Given the large uncertainties characterizing each component of the climate change problem, it is difficult for decision makers to establish a globally acceptable level of stabilizing GHG concentrations today. Studies appraised in Chapter 10 support the obvious expectations that lower stabilization targets involve substantially higher mitigation costs and relatively more ambitious near-term emission reductions on the one hand, but, as reported by WGII, lower targets induce significantly smaller bio/geophysical impacts and thus induce smaller damages and adaptation costs.

11 Gaps in Knowledge

Important gaps in own knowledge on which additional research could be useful to support future assessments include:

- *Further exploration of the regional, country, and sector specific potentials of technological and social innovation options, including:*
  - The short, medium, and long-term potential and costs of both CO_2 and non-CO_2, non-energy mitigation options;
  - Understanding of technology diffusion across different regions;
  - Identifying opportunities in the area of social innovation leading to decreased greenhouse gas emissions;
  - Comprehensive analysis of the impact of mitigation measures on C flows in and out of the terrestrial system; and
  - Some basic inquiry in the area of geo-engineering.

- *Economic, social, and institutional issues related to climate change mitigation in all countries. Priority areas include:*
  - Much more analysis of regionally specific mitigation options, barriers, and policies is recommended as these are conditioned by the regions' mitigative capacity;
  - The implications of mitigation on equity;
  - Appropriate methodologies and improved data sources for climate change mitigation and capacity building in the area of integrated assessment;
  - Strengthening future research and assessments, especially in developing countries.

- *Methodologies for analysis of the potential of mitigation options and their cost, with special attention to comparability of results. Examples include:*
  - Characterizing and measuring barriers that inhibit greenhouse gas-reducing action;
  - Make mitigation modelling techniques more consistent, reproducible, and accessible;
  - Modelling technology learning; improving analytical tools for evaluating ancillary benefits, e.g. assigning the costs of abatement to greenhouse gases and to other pollutants;
  - Systematically analyzing the dependency of costs on baseline assumptions for various greenhouse gas stabilization scenarios;

- Developing decision analytical frameworks for dealing with uncertainty as well as socio-economic and ecological risk in climate policymaking;
- Improving global models and studies, their assumptions, and their consistency in the treatment and reporting of non-Annex I countries and regions.

- *Evaluating climate mitigation options in the context of development, sustainability, and equity. Examples include:*
  - More research is needed on the balance of options in the areas of mitigation and adaptation and of the mitigative and adaptive capacity in the context of DES;
  - Exploration of alternative development paths including sustainable consumption patterns in all sectors, including the transportation sector, and integrated analysis of mitigation and adaptation;
  - Identifying opportunities for synergy between explicit climate policies and general policies promoting sustainable development;
  - Integration of inter- and intragenerational equity in climate change mitigation studies; Implications of equity assessments;
  - Analysis of scientific, technical, and economic aspects of implications of options under a wide variety of stabilization regimes;
  - Determining what kinds of policies interact with what sorts of socio-economic conditions to result in futures characterized by low CO_2 emissions;
  - Investigation on how changes in societal values may be encouraged to promote sustainable development; and
  - Evaluating climate mitigation options in the context of and for synergy with potential or actual adaptive measures.

- *Development of engineering-economic, end-use, and sectoral studies of GHG emissions mitigation potentials for specific regions and/or countries of the world, focusing on:*
  - Identification and assessment of mitigation technologies and measures that are required to deviate from "business-as-usual" in the short term (2010, 2020); Development of standardized methodologies for
  - quantifying emissions reductions and costs of mitigation technologies and measures;
  - Identification of barriers to the implementation of the mitigation technologies and measures;
  - Identification of opportunities to increase adoption of GHG emissions mitigation technologies and measures through connections with ancillary benefits as well as furtherance of the DES goals; and
  - Linking the results of the assessments to specific policies and programmes that can overcome the identified barriers as well as leverage the identified ancillary benefits.

Climate Change 2001:
Synthesis Report

Annexes

An Assessment of the Intergovernmental Panel on Climate Change

The glossary and companion annexes were not submitted to the Panel for action, consistent with normal practice within the IPCC.

Annex A. Authors and Expert Reviewers

Argentina
Daniel Bouille   Fundecion Bariloche
Marcelo Cabido   IMBIV, University of Cordoba
Osvaldo F. Canziani   Co-Chair, WGII
Rodolfo Carcavallo   Department of Entomology
Jorge O. Codignotto   Laboratorio Geologia y Dinamica Costera
Martin de Zuviria   Aeroterra S.A.
Sandra Myrna Diaz   Instituto Multidisciplinario de Biologia Vegetal
Jorge Frangi   Universidad Nacional de la Plata
Hector Ginzo   Instituto de Neurobiologia
Osvaldo Girardin   Fundacion Bariloche
Carlos Labraga   Consejo Nacional de Investigaciones Cientificas y Tecnicas, Centro Nactional Patagonico
Gabriel Soler   Fundacion Instituto   Latinoamericano de Politicas Sociales (ILAPS)
Walter Vargas   University of Buenos Aires - IEIMA
Ernesto F. Viglizzo   PROCISUR/INTO/CONICET

Australia
Susan Barrell   Bureau of Meteorology
Bryson Bates   CSIRO
Ian Carruthers   Australian Greenhouse Office
Habiba Gitay   Australian National University
John A. Church   CSIRO Division of Oceanography
Ove Hoegh-Guldberg   The University of Queensland
Roger Jones   CSIRO Atmospheric Research
Bryant McAvaney   Bureau of Meteorology Research Centre
Chris Mitchell   CSIRO Atmosphere Research
Ian Noble   Australian National University
Barrie Pittock   CSIRO (Climate Impact Group)
Andy Reisinger   Ministry for the Environment
B. Soderbaum   Greenhouse Policy Office, Australian Greenhouse Office
Greg Terrill   Australian Greenhouse Office
Kevin Walsh   Principal Research Scientist CSIRO Atmospheric Research
John Zillman   Vice-Chair, WGI

Austria
Renate Christ   IPCC Secretariat
Helmut Hojesky   Federal Ministry for Environment
K. Radunsky   Federal Environment Agency

Bangladesh
Q.K. Ahmad   Bangladesh Unnayan Parishad

Barbados
Leonard Nurse   Coastal Zone Management Unit

Belgium
Philippe Huybrechts   Vrije Universiteit Brussel
C. Vinckier   Department of Chemistry, KULeuven
R. Zander   University of Liege

Benin
Epiphane Dotou Ahlonsou   Service Météorologique National
Michel Boko   Universite de Bourgogne

Bosnia
Permanent Mission of Bosnia & Herzegovina

Botswana
Pauline O. Dube   University of Botswana

Brazil
Gylvan Meira Filho   Vice-Chair, IPCC
Jose Roberto Moreira   Biomass User Network (BUN)

Canada
Brad Bass   Environment Canada
James P. Bruce   Canadian Climate Program Board
Margo Burgess   Natural Resources Canada
Wenjun Chen   Natural Resources Canada
Jing Chen   University of Toronto
Stewart J. Cohen   Environment Canada
Patti Edwards   Environment Canada
David Etkin   Environment Canada
Darren Goetze   Environment Canada
J. Peter Hall   Canadian Forest Service
H. Hengeveld   Environment Canada
Pamela Kertland   Natural Resources Canada
Abdel Maaroud   Environment Canada
Joan Masterton   Environment Canada
Chris McDermott   Environment Canada
Brian Mills   Environment Canada
Linda Mortsch   Environment Canada
Tad Murty   Baird and Associates Coastal Engineers
Paul Parker   University of Waterloo
John Robinson   University of British Columbia
Hans-Holger Rogner   University of Victoria
Daniel Scott   Environment Canada
Sharon Smith   Natural Resources Canada
Barry Smit   University of Guelph
John Stone   Vice-Chair, WGI
Tana Lowen Stratton   Dept. Foreign Affairs and International Trade
Roger Street   Environment Canada
Eric Taylor   Natural Resources Canada
G. Daniel Williams   Environment Canada (retired)

Chile
E. Basso   Independent Consultant

China
Du Bilan   China Institute for Marine Development Strategy
Z. Chen   China Meteorological Administration
Liu Chunzhen   Hydrological Forecasting & Water Control Center
Zhou Dadi   Energy Research Institute
Qin Dahe   China Meteorological Administration
Xiaosu Dai   IPCC WGI TSU
Lin Erda   Chinese Academy of Agricultural Science
Mingshan Su   Tsinghua University
Yihui Ding   Co-Chair, WGI
Guangsheng Zhou   Chinese Academy of Sciences
Z.C. Zhao   National Climate Center

Cuba
Ramon Pichs-Madruga — Vice-Chair, WGIII
A.G. Suarez — Cuban Environmental Agency

Czech Republic
Jan Pretel — Vice-Chair, WGII

Denmark
Jesper Gundermann — Danish Energy Agency
Kirsten Halsnaes — Riso International Laboratory
Erik Rasmussen — Danish Energy Agency
Martin Stendel — Danish Meteorological Institute

Finland
Timothy Carter — Finnish Environment Institute
P. Heikinheimo — Ministry of Environment
Raino Heino — Finnish Meteorological Institute
Pekka E. Kauppi — University of Helsinki
R. Korhonen — VTT Energy
A. Lampinen — University of Jyväskyla
I. Savolainen — VTT Energy

France
Olivier Boucher — Universite de Lille I
Marc Darras — Gaz de France
Jane Ellis — OECD
Jean-Charles Hourcade — CIRED/CNRS
J.C. Morlot — Environment Department
M. Petit — Ecole Polytechnique

Gambia
B.E. Gomez — Department of Water Resources
M. Njie — Department of Water Resources

Germany
Heinz-Jurgen Ahlgrimm — Institute for Technology & Biosystems
Rosemarie Benndorf — Umweltbundesamt
Peter Burschel — Technische Universitat Munchen
Ulrich Cubasch — Max Planck Institut für Meteorologie
U. Fuentes — German Advisory Council on Global Change
Joanna House Max — Planck Inst. Biogeochemie
Jucundus Jacobeit — Universitaet Wuerzburg
Eberhard Jochem — Vice-Chair, WGIII
Harald Kohl — Federal Ministry of the Environment
Petra Mahrenholz — Federal Environmental Agency of Germany
I. Colin Prentice — Max Planck Institute for Biogeochemistry
C. le Quéré — Max Planck Institute for Biogeochemistry
Sarah Raper — University of East Anglia
Ferenc Toth — Potsdam Institute for Climate Impact Research
Manfred Treber — Germanwatch
R. Sartorius — Umweltbundesamt
Michael Weber — Ludwig-Maximilians Universität München
Gerd-Rainer Weber — Gesamtverband des Deutschen Steinkohlenberghaus

Hungary
G. Koppany   University of Szeged
Halldor Thorgeirsson   Ministry for the Environment

India
Murari Lal   Indian Institute of Technology
Rajendra K. Pachauri   Tata Energy Research Institute
N.H. Ravindranath   Indian Institute of Sciences
Priyadarshi Shukla   Indian Institute of Management
Leena Srivastava   Tata Energy Research Institute

Indonesia
R.T.M. Sutamihardja   Vice-Chair, WGIII

Israel
Simon Krichak   Tel Aviv University

Italy
Filippo Giorgi   Abdus Salam International Centre for Theoretical Physics (ICTP)
Annarita Mariotti   ENEA Climate Section

Japan
Kazuo Asakura   Central Research Institute (CRIEPI)
Noriyuki Goto   University of Tokyo, Komaba
Mariko Handa   Organization for Landscape and Urban Greenery Technology Development
Hideo Harasawa   Social and Environmental Systems Division
Yasuo Hosoya   Tokyo Electric Power Company
Y. Igarashi   Ministry of Foreign Affairs
Takeshi Imai   The Kansai Electric Power Co., Inc.
M. Inoue   Ministry of Economy, Trade and Industry
Hisashi Kato   Central Research Institute of Electric Power Industry
Naoki Matsuo   Global Industrial and Social Progress Research Institute (GISPRI)
Hisayoshi Morisugi   Tohoku University
Tsuneyuki Morita   National Institute for Environmental Studies
Shinichi Nagata   Environment Agency
S. Nakagawa   Japan Meteorological Agency
Yoshiaki Nishimura   Central Research Institute of Electric Power Industry
Ichiro Sadamori   Global Industrial and Social Progress Research Institute (GISPRI)
Akihiko Sasaki   National Institute of Public Health
Shojiro Sato   Chuba Electric Power Co.
A. Takeuchi   Japan Meteorological Agency
Kanako Tanaka   Global Industrial and Social Progress
Tomihiro Taniguchi   Vice-Chair, IPCC

Kenya
Richard S. Odingo   Vice-Chair, WGIII
Kingiri Senelwa   Moi University

Malawi
Paul Desanker   University of Virginia

Mexico
Gustavo Albin   Permanent Representative Mission of Mexico

Morocco
Abdelkader Allali   Ministry of Agriculture, Rural Development and Fishing
Abdalah Mokssit   Centre National du Climat et de Recherchco Meteorologiques

Netherlands
Alphonsus P.M. Baede   Royal Netherlands Meteorological Institute (KNMI)
T.A. Buishand   Royal Netherlands Meteorological Institute
W.L. Hare   Greenpeace International
Catrinus J. Jepma   University of Groningen
E. Koekkoek   Ministry of Housing, Spacial Planning and the Environment
Rik Leemans   National Institute of Public Health and Environmental Protection
K. McKullen   Greenpeace International
Bert Metz   Co-Chair, WGIII
Leo Meyer   Ministry of the Environment
Maresa Oosterman   Ministerie van Buitenlandse Zaken
M.B.A.M. Scheffers   National Institute for Coastal and Marine Management
Rob Swart   Head, WGIII TSU
H.M. ten Brink   ECN
Aad P. van Ulden   Royal Netherlands Meteorological Institute
J. Verbeek   Ministry of Transport, Public Works and Water Management

New Zealand
Jon Barnett   Macmillan Brown Centre for Pacific Studies, University of Canterbury
Vincent Gray   Climate Consultant
Wayne Hennessy   Coal Research Association of New Zealand, Inc.
Piers Maclaren   NZ Forest Research Institute
Martin Manning   Vice-Chair, WGII
Helen Plume   Ministry for the Environment
A. Reisinger   Ministry for the Environment
J. Salinger   National Institute of Water and Atmospheric Research Ltd (NIWA)
Ralph Sims   Massey University

Niger
Garba Goudou Dieudonne   Office of the Prime Minister

Nigeria
Sani Sambo   Abubakar Tafawa Balewa University

Norway
Torgrim Aspjell   The Norwegian Pollution Control Authorities
Oyvind Christophersen   Ministry of Environment
Eirik J. Forland   Norwegian Meteorological Institute
S. Gornas   University of Bergen
Jarle Inge Holten   Terrestrial Ecology Research
Snorre Kverndokk   Frischsenteret/Frisch Centre
A. Moene   The Norwegian Meteorological Institute
Audun Rossland   The Norwegian Pollution Control Authorities
Nils R. Saelthun   Norwegian Water Resources and Energy Administration
Tom Segalstad   University of Oslom Norway
S. Sundby   Institute of Marine Research
Kristian Tangen   The Fridtjof Nansen Institute

Oman
Mohammed bin Ali Al-Hakmani   Ministry of Regional Municipalities, Environment & Water Resources

Pakistan
Tariq Banuri   Sustainable Development Policy Institute

Peru
Eduardo Calvo   Vice-Chair, WGIII
Nadia Gamboa   Pontificia Universidad Catolica del Peru

Phillipines
Lewis H. Ziska   International Rice Research Institute

Poland
Jan Dobrowolski   Goetel's School of Environmental Protection & Engineering
Zbyszek Kundzewicz   Polish Academy of Sciences
Miroslaw Mietus   Institute of Meteorology & Water Management
A. Olecka   National Fund for Environmental Protection and Water Management
M. Sadowski   National Fund for Environmental Protection and Water Management
Wojciech Suchorzewski   Warsaw University of Technology

Romania
Vasile Cuculeanu   National Institute of Meteorology and Hydrology
Adriana Marica   National Institute of Meteorology and Hydrology

Russia
Yurij Anokhin   Institute of Global Climate & Ecology
Oleg Anisimov   State Hydrological Institute
Igor Bashmakov   Centre for Energy Efficiency (CENEF)
Igor Karol   Main Geophysical Observatory
Alla Tsyban   Institute of Global Climate and Ecology
Yuri Izrael   Vice-Chair, IPCC

Senegal
Alioune Ndiaye   Vice-Chair, WGII

Sierra Leone
Ogunlade R. Davidson   Co-chair, WGIII

Slovak Republic
Milan Lapin   Comenius University

South Africa
Gerrie Coetzee   Department of Environmental Affairs and Tourism
Bruce Hewitson   University of Capetown
Steve Lennon   Eskom
Robert J. Scholes   CSIR

Spain
Sergio Alonso   University of the Balearic Islands
Francisco Ayala-Carcedo   Geomining Technological Institute of Spain
Luis Balairon   National Meteorological Institute
Felix Hernandez   CSIC
Don Antonio Labajo Salazar   Government of Spain
Maria-Carmen Llasat   Botija University of Barcelona
Josep Penuelas   Center for Ecological Research & Forestry Applications
Ana Yaber   University, Complutense of Madrid

Sri Lanka
Mohan Munasinghe — Vice-Chair, WGIII
B. Punyawardena — Department of Agriculture

Sudan
Nagmeldin Elhassan — Higher Coucil for Environment & Natural Resources

Sweden
Marianne Lillieskold — Swedish Environmental Protection Agency
Ulf Molau — University of Gothenburg
Nils-Axel Morner — Paleogeophysics & Geodynamics Stockholm University
Markku Rummukainen — Swedish Meterorological and Hydrological Institute

Switzerland
Christof Appenzeller — Federal Office of Meteorology and Climatology (MetroSwiss)
Fortunat Joos — Vice-Chair, WGI
Herbert Lang — Swiss Federal Institute of Technology Zurich (ETH)
José Romero — Office Federal de l'Environnement, des Forets et du Paysage
T. Stocker — University of Bern

Tanzania
M.J. Mwandosya — Centre for Energy, Environment, Science, and Technology
Buruhani S. Nyenzi — Vice-Chair, WGI

United Kingdom
Nigel Arnell — University of Southampton
C. Baker — Natural Environment Research Council
Terry Barker — University of Cambridge
K. G. Begg — University of Surrey
S.A. Boehmer-Christiansen — University of Hull
Richard Courtney — The Libert
K. Deyes — Department for Environment, Food and Rural Affairs
Thomas E. Downing — Environmental Change Institute University of Oxford
Caroline Fish — Global Atmosphere Division
Chris Folland — Met Office, Hadley Centre
Jonathan Gregory — Hadley Climate Research Centre
Steve Gregory — Forestry Commission
David Griggs — Head, WG-I TSU
Joanna Haigh — Imperial College
M. Harley — English Nature
Susan Haseldine — Department for Environment, Food and Rural Affairs (DEFRA)
John Houghton — Co-Chair, WG-I
Mike Hulme — University of East Anglia
Michael Jefferson — World Energy Council
Cathy Johnson — IPCC, Working Group I
Sari Kovats — London School of Hygiene and Tropical Medicine
David Mansell-Moullin — International Petroleum Industry Environmental Conservation Association (IPIECA)
Anil Markandya — University of Bath
A. McCulloch — ICI Chemicals & Polymers Limited
Gordon McFadyen — Global Atmospheric Division Deparment of the Environment, Transport and the Regions
A.J. McMichael — London School of Hygiene and Tropical Medicine
Aubrey Meyer — Global Commons Institute
John Mitchell — Hadley Center
Martin Parry — Jackson Environment Institute

J.M. Penman	Department of the Environment, Transport and the Regions
S. Raper	University of East Anglia
Keith Shine	Department of Meteorology, University of Reading
P. Singleton	Scottish Environment Protection Agency
Peter Smith	IACR-Rothamsted
P. Smithson	University of Sheffield
Peter Thorne	School of Environmental Sciences, University of East Anglia
P. van der Linden	Met Office Hadley Centre for Climate Prediction and Research
David Warrilow	Department of the Environment, Food and Rural Affairs
Philip L. Woodworth	Bidston Observatory

United States

Dilip Ahuja	National Institute of Advanced Studies
Dan Albritton	NOAA Aeronomy Laboratory
Jeffrey S. Amthor	Oak Ridge National Laboratory
Peter Backlund	Office of Science and Technology Policy/Environment Division
Lee Beck	U.S. Environmental Protection Agency
Leonard Bernstein	IPIECA
Daniel Bodansky	U.S. Department of State
Rick Bradley	US Department of Energy
James L. Buizer	National Oceanic & Amtospheric Administration
John Christy	University of Alabama
Susan Conard	Office of Science and Technology Policy/Environment Division
Curt Covey	Lawrence Livermore National Laboratory
Benjamin DeAngelo	U.S. Environmental Protection Agency
Robert Dickinson	University of Arizona
David Dokken	University Corporation for Atmospheric Research
Rayola Dougher	American Petroleum Institute
William Easterling	Pennsylvania State University
Jerry Elwood	Department of Enegry
Paul R. Epstein	Harvard Medical School
Paul D. Farrar	Naval Oceanographic Office
Howard Feldman	American Petroleum Institute
Josh Foster	NOAA Office of Global Programs
Laurie Geller	National Research Council
Michael Ghil	University of California, Los Angeles
Vivien Gornitz	Columbia University
Kenneth Green	Reason Public Policy Institute
David Harrison	National Economic Research Associates
David D. Houghton	University of Wisconsin-Madison
Malcolm Hughes	University of Arizona
Stanley Jacobs	Lamont-Doherty Earth Observatory of Columbia University
Henry D. Jacoby	Massachusetts Institute of Technology
Judson Jaffe	Council of Economic Advisers
Steven M. Japar	Ford Motor Company
Russell O. Jones	American Petroleum Institute
Sally Kane	NOAA
T. Karl	NOAA National Climatic Data Center
Charles Keller	IGPP.SIO.UCSD
Haroon Kheshgi	Exxon Research & Engineering Company
Ann Kinzig	Arizona State University
Maureen T. Koetz	Nuclear Energy Institute
Rattan Lal	Ohio State Universtiy
Chris Landsea	NOAA AOML/Hurricane Research Division
Neil Leary	Head, WGII TSU

Sven B. Lundstedt	The Ohio State University
Anthony Lupo	University of Missouri - Columbia
Michael C. MacCracken	U.S. Global Change Research Program
James J. McCarthy	Co-Chair, WGII
Gerald Meehl	NCAR
Robert Mendelsohn	Yale University
Patrick Michaels	University of Virginia
Evan Mills	Lawrence Berkeley National Laboratory
William Moomaw	The Fletcher School of Law and Diplomacy, Tufts University
Berrien Moore	University of New Hampshire
James Morison	University of Washington
Jennifer Orme-Zavaleta	USEP/NHEERL/WED
Camille Parmesan	University of Texas
J.A. Patz	Johns Hopkins University
Joyce Penner	University of Michigan
Roger A. Pielke	Colorado State University
Michael Prather	University of California Irvine
Lynn K. Price Lawrence	Berkeley National Laboratory
V. Ramaswamy	NOAA
Robert L. Randall	The RainForest ReGeneration Institute
Richard Richels	Electric Power Research Institute
David Rind	National Aeronautics and Space Agency
Catriona Rogers	U.S. Global Change Research Program
Matthias Ruth	University of Maryland
Jayant Sathaye	Lawrence Berkeley National Laboratory
Michael Schlesinger	University of Illinois-Urbana-Champaign
Stephen Schneider	Stanford University
Michael J. Scott	Battelle Pacific Northwest Nat'l Laboratory
Roger Sedjo	Resources for the Future
Walter Short	National Renewable Energy Laboratory
Joel B. Smith	Stratus Consulting Inc.
Robert N. Stavins	John F. Kennedy School of Government, Harvard University
Ron Stouffer	US Dept of Commerce/NOAA
T. Talley	Office of Global Change, U.S. Department of State
Kevin Trenberth	NCAR
Edward Vine	Lawrence Berkeley National Laboratory
Henry Walker	U.S. Environmental Protection Agency
Robert Watson	Chair, IPCC
Howard Wesoky	Federal Aviation Administration
John P. Weyant	Energy Modeling Forum, Stanford University
Tom Wilbanks	Oak Ridge National Laboratory

Venezuela

Armando Ramirez Rojas	Vice-Chair, WGI

Zimbabwe

Chris Magadza	University of Zimbabwe
M.C. Zinyowera	MSU Zimbabwe Gvt

Annex B. Glossary of Terms

This Glossary is based on the glossaries published in the IPCC Third Assessment Report (IPCC, 2001a,b,c); however, additional work has been undertaken on consistency and refinement of some of the terms. The terms that are independent entries in this glossary are highlighted in *italics*.

Acclimatization
The physiological adaptation to climatic variations.

Activities Implemented Jointly (AIJ)
The pilot phase for *Joint Implementation*, as defined in Article 4.2(a) of the *United Nations Framework Convention on Climate Change*, that allows for project activity among developed countries (and their companies) and between developed and developing countries (and their companies). AIJ is intended to allow Parties to the United Nations Framework Convention on Climate Change to gain experience in jointly implemented project activities. There is no crediting for AIJ activity during the pilot phase. A decision remains to be taken on the future of AIJ projects and how they may relate to the *Kyoto Mechanisms*. As a simple form of tradable permits, AIJ and other market-based schemes represent important potential mechanisms for stimulating additional resource flows for the global environmental good. See also *Clean Development Mechanism* and *emissions trading*.

Adaptability
See *Adaptive capacity*.

Adaptation
Adjustment in natural or *human systems* to a new or changing environment. Adaptation to *climate change* refers to adjustment in natural or human systems in response to actual or expected climatic *stimuli* or their effects, which moderates harm or exploits beneficial opportunities. Various types of adaptation can be distinguished, including anticipatory and reactive adaptation, private and public adaptation, and autonomous and planned adaptation.

Adaptation assessment
The practice of identifying options to adapt to *climate change* and evaluating them in terms of criteria such as availability, benefits, costs, effectiveness, efficiency, and feasibility.

Adaptation benefits
The avoided damage costs or the accrued benefits following the adoption and *implementation* of *adaptation* measures.

Adaptation costs
Costs of planning, preparing for, facilitating, and implementing *adaptation* measures, including transition costs.

Adaptive capacity
The ability of a system to adjust to *climate change* (including *climate variability* and extremes) to moderate potential damages, to take advantage of opportunities, or to cope with the consequences.

Additionality
Reduction in *emissions* by *sources* or enhancement of removals by *sinks* that is additional to any that would occur in the absence of a *Joint Implementation* or a *Clean Development Mechanism* project activity as defined in the *Kyoto Protocol* Articles on Joint Implementation and the Clean Development Mechanism. This definition may be further broadened to include financial, investment, and *technology* additionality. Under "financial additionality," the project activity funding shall be additional to existing Global Environmental Facility, other financial commitments of Parties included in Annex I, Official Development Assistance, and other systems of cooperation. Under "investment additionality," the value of the *Emissions Reduction Unit/Certified Emission Reduction Unit* shall significantly improve the financial and/or commercial viability of the project activity. Under "technology additionality," the technology used for the project activity shall be the best available for the circumstances of the host Party.

Adjustment time
See *Lifetime*; see also *Response time*.

Aerosols
A collection of airborne solid or liquid particles, with a typical size between 0.01 and 10 mm that reside in the *atmosphere* for at least several hours. Aerosols may be of either natural or *anthropogenic* origin. Aerosols may influence *climate* in two ways: directly through scattering and absorbing radiation, and indirectly through acting as condensation nuclei for cloud formation or modifying the optical properties and lifetime of clouds. See *indirect aerosol effect*.

Afforestation
Planting of new *forests* on lands that historically have not contained forests. For a discussion of the term forest and related terms such as *afforestation*, *reforestation*, and *deforestation*, see the IPCC Special Report on Land Use, Land-Use Change, and Forestry (IPCC, 2000b).

Aggregate impacts
Total impacts summed up across sectors and/or regions. The aggregation of impacts requires knowledge of (or assumptions about) the relative importance of impacts in different sectors and regions. Measures of aggregate impacts include, for example, the total number of people affected, change in net primary productivity, number of systems undergoing change, or total economic costs.

Albedo
The fraction of *solar radiation* reflected by a surface or object, often expressed as a percentage. Snow covered surfaces have

a high albedo; the albedo of soils ranges from high to low; vegetation covered surfaces and oceans have a low albedo. The Earth's albedo varies mainly through varying cloudiness, snow, ice, leaf area, and land cover changes.

Algal blooms
A reproductive explosion of algae in a lake, river, or ocean.

Alpine
The biogeographic zone made up of slopes above timberline and characterized by the presence of rosette-forming herbaceous plants and low shrubby slow-growing woody plants.

Alternative development paths
Refer to a variety of possible *scenarios* for societal *values* and consumption and production patterns in all countries, including, but not limited to, a continuation of today's trends. In this report, these paths do not include additional *climate* initiatives which means that no scenarios are included that explicitly assume *implementation* of the *United Nations Framework Convention on Climate Change* or the emission targets of the *Kyoto Protocol*, but do include assumptions about other policies that influence *greenhouse gas emissions* indirectly.

Alternative energy
Energy derived from non-fossil-fuel sources.

Ancillary benefits
The ancillary, or side effects, of policies aimed exclusively at *climate change mitigation*. Such policies have an impact not only on *greenhouse gas emissions*, but also on resource use efficiency, like reduction in emissions of local and regional air pollutants associated with *fossil-fuel* use, and on issues such as transportation, agriculture, *land-use* practices, employment, and fuel security. Sometimes these benefits are referred to as "ancillary impacts" to reflect that in some cases the benefits may be negative. From the perspective of policies directed at abating local air pollution, greenhouse gas mitigation may also be considered an ancillary benefit, but these relationships are not considered in this assessment.

Annex I countries/Parties
Group of countries included in Annex I (as amended in 1998) to the *United Nations Framework Convention on Climate Change*, including all the developed countries in the Organisation for Economic Cooperation and Development, and *economies in transition*. By default, the other countries are referred to as *non-Annex I countries*. Under Articles 4.2(a) and 4.2(b) of the Convention, Annex I countries commit themselves specifically to the aim of returning individually or jointly to their 1990 levels of *greenhouse gas emissions* by the year 2000. See also *Annex II, Annex B*, and *non-Annex B countries*.

Annex II countries
Group of countries included in Annex II to the *United Nations Framework Convention on Climate Change*, including all developed countries in the Organisation for Economic Cooperation and Development. Under Article 4.2(g) of the Convention, these countries are expected to provide financial resources to assist developing countries to comply with their obligations, such as preparing national reports. Annex II countries are also expected to promote the transfer of *environmentally sound technologies* to developing countries. See also *Annex I, Annex B, non-Annex I*, and *non-Annex B countries/Parties*.

Annex B countries/Parties
Group of countries included in Annex B in the *Kyoto Protocol* that have agreed to a target for their *greenhouse gas emissions*, including all the *Annex I countries* (as amended in 1998) but Turkey and Belarus. See also *Annex II, non-Annex I*, and *non-Annex B countries/Parties*.

Anthropogenic
Resulting from or produced by human beings.

Anthropogenic emissions
Emissions of *greenhouse gases*, greenhouse gas *precursors*, and *aerosols* associated with human activities. These include burning of *fossil fuels* for energy, *deforestation*, and *land-use* changes that result in net increase in emissions.

Aquaculture
Breeding and rearing fish, shellfish, etc., or growing plants for food in special ponds.

Aquifer
A stratum of permeable rock that bears water. An unconfined aquifer is recharged directly by local rainfall, rivers, and lakes, and the rate of recharge will be influenced by the permeability of the overlying rocks and soils. A confined aquifer is characterized by an overlying bed that is impermeable and the local rainfall does not influence the aquifer.

Arid regions
Ecosystems with less than 250 mm precipitation per year.

Assigned amounts (AAs)
Under the *Kyoto Protocol*, the total amount of *greenhouse gas emissions* that each *Annex B country* has agreed that its emissions will not exceed in the first commitment period (2008 to 2012) is the assigned amount. This is calculated by multiplying the country's total greenhouse gas emissions in 1990 by five (for the 5-year commitment period) and then by the percentage it agreed to as listed in Annex B of the Kyoto Protocol (e.g., 92% for the European Union, 93% for the USA).

Assigned amount unit (AAU)
Equal to 1 tonne (metric ton) of *CO₂-equivalent emissions* calculated using the *Global Warming Potential*.

Atmosphere
The gaseous envelop surrounding the Earth. The dry atmosphere consists almost entirely of nitrogen (78.1% *volume mixing ratio*) and oxygen (20.9% volume mixing ratio), together with a number of trace gases, such as argon (0.93% volume mixing ratio), helium, and radiatively active *greenhouse gases* such as *carbon dioxide* (0.035% volume mixing ratio) and *ozone*. In addition, the atmosphere contains water vapor, whose amount is highly variable but typically 1% volume mixing ratio. The atmosphere also contains clouds and *aerosols*.

Attribution
See *detection and attribution*.

Banking
According to the *Kyoto Protocol* [Article 3(13)], Parties included in Annex I to the *United Nations Framework Convention on Climate Change* may save excess *emissions* allowances or credits from the first commitment period for use in subsequent commitment periods (post-2012).

Barrier
A barrier is any obstacle to reaching a potential that can be overcome by a policy, program, or measure.

Baseline
The baseline (or reference) is any datum against which change is measured. It might be a "current baseline," in which case it represents observable, present-day conditions. It might also be a "future baseline," which is a projected future set of conditions excluding the driving factor of interest. Alternative interpretations of the reference conditions can give rise to multiple baselines.

Basin
The drainage area of a stream, river, or lake.

Biodiversity
The numbers and relative abundances of different genes (genetic diversity), species, and *ecosystems* (communities) in a particular area.

Biofuel
A fuel produced from dry organic matter or combustible oils produced by plants. Examples of biofuel include alcohol (from fermented sugar), black liquor from the paper manufacturing process, wood, and soybean oil.

Biomass
The total mass of living organisms in a given area or volume; recently dead plant material is often included as dead biomass.

Biome
A grouping of similar plant and animal communities into broad landscape units that occur under similar environmental conditions.

Biosphere (terrestrial and marine)
The part of the Earth system comprising all *ecosystems* and living organisms in the *atmosphere*, on land (terrestrial biosphere), or in the oceans (marine biosphere), including derived dead organic matter such as litter, soil organic matter, and oceanic detritus.

Biota
All living organisms of an area; the flora and fauna considered as a unit.

Black carbon
Operationally defined species based on measurement of light absorption and chemical reactivity and/or thermal stability; consists of soot, charcoal, and/or possible light-absorbing refractory organic matter (Charlson and Heintzenberg, 1995).

Bog
A poorly drained area rich in accumulated plant material, frequently surrounding a body of open water and having a characteristic flora (such as sedges, heaths, and sphagnum).

Boreal forest
Forests of pine, spruce, fir, and larch stretching from the east coast of Canada westward to Alaska and continuing from Siberia westward across the entire extent of Russia to the European Plain.

Bottom-up models
A modeling approach that includes technological and engineering details in the analysis. See also *top-down models*.

Burden
The total mass of a gaseous substance of concern in the atmosphere.

Capacity building
In the context of *climate change*, capacity building is a process of developing the technical skills and institutional capability in developing countries and *economies in transition* to enable them to participate in all aspects of *adaptation* to, *mitigation* of, and research on climate change, and the *implementation* of the *Kyoto Mechanisms*, etc.

Carbonaceous aerosol
Aerosol consisting predominantly of organic substances and various forms of black carbon (Charlson and Heintzenberg, 1995).

Carbon cycle

The term used to describe the flow of carbon (in various forms such as as *carbon dioxide*) through the *atmosphere*, ocean, terrestrial *biosphere*, and *lithosphere*.

Carbon dioxide (CO_2)

A naturally occurring gas, and also a by-product of burning *fossil fuels* and *biomass*, as well as *land-use changes* and other industrial processes. It is the principal *anthropogenic greenhouse gas* that affects the Earth's *radiative balance*. It is the reference gas against which other greenhouse gases are measured and therefore has a *Global Warming Potential* of 1.

Carbon dioxide (CO_2) fertilization

The enhancement of the growth of plants as a result of increased atmospheric *carbon dioxide* concentration. Depending on their mechanism of *photosynthesis*, certain types of plants are more sensitive to changes in atmospheric carbon dioxide concentration. In particular, plants that produce a three-carbon compound (C_3) during photosynthesis—including most trees and agricultural crops such as rice, wheat, soybeans, potatoes, and vegetables—generally show a larger response than plants that produce a four-carbon compound (C_4) during photosynthesis—mainly of tropical origin, including grasses and the agriculturally important crops maize, sugar cane, millet, and sorghum.

Carbon leakage

See *leakage*.

Carbon taxes

See *emissions tax*.

Catchment

An area that collects and drains rainwater.

Certified Emission Reduction (CER) Unit

Equal to 1 tonne (metric ton) of *CO_2-equivalent emissions* reduced or sequestered through a *Clean Development Mechanism* project, calculated using *Global Warming Potentials*. See also *Emissions Reduction Unit*.

Chlorofluorocarbons (CFCs)

Greenhouse gases covered under the 1987 *Montreal Protocol* and used for refrigeration, air conditioning, packaging, insulation, solvents, or aerosol propellants. Since they are not destroyed in the lower *atmosphere*, CFCs drift into the upper atmosphere where, given suitable conditions, they break down *ozone*. These gases are being replaced by other compounds, including hydrochlorofluorocarbons and *hydrofluorocarbons*, which are greenhouse gases covered under the *Kyoto Protocol*.

Cholera

An intestinal infection that results in frequent watery stools, cramping abdominal pain, and eventual collapse from dehydration.

Clean Development Mechanism (CDM)

Defined in Article 12 of the *Kyoto Protocol*, the Clean Development Mechanism is intended to meet two objectives: (1) to assist Parties not included in *Annex I* in achieving *sustainable development* and in contributing to the ultimate objective of the convention; and (2) to assist Parties included in Annex I in achieving *compliance* with their quantified emission limitation and reduction commitments. *Certified Emission Reduction Units* from Clean Development Mechanism projects undertaken in *non-Annex I countries* that limit or reduce *greenhouse gas emissions*, when certified by operational entities designated by *Conference of the Parties/Meeting of the Parties,* can be accrued to the investor (government or industry) from Parties in *Annex B*. A share of the proceeds from the certified project activities is used to cover administrative expenses as well as to assist developing country Parties that are particularly vulnerable to the adverse effects of *climate change* to meet the costs of *adaptation*.

Climate

Climate in a narrow sense is usually defined as the "average weather" or more rigorously as the statistical description in terms of the mean and variability of relevant quantities over a period of time ranging from months to thousands or millions of years. The classical period is 30 years, as defined by the World Meteorological Organization (WMO). These relevant quantities are most often surface variables such as temperature, precipitation, and wind. Climate in a wider sense is the state, including a statistical description, of the *climate system*.

Climate change

Climate change refers to a statistically significant variation in either the mean state of the *climate* or in its variability, persisting for an extended period (typically decades or longer). Climate change may be due to natural internal processes or *external forcings*, or to persistent *anthropogenic* changes in the composition of the *atmosphere* or in *land use*. Note that the *United Nations Framework Convention on Climate Change* (UNFCCC), in its Article 1, defines "climate change" as: "a change of climate which is attributed directly or indirectly to human activity that alters the composition of the global atmosphere and which is in addition to natural climate variability observed over comparable time periods." The UNFCCC thus makes a distinction between "climate change" attributable to human activities altering the atmospheric composition, and "climate variability" attributable to natural causes. See also *climate variability*.

Climate feedback

An interaction mechanism between processes in the *climate system* is called a climate feedback, when the result of an initial process triggers changes in a second process that in turn influences the initial one. A positive feedback intensifies the original process, and a negative feedback reduces it.

Climate model (hierarchy)

A numerical representation of the *climate system* based on the physical, chemical, and biological properties of its components, their interactions and *feedback* processes, and accounting for all or some of its known properties. The climate system can be represented by models of varying complexity—that is, for any one component or combination of components a "hierarchy" of models can be identified, differing in such aspects as the number of spatial dimensions, the extent to which physical, chemical or biological processes are explicitly represented, or the level at which empirical *parametrizations* are involved. Coupled atmosphere/ocean/sea-ice *general circulation models* (AOGCMs) provide a comprehensive representation of the climate system. There is an evolution towards more complex models with active chemistry and biology. Climate models are applied, as a research tool, to study and simulate the climate, but also for operational purposes, including monthly, seasonal, and interannual *climate predictions*.

Climate prediction

A climate prediction or climate forecast is the result of an attempt to produce a most likely description or estimate of the actual evolution of the *climate* in the future (e.g., at seasonal, interannual, or long-term *time-scales*). See also *climate projection* and *climate (change) scenario*.

Climate projection

A *projection* of the response of the *climate system* to *emission* or concentration *scenarios* of *greenhouse gases* and *aerosols*, or *radiative forcing scenarios*, often based upon simulations by *climate models*. Climate projections are distinguished from *climate predictions* in order to emphasize that climate projections depend upon the emission/concentration/radiative forcing scenario used, which are based on assumptions, concerning, for example, future socio-economic and technological developments that may or may not be realized, and are therefore subject to substantial *uncertainty*.

Climate scenario

A plausible and often simplified representation of the future *climate*, based on an internally consistent set of climatological relationships, that has been constructed for explicit use in investigating the potential consequences of *anthropogenic climate change*, often serving as input to impact models. *Climate projections* often serve as the raw material for constructing climate scenarios, but climate scenarios usually require additional information such as about the observed current climate. A "climate change scenario" is the difference between a climate scenario and the current climate.

Climate sensitivity

In IPCC assessments, "equilibrium climate sensitivity" refers to the equilibrium change in global mean surface temperature following a doubling of the atmospheric (*equivalent*) CO_2 concentration. More generally, equilibrium climate sensitivity refers to the equilibrium change in surface air temperature following a unit change in *radiative forcing* ($°C/Wm^{-2}$). In practice, the evaluation of the equilibrium climate sensitivity requires very long simulations with coupled *general circulation models*. The "effective climate sensitivity" is a related measure that circumvents this requirement. It is evaluated from model output for evolving non-equilibrium conditions. It is a measure of the strengths of the *feedbacks* at a particular time and may vary with forcing history and climate state. See *climate model*.

Climate system

The climate system is the highly complex system consisting of five major components: the *atmosphere*, the *hydrosphere*, the *cryosphere*, the land surface and the *biosphere*, and the interactions between them. The climate system evolves in time under the influence of its own internal dynamics and because of external forcings such as volcanic eruptions, solar variations, and human-induced forcings such as the changing composition of the atmosphere and *land-use change*.

Climate variability

Climate variability refers to variations in the mean state and other statistics (such as standard deviations, the occurrence of extremes, etc.) of the *climate* on all *temporal and spatial scales* beyond that of individual weather events. Variability may be due to natural internal processes within the *climate system* (internal variability), or to variations in natural or *anthropogenic external forcing* (external variability). See also *climate change*.

CO_2-equivalent

See *equivalent CO_2*.

CO_2 fertilization

See *carbon dioxide (CO_2) fertilization*.

Co-benefits

The benefits of policies that are implemented for various reasons at the same time—including *climate change mitigation*—acknowledging that most policies designed to address *greenhouse gas mitigation* also have other, often at least equally important, rationales (e.g., related to objectives of development, sustainability, and equity). The term co-impact is also used in a more generic sense to cover both the positive and negative sides of the benefits. See also *ancillary benefits*.

Co-generation

The use of waste heat from electric generation, such as exhaust from gas turbines, for either industrial purposes or district heating.

Compliance

See *implementation*.

Conference of the Parties (COP)

The supreme body of the *United Nations Framework Convention on Climate Change (UNFCCC)*, comprising countries that have ratified or acceded to the UNFCCC. The first session of the

Conference of the Parties (COP-1) was held in Berlin in 1995, followed by COP-2 in Geneva 1996, COP-3 in Kyoto 1997, COP-4 in Buenos Aires 1998, COP-5 in Bonn 1999, COP-6 Part 1 in The Hague 2000, and COP-6 Part 2 in Bonn 2001. COP-7 is scheduled for November 2001 in Marrakech. See also *Meeting of the Parties (MOP)*.

Cooling degree days
The integral over a day of the temperature above 18°C (e.g., a day with an average temperature of 20°C counts as 2 cooling degree days). See also *heating degree days*.

Coping range
The variation in climatic *stimuli* that a system can absorb without producing significant impacts.

Coral bleaching
The paling in color of corals resulting from a loss of symbiotic algae. Bleaching occurs in response to physiological shock in response to abrupt changes in temperature, salinity, and turbidity.

Cost-effective
A criterion that specifies that a *technology* or measure delivers a good or service at equal or lower cost than current practice, or the least-cost alternative for the achievement of a given target.

Cryosphere
The component of the *climate system* consisting of all snow, ice, and *permafrost* on and beneath the surface of the earth and ocean. See also *glacier* and *ice sheet*.

Deepwater formation
Occurs when seawater freezes to form sea ice. The local release of salt and consequent increase in water density leads to the formation of saline coldwater that sinks to the ocean floor.

Deforestation
Conversion of *forest* to non-forest. For a discussion of the term forest and related terms such as *afforestation*, *reforestation*, and *deforestation*, see the IPCC Special Report on Land Use, Land-Use Change, and Forestry (IPCC, 2000b).

Demand-side management
Policies and programs designed for a specific purpose to influence consumer demand for goods and/or services. In the energy sector, for instance, it refers to policies and programs designed to reduce consumer demand for electricity and other energy sources. It helps to reduce *greenhouse gas emissions*.

Dengue Fever
An infectious viral disease spread by mosquitoes often called breakbone fever because it is characterized by severe pain in joints and back. Subsequent infections of the virus may lead to dengue haemorrhagic fever (DHF) and dengue shock syndrome (DSS), which may be fatal.

Deposit–refund system
Combines a deposit or fee (tax) on a commodity with a refund or rebate (*subsidy*) for *implementation* of a specified action. Se also *emissions tax*.

Desert
An *ecosystem* with less than 100 mm precipitation per year.

Desertification
Land degradation in arid, *semi-arid*, and dry sub-humid areas resulting from various factors, including climatic variations and human activities. Further, the United Nations Convention to Combat Desertification defines land degradation as a reduction or loss in arid, semi-arid, and dry sub-humid areas of the biological or economic productivity and complexity of rain-fed cropland, irrigated cropland, or range, pasture, *forest*, and woodlands resulting from *land uses* or from a process or combination of processes, including processes arising from human activities and habitation patterns, such as: (i) soil *erosion* caused by wind and/or water; (ii) deterioration of the physical, chemical, and biological or economic properties of soil; and (iii) long-term loss of natural vegetation.

Detection and attribution
Climate varies continually on all *time scales*. Detection of *climate change* is the process of demonstrating that climate has changed in some defined statistical sense, without providing a reason for that change. Attribution of causes of climate change is the process of establishing the most likely causes for the detected change with some defined level of confidence.

Disturbance regime
Frequency, intensity, and types of disturbances, such as fires, inspect or pest outbreaks, floods, and *droughts*.

Diurnal temperature range
The difference between the maximum and minimum temperature during a day.

Double dividend
The effect that revenue-generating instruments, such as *carbon taxes* or auctioned (tradable) carbon emission permits, can (i) limit or reduce *greenhouse gas emissions* and (ii) offset at least part of the potential welfare losses of climate policies through recycling the revenue in the economy to reduce other taxes likely to be distortionary. In a world with involuntary unemployment, the *climate change* policy adopted may have an effect (a positive or negative "third dividend") on employment. Weak double dividend occurs as long as there is a *revenue recycling* effect—that is, as long as revenues are recycled through reductions in the marginal rates of distortionary taxes. Strong double dividend requires that the (beneficial) revenue recycling effect more than offset the combination of the primary cost and, in this case, the net cost of abatement is negative.

Drought
The phenomenon that exists when precipitation has been significantly below normal recorded levels, causing serious hydrological imbalances that adversely affect land resource production systems.

Economic potential
Economic potential is the portion of *technological potential* for *greenhouse gas emissions* reductions or *energy efficiency* improvements that could be achieved *cost-effectively* through the creation of markets, reduction of market failures, or increased financial and technological transfers. The achievement of economic potential requires additional *policies and measures* to break down *market barriers*. See also *market potential*, *socio-economic potential*, and *technological potential*.

Economies in transition (EITs)
Countries with national economies in the process of changing from a planned economic system to a market economy.

Ecosystem
A system of interacting living organisms together with their physical environment. The boundaries of what could be called an ecosystem are somewhat arbitrary, depending on the focus of interest or study. Thus, the extent of an ecosystem may range from very small *spatial scales* to, ultimately, the entire Earth.

Ecosystem services
Ecological processes or functions that have *value* to individuals or society.

El Niño Southern Oscillation (ENSO)
El Niño, in its original sense, is a warmwater current that periodically flows along the coast of Ecuador and Peru, disrupting the local fishery. This oceanic event is associated with a fluctuation of the intertropical surface pressure pattern and circulation in the Indian and Pacific Oceans, called the Southern Oscillation. This coupled atmosphere-ocean phenomenon is collectively known as El Niño Southern Oscillation, or ENSO. During an El Niño event, the prevailing trade winds weaken and the equatorial countercurrent strengthens, causing warm surface waters in the Indonesian area to flow eastward to overlie the cold waters of the Peru current. This event has great impact on the wind, sea surface temperature, and precipitation patterns in the tropical Pacific. It has climatic effects throughout the Pacific region and in many other parts of the world. The opposite of an El Niño event is called *La Niña*.

Emissions
In the *climate change* context, emissions refer to the release of *greenhouse gases* and/or their *precursors* and *aerosols* into the *atmosphere* over a specified area and period of time.

Emissions permit
An emissions permit is the non-transferable or tradable allocation of entitlements by an administrative authority (intergovernmental organization, central or local government agency) to a regional (country, sub-national) or a sectoral (an individual firm) entity to emit a specified amount of a substance.

Emissions quota
The portion or share of total allowable *emissions* assigned to a country or group of countries within a framework of maximum total emissions and mandatory allocations of resources.

Emissions Reduction Unit (ERU)
Equal to 1 tonne (metric ton) of *carbon dioxide emissions* reduced or sequestered arising from a *Joint Implementation* (defined in Article 6 of the *Kyoto Protocol*) project calculated using *Global Warming Potential*. See also *Certified Emission Reduction Unit* and *emissions trading*.

Emissions tax
Levy imposed by a government on each unit of CO_2-equivalent *emissions* by a *source* subject to the tax. Since virtually all of the carbon in *fossil fuels* is ultimately emitted as *carbon dioxide*, a levy on the carbon content of fossil fuels—a *carbon tax*—is equivalent to an emissions tax for emissions caused by fossil-fuel combustion. An *energy tax*—a levy on the energy content of fuels—reduces demand for energy and so reduces carbon dioxide emissions from fossil-fuel use. An ecotax is designated for the purpose of influencing human behavior (specifically economic behavior) to follow an ecologically benign path. International emissions/carbon/energy tax is a tax imposed on specified sources in participating countries by an international agency. The revenue is distributed or used as specified by participating countries or the international agency.

Emissions trading
A market-based approach to achieving environmental objectives that allows, those reducing *greenhouse gas emissions* below what is required, to use or trade the excess reductions to offset emissions at another source inside or outside the country. In general, trading can occur at the intracompany, domestic, and international levels. The IPCC Second Assessment Report adopted the convention of using "permits" for domestic trading systems and "quotas" for international trading systems. Emissions trading under Article 17 of the *Kyoto Protocol* is a tradable quota system based on the *assigned amounts* calculated from the emission reduction and limitation commitments listed in *Annex B* of the Protocol. See also *Certified Emission Reduction Unit* and *Clean Development Mechanism*.

Emissions scenario
A plausible representation of the future development of *emissions* of substances that are potentially radiatively active (e.g., *greenhouse gases*, *aerosols*), based on a coherent and internally consistent set of assumptions about driving forces (such as demographic and socio-economic development, technological change) and their key relationships. Concentration scenarios, derived from emissions scenarios, are used as input

into a *climate model* to compute *climate projections*. In IPCC (1992), a set of emissions scenarios were used as a basis for the climate projections in IPCC (1996). These emissions scenarios are referred to as the IS92 scenarios. In the IPCC Special Report on Emissions Scenarios (Nakicenovic *et al.*, 2000), new emissions scenarios—the so-called *SRES scenarios*—were published. For the meaning of some terms related to these scenarios, see *SRES scenarios*.

Endemic
Restricted or peculiar to a locality or region. With regard to human health, endemic can refer to a disease or agent present or usually prevalent in a population or geographical area at all times.

Energy balance
Averaged over the globe and over longer time periods, the energy budget of the *climate system* must be in balance. Because the climate system derives all its energy from the Sun, this balance implies that, globally, the amount of incoming *solar radiation* must on average be equal to the sum of the outgoing reflected solar radiation and the outgoing *infrared radiation* emitted by the climate system. A perturbation of this global radiation balance, be it human-induced or natural, is called *radiative forcing*.

Energy conversion
See *energy transformation*.

Energy efficiency
Ratio of energy output of a conversion process or of a system to its energy input.

Energy intensity
Energy intensity is the ratio of energy consumption to economic or physical output. At the national level, energy intensity is the ratio of total domestic *primary energy* consumption or *final energy* consumption to *Gross Domestic Product* or physical output.

Energy service
The application of useful energy to tasks desired by the consumer such as transportation, a warm room, or light.

Energy tax
See *emissions tax*.

Energy transformation
The change from one form of energy, such as the energy embodied in *fossil fuels*, to another, such as electricity.

Environmentally Sound Technologies (ESTs)
Technologies that protect the environment, are less polluting, use all resources in a more sustainable manner, recycle more of their wastes and products, and handle residual wastes in a more acceptable manner than the technologies for which they were substitutes and are compatible with nationally determined socio-economic, cultural, and environmental priorities. ESTs in this report imply mitigation and adaptation technologies, hard and soft technologies.

Epidemic
Occurring suddenly in numbers clearly in excess of normal expectancy, said especially of *infectious diseases* but applied also to any disease, injury, or other health-related event occurring in such outbreaks.

Equilibrium and transient climate experiment
An "equilibrium climate experiment" is an experiment in which a *climate model* is allowed to fully adjust to a change in *radiative forcing*. Such experiments provide information on the difference between the initial and final states of the model, but not on the time-dependent response. If the forcing is allowed to evolve gradually according to a prescribed *emission scenario*, the time-dependent response of a climate model may be analyzed. Such an experiment is called a "transient climate experiment." See also *climate projection*.

Equivalent CO₂ (carbon dioxide)
The concentration of *carbon dioxide* that would cause the same amount of *radiative forcing* as a given mixture of carbon dioxide and other *greenhouse gases*.

Erosion
The process of removal and transport of soil and rock by weathering, mass wasting, and the action of streams, *glaciers*, waves, winds, and underground water.

Eustatic sea-level change
A change in global average sea level brought about by an alteration to the volume of the world ocean. This may be caused by changes in water density or in the total mass of water. In discussions of changes on geological time scales, this term sometimes also includes changes in global average sea level caused by an alteration to the shape of the ocean basins. In this report, the term is not used in that sense.

Eutrophication
The process by which a body of water (often shallow) becomes (either naturally or by pollution) rich in dissolved nutrients with a seasonal deficiency in dissolved oxygen.

Evaporation
The process by which a liquid becomes a gas.

Evapotranspiration
The combined process of *evaporation* from the Earth's surface and *transpiration* from vegetation.

Exotic species
See *introduced species*.

Exposure
The nature and degree to which a system is exposed to significant climatic variations.

Externality
See *external cost*.

External cost
Used to define the costs arising from any human activity, when the agent responsible for the activity does not take full account of the impacts on others of his or her actions. Equally, when the impacts are positive and not accounted for in the actions of the agent responsible they are referred to as external benefits. *Emissions* of particulate pollution from a power station affect the health of people in the vicinity, but this is not often considered, or is given inadequate weight, in private decision making and there is no market for such impacts. Such a phenomenon is referred to as an "externality," and the costs it imposes are referred to as the external costs.

External forcing
See *climate system*.

Extinction
The complete disappearance of an entire species.

Extirpation
The disappearance of a species from part of its range; local extinction.

Extreme weather event
An extreme weather event is an event that is rare within its statistical reference distribution at a particular place. Definitions of "rare" vary, but an extreme weather event would normally be as rare as or rarer than the 10th or 90th percentile. By definition, the characteristics of what is called extreme weather may vary from place to place. An extreme *climate* event is an average of a number of weather events over a certain period of time, an average which is itself extreme (e.g., rainfall over a season).

Feedback
See *climate feedback*.

Fiber
Wood, fuelwood (either woody or non-woody).

Final energy
Energy supplied that is available to the consumer to be converted into usable energy (e.g., electricity at the wall outlet).

Flexibility mechanisms
See *Kyoto Mechanisms*.

Flux adjustment
To avoid the problem of coupled atmosphere-ocean general circulation models drifting into some unrealistic climate state, adjustment terms can be applied to the atmosphere-ocean fluxes of heat and moisture (and sometimes the surface stresses resulting from the effect of the wind on the ocean surface) before these fluxes are imposed on the model ocean and atmosphere. Because these adjustments are pre-computed and therefore independent of the coupled model integration, they are uncorrelated to the anomalies that develop during the integration.

Food insecurity
A situation that exists when people lack secure access to sufficient amounts of safe and nutritious food for normal growth and development and an active and healthy life. It may be caused by the unavailability of food, insufficient purchasing power, inappropriate distribution, or inadequate use of food at the household level. Food insecurity may be chronic, seasonal, or transitory.

Forest
A vegetation type dominated by trees. Many definitions of the term forest are in use throughout the world, reflecting wide differences in bio-geophysical conditions, social structure, and economics. For a discussion of the term forest and related terms such as *afforestation*, *reforestation*, and *deforestation*: see the IPCC Special Report on Land Use, Land-Use Change, and Forestry (IPCC, 2000b).

Fossil CO_2 (carbon dioxide) emissions
Emissions of *carbon dioxide* resulting from the combustion of fuels from fossil carbon deposits such as oil, natural gas, and coal.

Fossil fuels
Carbon-based fuels from fossil carbon deposits, including coal, oil, and natural gas.

Freshwater lens
A lenticular fresh groundwater body that underlies an oceanic island. It is underlain by saline water.

Fuel switching
Policy designed to reduce *carbon dioxide emissions* by switching to lower carbon-content fuels, such as from coal to natural gas.

Full-cost pricing
The pricing of commercial goods—such as electric power—that includes in the final prices faced by the end user not only the private costs of inputs, but also the costs of externalities created by their production and use.

Framework Convention on Climate Change
See *United Nations Framework Convention on Climate Change*.

General circulation

The large scale motions of the *atmosphere* and the ocean as a consequence of differential heating on a rotating Earth, aiming to restore the *energy balance* of the system through transport of heat and momentum.

General Circulation Model (GCM)

See *climate model*.

Geo-engineering

Efforts to stabilize the climate system by directly managing the energy balance of the Earth, thereby overcoming the enhanced *greenhouse effect*.

Glacier

A mass of land ice flowing downhill (by internal deformation and sliding at the base) and constrained by the surrounding topography (e.g., the sides of a valley or surrounding peaks); the bedrock topography is the major influence on the dynamics and surface slope of a glacier. A glacier is maintained by accumulation of snow at high altitudes, balanced by melting at low altitudes or discharge into the sea.

Global surface temperature

The global surface temperature is the area-weighted global average of (i) the sea surface temperature over the oceans (i.e., the sub-surface bulk temperature in the first few meters of the ocean), and (ii) the surface air temperature over land at 1.5 m above the ground.

Global Warming Potential (GWP)

An index, describing the radiative characteristics of well-mixed *greenhouse gases*, that represents the combined effect of the differing times these gases remain in the *atmosphere* and their relative effectiveness in absorbing outgoing *infrared radiation*. This index approximates the time-integrated warming effect of a unit mass of a given greenhouse gas in today's atmosphere, relative to that of *carbon dioxide*.

Greenhouse effect

Greenhouse gases effectively absorb *infrared radiation*, emitted by the Earth's surface, by the *atmosphere* itself due to the same gases, and by clouds. Atmospheric radiation is emitted to all sides, including downward to the Earth's surface. Thus greenhouse gases trap heat within the surface-troposphere system. This is called the "natural greenhouse effect." Atmospheric radiation is strongly coupled to the temperature of the level at which it is emitted. In the *troposphere,* the temperature generally decreases with height. Effectively, infrared radiation emitted to space originates from an altitude with a temperature of, on average, -19° C, in balance with the net incoming *solar radiation*, whereas the Earth's surface is kept at a much higher temperature of, on average, +14° C. An increase in the concentration of greenhouse gases leads to an increased infrared opacity of the atmosphere, and therefore to an effective radiation into space from a higher altitude at a lower temperature. This causes a *radiative forcing*, an imbalance that can only be compensated for by an increase of the temperature of the surface-troposphere system. This is the "enhanced greenhouse effect."

Greenhouse gas

Greenhouse gases are those gaseous constituents of the *atmosphere*, both natural and *anthropogenic*, that absorb and emit radiation at specific wavelengths within the spectrum of *infrared radiation* emitted by the Earth's surface, the atmosphere, and clouds. This property causes the *greenhouse effect*. Water vapor (H_2O), *carbon dioxide* (CO_2), *nitrous oxide* (N_2O), *methane* (CH_4), and *ozone* (O_3) are the primary greenhouse gases in the Earth's atmosphere. Moreover there are a number of entirely human-made greenhouse gases in the atmosphere, such as the *halocarbons* and other chlorine- and bromine-containing substances, dealt with under the *Montreal Protocol*. Besides CO_2, N_2O, and CH_4, the *Kyoto Protocol* deals with the greenhouse gases *sulfur hexafluoride* (SF_6), *hydrofluorocarbons* (HFCs), and *perfluorocarbons* (PFCs).

Groin

A low, narrow jetty, usually extending roughly perpendicular to the shoreline, designed to protect the shore from *erosion* by currents, tides, or waves, or to trap sand for the purpose of building up or making a beach.

Gross Domestic Product (GDP)

The sum of gross *value added*, at purchasers' prices, by all resident and non-resident producers in the economy, plus any taxes and minus any subsidies not included in the value of the products in a country or a geographic region for a given period of time, normally 1 year. It is calculated without deducting for depreciation of fabricated assets or depletion and degradation of natural *resources*. GDP is an often used but incomplete measure of welfare.

Gross Primary Production (GPP)

The amount of carbon fixed from the *atmosphere* through *photosynthesis*.

Groundwater recharge

The process by which external water is added to the zone of saturation of an *aquifer*, either directly into a formation or indirectly by way of another formation.

Habitat

The particular environment or place where an organism or species tend to live; a more locally circumscribed portion of the total environment.

Halocarbons

Compounds containing carbon and either chlorine, bromine, or fluorine. Such compounds can act as powerful *greenhouse gases* in the *atmosphere*. The chlorine- and bromine-containing halocarbons are also involved in the depletion of the *ozone layer*.

Harmonized emissions/carbon/energy tax

Commits participating countries to impose a tax at a common rate on the same *sources*. Each country can retain the tax revenue it collects. A harmonized tax would not necessarily require countries to impose a tax at the same rate, but imposing different rates across countries would not be *cost-effective*. See also *emissions tax*.

Heat island

An area within an urban area characterized by ambient temperatures higher than those of the surrounding area because of the absorption of solar energy by materials like asphalt.

Heating degree days

The integral over a day of the temperature below $18°C$ (e.g., a day with an average temperature of $16°C$ counts as 2 heating degree days). See also *cooling degree days*.

Hedging

In the context of climate change mitigation, hedging is defined as balancing the risks of acting too slowly against acting too quickly, and it depends on society's attitude towards risks.

Heterotrophic respiration

The conversion of organic matter to CO_2 by organisms other than plants.

Human settlement

A place or area occupied by settlers.

Human system

Any system in which human organizations play a major role. Often, but not always, the term is synonymous with "society" or "social system" (e.g., agricultural system, political system, technological system, economic system).

Hydrofluorocarbons (HFCs)

Among the six *greenhouse gases* to be curbed under the *Kyoto Protocol*. They are produced commercially as a substitute for *chlorofluorocarbons*. HFCs largely are used in refrigeration and semiconductor manufacturing. Their *Global Warming Potentials* range from 1,300 to 11,700.

Hydrosphere

The component of the *climate system* composed of liquid surface and subterranean water, such as oceans, seas, rivers, freshwater lakes, underground water, etc.

Ice cap

A dome shaped ice mass covering a highland area that is considerably smaller in extent than an *ice sheet*.

Ice sheet

A mass of land ice that is sufficiently deep to cover most of the underlying bedrock topography, so that its shape is mainly determined by its internal dynamics (the flow of the ice as it deforms internally and slides at its base). An ice sheet flows outward from a high central plateau with a small average surface slope. The margins slope steeply, and the ice is discharged through fast-flowing ice streams or outlet *glaciers*, in some cases into the sea or into *ice shelves* floating on the sea. There are only two large ice sheets in the modern world, on Greenland and Antarctica, the Antarctic ice sheet being divided into East and West by the Transantarctic Mountains; during glacial periods there were others.

Ice shelf

A floating *ice sheet* of considerable thickness attached to a coast (usually of great horizontal extent with a level or gently undulating surface); often a seaward extension of ice sheets.

(Climate) Impact assessment

The practice of identifying and evaluating the detrimental and beneficial consequences of *climate change* on natural and *human systems*.

(Climate) Impacts

Consequences of *climate change* on natural and *human systems*. Depending on the consideration of *adaptation*, one can distinguish between potential impacts and residual impacts.

- Potential impacts: All impacts that may occur given a projected change in *climate*, without considering adaptation.
- Residual impacts: The impacts of climate change that would occur after adaptation.

See also a*ggregate impacts*, *market impacts*, and *non-market impacts*.

Implementation

Implementation refers to the actions (legislation or regulations, judicial decrees, or other actions) that governments take to translate international accords into domestic law and policy. It includes those events and activities that occur after the issuing of authoritative public policy directives, which include the effort to administer and the substantive impacts on people and events. It is important to distinguish between the legal implementation of international commitments (in national law) and the effective implementation (measures that induce changes in the behavior of target groups). Compliance is a matter of whether and to what extent countries do adhere to the provisions of the accord. Compliance focuses on not only whether implementing measures are in effect, but also on whether there is compliance with the implementing actions. Compliance measures the degree to which the actors whose behavior is targeted by the agreement, whether they are local government units, corporations, organizations, or individuals, conform to the implementing measures and obligations.

Implementation costs

Costs involved in the *implementation* of *mitigation* options. These costs are associated with the necessary institutional changes, information requirements, market size, *opportunities* for *technology* gain and learning, and economic incentives needed (grants, subsidies, and taxes).

Indigenous peoples
People whose ancestors inhabited a place or a country when persons from another culture or ethnic background arrived on the scene and dominated them through conquest, settlement, or other means and who today live more in conformity with their own social, economic, and cultural customs and traditions than those of the country of which they now form a part (also referred to as "native," "aboriginal," or "tribal" peoples).

Indirect aerosol effect
Aerosols may lead to an indirect *radiative forcing* of the *climate system* through acting as condensation nuclei or modifying the optical properties and lifetime of clouds. Two indirect effects are distinguished:
- First indirect effect: A radiative forcing induced by an increase in *anthropogenic* aerosols which cause an initial increase in droplet concentration and a decrease in droplet size for fixed liquid water content, leading to an increase of cloud *albedo*. This effect is also known as the "Twomey effect." This is sometimes referred to as the cloud albedo effect. However this is highly misleading since the second indirect effect also alters cloud albedo.
- Second indirect effect: A radiative forcing induced by an increase in anthropogenic aerosols which cause a decrease in droplet size, reducing the precipitation efficiency, thereby modifying the liquid water content, cloud thickness, and cloud lifetime. This effect is also known as the "cloud lifetime effect" or "Albrecht effect."

Industrial Revolution
A period of rapid industrial growth with far-reaching social and economic consequences, beginning in England during the second half of the 18th century and spreading to Europe and later to other countries including the United States. The invention of the steam engine was an important trigger of this development. The Industrial Revolution marks the beginning of a strong increase in the use of *fossil fuels* and emission of, in particular, fossil *carbon dioxide*. In this report, the terms "pre-industrial" and "industrial" refer, somewhat arbitrarily, to the periods before and after the year 1750, respectively.

Inertia
Delay, slowness, or resistance in the response of the *climate*, biological, or *human systems* to factors that alter their rate of change, including continuation of change in the system after the cause of that change has been removed.

Infectious diseases
Any disease that can be transmitted from one person to another. This may occur by direct physical contact, by common handling of an object that has picked up infective organisms, through a disease carrier, or by spread of infected droplets coughed or exhaled into the air.

Infrared radiation
Radiation emitted by the Earth's surface, the *atmosphere*, and clouds. It is also known as terrestrial or long-wave radiation. Infrared radiation has a distinctive range of wavelengths ("spectrum") longer than the wavelength of the red color in the visible part of the spectrum. The spectrum of infrared radiation is practically distinct from that of solar or short-wave radiation because of the difference in temperature between the Sun and the Earth-atmosphere system.

Infrastructure
The basic equipment, utilities, productive enterprises, installations, institutions, and services essential for the development, operation, and growth of an organization, city, or nation. For example, roads; schools; electric, gas, and water utilities; transportation; communication; and legal systems would be all considered as infrastructure.

Integrated assessment
A method of analysis that combines results and models from the physical, biological, economic, and social sciences, and the interactions between these components, in a consistent framework, to evaluate the status and the consequences of environmental change and the policy responses to it.

Interaction effect
The result or consequence of the interaction of *climate change* policy instruments with existing domestic tax systems, including both cost-increasing tax interaction and cost-reducing revenue-recycling effect. The former reflects the impact that *greenhouse gas* policies can have on the functioning of labor and capital markets through their effects on real wages and the real return to capital. By restricting the allowable greenhouse gas *emissions*, permits, regulations, or a *carbon tax* raise the costs of production and the prices of output, thus reducing the real return to labor and capital. For policies that raise revenue for the government—carbon taxes and auctioned permits—the revenues can be recycled to reduce existing distortionary taxes. See also *double dividend*.

Internal variability
See *climate variability*.

International emissions/carbon/energy tax
See *emissions tax*.

International Energy Agency (IEA)
Paris-based energy forum established in 1974. It is linked with the Organisation for Economic Cooperation and Development to enable member countries to take joint measures to meet oil supply emergencies, to share energy information, to coordinate their energy policies, and to cooperate in the development of rational energy programs.

International product and/or technology standards
See *standards*.

Introduced species
A species occurring in an area outside its historically known natural range as a result of accidental dispersal by humans (also referred to as "*exotic species*" or "alien species").

Invasive species
An *introduced species* that invades natural *habitats*.

Isostatic land movements
Isostasy refers to the way in which the *lithosphere* and mantle respond to changes in surface loads. When the loading of the lithosphere is changed by alterations in land ice mass, ocean mass, sedimentation, erosion, or mountain building, vertical isostatic adjustment results, in order to balance the new load.

Joint Implementation (JI)
A market-based implementation mechanism defined in Article 6 of the *Kyoto Protocol*, allowing *Annex I countries* or companies from these countries to implement projects jointly that limit or reduce *emissions*, or enhance *sinks*, and to share the *Emissions Reduction Units*. JI activity is also permitted in Article 4.2(a) of the *United Nations Framework Convention on Climate Change*. See also *Activities Implemented Jointly* and *Kyoto Mechanisms*.

Known technological options
Refer to technologies that exist in operation or pilot plant stage today. It does not include any new technologies that will require drastic technological breakthroughs.

Kyoto Mechanisms
Economic mechanisms based on market principles that Parties to the *Kyoto Protocol* can use in an attempt to lessen the potential economic impacts of *greenhouse gas* emission-reduction requirements. They include *Joint Implementation* (Article 6), the *Clean Development Mechanism* (Article 12), and *Emissions Trading* (Article 17).

Kyoto Protocol
The Kyoto Protocol to the *United Nations Framework Convention on Climate Change* (UNFCCC) was adopted at the Third Session of the *Conference of the Parties* to the UNFCCC in 1997 in Kyoto, Japan. It contains legally binding commitments, in addition to those included in the UNFCCC. Countries included in *Annex B* of the Protocol (most countries in the Organisation for Economic Cooperation and Development, and countries with *economies in transition*) agreed to reduce their *anthropogenic greenhouse gas emissions* (*carbon dioxide, methane, nitrous oxide, hydrofluorocarbons, perfluorocarbons, and sulfur hexafluoride*) by at least 5% below 1990 levels in the commitment period 2008 to 2012. The Kyoto Protocol has not yet entered into force (September 2001).

La Niña
See *El Niño Southern Oscillation*.

Land use
The total of arrangements, activities, and inputs undertaken in a certain land cover type (a set of human actions). The social and economic purposes for which land is managed (e.g., grazing, timber extraction, and conservation).

Land-use change
A change in the use or management of land by humans, which may lead to a change in land cover. Land cover and land-use change may have an impact on the *albedo, evapotranspiration, sources,* and *sinks* of *greenhouse gases*, or other properties of the *climate system*, and may thus have an impact on *climate*, locally or globally. See also the IPCC Special Report on Land Use, Land-Use Change, and Forestry (IPCC, 2000b).

Landslide
A mass of material that has slipped downhill by gravity, often assisted by water when the material is saturated; rapid movement of a mass of soil, rock, or debris down a slope.

Leakage
The part of *emissions* reductions in *Annex B countries* that may be offset by an increase of the emission in the non-constrained countries above their *baseline* levels. This can occur through (i) relocation of energy-intensive production in non-constrained regions; (ii) increased consumption of *fossil fuels* in these regions through decline in the international price of oil and gas triggered by lower demand for these energies; and (iii) changes in incomes (thus in energy demand) because of better terms of trade. Leakage also refers to the situation in which a carbon *sequestration* activity (e.g., tree planting) on one piece of land inadvertently, directly or indirectly, triggers an activity, which in whole or part, counteracts the carbon effects of the initial activity.

Lifetime
Lifetime is a general term used for various *time scales* characterizing the rate of processes affecting the concentration of trace gases. In general, lifetime denotes the average length of time that an atom or molecule spends in a given *reservoir*, such as the *atmosphere* or oceans. The following lifetimes may be distinguished:
· "Turnover time" (T) or "atmospheric lifetime" is the ratio of the mass M of a reservoir (e.g., a gaseous compound in the *atmosphere*) and the total rate of removal S from the reservoir: $T = M/S$. For each removal process separate turnover times can be defined. In soil carbon biology, this is referred to as Mean Residence Time.
· "Adjustment time," "response time," or "perturbation lifetime" (T_a) is the time scale characterizing the decay of an instantaneous pulse input into the reservoir. The term adjustment time is also used to characterize the adjustment

of the mass of a reservoir following a step change in the source strength. Half-life or decay constant is used to quantify a first-order exponential decay process. See *response time* for a different definition pertinent to *climate* variations. The term "lifetime" is sometimes used, for simplicity, as a surrogate for "adjustment time."

In simple cases, where the global removal of the compound is directly proportional to the total mass of the reservoir, the adjustment time equals the turnover time: $T = T_a$. An example is CFC-11 which is removed from the atmosphere only by photochemical processes in the *stratosphere*. In more complicated cases, where several reservoirs are involved or where the removal is not proportional to the total mass, the equality $T = T_a$ no longer holds. *Carbon dioxide* is an extreme example. Its turnover time is only about 4 years because of the rapid exchange between atmosphere and the ocean and terrestrial biota. However, a large part of that CO_2 is returned to the atmosphere within a few years. Thus, the adjustment time of CO_2 in the atmosphere is actually determined by the rate of removal of carbon from the surface layer of the oceans into its deeper layers. Although an approximate value of 100 years may be given for the adjustment time of CO_2 in the atmosphere, the actual adjustment is faster initially and slower later on. In the case of *methane*, the adjustment time is different from the turnover time, because the removal is mainly through a chemical reaction with the hydroxyl radical OH, the concentration of which itself depends on the CH_4 concentration. Therefore the CH_4 removal S is not proportional to its total mass M.

Lithosphere

The upper layer of the solid Earth, both continental and oceanic, which is composed of all crustal rocks and the cold, mainly elastic, part of the uppermost mantle. Volcanic activity, although part of the lithosphere, is not considered as part of the *climate system*, but acts as an *external forcing* factor.

Leapfrogging

Leapfrogging (or technological leapfrogging) refers to the opportunities in developing countries to bypass several stages of technology development, historically observed in industrialized countries, and apply the most advanced presently available technologies in the energy and other economic sectors, through investments in technological development and capacity building.

Level of scientific understanding

This is an index on a 4-step scale (High, Medium, Low, and Very Low) designed to characterize the degree of scientific understanding of the *radiative forcing* agents that affect *climate change*. For each agent, the index represents a subjective judgement about the reliability of the estimate of its forcing, involving such factors as the assumptions necessary to evaluate the forcing, the degree of knowledge of the physical/chemical mechanisms determining the forcing, and the uncertainties surrounding the quantitative estimate.

Local Agenda 21

Local Agenda 21s are the local plans for environment and development that each local authority is meant to develop through a consultative process with their populations, with particular attention paid to involving women and youth. Many local authorities have developed Local Agenda 21s through consultative processes as a means of reorienting their policies, plans, and operations towards the achievement of *sustainable development* goals. The term comes from Chapter 28 of Agenda 21—the document formally endorsed by all government representatives attending the United Nations Conference on Environment and Development (also known as the Earth Summit) in Rio de Janeiro in 1992.

Lock-in technologies and practices

Technologies and practices that have market advantages arising from existing institutions, services, infrastructure, and available resources; they are very difficult to change because of their widespread use and the presence of associated infrastructure and socio-cultural patterns.

Maladaptation

Any changes in natural or *human systems* that inadvertently increase *vulnerability* to climatic *stimuli*; an *adaptation* that does not succeed in reducing vulnerability but increases it instead.

Malaria

Endemic or *epidemic* parasitic disease caused by species of the genus Plasmodium (protozoa) and transmitted by mosquitoes of the genus Anopheles; produces high fever attacks and systemic disorders, and kills approximately 2 million people every year.

Marginal cost pricing

The pricing of commercial goods and services such that the price equals the additional cost that arises from the expansion of production by one additional unit.

Market barriers

In the context of *mitigation* of *climate change*, conditions that prevent or impede the diffusion of *cost-effective* technologies or practices that would mitigate *greenhouse gas emissions*.

Market-based incentives

Measures intended to use price mechanisms (e.g., taxes and tradable permits) to reduce *greenhouse gas emissions*.

Market impacts

Impacts that are linked to market transactions and directly affect *Gross Domestic Product* (a country's national accounts)—for example, changes in the supply and price of agricultural goods. See also *non-market impacts*.

Market penetration

Market penetration is the share of a given market that is provided by a particular good or service at a given time.

Market potential

The portion of the *economic potential* for *greenhouse gas emissions* reductions or *energy-efficiency* improvements that could be achieved under forecast market conditions, assuming no new *policies and measures*. See also *economic potential*, *socio-economic potential*, and *technological potential*.

Mass movement

Applies to all unit movements of land material propelled and controlled by gravity.

Mean Sea Level (MSL)

Mean Sea Level is normally defined as the average r*elative sea level* over a period, such as a month or a year, long enough to average out transients such as waves. See also *sea-level rise*.

Methane (CH$_4$)

A hydrocarbon that is a *greenhouse gas* produced through anaerobic (without oxygen) decomposition of waste in landfills, animal digestion, decomposition of animal wastes, production and distribution of natural gas and oil, coal production, and incomplete fossil-fuel combustion. *Methane* is one of the six *greenhouse gases* to be mitigated under the *Kyoto Protocol*.

Methane recovery

Method by which *methane emissions* (e.g., from coal mines or waste sites) are captured and then reused either as a fuel or for some other economic purpose (e.g., reinjection in oil or gas *reserves*).

Meeting of the Parties (to the Kyoto Protocol) (MOP)

The *Conference of the Parties* of the *United Nations Framework Convention on Climate Change* will serve as the *Meeting of the Parties (MOP)*, the supreme body of the *Kyoto Protocol*, but only Parties to the Kyoto Protocol may participate in deliberations and make decisions. Until the Protocol enters into force, MOP cannot meet.

Mitigation

An *anthropogenic* intervention to reduce the *sources* or enhance the *sinks* of *greenhouse gases*.

Mitigative capacity

The social, political, and economic structures and conditions that are required for effective *mitigation*.

Mixed layer

The upper region of the ocean well-mixed by interaction with the overlying *atmosphere*.

Mixing ratio

See *mole fraction*.

Model hierarchy

See *climate model*.

Mole fraction

Mole fraction, or mixing ratio, is the ratio of the number of moles of a constituent in a given volume to the total number of moles of all constituents in that volume. It is usually reported for dry air. Typical values for long-lived *greenhouse gases* are in the order of mmol/mol (parts per million: ppm), nmol/mol (parts per billion: ppb), and fmol/mol (parts per trillion: ppt). Mole fraction differs from volume mixing ratio, often expressed in ppmv, etc., by the corrections for non-ideality of gases. This correction is significant relative to measurement precision for many greenhouse gases (Schwartz and Warneck, 1995).

Monsoon

Wind in the general atmospheric circulation typified by a seasonal persistent wind direction and by a pronounced change in direction from one season to the next.

Montane

The biogeographic zone made up of relatively moist, cool upland slopes below timberline and characterized by the presence of large evergreen trees as a dominant life form.

Montreal Protocol

The Montreal Protocol on substances that deplete the *ozone layer* was adopted in Montreal in 1987, and subsequently adjusted and amended in London (1990), Copenhagen (1992), Vienna (1995), Montreal (1997), and Beijing (1999). It controls the consumption and production of chlorine- and bromine-containing chemicals that destroy stratospheric ozone, such as *chlorofluorocarbons* (CFCs), methyl chloroform, carbon tetrachloride, and many others.

Morbidity

Rate of occurrence of disease or other health disorder within a population, taking account of the age-specific morbidity rates. Health outcomes include chronic disease incidence/prevalence, rates of hospitalization, primary care consultations, disability-days (i.e., days when absent from work), and prevalence of symptoms.

Mortality

Rate of occurrence of death within a population within a specified time period; calculation of mortality takes account of age-specific death rates, and can thus yield measures of life expectancy and the extent of premature death.

Net Biome Production (NBP)

Net gain or loss of carbon from a region. NBP is equal to the *Net Ecosystem Production* minus the carbon lost due to a disturbance (e.g., a *forest* fire or a forest harvest).

Net carbon dioxide emissions
Difference between sources and sinks of carbon dioxide in a given period and specific area or region.

Net Ecosystem Production (NEP)
Net gain or loss of carbon from an *ecosystem*. NEP is equal to the *Net Primary Production* minus the carbon lost through heterotrophic *respiration*.

Net Primary Production (NPP)
The increase in plant *biomass* or carbon of a unit of a landscape. NPP is equal to the *Gross Primary Production* minus carbon lost through autotrophic *respiration*.

Nitrogen fertilization
Enhancement of plant growth through the addition of nitrogen compounds. In IPCC assessments, this typically refers to fertilization from *anthropogenic sources* of nitrogen such as human-made fertilizers and *nitrogen oxides* released from burning *fossil fuels*.

Nitrogen oxides (NO_x)
Any of several oxides of nitrogen.

Nitrous oxide (N_2O)
A powerful greenhouse gas emitted through soil cultivation practices, especially the use of commercial and organic fertilizers, fossil-fuel combustion, nitric acid production, and biomass burning. One of the six *greenhouse gases* to be curbed under the *Kyoto Protocol*.

Non-point-source pollution
Pollution from *sources* that cannot be defined as discrete points, such as areas of crop production, timber, surface mining, disposal of refuse, and construction. See also *point-source pollution*.

No-regrets opportunities
See *no-regrets policy*.

No-regret options
See *no-regrets policy*.

No-regrets policy
One that would generate net social benefits whether or not there is *climate change*. No-regrets opportunities for *greenhouse gas emissions* reduction are defined as those options whose benefits such as reduced energy costs and reduced emissions of local/regional pollutants equal or exceed their costs to society, excluding the benefits of avoided climate change. No-regrets potential is defined as the gap between the *market potential* and the *socio-economic potential*.

No-regrets potential
See *no-regrets policy*.

Non-Annex B countries/Parties
The countries that are not included in Annex B in the *Kyoto Protocol*. See also *Annex B countries*.

Non-Annex I countries/Parties
The countries that have ratified or acceded to the *United Nations Framework Convention on Climate Change* that are not included in Annex I of the Climate Convention. See also *Annex I countries*.

Non-linearity
A process is called "non-linear" when there is no simple proportional relation between cause and effect. The *climate system* contains many such non-linear processes, resulting in a system with a potentially very complex behavior. Such complexity may lead to *rapid climate change*.

Non-market impacts
Impacts that affect *ecosystems* or human welfare, but that are not directly linked to market transactions—for example, an increased risk of premature death. See also *market impacts*.

North Atlantic Oscillation (NAO)
The North Atlantic Oscillation consists of opposing variations of barometric pressure near Iceland and near the Azores. On average, a westerly current, between the Icelandic low pressure area and the Azores high pressure area, carries cyclones with their associated frontal systems towards Europe. However, the pressure difference between Iceland and the Azores fluctuates on *time scales* of days to decades, and can be reversed at times. It is the dominant mode of winter *climate variability* in the North Atlantic region, ranging from central North America to Europe.

Ocean conveyor belt
The theoretical route by which water circulates around the entire global ocean, driven by wind and the *thermohaline circulation*.

Opportunity
An opportunity is a situation or circumstance to decrease the gap between the *market potential* of any *technology* or practice and the *economic potential*, *socio-economic potential*, or *technological potential*.

Opportunity costs
The cost of an economic activity forgone by the choice of another activity.

Optimal policy
A policy is assumed to be "optimal" if marginal abatement costs are equalized across countries, thereby minimizing *total costs*.

Organic aerosol
Aerosol particles consisting predominantly of organic compounds, mainly C, H, and O, and lesser amounts of other elements (Charlson and Heintzenberg, 1995). See *carbonaceous aerosol*.

Ozone (O_3)

Ozone, the triatomic form of oxygen (O_3), is a gaseous atmospheric constituent. In the *troposphere* it is created both naturally and by photochemical reactions involving gases resulting from human activities (photochemical "smog"). In high concentrations, tropospheric ozone can be harmful to a wide-range of living organisms. Tropospheric ozone acts as a *greenhouse gas*. In the *stratosphere*, ozone is created by the interaction between solar ultraviolet radiation and molecular oxygen (O_2). Stratospheric ozone plays a decisive role in the stratospheric *radiative balance*. Its concentration is highest in the *ozone layer*. Depletion of stratospheric ozone, due to chemical reactions that may be enhanced by *climate change*, results in an increased ground-level flux of *ultraviolet-B radiation*. See also *Montreal Protocol* and *ozone layer*.

Ozone hole

See *ozone layer*.

Ozone layer

The *stratosphere* contains a layer in which the concentration of *ozone* is greatest, the so-called ozone layer. The layer extends from about 12 to 40 km. The ozone concentration reaches a maximum between about 20 and 25 km. This layer is being depleted by human *emissions* of chlorine and bromine compounds. Every year, during the Southern Hemisphere spring, a very strong depletion of the ozone layer takes place over the Antarctic region, also caused by human-made chlorine and bromine compounds in combination with the specific meteorological conditions of that region. This phenomenon is called the *ozone hole*.

Parameterization

In *climate models*, this term refers to the technique of representing processes, that cannot be explicitly resolved at the spatial or temporal resolution of the model (sub-grid scale processes), by relationships between the area- or time-averaged effect of such sub-grid-scale processes and the larger scale flow.

Pareto criterion/Pareto optimum

A requirement or status that an individual's welfare could not be further improved without making others in the society worse off.

Perfluorocarbons (PFCs)

Among the six *greenhouse gases* to be abated under the *Kyoto Protocol*. These are by-products of aluminum smelting and uranium enrichment. They also replace *chlorofluorocarbons* in manufacturing semiconductors. The *Global Warming Potential* of PFCs is 6,500–9,200 times that of *carbon dioxide*.

Permafrost

Perennially frozen ground that occurs wherever the temperature remains below 0°C for several years.

Perturbation lifetime

See *lifetime*.

Photosynthesis

The process by which plants take *carbon dioxide* (CO_2) from the air (or bicarbonate in water) to build carbohydrates, releasing oxygen (O_2) in the process. There are several pathways of photosynthesis with different responses to atmospheric CO_2 concentrations. See also *carbon dioxide fertilization*.

Phytoplankton

The plant forms of *plankton* (e.g., diatoms). Phytoplankton are the dominant plants in the sea, and are the bast of the entire marine food web. These single-celled organisms are the principal agents for photosynthetic carbon fixation in the ocean. See also *zooplankton*.

Plankton

Aquatic organisms that drift or swim weakly. See also *phytoplankton* and *zooplankton*.

Point-source pollution

Pollution resulting from any confined, discrete source, such as a pipe, ditch, tunnel, well, container, concentrated animal-feeding operation, or floating craft. See also *non-point-source pollution*.

Policies and measures

In *United Nations Framework Convention on Climate Change* parlance, "policies" are actions that can be taken and/or mandated by a government—often in conjunction with business and industry within its own country, as well as with other countries—to accelerate the application and use of measures to curb *greenhouse gas emissions*. "Measures" are technologies, processes, and practices used to implement policies, which, if employed, would reduce greenhouse gas emissions below anticipated future levels. Examples might include carbon or other *energy taxes*, standardized fuel-efficiency *standards* for automobiles, etc. "Common and coordinated" or "harmonized" policies refer to those adopted jointly by Parties.

Pool

See *reservoir*.

Post-glacial rebound

The vertical movement of the continents and sea floor following the disappearance and shrinking of *ice sheets*—for example, since the Last Glacial Maximum (21 ky BP). The rebound is an *isostatic land movement*.

Precursors

Atmospheric compounds which themselves are not *greenhouse gases* or *aerosols*, but which have an effect on greenhouse gas or aerosol concentrations by taking part in physical or chemical processes regulating their production or destruction rates.

Pre-industrial

See *Industrial Revolution*.

Present value cost

The sum of all costs over all time periods, with future costs discounted.

Primary energy

Energy embodied in natural *resources* (e.g., coal, crude oil, sunlight, uranium) that has not undergone any *anthropogenic* conversion or transformation.

Private cost

Categories of costs influencing an individual's decision making are referred to as private costs. See also *social cost* and *total cost*.

Profile

A smoothly changing set of concentrations representing a possible pathway towards stabilization. The word "profile"is used to distinguish such pathways from emissions pathways, which are usually referred to as "*scenarios.*"

Projection (generic)

A projection is a potential future evolution of a quantity or set of quantities, often computed with the aid of a model. Projections are distinguished from "predictions" in order to emphasize that projections involve assumptions concerning, for example, future socio-economic and technological developments that may or may not be realized, and are therefore subject to substantial *uncertainty*. See also *climate projection* and *climate prediction*.

Proxy

A proxy *climate* indicator is a local record that is interpreted, using physical and biophysical principles, to represent some combination of climate-related variations back in time. Climate-related data derived in this way are referred to as proxy data. Examples of proxies are tree ring records, characteristics of corals, and various data derived from ice cores.

Purchasing Power Parity (PPP)

Estimates of *Gross Domestic Product* based on the purchasing power of currencies rather than on current exchange rates. Such estimates are a blend of extrapolated and regression-based numbers, using the results of the International Comparison Program. PPP estimates tend to lower per capita GDPs in industrialized countries and raise per capita GDPs in developing countries. PPP is also an acronym for polluter-pays-principle.

Radiative balance

See *energy balance*.

Radiative forcing

Radiative forcing is the change in the net vertical irradiance (expressed in Wm^{-2}) at the *tropopause* due to an internal change or a change in the external forcing of the *climate system*, such as, for example, a change in the concentration of *carbon dioxide* or the output of the Sun. Usually radiative forcing is computed after allowing for stratospheric temperatures to readjust to radiative equilibrium, but with all tropospheric properties held fixed at their unperturbed values.

Radiative forcing scenario

A plausible representation of the future development of *radiative forcing* associated, for example, with changes in atmospheric composition or *land-use* change, or with external factors such as variations in *solar activity*. Radiative forcing scenarios can be used as input into simplified *climate models* to compute *climate projections*.

Rangeland

Unimproved grasslands, shrublands, savannahs, and tundra.

Regeneration

The renewal of a stand of trees through either natural means (seeded onsite or adjacent stands or deposited by wind, birds, or animals) or artificial means (by planting seedlings or direct seeding).

Rapid climate change

The *non-linearity* of the *climate system* may lead to rapid *climate change*, sometimes called abrupt events or even surprises. Some such abrupt events may be imaginable, such as a dramatic reorganization of the *thermohaline circulation*, rapid deglaciation, or massive melting of *permafrost* leading to fast changes in the *carbon cycle*. Others may be truly unexpected, as a consequence of a strong, rapidly changing, forcing of a non-linear system.

Rebound effect

Occurs because, for example, an improvement in motor efficiency lowers the cost per kilometer driven; it has the perverse effect of encouraging more trips.

Reference scenario

See *baseline*.

Reforestation

Planting of *forests* on lands that have previously contained forests but that have been converted to some other use. For a discussion of the term forest and related terms such as *afforestation*, *reforestation*, and *deforestation*, see the IPCC Special Report on Land Use, Land-Use Change, and Forestry (IPCC, 2000b).

Regulatory measures

Rules or codes enacted by governments that mandate product specifications or process performance characteristics. See also *standards*.

Reinsurance

The transfer of a portion of primary insurance risks to a secondary tier of insurers (reinsurers); essentially "insurance for insurers."

Relative sea level

Sea level measured by a *tide gauge* with respect to the land upon which it is situated. See also *Mean Sea Level*.

(Relative) Sea level secular change

Long-term changes in relative sea level caused by either eustatic changes (e.g., brought about by *thermal expansion*) or changes in vertical land movements.

Renewables

Energy sources that are, within a short time frame relative to the Earth's natural cycles, sustainable, and include non-carbon technologies such as solar energy, hydropower, and wind, as well as carbon-neutral technologies such as *biomass*.

Research, development, and demonstration

Scientific and/or technical research and development of new production processes or products, coupled with analysis and measures that provide information to potential users regarding the application of the new product or process; demonstration tests; and feasibility of applying these products processes via pilot plants and other pre-commercial applications.

Reserves

Refer to those occurrences that are identified and measured as economically and technically recoverable with current technologies and prices. See also *resources*.

Reservoir

A component of the *climate system*, other than the *atmosphere*, which has the capacity to store, accumulate, or release a substance of concern (e.g., carbon, a *greenhouse gas*, or a *precursor*). Oceans, soils, and *forests* are examples of reservoirs of carbon. *Pool* is an equivalent term (note that the definition of pool often includes the atmosphere). The absolute quantity of substance of concerns, held within a reservoir at a specified time, is called the stock. The term also means an artificial or natural storage place for water, such as a lake, pond, or *aquifer*, from which the water may be withdrawn for such purposes as irrigation, water supply, or irrigation.

Resilience

Amount of change a system can undergo without changing state.

Resource base

Resource base includes both *reserves* and *resources*.

Resources

Resources are those occurrences with less certain geological and/or economic characteristics, but which are considered potentially recoverable with foreseeable technological and economic developments.

Respiration

The process whereby living organisms converts organic matter to *carbon dioxide*, releasing energy and consuming oxygen.

Response time

The response time or adjustment time is the time needed for the *climate system* or its components to re-equilibrate to a new state, following a forcing resulting from external and internal processes or *feedbacks*. It is very different for various components of the climate system. The response time of the *troposphere* is relatively short, from days to weeks, whereas the *stratosphere* comes into equilibrium on a *time scale* of typically a few months. Due to their large heat capacity, the oceans have a much longer response time, typically decades, but up to centuries or millennia. The response time of the strongly coupled surface-troposphere system is, therefore, slow compared to that of the stratosphere, and mainly determined by the oceans. The *biosphere* may respond fast (e.g., to *droughts*), but also very slowly to imposed changes. See *lifetime* for a different definition of response time pertinent to the rate of processes affecting the concentration of trace gases.

Revenue recycling

See *interaction effect*.

Runoff

That part of precipitation that does not evaporate. In some countries, runoff implies *surface runoff* only.

S profiles

The carbon dioxide concentration *profiles* leading to stabilization defined in the IPCC 1994 assessment (Enting *et al.*, 1994; Schimel *et al.*, 1995). For any given stabilization level, these profiles span a wide range of possibilities. The S stands for "Stabilization." See also *WRE profiles*.

Safe-landing approach

See *tolerable windows approach*.

Salinization

The accumulation of salts in soils.

Saltwater intrusion/encroachment

Displacement of fresh surfacewater or groundwater by the advance of saltwater due to its greater density, usually in coastal and estuarine areas.

Scenario (generic)

A plausible and often simplified description of how the future may develop, based on a coherent and internally consistent set of assumptions about key driving forces (e.g., rate of *technology* change, prices) and relationships. Scenarios are neither predictions nor forecasts and sometimes may be based on a "narrative storyline." Scenarios may be derived from *projections*, but are often based on additional information from other sources. See also *SRES scenarios*, *climate scenario*, and *emission scenarios*.

Sea-level rise
An increase in the mean level of the ocean. Eustatic sea-level rise is a change in global average sea level brought about by an alteration to the volume of the world ocean. *Relative sea-level* rise occurs where there is a net increase in the level of the ocean relative to local land movements. Climate modelers largely concentrate on estimating eustatic sea-level change. *Impact* researchers focus on relative sea-level change.

Seawall
A human-made wall or embankment along a shore to prevent wave *erosion*.

Semi-arid regions
Ecosystems that have more than 250 mm precipitation per year but are not highly productive; usually classified as *rangelands*.

Sensitivity
Sensitivity is the degree to which a system is affected, either adversely or beneficially, by climate-related *stimuli*. The effect may be direct (e.g., a change in crop yield in response to a change in the mean, range, or variability of temperature) or indirect (e.g., damages caused by an increase in the frequency of coastal flooding due to *sea-level rise*). See also *climate sensitivity*.

Sequential decision making
Stepwise decision making aiming to identify short-term strategies in the face of long-term uncertainties, by incorporating additional information over time and making mid-course corrections.

Sequestration
The process of increasing the carbon content of a carbon *reservoir* other than the *atmosphere*. Biological approaches to sequestration include direct removal of *carbon dioxide* from the atmosphere through *land-use change*, *afforestation*, *reforestation*, and practices that enhance soil carbon in agriculture. Physical approaches include separation and disposal of carbon dioxide from flue gases or from processing *fossil fuels* to produce hydrogen- and carbon dioxide-rich fractions and long-term storage in underground in depleted oil and gas reservoirs, coal seams, and saline *aquifers*. See also *uptake*.

Silt
Unconsolidated or loose sedimentary material whose constituent rock particles are finer than grains of sand and larger than clay particles.

Silviculture
Development and care of forests.

Sink
Any process, activity or mechanism that removes a *greenhouse gas*, an *aerosol*, or a *precursor* of a greenhouse gas or aerosol from the *atmosphere*.

Snowpacks
A seasonal accumulation of slow-melting snow.

Social cost
The social cost of an activity includes the *value* of all the *resources* used in its provision. Some of these are priced and others are not. Non-priced resources are referred to as externalities. It is the sum of the costs of these externalities and the priced resources that makes up the social cost. See also *private cost* and *total cost*.

Socio-economic potential
The socio-economic potential represents the level of greenhouse gas *mitigation* that would be approached by overcoming social and cultural obstacles to the use of technologies that are *cost-effective*. See also *economic potential*, *market potential*, and *technology potential*.

Soil moisture
Water stored in or at the land surface and available for evaporation.

Solar activity
The Sun exhibits periods of high activity observed in numbers of *sunspots*, as well as radiative output, magnetic activity, and emission of high energy particles. These variations take place on a range of *time scales* from millions of years to minutes. See also *solar cycle*.

Solar ("11 year") cycle
A quasi-regular modulation of *solar activity* with varying amplitude and a period of between 9 and 13 years.

Solar radiation
Radiation emitted by the Sun. It is also referred to as short-wave radiation. Solar radiation has a distinctive range of wavelengths (spectrum) determined by the temperature of the Sun. See also *infrared radiation*.

Soot particles
Particles formed during the quenching of gases at the outer edge of flames of organic vapors, consisting predominantly of carbon, with lesser amounts of oxygen and hydrogen present as carboxyl and phenolic groups and exhibiting an imperfect graphitic structure (Charlson and Heintzenberg, 1995). See also *black carbon*.

Source
Any process, activity, or mechanism that releases a *greenhouse gas*, an *aerosol*, or a *precursor* of a greenhouse gas or aerosol into the *atmosphere*.

Southern Oscillation
See *El Niño Southern Oscillation*.

Spatial and temporal scales
Climate may vary on a large range of spatial and temporal scales. Spatial scales may range from local (less than 100,000 km²),

through regional (100,000 to 10 million km²) to continental (10 to 100 million km²). Temporal scales may range from seasonal to geological (up to hundreds of millions of years).

Spill-over effect
The economic effects of domestic or sectoral *mitigation* measures on other countries or sectors. In this report, no assessment is made on environmental spillover effects. Spillover effects can be positive or negative and include effects on trade, carbon *leakage*, transfer, and diffusion of *environmentally sound technology* and other issues.

SRES scenarios
SRES scenarios are *emissions scenarios* developed by Nakicenovic *et al.* (2000) and used, among others, as a basis for the *climate projections* in the IPCC WGI contribution to the Third Assessment Report (IPCC, 2001a). The following terms are relevant for a better understanding of the structure and use of the set of SRES scenarios:

· *(Scenario) Family*: Scenarios that have a similar demographic, societal, economic, and technical-change *storyline*. Four scenario families comprise the SRES scenario set: A1, A2, B1, and B2.
· *(Scenario) Group*: Scenarios within a family that reflect a consistent variation of the storyline. The A1 scenario family includes four groups designated as A1T, A1C, A1G, and A1B that explore alternative structures of future energy systems. In the Summary for Policymakers of Nakicenovic *et al.* (2000), the A1C and A1G groups have been combined into one "Fossil-Intensive" A1FI scenario group. The other three scenario families consist of one group each. The SRES scenario set reflected in the Summary for Policymakers of Nakicenovic *et al.* (2000) thus consist of six distinct *scenario groups*, all of which are equally sound and together capture the range of uncertainties associated with driving forces and emissions.
· *Illustrative Scenario*: A scenario that is illustrative for each of the six *scenario groups* reflected in the Summary for Policymakers of Nakicenovic *et al.* (2000). They include four revised *scenario markers* for the *scenario groups* A1B, A2, B1, B2, and two additional scenarios for the A1FI and A1T groups. All *scenario groups* are equally sound.
· *(Scenario) Marker*: A scenario that was originally posted in draft form on the SRES website to represent a given *scenario family*. The choice of markers was based on which of the initial quantifications best reflected the storyline, and the features of specific models. Markers are no more likely than other scenarios, but are considered by the SRES writing team as illustrative of a particular storyline. They are included in revised form in Nakicenovic *et al.* (2000). These scenarios have received the closest scrutiny of the entire writing team and via the SRES open process. Scenarios have also been selected to illustrate the other two *scenario groups*.
· *(Scenario) Storyline*: A narrative description of a scenario (or family of scenarios) highlighting the main scenario characteristics, relationships between key driving forces, and the dynamics of their evolution.

Stabilization
The achievement of stabilization of atmospheric concentrations of one or more *greenhouse gases* (e.g., *carbon dioxide* or a CO_2-equivalent basket of greenhouse gases).

Stabilization analysis
In this report, this refers to analyses or *scenarios* that address the *stabilization* of the concentration of *greenhouse gases*.

Stabilization scenarios
See s*tabilization analysis*.

Stakeholders
Person or entity holding grants, concessions, or any other type of *value* that would be affected by a particular action or policy.

Standards
Set of rules or codes mandating or defining product performance (e.g., grades, dimensions, characteristics, test methods, and rules for use). International product and/or *technology* or performance standards establish minimum requirements for affected products and/or technologies in countries where they are adopted. The standards reduce *greenhouse gas emissions* associated with the manufacture or use of the products and/or application of the technology. See also *regulatory measures*.

Stimuli (climate-related)
All the elements of *climate change*, including mean *climate* characteristics, *climate variability*, and the frequency and magnitude of extremes.

Stock
See *reservoir*.

Storm surge
The temporary increase, at a particular locality, in the height of the sea due to extreme meteorological conditions (low atmospheric pressure and/or strong winds). The storm surge is defined as being the excess above the level expected from the tidal variation alone at that time and place.

Storyline
See *SRES scenarios*.

Streamflow
Water within a river channel, usually expressed in $m^3 sec^{-1}$.

Stratosphere
The highly stratified region of the *atmosphere* above the *troposphere* extending from about 10 km (ranging from 9 km in high latitudes to 16 km in the tropics on average) to about 50 km.

Structural change
Changes, for example, in the relative share of *Gross Domestic Product* produced by the industrial, agricultural, or services sectors

of an economy; or more generally, systems transformations whereby some components are either replaced or potentially substituted by other ones.

Submergence
A rise in the water level in relation to the land, so that areas of formerly dry land become inundated; it results either from a sinking of the land or from a rise of the water level.

Subsidence
The sudden sinking or gradual downward settling of the Earth's surface with little or no horizontal motion.

Subsidy
Direct payment from the government to an entity, or a tax reduction to that entity, for implementing a practice the government wishes to encourage. *Greenhouse gas emissions* can be reduced by lowering existing subsidies that have the effect of raising emissions, such as subsidies to *fossil-fuel* use, or by providing subsidies for practices that reduce emissions or enhance *sinks* (e.g., for insulation of buildings or planting trees).

Sulfur hexafluoride (SF_6)
One of the six *greenhouse gases* to be curbed under the *Kyoto Protocol*. It is largely used in heavy industry to insulate high-voltage equipment and to assist in the manufacturing of cable-cooling systems. Its *Global Warming Potential* is 23,900.

Sunspots
Small dark areas on the Sun. The number of sunspots is higher during periods of high *solar activity*, and varies in particular with the *solar cycle*.

Surface runoff
The water that travels over the soil surface to the nearest surface stream; runoff of a drainage *basin* that has not passed beneath the surface since precipitation.

Sustainable development
Development that meets the needs of the present without compromising the ability of future generations to meet their own needs.

Targets and time tables
A target is the reduction of a specific percentage of *greenhouse gas emissions* from a *baseline* date (e.g., "below 1990 levels") to be achieved by a set date or time table (e.g., 2008 to 2012). For example, under the *Kyoto Protocol's* formula, the European Union has agreed to reduce its *greenhouse gas emissions* by 8% below 1990 levels by the 2008 to 2012 commitment period. These targets and time tables are, in effect, an emissions cap on the total amount of greenhouse gas emissions that can be emitted by a country or region in a given time period.

Tax-interaction effect
See *interaction effect*.

Technological potential
The amount by which it is possible to reduce *greenhouse gas emissions* or improve *energy efficiency* by implementing a *technology* or practice that has already been demonstrated. See also *economic potential*, *market potential*, and *socio-economic potential*.

Technology
A piece of equipment or a technique for performing a particular activity.

Technology or performance standard
See *standards*.

Technology transfer
The broad set of processes that cover the exchange of knowledge, money, and goods among different *stakeholders* that lead to the spreading of *technology* for adapting to or mitigating *climate change*. As a generic concept, the term is used to encompass both diffusion of technologies and technological cooperation across and within countries.

Thermal erosion
The *erosion* of ice-rich *permafrost* by the combined thermal and mechanical action of moving water.

Thermal expansion
In connection with sea level, this refers to the increase in volume (and decrease in density) that results from warming water. A warming of the ocean leads to an expansion of the ocean volume and hence an increase in sea level.

Thermohaline circulation
Large-scale density-driven circulation in the ocean, caused by differences in temperature and salinity. In the North Atlantic, the thermohaline circulation consists of warm surface water flowing northward and cold deepwater flowing southward, resulting in a net poleward transport of heat. The surface water sinks in highly restricted sinking regions located in high latitudes.

Thermokarst
Irregular, hummocky topography in frozen ground caused by melting of ice.

Tide gauge
A device at a coastal location (and some deep sea locations) which continuously measures the level of the sea with respect to the adjacent land. Time-averaging of the sea level so recorded gives the observed *relative sea level secular changes*.

Time scale
Characteristic time for a process to be expressed. Since many processes exibit most of their effects early, and then have a long

period during which they gradually approach full expression, for the purpose of this report the time scale is numerically defined as the time required for a perturbation in a process to show at least half of its final effect.

Tolerable-windows approach

These approaches analyze *greenhouse gas emissions* as they would be constrained by adopting a long-term *climate*—rather than greenhouse gas concentration *stabilization*—target (e.g., expressed in terms of temperature or sea level changes or the rate of such changes). The main objective of these approaches is to evaluate the implications of such long-term targets for short- or medium-term "tolerable" ranges of global *greenhouse gas emissions*. Also referred to as safe-landing approaches.

Top-down models

The terms "top" and "bottom" are shorthand for aggregate and disaggregated models. The top-down label derives from how modelers applied macro-economic theory and econometric techniques to historical data on consumption, prices, incomes, and factor costs to model final demand for goods and services, and supply from main sectors, like the energy sector, transportation, agriculture, and industry. Therefore, top-down models evaluate the system from aggregate economic variables, as compared to *bottom-up models* that consider technological options or project specific *climate change mitigation* policies. Some technology data were, however, integrated into top-down analysis and so the distinction is not that clear-cut.

Total cost

All items of cost added together. The total cost to society is made up of both the *external cost* and the *private cost*, which together are defined as *social cost*.

Trade effects

Economic impacts of changes in the purchasing power of a bundle of exported goods of a country for bundles of goods imported from its trade partners. Climate policies change the relative production costs and may change terms of trade substantially enough to change the ultimate economic balance.

Transient climate response

The globally averaged surface air temperature increase, averaged over a 20-year period, centered at the time of CO_2 doubling (i.e., at year 70 in a 1% per year compound CO_2 increase experiment with a global coupled *climate model*).

Tropopause

The boundary between the *troposphere* and the *stratosphere*.

Troposphere

The lowest part of the *atmosphere* from the surface to about 10 km in altitude in mid-latitudes (ranging from 9 km in high latitudes to 16 km in the tropics on average) where clouds and "weather" phenomena occur. In the troposphere, temperatures generally decrease with height.

Tundra

A treeless, level, or gently undulating plain characteristic of arctic and subarctic regions.

Turnover time

See *lifetime*.

Ultraviolet (UV)-B radiation

Solar radiation within a wavelength range of 280-320 nm, the greater part of which is absorbed by stratospheric *ozone*. Enhanced UV-B radiation suppresses the immune system and can have other adverse effects on living organisms.

Uncertainty

An expression of the degree to which a value (e.g., the future state of the *climate system*) is unknown. Uncertainty can result from lack of information or from disagreement about what is known or even knowable. It may have many types of sources, from quantifiable errors in the data to ambiguously defined concepts or terminology, or uncertain *projections* of human behavior. Uncertainty can therefore be represented by quantitative measures (e.g., a range of values calculated by various models) or by qualitative statements (e.g., reflecting the judgment of a team of experts). See Moss and Schneider (2000).

Undernutrition

The result of food intake that is insufficient to meet dietary energy requirements continuously, poor absorption, and/or poor biological use of nutrients consumed.

Unique and threatened systems

Entities that are confined to a relatively narrow geographical range but can affect other, often larger entities beyond their range; narrow geographical range points to *sensitivity* to environmental variables, including *climate*, and therefore attests to potential *vulnerability* to *climate change*.

United Nations Framework Convention on Climate Change (UNFCCC)

The Convention was adopted on 9 May 1992 in New York and signed at the 1992 Earth Summit in Rio de Janeiro by more than 150 countries and the European Community. Its ultimate objective is the "stabilization of greenhouse gas concentrations in the atmosphere at a level that would prevent dangerous anthropogenic interference with the climate system." It contains commitments for all Parties. Under the Convention, Parties included in *Annex I* aim to return *greenhouse gas emissions* not controlled by the *Montreal Protocol* to 1990 levels by the year 2000. The Convention entered into force in March 1994. See also *Kyoto Protocol* and *Conference of the Parties (COP)*.

Uptake
The addition of a substance of concern to a *reservoir*. The uptake of carbon-containing substances, in particular *carbon dioxide*, is often called (carbon) *sequestration*. See also *sequestration*.

Upwelling
Transport of deeper water to the surface, usually caused by horizontal movements of surface water.

Urbanization
The conversion of land from a natural state or managed natural state (such as agriculture) to cities; a process driven by net rural-to-urban migration through which an increasing percentage of the population in any nation or region come to live in settlements that are defined as "urban centres."

Value added
The net output of a sector after adding up all outputs and subtracting intermediate inputs.

Values
Worth, desirability, or utility based on individual preferences. The total value of any resource is the sum of the values of the different individuals involved in the use of the resource. The values, which are the foundation of the estimation of costs, are measured in terms of the willingness to pay (WTP) by individuals to receive the resource or by the willingness of individuals to accept payment (WTA) to part with the resource.

Vector
An organism, such as an insect, that transmits a pathogen from one host to another. See also *vector-borne diseases*.

Vector-borne diseases
Disease that is transmitted between hosts by a *vector* organism such as a mosquito or tick (e.g., *malaria, dengue fever,* and leishmaniasis).

Volume mixing ratio
See *mole fraction*.

Voluntary agreement
An agreement between a government authority and one or more private parties, as well as a unilateral commitment that is recognized by the public authority, to achieve environmental objectives or to improve environmental performance beyond *compliance*.

Vulnerability
The degree to which a system is susceptible to, or unable to cope with, adverse effects of *climate change*, including *climate variability* and extremes. Vulnerability is a function of the character, magnitude, and rate of climate variation to which a system is exposed, its *sensitivity*, and its *adaptive capacity*.

Water stress
A country is water-stressed if the available freshwater supply relative to water withdrawals acts as an important constraint on development. Withdrawals exceeding 20% of renewable water supply has been used as an indicator of water stress.

Water-use efficiency
Carbon gain in *photosynthesis* per unit water lost in *evapotranspiration*. It can be expressed on a short-term basis as the ratio of photosynthetic carbon gain per unit transpirational water loss, or on a seasonal basis as the ratio of *net primary production* or agricultural yield to the amount of available water.

Water withdrawal
Amount of water extracted from water bodies.

WRE profiles
The carbon dioxide concentration *profiles* leading to stabilization defined by Wigley, Richels, and Edmonds (1996) whose initials provide the acronym. For any given stabilization level, these profiles span a wide range of possibilities. See also *S profiles*.

Zooplankton
The animal forms of *plankton*. They consume *phytoplankton* or other *zooplankton*. See also *phytoplankton*.

Sources

Charlson, R.J., and J. Heintzenberg (eds.), 1995: *Aerosol Forcing of Climate*. John Wiley and Sons Limited, Chichester, United Kingdom, pp. 91–108 (reproduced with permission).

Enting, I.G., T.M.L. Wigley, and M. Heimann, 1994: Future emissions and concentrations of carbon dioxide: key ocean/atmosphere/land analyses. *CSIRO Division of Atmospheric Research Technical Paper 31*, Mordialloc, Australia, 120 pp.

IPCC, 1992: *Climate Change 1992: The Supplementary Report to the IPCC Scientific Assessment* [Houghton, J.T., B.A. Callander, and S.K. Varney (eds.)]. Cambridge University Press, Cambridge, UK, xi + 116 pp.

IPCC, 1994: *Climate Change 1994: Radiative Forcing of Climate Change and an Evaluation of the IPCC IS92 Emission Scenarios*, [Houghton, J.T., L.G. Meira Filho, J. Bruce, Hoesung Lee, B.A. Callander, E. Haites, N. Harris, and K. Maskell (eds.)]. Cambridge University Press, Cambridge, UK and New York, NY, USA, 339 pp.

IPCC, 1996: *Climate Change 1995: The Science of Climate Change. Contribution of Working Group I to the Second Assessment Report of the Intergovernmental Panel on Climate Change* [Houghton., J.T., L.G. Meira Filho, B.A. Callander, N. Harris, A. Kattenberg, and K. Maskell (eds.)]. Cambridge University Press, Cambridge, United Kingdom and New York, NY, USA, 572 pp.

IPCC, 1997a: *IPCC Technical Paper 2: An Introduction to Simple Climate Models used in the IPCC Second Assessment Report* [Houghton, J.T., L.G. Meira Filho, D.J. Griggs, and K. Maskell (eds.)]. Intergovernmental Panel on Climate Change, World Meteorological Organization, Geneva, Switzerland, 51 pp.

IPCC, 1997b: *Revised 1996 IPCC Guidelines for National Greenhouse Gas Inventories* (3 volumes) [Houghton, J.T., L.G. Meira Filho, B. Lim, K. Tréanton, I. Mamaty, Y. Bonduki, D.J. Griggs, and B.A. Callander (eds.)]. Intergovernmental Panel on Climate Change, World Meteorological Organization, Geneva, Switzerland.

IPCC, 1997c: *IPCC Technical Paper 4: Implications of Proposed CO_2 Emissions Limitations*. [Houghton, J.T., L.G. Meira Filho, D.J. Griggs, and M. Noguer (eds.)]. Intergovernmental Panel on Climate Change, World Meteorological Organization, Geneva, Switzerland, 41 pp.

IPCC, 1998: *The Regional Impacts of Climate Change: An Assessment of Vulnerability. A Special Report of IPCC Working Group II* [Watson, R.T., M.C. Zinyowera, and R.H. Moss (eds.)]. Cambridge University Press, Cambridge, United Kingdom and New York, NY, USA, 517 pp.

IPCC, 2000a: *Methodological and Technical Issues in Technology Transfer. A Special Report of IPCC Working Group III* [Metz, B., O.R. Davidson, J.-W. Martens, S.N.M. van Rooijen, and L. van Wie McGrory (eds.)] Cambridge University Press, Cambridge, United Kingdom and New York, NY, USA, 466 pp.

IPCC, 2000b: *Land Use, Land-Use Change, and Forestry. A Special Report of the IPCC* [Watson, R.T., I.R. Noble, B. Bolin, N.H. Ravindranath, D.J. Verardo, and D.J. Dokken (eds.)] Cambridge University Press, Cambridge, United Kingdom and New York, NY, USA, 377 pp.

IPCC, 2001a: *Climate Change 2001: The Scientific Basis. Contribution of Working Group I to the Third Assessment Report of the Intergovernmental Panel on Climate Change* [Houghton, J.T., Y. Ding, D.G. Griggs, M. Noguer, P.J. van der Linden, X. Dai, K. Maskell, and C.A. Johnson (eds.)]. Cambridge University Press, Cambridge, United Kingdom and New York, NY, USA, 881 pp.

IPCC, 2001b: *Climate Change 2001: Impacts, Adaptation, and Vulnerability. Contribution of Working Group II to the Third Assessment Report of the Intergovernmental Panel on Climate Change* [McCarthy, J.J., O.F. Canziani, N.A. Leary, D.J. Dokken, and K.S. White (eds.)]. Cambridge University Press, Cambridge, United Kingdom and New York, NY, USA, 1031 pp.

IPCC, 2001c: *Climate Change 2001: Mitigation. Contribution of Working Group III to the Third Assessment Report of the Intergovernmental Panel on Climate Change* [Metz, B., O.R. Davidson, R. Swart, and J. Pan (eds.)]. Cambridge University Press, Cambridge, United Kingdom and New York, NY, USA, 752 pp.

Jackson, J. (ed.), 1997: *Glossary of Geology*. American Geological Institute, Alexandria, Virginia.

Maunder, W.J., 1992: *Dictionary of Global Climate Change*, UCL Press Ltd.

Moss, R. and S. Schneider, 2000: Uncertainties in the IPCC TAR: recommendations to Lead Authors for more consistent assessment and reporting. In: *Guidance Papers on the Cross-Cutting Issues of the Third Assessment Report of the IPCC* [Pachauri, R., T. Taniguchi, and K. Tanaka (eds.)]. Intergovernmental Panel on Climate Change, World Meteorological Organization, Geneva, Switzerland, pp. 33–51. Available online at http://www.gispri.or.jp.

Nakicenovic, N., J. Alcamo, G. Davis, B. de Vries, J. Fenhann, S. Gaffin, K. Gregory, A. Grübler, T.Y. Jung, T. Kram, E.L. La Rovere, L. Michaelis, S. Mori, T. Morita, W. Pepper, H. Pitcher, L. Price, K. Raihi, A. Roehrl, H.-H. Rogner, A. Sankovski, M. Schlesinger, P. Shukla, S. Smith, R. Swart, S. van Rooijen, N. Victor, and Z. Dadi, 2000: *Emissions Scenarios. A Special Report of Working Group III of the Intergovernmental Panel on Climate Change*. Cambridge University Press, Cambridge, United Kingdom and New York, NY, USA, 599 pp.

Schwartz, S. E. and P. Warneck, 1995: Units for use in atmospheric chemistry, *Pure & Appl. Chem.*, **67**, 1377–1406.

UNEP, 1995: *Global Biodiversity Assessment* [Heywood, V.H. and R.T. Watson (eds.)]. Cambridge Unive rsity Press, Cambridge, United Kingdom and New York, NY, USA, 1140 pp.

Wigley, T.M.L., R. Richels, and J.A. Edmonds, 1996: Economic and environmental choices in the stabilization of atmospheric CO_2 concentrations. *Nature*, **379**, 242–245.

Annex C. Acronyms, Abbreviations, and Units

Acronyms and Abbreviations

AA	Assigned Amount
AAU	Assigned Amount Unit
AD	*Anno Domini*
AIJ	Activities Implemented Jointly
A-O	Atmosphere-Ocean
AO	Arctic Oscillation
AOGCM	Atmosphere-Ocean General Circulation Model
Bern-CC	Bern Carbon Cycle
BP	Before Present
C_2F_6	Perfluoroethane / Hexafluoroethane
C_3	Three-Carbon Compound
C_4	Four-Carbon Compound
CANZ	Canada, Australia, and New Zealand
CBA	Cost-Benefit Analysis
CCC(ma)	Canadian Centre for Climate (Modeling and Analysis) (Canada)
CCGT	Combined Cycle Gas Turbine
CDM	Clean Development Mechanism
CEA	Cost-Effectiveness Analysis
CER	Certified Emission Reduction
CF_4	Perfluoromethane / Tetrafluoromethane
CFC	Chlorofluorocarbon
CGCM	Coupled GCM from CCC(ma)
CGE	Computable General Equilibrium
CGIAR	Consultative Group on International Agricultural Research
CH_4	Methane
CHP	Combined Heat and Power
CMIP	Coupled Model Intercomparison Project
CO_2	Carbon Dioxide
COP	Conference of the Parties
DAF	Decision Analysis Framework
DES	Development, Equity, and Sustainability
DES GP	Guidance Paper on Development, Equity, and Sustainability
DHF	Dengue Haemorrhagic Fever
DMF	Decision Making Framework
DSS	Dengue Shock Syndrome
ECE	Economic Commission for Europe
EIT	Economy in Transition
ENSO	El Niño Southern Oscillation
ERU	Emissions Reduction Unit
ES	Executive Summary
ESCO	Energy Service Company
EST	Environmentally Sound Technology
FCCC	Framework Convention on Climate Change
FSU	Former Soviet Union
GCM	General Circulation Model
GDP	Gross Domestic Product
GFDL	Geophysical Fluid Dynamics Laboratory (USA)
GHG	Greenhouse Gas
GNP	Gross National Product
GP	Guidance Paper
GPP	Gross Primary Production
GWP	Global Warming Potential

H_2O	Water Vapor
HadCM	Hadley Centre Coupled Model
HFC	Hydrofluorocarbon
IAM	Integrated Assessment Model
ICSU	International Council of Scientific Unions
IEA	International Energy Agency
IET	International Emissions Trading
IGCCS	Integrated Gasification Combined Cycle or Supercritical
IPCC	Intergovernmental Panel on Climate Change
IPCC TP3	Technical Paper on Stabilization of Atmospheric Greenhouse Gases: Physical, Biological, and Socio-Economic Implications
IPCC TP4	Technical Paper on Implications of Proposed CO_2 Emissions Limitations
ISAM	Integrated Science Assessment Model
JI	Joint Implementation
LCC	Land-Cover Change
LSG	Large-Scale Geostrophic Ocean Model
LUC	Land-Use Change
MAC	Marginal Abatement Cost
MOP	Meeting of the Parties
MSL	Mean Sea Level
MSU	Microwave Sounding Unit
N_2O	Nitrous Oxide
NAO	North Atlantic Oscillation
NBP	Net Biome Production
NEP	Net Ecosystem Production
NGOs	Non-Governmental Organization
NO_x	Nitrogen Oxides
NPP	Net Primary Production
NSI	National Systems of Innovation
O_2	Molecular Oxygen
O_3	Ozone
ODS	Ozone-Depleting Substance
OECD	Organisation for Economic Cooperation and Development
OPEC	Organization of Petroleum-Exporting Countries
OPYC	Ocean Isopycnal GCM
PFC	Perfluorocarbon
PMIP	Paleoclimate Model Intercomparison Project
PPM	Processes and Production Method
PPP	Purchasing Power Parity
R&D	Research and Development
RCM	Regional Climate Model
SAR	Second Assessment Report
SF_6	Sulfur Hexafluoride
SME	Small and Medium Sized Enterprise
SO_2	Sulfur Dioxide
SPM	Summary for Policymakers
SRAGA	Special Report on Aviation and the Global Atmosphere
SRES	Special Report on Emissions Scenarios
SRLULUCF	Special Report on Land Use, Land-Use Change, and Forestry
SRTT	Special Report on the Methodological and Technological Issues in Technology Transfer
SST	Sea Surface Temperature
TAR	Third Assessment Report
TCR	Transient Climate Response
THC	Thermohaline Circulation
TP	Technical Paper
TS	Technical Summary

TSI	Total Solar Irradiance
UNEP	United Nations Environment Programme
UNESCO	United Nations Education, Scientific and Cultural Organisation
UNFCCC	United Nations Framework Convention on Climate Change
UV	Ultraviolet
VA	Voluntary Agreement or Value-Added
VOC	Volatile Organic Compounds
WAIS	West Antarctic Ice Sheet
WGI TAR	Working Group I Contribution to the Third Assessment Report
WGII SAR	Working Group II Contribution to the Second Assessment Report
WGII TAR	Working Group II Contribution to the Third Assessment Report
WGIII TAR	Working Group III Contribution to the Third Assessment Report
WMO	World Meteorological Organization
WRE	Wigley, Richels, and Edmonds
WTA	Willingness to Accept
WTP	Willingness to Pay
WUE	Water-Use Efficiency

Units

SI (Systeme Internationale) Units

Physical Quantity	Name of Unit	Symbol
length	meter	m
mass	kilogram	kg
time	second	s
thermodynamic temperature	kelvin	K
amount of substance	mole	mol

Fraction	Prefix	Symbol	Multiple	Prefix	Symbol
10^{-1}	deci	d	10	deca	da
10^{-2}	centi	c	10^2	hecto	h
10^{-3}	milli	m	10^3	kilo	k
10^{-6}	micro	μ	10^6	mega	M
10^{-9}	nano	n	10^9	giga	G
10^{-12}	pico	p	10^{12}	tera	T
10^{-15}	femto	t	10^{15}	peta	P

Special Names and Symbols for Certain SI-Derived Units

Physical Quantity	Name of SI Unit	Symbol for SI Unit	Definition of Unit
force	newton	N	$\mathrm{kg\ m\ s^{-2}}$
pressure	pascal	Pa	$\mathrm{kg\ m^{-1}\ s^{-2}}\ (=\mathrm{N\ m^{-2}})$
energy	joule	J	$\mathrm{kg\ m^2\ s^{-2}}$
power	watt	W	$\mathrm{kg\ m^2\ s^{-3}}\ (=\mathrm{J\ s^{-1}})$
frequency	hertz	Hz	$\mathrm{s^{-1}}$ (cycles per second)

Decimal Fractions and Multiples of SI Units having Special Names

Physical Quantity	Name of SI Unit	Symbol for SI Unit	Definition of Unit
length	Ångstrom	Å	$10^{-10}\ \mathrm{m} = 10^{-8}\ \mathrm{cm}$
length	micron	μm	$10^{-6}\ \mathrm{m}$
area	hectare	ha	$10^4\ \mathrm{m^2}$
force	dyne	dyn	$10^{-5}\ \mathrm{N}$
pressure	bar	bar	$10^5\ \mathrm{N\ m^{-2}} = 10^5\ \mathrm{Pa}$
pressure	millibar	mb	$10^2\ \mathrm{N\ m^{-2}} = 1\ \mathrm{hPa}$
mass	tonne	t	$10^3\ \mathrm{kg}$
mass	gram	g	$10^{-3}\ \mathrm{kg}$
column density	Dobson units	DU	$2.687 \times 10^{16}\ \mathrm{molecules\ cm^{-2}}$
streamfunction	Sverdrup	Sv	$10^6\ \mathrm{m^3\ s^{-1}}$

Non-SI Units

°C	degree Celsius (0° C = 273 K approximately)
	Temperature differences are also given in ° C (=K) rather than the more correct form of "Celsius degrees"
ppmv	parts per million (10^6) by volume
ppbv	parts per billion (10^9) by volume
pptv	parts per trillion (10^{12}) by volume
yr	year
ky	thousands of years
bp	before present

Annex D. Scientific, Technical, and Socio-Economic Questions Selected by the Panel

Question 1

What can scientific, technical, and socio-economic analyses contribute to the determination of what constitutes dangerous anthropogenic interference with the climate system as referred to in Article 2 of the Framework Convention on Climate Change?

Question 2

What is the evidence for, causes of, and consequences of changes in the Earth's climate since the pre-industrial era?

a) Has the Earth's climate changed since the pre-industrial era at the regional and/or global scale? If so, what part, if any, of the observed changes can be attributed to human influence and what part, if any, can be attributed to natural phenomena? What is the basis for that attribution?

b) What is known about the environmental, social, and economic consequences of climate changes since the pre-industrial era with an emphasis on the last 50 years?

Question 3

What is known about the regional and global climatic, environmental, and socio-economic consequences in the next 25, 50, and 100 years associated with a range of greenhouse gas emissions arising from scenarios used in the TAR (projections which involve no climate policy intervention)?

To the extent possible evaluate the:
- Projected changes in atmospheric concentrations, climate, and sea level
- Impacts and economic costs and benefits of changes in climate and atmospheric composition on human health, diversity and productivity of ecological systems, and socio-economic sectors (particularly agriculture and water)
- The range of options for adaptation, including the costs, benefits, and challenges
- Development, sustainability, and equity issues associated with impacts and adaptation at a regional and global level.

Question 4

What is known about the influence of the increasing atmospheric concentrations of greenhouse gases and aerosols, and the projected human-induced change in climate regionally and globally on:

a. The frequency and magnitude of climate fluctuations, including daily, seasonal, inter-annual, and decadal variability, such as the El Niño Southern Oscillation cycles and others?

b. The duration, location, frequency, and intensity of extreme events such as heat waves, droughts, floods, heavy precipitation, avalanches, storms, tornadoes, and tropical cyclones?

c. The risk of abrupt/non-linear changes in, among others, the sources and sinks of greenhouse gases, ocean circulation, and the extent of polar ice and permafrost? If so, can the risk be quantified?

d. The risk of abrupt or non-linear changes in ecological systems?

Question 5

What is known about the inertia and time scales associated with the changes in the climate system, ecological systems, and socio-economic sectors and their interactions?

Question 6

a) How does the extent and timing of the introduction of a range of emissions reduction actions determine and affect the rate, magnitude, and impacts of climate change, and affect the global and regional economy, taking into account the historical and current emissions?

b) What is known from sensitivity studies about regional and global climatic, environmental, and socio-economic consequences of stabilizing the atmospheric concentrations of greenhouse gases (in carbon dioxide equivalents), at a range of levels from today's to double that level or more, taking into account to the extent possible the effects of aerosols? For each stabilization scenario, including different pathways to stabilization, evaluate the range of costs and benefits, relative to the range of scenarios considered in Question 3, in terms of:
- Projected changes in atmospheric concentrations, climate, and sea level, including changes beyond 100 years
- Impacts and economic costs and benefits of changes in climate and atmospheric composition on human health, diversity and productivity of ecological systems, and socio-economic sectors (particularly agriculture and water)
- The range of options for adaptation, including the costs, benefits, and challenges
- The range of technologies, policies, and practices that could be used to achieve each of the stabilization levels, with an evaluation of the national and global costs and benefits, and an assessment of how these costs and benefits would compare, either qualitatively or quantitatively, to the avoided environmental harm that would be achieved by the emissions reductions
- Development, sustainability, and equity issues associated with impacts, adaptation, and mitigation at a regional and global level.

Question 7

What is known about the potential for, and costs and benefits of, and time frame for reducing greenhouse gas emissions?
- What would be the economic and social costs and benefits and equity implications of options for policies and measures,

and the mechanisms of the Kyoto Protocol, that might be considered to address climate change regionally and globally?

- What portfolios of options of research and development, investments, and other policies might be considered that would be most effective to enhance the development and deployment of technologies that address climate change?
- What kind of economic and other policy options might be considered to remove existing and potential barriers and to stimulate private- and public-sector technology transfer and deployment among countries, and what effect might these have on projected emissions?
- How does the timing of the options contained in the above affect associated economic costs and benefits, and the
- atmospheric concentrations of greenhouse gases over the next century and beyond?

Question 8

What is known about the interactions between projected human-induced changes in climate and other environmental issues (e.g., urban air pollution, regional acid deposition, loss of biological diversity, stratospheric ozone depletion, and desertification and land degradation)? What is known about environmental, social, and economic costs and benefits and implications of these interactions for integrating climate change response strategies in an equitable manner into broad sustainable development strategies at the local, regional, and global scales?

Question 9

What are the most robust findings and key uncertainties regarding attribution of climate change and regarding model projections of:

- Future emissions of greenhouse gases and aerosols?
- Future concentrations of greenhouse gases and aerosols?
- Future changes in regional and global climate?
- Regional and global impacts of climate change?
- Costs and benefits of mitigation and adaptation options?

Annex E. List of Major IPCC Reports

Climate Change—The IPCC Scientific Assessment
The 1990 Report of the IPCC Scientific Assessment Working Group (also in Chinese, French, Russian, and Spanish)

Climate Change—The IPCC Impacts Assessment
The 1990 Report of the IPCC Impacts Assessment Working Group (also in Chinese, French, Russian, and Spanish)

Climate Change—The IPCC Response Strategies
The 1990 Report of the IPCC Response Strategies Working Group (also in Chinese, French, Russian, and Spanish)

Emissions Scenarios
Prepared for the IPCC Response Strategies Working Group, 1990

Assessment of the Vulnerability of Coastal Areas to Sea Level Rise—A Common Methodology
1991 (also in Arabic and French)

Climate Change 1992—The Supplementary Report to the IPCC Scientific Assessment
The 1992 Report of the IPCC Scientific Assessment Working Group

Climate Change 1992—The Supplementary Report to the IPCC Impacts Assessment
The 1992 Report of the IPCC Impacts Assessment Working Group

Climate Change: The IPCC 1990 and 1992 Assessments
IPCC First Assessment Report Overview and Policymaker Summaries, and 1992 IPCC Supplement

Global Climate Change and the Rising Challenge of the Sea
Coastal Zone Management Subgroup of the IPCC Response Strategies Working Group, 1992

Report of the IPCC Country Studies Workshop
1992

Preliminary Guidelines for Assessing Impacts of Climate Change
1992

IPCC Guidelines for National Greenhouse Gas Inventories
Three volumes, 1994 (also in French, Russian, and Spanish)

IPCC Technical Guidelines for Assessing Climate Change Impacts and Adaptations
1995 (also in Arabic, Chinese, French, Russian, and Spanish)

Climate Change 1994—Radiative Forcing of Climate Change and an Evaluation of the IPCC IS92 Emission Scenarios
1995

Climate Change 1995—The Science of Climate Change — Contribution of Working Group I to the IPCC Second Assessment Report
1996

Climate Change 1995—Impacts, Adaptations, and Mitigation of Climate Change: Scientific-Technical Analyses — Contribution of Working Group II to the IPCC Second Assessment Report
1996

Climate Change 1995—Economic and Social Dimensions of Climate Change — Contribution of Working Group III to the IPCC Second Assessment Report
1996

Climate Change 1995—IPCC Second Assessment Synthesis of Scientific-Technical Information Relevant to Interpreting Article 2 of the UN Framework Convention on Climate Change
1996 (also in Arabic, Chinese, French, Russian, and Spanish)

Technologies, Policies, and Measures for Mitigating Climate Change — IPCC Technical Paper I
1996 (also in French and Spanish)

An Introduction to Simple Climate Models used in the IPCC Second Assessment Report — IPCC Technical Paper II
1997 (also in French and Spanish)

Stabilization of Atmospheric Greenhouse Gases: Physical, Biological and Socio-economic Implications — IPCC Technical Paper III
1997 (also in French and Spanish)

Implications of Proposed CO_2 Emissions Limitations — IPCC Technical Paper IV
1997 (also in French and Spanish)

The Regional Impacts of Climate Change: An Assessment of Vulnerability — IPCC Special Report
1998

Aviation and the Global Atmosphere — IPCC Special Report
1999

Methodological and Technological Issues in Technology Transfer — IPCC Special Report
2000

Land Use, Land-Use Change, and Forestry — IPCC Special Report
2000

Emission Scenarios — IPCC Special Report
2000

**Good Practice Guidance and Uncertainty Management
in National Greenhouse Gas Inventories**
2000

**Climate Change 2001: The Scientific Basis − Contribution
of Working Group I to the IPCC Third Assessment Report**
2001

**Climate Change 2001: Impacts, Adaptation, and
Vulnerability − Contribution of Working Group II to the
IPCC Third Assessment Report**
2001

**Climate Change 2001: Mitigation − Contribution of
Working Group III to the IPCC Third Assessment Report**
2001

Enquiries: IPCC Secretariat, c/o World Meteorological
Organization, 7 bis, Avenue de la Paix, Case Postale 2300,
1211 Geneva 2, Switzerland